P9-EIF-046

No. 1139
$24.95

The Complete Powerplant Efficiency Manual

by Frank G. McAleese

TAB BOOKS Inc.
BLUE RIDGE SUMMIT, PA. 17214

FIRST EDITION

FIRST PRINTING

Printed in the United States of America

Library of Congress Cataloging in Publication Data

McAleese, Frank G
The complete powerplant efficiency manual.

Includes index.
1. Power-plants—Handbooks, manuals, etc. I. Title.
TJ164.M32 621.4 80-19903
ISBN 0-8306-9920-1

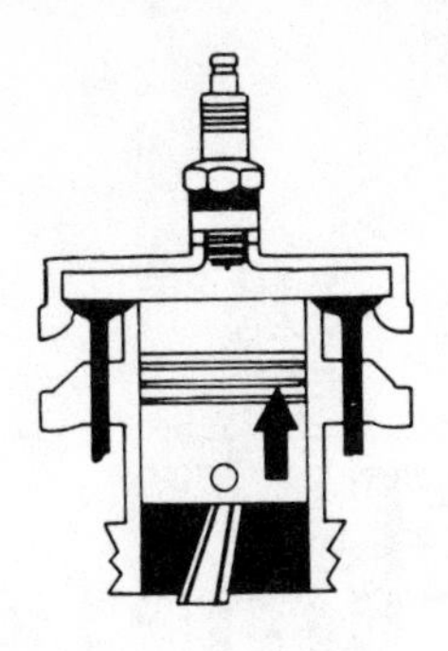

Acknowledgments

I wish to thank the following organizations and individuals for their contributions and assistance: Fairbanks Morse Engine Division, Colt Industries, Beloit, WI; Westinghouse Electric Corporation, Pittsburgh, PA; Weyerhaeuser Company, Tacoma, WA; Babcock and Wilcox Company, Barberton, OH, and Philadelphia, PA; E. Keeler Company, Williamsport, PA; Cleaver-Brooks, Division of Aqua-Chem, Inc., Milwaukee, WI; North American Manufacturing Company, Cleveland, OH; Dresser Industries, Dresser Clark Division, Olean, NY; Milton Roy Company, Hays-Republic Division, Michigan City, IN; Sprague Electric Company, North Adams, MA; United Technologies, Power Sytems Division, Farmington, CT; Alco Power Inc., Auburn, NY; AMPROBE Instrument, Division of Core Industries, Lynbrook, NY; Kewanee Boiler Corporation, Kewanee, IL; Caterpillar Tractor Company, Peoria, IL; Atlantic City Electric Company, Atlantic City, NJ; Krop Boiler Cleaning Company, Blue Bell, PA; American Forest Institute, Washington, D.C.; United States Department of the Interior, Bureau of Mines, Washington, D.C.; United States Department of Energy, Energy, Research and Development Administration, Washington, D.C.; United States Department of Agriculture, Forest Service, Washington, D.C.; United Sterling Company, Sweden; Philips Research Laboratories, Erndhover, Holland; Dr. Wellington Woods Jr., Glassboro State Teachers College, Glassboro, NJ; Inspector Joseph Czop, Hartford Steam Boiler Inspection and Insurance Company, King of Prussia, PA; Walter "Lodge" Czop, Blue Bell, PA; Donna Brown, Cumberland County College, Vineland, NJ; Robert E. Kahler, Fidelity Corporation, Philadelphia, PA; Chief Engineer George Ingram, Philadelphia, PA.

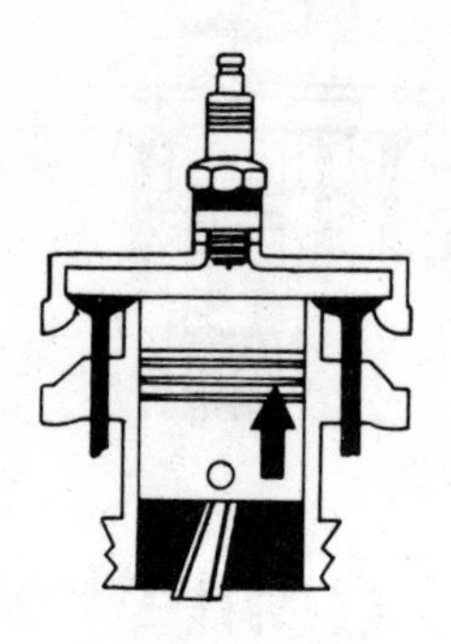

Dedication:

Dedicated to Mr. and Mrs. Joseph McAleese,
My mother and father.
Granted to me by concern and benefaction,
through the providence of God's certain Love.

Other TAB books by the author:

No. 1123 *The Laser Experimenter's Handbook*

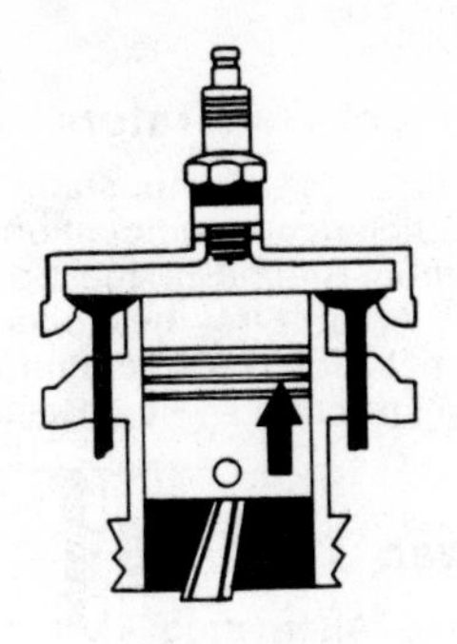

Contents

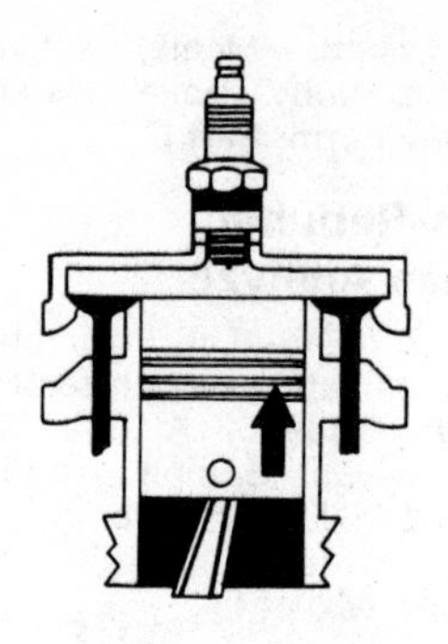

Introduction

With today's justified concern for the efficient and economic utilization of our ever depleting natural fuel resources and environmentally damaging effects to our ecological life support systems, the emphasis of conservation efforts has expanded from The Scientific Community into the power plants and homes of our nation. It is my feeling that a truly meaningful effort for conservation and ecological safety must be launched from the extreme end of the supply chain: the management and staff of our nation's power plant.

The essence of conservation could best be defined as utilizing the most of what one has available by extracting the greatest output from the least input. The viable truth is that conservation is an inspired attitude which, to be effective, must be embraced by our administrative and industrial leaders on down and enacted supportively by our power plant personnel on up.

It stands imperative for us to participate on a national level in a concerted effort for conservation. We must be equal to adopting a totally mindful perspective of the interrelated benefits and effects afforded by conservation, both in economical as well as ecological terms. Additionally, we must be willing through education and research to attain improvement through experimentation with new and perhaps more efficient means of energy conversion devices.

Conservation of our natural resources, be they petroleum, gas, wood forests, coal, water or any other commodities, is a categorical "must." We cannot return the resources we have to date used or even misused. It certainly doesn't require a graduate ecologist to certify the truism of the irreversible fact of supply and demand: the more heavier our demand, the more hastened will be the eventual exhaustion of that supply.

If there is no use in crying over spilled milk, there is even less to profit ourselves with criticism over wasted fuel supplies. We may however profit materially by resolving that those resources yet on hand will be seriously and wisely used from this day forward.

This resolution can best be spearheaded by front line technicians — the power plant team. Appropriately, it is to this level of personnel to whom this book is dedicated, particularly the chief engineer. The chief engineer has both the authority to enforce and demand efficiency and the first hand expertise to educationally equip his subordinate engineers, firemen and maintenance mechanics with the competency to work and to aspire for efficient results.

This manual has been compiled as a practical working guide for the chief engineer to assist him in conducting a comprehensive and total plant survey in order to increase the plant's efficiency. It is assumed that the reader, being a member of the power plant team, has at least a conversant familiarity with efficiency, the use of the respective measuring instruments necessary in determining the values required for making such evaluations, and the technical perspective needed to appreciate the values obtained.

The reader should acquaint himself with the data, specifications and instructions as provided by the manufacturer or builder for each piece of equipment being tested. Additionally, he should be capable of taking accurate measurements and indications with the appropriate instruments and gauges, consulting instructions where necessary. This manual does not provide a course in instrumentation. Rather, it is intended to provide a systematic approach as to just what factual measurements are necessary and how they will be evaluated in depicting the possible deficiencies of the systems.

The author has incorporated in this manual forms for the reader to employ in his fact finding excursion of collected data and measurements. The forms which the author has provided are a result of several years of trials and applications and represent a logical and comprehensive system of collecting, recording and evaluating the information which will ultimately determine how efficient the power plant is.

It must be understood that what we are seeking is the greatest possible total plant efficiency. Occasionally we may find ourselves in the predicament where we must sacrifice a component's individual efficiency in preference to the more desirable total effect it would alternately yield. By carefully studying the logical and collective evaluation as it will present itself, we should be able to negotiate such a problem with no great catastrophies.

Only real and physical action is going to stretch out our depleting reserves. That action is going to have to come from the American engineer in concert with the scientific community in its entirety. Power plant engineers must take a leading role in energy conservation; that is the theme of this book.

Frank G. McAleese

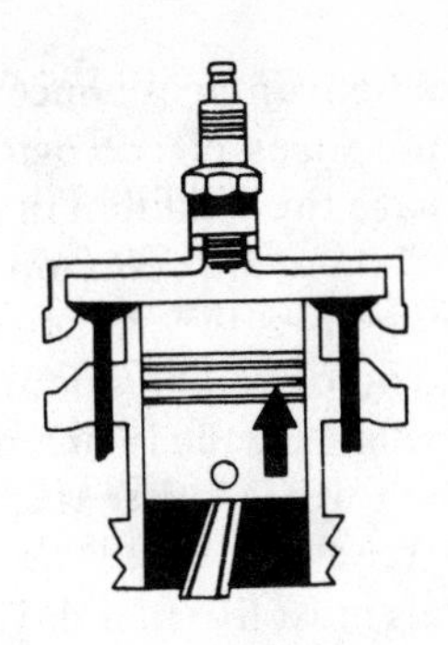

Chapter 1
Natural Fuel Resources - Coal, Petroleum and Natural Gas

Because this book deals with the energy provided by our earth's resources, we will confine our interests to those resources of energy bearing characteristics—more specifically, to those resources that may directly or indirectly be translated in terms of available heat values. Fundamentally, all energy *per se* comes from the sun in one form or another, whether it be in plants (by photosynthesis of cellular action) or the fossilized products of plants and vegetation. The woods we cut from the forest are a direct offspring of solar-biochemical relation.

THE NEED TO CONSERVE

The following are ramifications of solar effects in one application or another: the energy stored in organic matter, in the form of coal or petroleum; the force of wind power; the very usable hydraulic forces of water falling from one potential level to a lower level; and the flow of water over or under a water wheel. Ironically, after having sought elsewhere for our energy from the Industrial Revolution until today, we return to the sun directly for its benefits by the use of solar cells and solar concentrate heating.

We're living now during the last leg of the journey which carbonaceous fuels have adequately carried us. Carbonaceous fuels are certainly not objectionable scientifically other than their contaminate potential in ecological terms.

There are manifold advantages concerning the transition to nuclear fuels for the purposes of ecological safeguards, which would or should eliminate the pollution in the atmosphere from industrial sources. Plus, there are economic benefits with respect to decreasing power expenses to virtually only the initial investment costs of nuclear reactor and components. By the reduction in power costs and having essentially unlimited power at negligible costs, projects which would otherwise prove too expensive seem more within our reach.

Since we as mortals must live in today's world, then we must make do with today's resources and deal with problems with the equipment we have at our disposal. And that means relying upon natural fuel supplies, basically comprising wood forests, coal deposits and petroleum pockets. We could include solar energy, and natural forces such as wind and water power. Since these three supplies are of limited application, we'll concentrate our study to the convertible fuels yielded by drill, shovel and saw.

Because we're going to launch an in-depth inquiry into the preferred methods of utilizing these primary fuels in manners which will return the greatest possible return for the least input, we'll find a synopsis of each fuel to be both an education and an enjoyable venture into the woodlands, the mines and the oil fields.

I'm not suggesting that we sentimentally love a tree or adopt a piece of anthracite coal into the family. I *am* advocating a properly due concern for the preservation of the resources in themselves.

The following discussions of our three main and primary fuels as available resources will acquaint the reader with the origins of the fuels and the nature of their respective geological-chemical formations. Once these resources are gone, they are permanently and irreplaceably lost. Petroleum deposits require from 70 to 250 million years to create, while coal veins take 250 to 500 million. For timber land to be worth harvesting, 40 to 90 years must pass.

I will limit the discussion of each fuel source to its respective rank as a comparative combustible, weighing the inherent advantages and disadvantages relative to the fuel's availability, the extent of refinement prior to its use as a fuel, and *its* compatibility as a cooperating phase in our ecological life support system.

We must be ever mindful in the deliberation of both the "end" and "by-product" potentials in the selection of our primary fuels that they conform to both engineering and ecological criteria in respect to their contaminate matter. We cannot in good conscience offer our dependent civilization a proposed energy source of ideal engineering endorsements, which would at the same instant prove detrimental to the environment or disruptive to our biological welfare.

A broader embrace of the life support system will be given later in the summation of natural fuel resources. Suffice it to say that we are bound by our professional commitments to investigate all contingencies of by-product reactions and interreactions resulting from the fuel selections, and fuel modulations, in respect to their products of combustion. Cheap toxins admitted to our already near saturated atmosphere are certainly no bargain at any price.

Further, the continued and unabated contamination of our atmosphere can hardly speak well of our professional credentials, or collective competency. Although we cannot directly imply that the air pollution index for any given day could be assumed as a qualification barometer of our practicing ability, it does nevertheless indicate that certain prompt attention is wanted in this critical area.

TIMBER RESERVES

Of the world's 13,037,063 square miles of forest and timber reserves, the United States has about 2,872,587 square miles of woods (Fig. 1-1). Until World War II, almost half of all the timber

Fig. 1-1. Typical commercial stand of Douglas fir and hemlock in Washington state. Mt. Rainier is in the background (courtesy of Weyerhaeuser Company).

Fig. 1-2. A chain saw makes tree cutting easy (courtesy of Weyerhaeuser Company).

cut was used as fuel (Figs. 1-2 through 1-5). This percentage has proportionately decreased as usage of oil and coal has increased. The chief displacement of wood as a fuel source has been due to the discovery of more oil fields, the receding forest lands (actually the depletion of them) from the proximity of the fuel burning plants, the higher heat content offered by oil, coal and gas, and the added expense in transporting and handling wood (Figs. 1-6 through 1-8). In less industrialized countries, wood fuels still offer economy.

Trees are divided into two main groups: the *coniferous* or pines and the *deciduous*, or the family of trees which shed their leaves seasonally. The conifers, which botanically are classified as

Fig. 1-3. This logger is in the process of falling a mature tree (courtesy of Weyerhaeuser Company).

softwoods, have a slightly higher heating value then do the hardwoods (deciduous). The higher heating value is a result of the resins which are peculiar to pine trees.

Heating Value

Table 1-1 gives the relative heat content per cord of wood versus a ton of anthracite coal. A cord of wood is a volumetric measurement, being a pile of cut lengths 4 feet wide by 4 feet high, by 8 feet in total length, which equals a required storage spce of 128 cubic feet. Because of the air spaces between the logs, only 70 percent is actually wood. Since moisture affects both the weight and consequent heating value, Table 1-1 is based upon "air dry" ready wood.

It will be evidenced by the comparative values from our table that wood, pound for pound, yields far less heat in Btus/pound as compared to coal. A far greater volume or mass of wood must be transported, fired and handled than coal. A much larger storage facility is needed to accommodate the extra bulk which wood requires by its lower Btu/pound ratio. Another factor is the protection wood requires. It must be kept as dry as possible. An increase in any amount of moisture can drastically decrease its calorific content per pound and proportionately increase its weight.

Even the closest wood to coal in terms of heat value, hickory, only yields 24,800,000 Btus/cord. When we consider that it takes a

Table 1-1. Fact on Types of Wood.

TYPE OF WOOD	WEIGHT PER CORD	Btus PER CORD	PERCENT EQUIV. TO TON OF COAL
White Ash	3800 pounds	20,500,000	0.79 ton
Beech	3900 " "	20,900,000	0.80 "
Yellow Birch	4000 " "	20,900,000	0.80 "
Chestnut	2700 " "	15,600,000	0.60 "
Cottonwood	2500 " "	15,000,000	0.58 "
White Elm	3100 " "	17,700,000	0.68 "
Hickory	4600 " "	24,800,000	0.95 "
Maple, Sugar	3900 " "	21,800,000	0.84 "
Maple, Red	3200 " "	19,100,000	0.73 "
Oak, Red	3900 " "	21,700,000	0.83 "
Oak, White	4300 " "	23,900,000	0.92 "
Yellow Pine	3900 " "	24,000,000	0.90 "
White Pine	3800 " "	15,000,000	0.58 "
Black Walnut	4000 " "	20,800,000	0.80 "
Willow	2300 " "	13,500,000	0.52 "

Fig. 1-4. The tree is being cut up into desirable lengths (courtesy of Weyerhaeuser Company).

cord of hickory, which weighs 4,600 pounds to equal the same 2,000 pounds of coal, we then appreciate the difference in the relative Btu/pound ratio. Hickory offers only 5,400 Btu/pound, compared to the 13,000 Btu/pound offered by anthracite coal. If we had a certain boiler requiring 20 tons of coal per hour, then to fire wood, in this case hickory, we would have to transport, prepare and fire almost 50 tons of hickory to equal the heat released by coal. We would also need a much larger combustion chamber and stoking equipment necessitated by the bulkier mass of wood.

One of the foremost reasons that railroads were so willing to change from wood to coal was because of the less bulk required in transporting their fuel. The less volume and weight required for fuel means that much more paying cargo can be realized, and thus more profit per mile traveled. Ideally, liquid fuels have answered the railroad and maritime quite well, as oil has an even greater Btu/pound ratio than coal and can very easily be accommodated in storage tanks of any design and location. Thus, fuel selection and ultimately resource preference were given primarily for economic factors, as oil serves better than coal and coal serves better than wood.

Sanctuaries

The reader musn't infer that I am biased towards fuels other than wood or that I have a prejudices against wood. I would prefer to see our timberlands remain sanctuaries rather than witness our forests having to be fired for fuel unnecessarily. While we're on this issue, I'll provide a direct correlation. If, in a wood burning plant, the carbon dioxide were to be increased from the typical 8 percent to the ideal 12 percent an additional saving of fuel would

Fig. 1-7. The tall pole in the background is a portable spar tree (courtesy of Weyerhaeuser Company).

Fig. 1-8. This loaded logging truck is headed for a mill in eastern Washington state (courtesy of Weyerhaeuser Company).

be realized in the neighborhood of 15 percent. Less wood would be needed to supply that particular boiler plant. For every 100 acres of timber land formerly cut, only 85 acres would then be required. The other 15 acres could be used for other worthy functions. It makes one wonder how many 15 acre "tracts" we could have saved during the hundreds of years we burned our timber lands, probably enough to equal a national wildlife refuge. This example, along with the other illustrations, is given to sensitize the engineer with the direct correlation of his efficiency minded attitude and

Fig. 1-5. Bucking a Douglas fir tree in Washington (courtesy of Weyerhaeuser Company).

Fig. 1-6. Logs are being loaded onto a truck (courtesy of Weyerhaeuser Company).

the savings potential related to it, both in terms of dollars and in natural resources.

THE COAL CYCLE

Coal is really a *sedimentary* rock. Because of its energy value as a primary heat source and its yielding of chemicals for medicines, nylon stockings, and synthetic compounds for gasoline substitutes, it is probably the most precious of minerals next to gold itself.

The sedimentary cycle of coal started in swamps millions of years ago, the immediate sources being plants and carbon dioxide drawn from the atmosphere and surface waters. The juvenile carbon dioxide, however, sprang from igneous sources, as did the original constituents of the soil that supported the plant life. Volcanic activity also played a part in the contribution of the carbon dioxide for the carbon matter in the coal bed. The carbon dioxide absorbed by plants and animals is largely released again; that locked up in limestone and coal is only partly returned. The trees and small plants that fall into swamp water undergo partial decay. The residue collects on the swamp bottom to form peat, the first stage of all coal. This gradually loses oxygen and hydrogen, and under compaction of overlaying beds, undergoes chemical and physical changes to become bituminous coal. By metamorphosis this coal may later evolve into anthracite coal, the last stage of the coal cycle.

TYPES OF COAL

In nontechnical terminology, four main types or variations of coal are recognized: *anthracite,* or hard coal; *bituminous,* or soft coal: *lignite* and *canned* coal. There are of course several interclassifications such as semi-bituminous, and subclassifications according to the areas where the coals are respectively mined. Peat lies below lignite and meta-anthracite, and graphite is above anthracite. Peat is not a coal, even through it is a fuel. It is an accumulation of partly decomposed vegetable matter that represents the first stage of all coals.

Lignite and Bituminous

Lignite and brown coal are composed of woody matter embedded in macerated and decomposed vegetable matter. The matter is banded and jointed and slacks or disintegrates upon drying. It has low heating value and is particularly subject to spontaneous combustion. Sub-bituminous coal is an intermediate coal that resembles bituminous. It is dull black, waxy, banded and splits parallel to the bedding. The coal is but of relatively low heating value. Bituminous coal is dense, black, brittle, banded and breaks into prismatic blocks. Its constituent vegetable matter is not ordinarily visible to the naked eye. Bituminous ignites easily and even under favorable conditions burns with a smoky yellow flame. Its moisture is comparatively low and its heating value is high. Bituminous is the most used coal in the world because of its abundance and is favored for steam generation, gas and coking.

Cannel

Cannel coal is a special variety of bituminous that is lusterless, does not soil the fingers, is not banded and breaks with a splintery fracture. It is made up of wind blown spores and pollen. The coal is clean, burns with a long flame and is sought for fireplace coal. The higher ranks of bituminous coals have the highest values of heat content, and thus are more desirable from engineering standpoints.

Anthracite

Anthracite is a jet black, hard coal of high luster and irregular fracture. It ignites slowly, is virtually smokeless and has the most desirable qualities of all naturally occurring coals. It is unfortunately limited in its geographical distribution, thus making it more expensive and less available economically.

CHEMICAL COMPOSITION OF COAL

Chemically, coals are made up of various proportions of carbon, hydrogen, oxygen, nitrogen and impurities. Toward the higher ranks (near anthracite) there is a progressive elimination of water, hydrogen and oxygen and an increase in carbon, which is present as fixed carbon, and is volatile matter. The volatile matter burns as a gas; it causes ready ignition and smoke. The fixed carbon is the steady lasting source of heat, producing a short, hot, smokeless flame.

The *fuel ratio* (composition), an important characteristic of coal, is the fixed carbon divided by the volatile matter. Ash is the residual in coal of noncombustible matter that comes from silt, clay, silica or other substances. Ash and nonconsumable matter are what cause the headaches associated with *clinker*, which must be cleaned from the fires regularly. Otherwise, it will fuse to form a surface or patch which will starve that area from air passage and retard combustion.

Physically, banded coals are made up of partly decomposed and macerated vegetable matter, mainly vascular land plants, of which the following have been recognized: *resins, waxes, cutin, lignose, gums, oils, fats, pigments, starches, cellulose,* and *protoplasm.* When the vegetable matter is attacked by bacteria, the most assailable substances are largely destroyed. The least assailable ones, such as resins and waxes, are present in most coals.

ORIGIN OF COAL

All ordinary coals are of vegetable parentage and the banded ones are now considered to have been formed in swamps, going through a peat stage. The microscope reveals that the raw material of banded coals was vascular swamp vegetation, near to that growing in present day peat-forming swamps. Well over 3,000 plant species have been identified from *carboniferous* coal beds. Roots and stumps found under clays beneath coal beds show that the vegetation grew and accumulated in place. Luxuriant vegetation flourished and consisted mainly of *ferns, lycopods* and flowering plants, with conifers and other varieties present. Ferns were tree-like, and rushes grew 90 feet high. Lycopods (small shrubs today) attained 100 feet in height. Bulbous and arched roots show that the trees lived in water. None of the plants examined were of the salt water species, and the same kinds of plants are found in coals of all ranks and degrees of evolution.

The extensive distribution of individual coal seams implies swamp accumulation on broad delta and coastal plains and broad interior lowlands, where shallow waters rest throughout the year. Most coals are underlain by carbonaceous shales of lake-bottom depositions. Low lying surrounding lands are necessary; otherwise, there would be too much inflow of silt. J.V. Lewis, a well-known authority, estimated that it would take 125 to 150 years to accumulate enough material for a bed of bituminous coal 1 foot thick, and 175 to 200 years for the accumulation of a bed of 1 foot of anthracite.

The climatic conditions were mild to subtropical, with moderate to heavy rainfall. Severe frosts were absent, but the climates were not without dry spells since some of the plants evidenced "water containers" in their trunks and roots. Probably the climate was like that of the Carolinas or Florida. Coals have been mined in places where they would not now be formed, as in Spitzbergen, Greenland and Antarctica, indicating severe and drastic climatic changes since their respective age of coal formation.

Biochemical Action

The change from plant debris to coal involves biochemical action, producing partical decay followed by preserval of this material from further decay and later processes. The type of coal depends largely upon the climate, the type of plants and particularly the duration of bacterial decay. When a tree falls on dry land, it decays. Its constituents are broken up into carbon dioxide and water with which the tree started, no coal thus accumulating. When vegetation falls into water, however, a similar but much slower decay sets in. An essential factor to coal formation is the arrest of bacterial decay before complete destruction takes place, so that there can be some residue remaining to accumulate. This is accomplished by the decay-prompting bacteria rendering the water toxic to themselves, preventing further decay to the vegetable tissues and thus permitting their preservation and accumulation.

The biochemical changes, as mentioned, eliminate oxygen and hydrogen and by so doing concentrate the carbon matter. The bacteria first attack the most decomposable constituents such as cellulose and starches. The resistant materials like resins and waxes, along with wood fragments, drop to the bottom of the swamp where the toxicity slows up or prevents further decay and humus accumulates.

This factor of duration of bacterial breakdown is most important in determining whether any coal will form and, as a consequence, what type of coal it will be characterized as. Under normal conditions of water supply and little surface agitation, all but the least resistant of the vegetable matter will be preserved. Heavy rains however will dilute the water, lessen the toxicity and foster further decay. Floods are catastrophic to coal formation because they lower the toxicity and wash away ingredient materials. Dry seasons are likewise destructive because the water level is lowered, the detritus becomes air exposed, and decay proceeds to total destruction. As you can visualize, water level plays an important role in the recipe. Accumulation of detritus in the swamp gives rise to peat, the initial stage of coal.

Physical and Chemical Changes

After the peat stage, bacteria seem to play little further part as most of the subsequent changes are chemical and physical. These changes are induced by pressure and slight increases in temperature, due to the exothermic reaction of the decomposition and the somewhat insulated conditions contributed by the overlying sediments. In peat, the oxygen content has already been reduced about 10 percent. Further progressive elimination is probably caused by the combination of oxygen with carbon to form carbon monoxide and carbon dioxide which escapes.

The important chemical changes from peat to the higher ranking coals are the progressive elimination of water, oxygen and *bitumens;* the conservation of hydrogen; the progressive increase of *ulmins*; dehydration; increased resistance to solvents, oxidation and heat; *and the development of heavy hydrocarbons.* The chief contemporaneous physical changes are compaction, drying and induration; jointing and cleavage; optical changes; dehydration: color change—brown to black: increase of density: change of luster; and fracture changes from bedding to cleavage to irregular.

The change in rank is mostly the combined effects of time pressure. The older the coals, the more likely they are to be found more deeply buried, thus increasing the pressure and accelerating the metamorphism. Folding increases the rank status because it escalates the application of pressure and temperature. This occurs in Pennsylvania, where the rank of the coal increases progressively with the intensity of the folding of the rocks. The anthracites are found in the most closely folded beds.

OCCURRENCE OF COAL

Coal beds occur in what are called coal measures which consist of alternating beds of sandstone, shale and clay, mostly of fresh water origin. These indicate alternating and fluctuating conditions of sedimentation between coal and ordinary water-borne sediments, since several coal seams generally occur within the coal measures. For example, in Pennsylvania there are 29 coal seams aggregating 106 feet of coal; in Alabama there are 55 coal seams.

Coal seams are huge flat lenses, although some are remarkably persistent. The Pittsburgh seam for example, underlays 15,000 square miles. The thickness of individual seams of coal ranges from a mere film up to as much as 100 feet. The thickest bed in the United States is 84 feet; a famous seam in Pennsylvania is 50 to 60 feet. Most coal seams range between 2 and 10 feet in thickness, and rarely exceed a thickness of 20 feet. Even these thicknesses represent a long period of quiescent sedimentation. Many coal seams contain thin partings of shale or clay called "bone," which represents a break in accumulation with influx of sediments. Other seams divide into splits toward the margin of the

Fig. 1-9. Unreclaimed mountainside scars winding through Appalachia show indelible marks of past contour strip mining for coal (courtesy of Bureau of Mines, United States Department of the Interior).

basin, indicating inroads of sedimentation from the side during accumulation. Other display bulges and rolls of the floor or roof.

Coals occurred in all of the geologic ages since the *Devonian* period, about 300 million years ago. The *Carboniferous* period, about 200 million years, received its name because of its world-wide distribution of coal. The *Cretaceous* period is the next most prolific one, and the *Tertiary* developed most of the world's lignite. In distribution, coal is world-wide, but it is rather phenomenal that coal is much more abundant in the northern hemisphere than in the southern half. There are few countries north of the equator that are entirely without coal. The countries most endowed with deposits are the United States, Russia, Canada, England, Germany, China and India. For some unexplainable reason, very little is to be found in Central and South America, Africa, Scandinavia and the Mediterranean countries.

METHODS OF COAL EXTRACTION

There are two basic methods of extracting, or in the case of coal, mining it from its natural vaults in the earth. These are *strip mining,* where the veins of coal are reasonably close to the surface; and *deep* or *underground mining,* where the coal veins are so deep as to require going in after them (Figs. 1-9 and 1-10). Approximately half of the coal mined in the United States is taken from deep mining techniques. In strip mining, giant shovels and earth-moving equipment are employed to remove the "over burden" from above and around the coal veins to expose the veins for extraction.

Strip Mining

Once the coal vein is exposed, the vein is then broken up into smaller more collectible chunks. Massive machines are used to gather the coal and load it into large trucks and transporting vehicles. The most common manner by which the coal seams are broken is by dynamiting the seams at predetermined points. Convenient-sized chunks of several tons may then be hauled out of the mouth of the excavation and delivered to the preparation plants. Although strip mining is fast and efficient, it also leaves very ugly scars on the land and contributes to the erosion of the soil. Many companies are reclaiming or rejuvenating the scarred land by planting trees, crops and shrubbery or converting the area to public recreation sites.

Because of the otherwise ruining effect from disreputable companies, many states have imposed restrictions upon the future application of strip mining. Figures 1-11 and 1-12 illustrate the relationship of the coal seams or veins with respect to the lands' surface. Modern techniques have yielded as much as 12 tons of coal per minute from each machine used in the operation of strip mining.

Deep Mining

Deep mining, which as the name suggests is the technique used where the coal seams are running at a substantial depth in the

Fig. 1-10. This giant shovel 20 stories high had to be assembled at the mining site. Coal is produced either from surface mines or from deep mines (courtesy of United States Department of Energy).

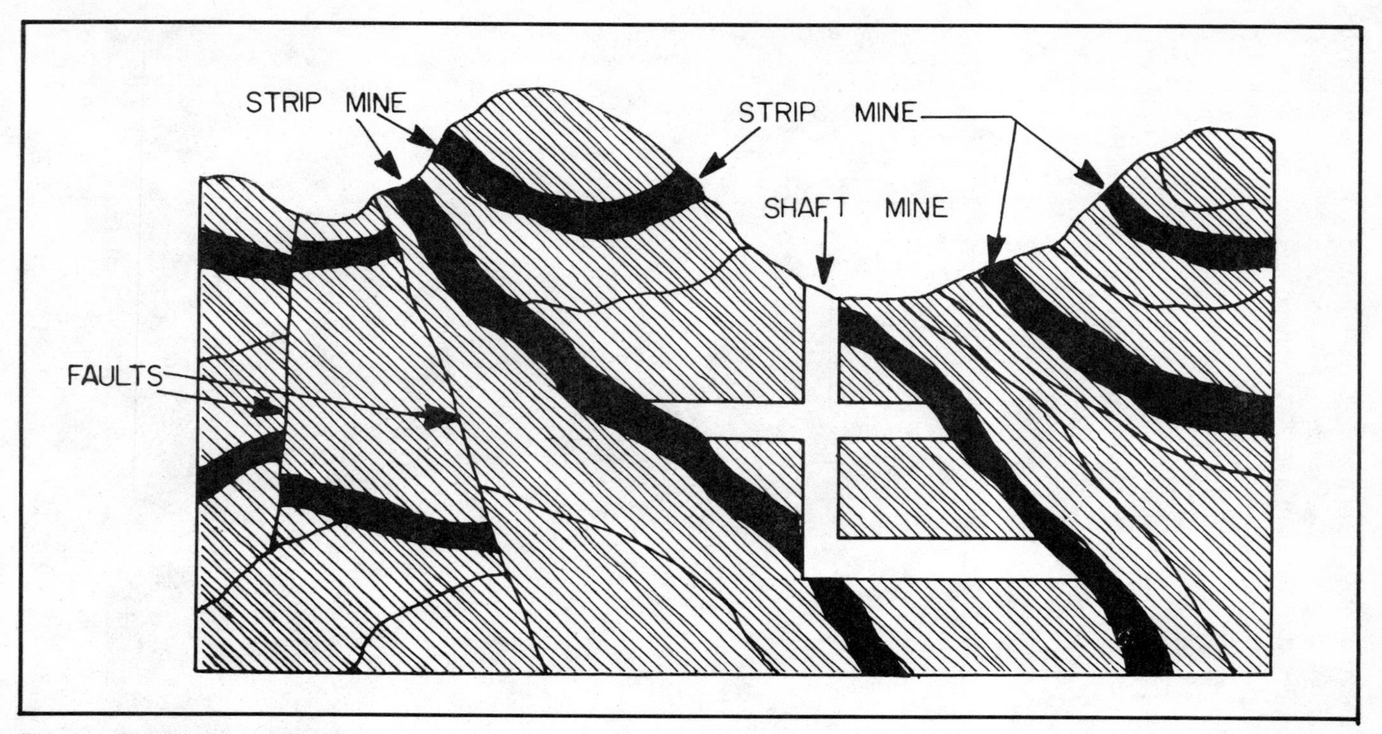

Fig. 1-11. The faults are plainly evident.

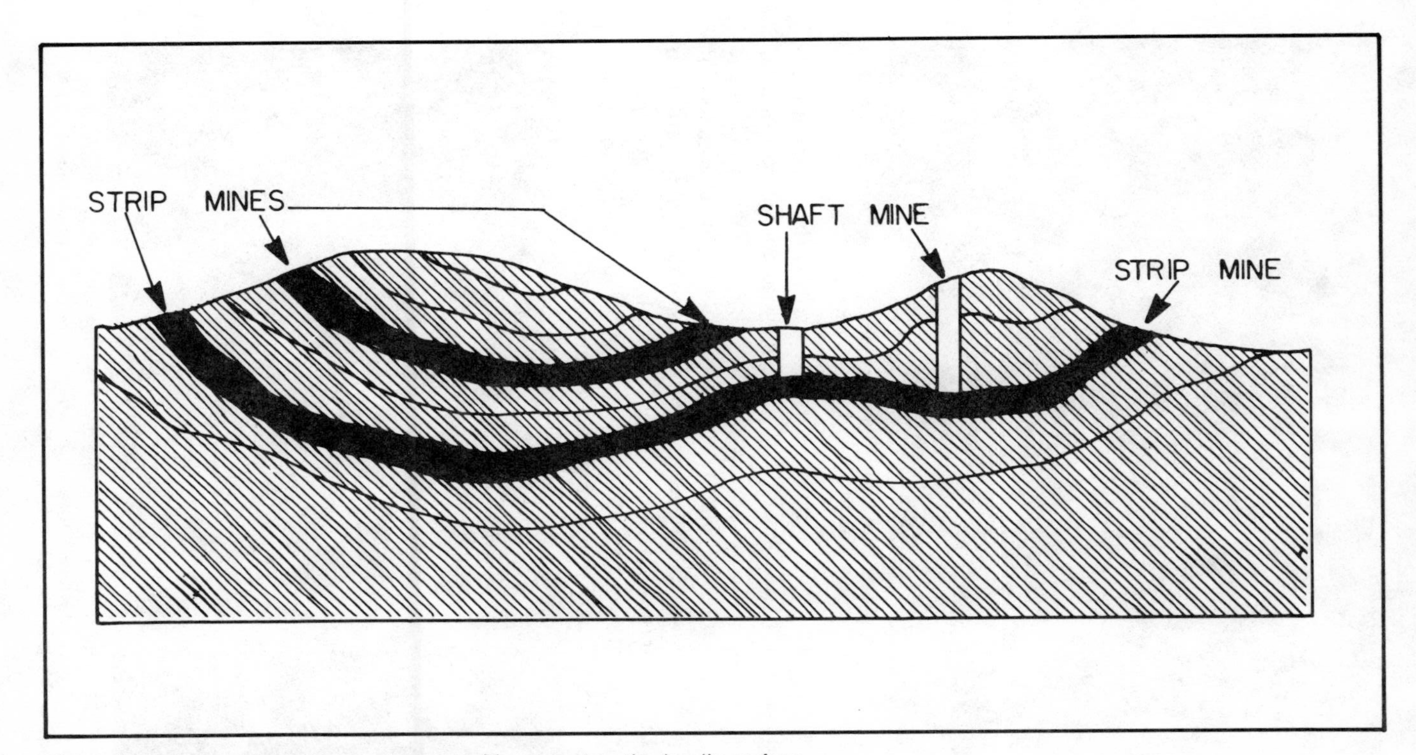

Fig. 1-12. Relationship of the coal seams with respect to the land's surface.

Fig. 1-13. A full-size continuous mining machine (courtesy of Bureau of Mines, United States Department of the Interior).

earth, must be reached by one of essentially three different techniques. The first is basic *shaft mining,* where a vertical hole or shaft is dug straight down until the seam is reached. Then the shaft is dug into the seam itself, following the entire seam until it expires. Separate shafts are dug at stages to facilitate ventilation as the extraction of the seam runs its length. The average depth of mines and the seams therein are about 280 feet in the United States, although in other countries seams as deep as 4,000 feet are

Fig. 1-14. A long row of hydraulically powered roof supports protects the coal cutting operation (courtesy of Bureau of Mines, United States Department of the Interior).

worked. The problem of cave-ins is reduced by leaving predetermined spots intact. Great advancements in mining technology have been made in recent years, but mining is still a hazardous occupation at best (Figs. 1-13 through 1-15). Miners earn every bit of their wages.

The two additional techniques of deep mining are slope and drift mining (Fig. 1-16). The actual terrain dictates the angle of

Fig. 1-15. Special tool for borehole mining (courtesy of Bureau of Mines, United States Department of the Interior).

Fig. 1-16. Researchers are working with concepts for inducing open pit slopes to fail, as a way of producing broken ore and waste rock at low cost, to allow recovery of ore that would normally be left in the pit walls (courtesy of Bureau of Mines, United States Department of the Interior).

entry to be made so that intersection of the coal seam can be arrived at such an angle that allows the equipment to be driven into the sloping entrance tunnel. Thus, tractive power can be used to drive the miners and equipment into the work area. The extracted coal can be removed without the inconvenience caused by the elevator system as required by the conventional shaft mine technique. Many times slope and drift techniques are used in strip mining. When surface digging equipment can no longer be adequately used to extract the coal, the operation then assumes the characteristics of a slope or drift mine.

After having extracted the coal, either by the surface or deep mining techniques, the coal is next transported to the preparation plant where the bulk is run through a continuous conveyor operation. The coal is screened according to its "as mined" size. The dust and carbon content is separated, delivering it directly to a waiting hopper which will retain it for pulverized systems and certain industrial uses. Medium-sized chunks are diverted to the washing process, where the coal is separated from any refuse and is further passed through sizing screens. The various sized pieces will go to respective hoppers based upon their dimensions as desired for specific requirements. The diverted coal is crushed to smaller pieces for use as stoker and nut size varieties. The coal, upon leaving the preparation plant, is now in sizes from pulverized, which is ¼-inch or smaller, up to and including lump coal, which is 5 inches and larger in size. The coal, having been separated from the greater part of its impurities through screening and washing, is now ready to be transported and be used as a combustible.

COAL STATISTICS

Basically, the operations as related are equally applied to both bituminous and anthracite coals. The extraction of anthracite is far less than the bituminous rank coals. In fact, for every 54 tons of bituminous mined, only one tone of anthracite is represented, primarily due to the limited distribution of anthracite deposits.

Statistically, in 1970 the total coal (both anthracite and bituminous) mined and processed exceeded 565,000,000 tons. Fifty-five percent used for electric power generation. 17 percent was used for coking purposes, and 16 percent was consumed by industry for general process and production. Ten percent was exported and the balance, approximately 2 percent, was burned for home heating. From every one ton of coal delivered for coking, which essentially consists of heating the coal in an airtight space and condensing its vapors, we can realize 1500 pounds of coke (used chiefly for heating and steel making), 8 to 10 gallons of coal tar (for further refinement and reduction to synthetic substances), 3 gallons of light oil, 5 to 6 pounds of ammonia and 10,000 to 12,000 cubic feet of coal gas.

Even more educational are the following facts, courtesy of the National Coal Association. America is second in the world's production of coal. The United States produces some 570,000,000 tons/year as compared to the estimated 670,000,000 tons/year of Russia. Of the United States' production, 97 percent was bituminous and only 2 percent was anthracite. West Virginia, Kentucky, and Pennsylvania mined 60 percent of the entire nation's output, and Pennsylvania supplies virtually all of the anthracite. anthractie.

UPDATE ON COAL AND ITS UTILIZATION

Through the courtesy of Jack Schneider of the Department of Energy, here is a compendium of information about the developmental status of our coal resources. The United States, as has been related, has vast resources of coal veins. Our reserves are equal to one-half of the world's known reserves. In addition, the United States enjoys five times the energy value from our domestic recoverable oil and natural gas.

Whether the United States can continue to maintain its advanced lifestyles that depend so much on the supply of energy will in the end be determined on how successfully and efficiently we utilize our available coal. Of a total of 1600 billion tons of identified coal resources (plus an equal amount believed to exit in uncharted areas), we have reserves of 434 billion tons which can be mined under today's economic and technological conditions. In 1975 the United States consumed and exported only 600 million tons.

Fig. 1-17. Coal hydrocarbonization facility (courtesy of Energy Research and Development Administration).

Fig. 1-18. This plant can process 72 tons of coal per day into 1.2 million cubic feet of gas, part of which is methanated to pipeline quality substitute natural gas (courtesy of United States Department of Energy).

At current rates of consumption, the United States has enough coal to last some 360 years, with present standard mining techniques and demand rates based upon current efficiency procedures. Several forecasts however predict that the demand will be tripled by the year 2000, if certain limiting technological and environmental problems are overcome. Of primary concern is the impact coal will have on our environmental situation. This situation has been only recently improving due to industry's adoption of more modern combustion and by-product techniques, which will adequately preserve the compromise of environmental protection and energy provision equal to our necessities.

Projections of our midterm (1975-2000) energy requirements will necessitate the establishment of an industry to manufacture

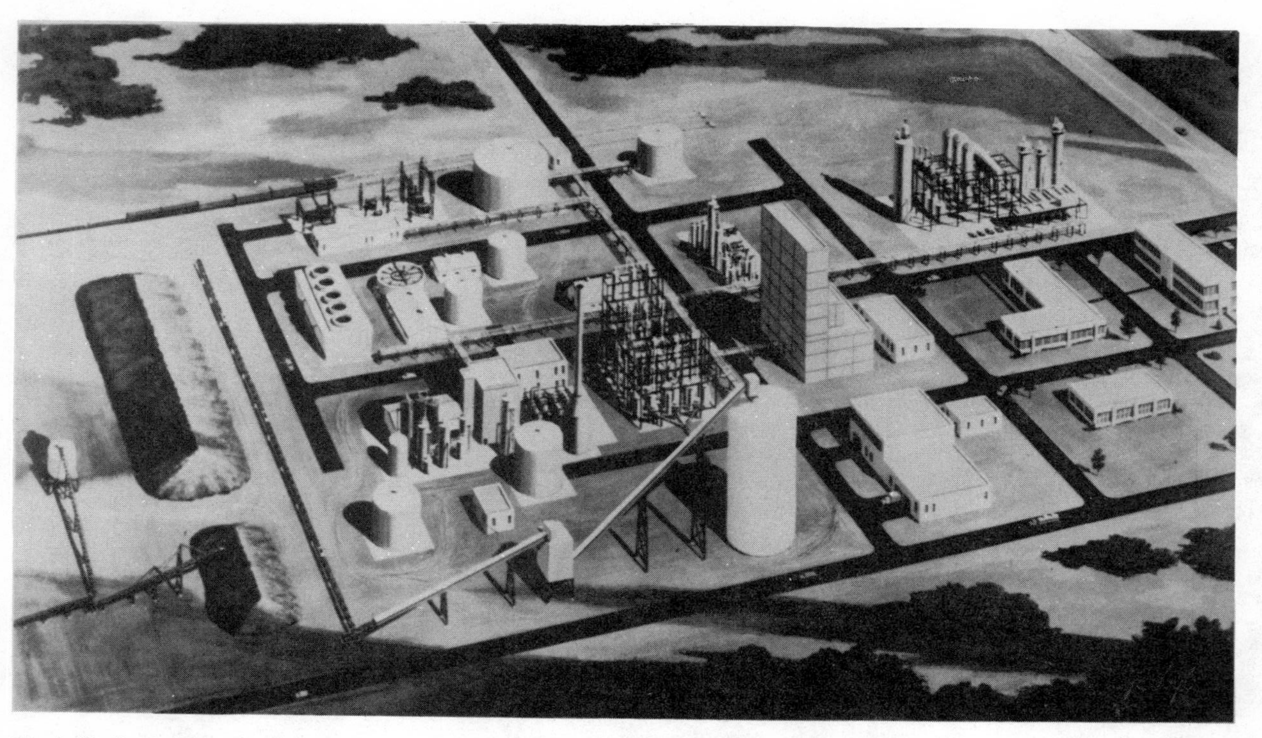

Fig. 1-19. Coal demonstration plant (courtesy of United States Department of Energy).

synthetic fuels from coal and oil shale (Fig. 1-17). Coal derived fuels, in both liquid and gaseous forms, would do much of the work previously done by oil and natural gas. Coal gasification will be needed to produce a high Btu substitute for natural gas (Methane) and also low Btu gas for boiler and other industrial fuel substitutes (Figs. 1-18 and 1-19). Liquid products made from coal and oil shale would also be refined into gasoline and heating fuels for domestic heat, industrial processes and electrical generation (Fig. 1-20). According to a 1976 energy survey, the United States will require a synthetic fuel process which will have a production capacity of not less than 5 million barrels a day by 1995, if oil imports are expected to be contained to current levels. This would then require the addition of 100 to 200 new synthetic fuel plants, each costing approximately $1 billion or more per plant.

Fig. 1-20. Solvent refined coal pilot plant (Courtesy of United States Department of Energy).

FLUIDIZED COAL BED COMBUSTION

Foremost in present development of coal combustion techniques, in addition to the advantageous benefits of coal pulverization and the cyclone furnace, are the *fluidized coal bed combustion process* and the manner of extracting gaseous coal from underground deposits. Many deposits of coal found east of the Mississippi River contain high levels of undesirable sulfur, which is a very objectionable contaminate. Its products of combustion yield sulfur dioxide with its related effects.

Nevertheless, combustion of high sulfur coals will be necessary to meet the estimates of future energy demands. Coal used for the generation of electricity alone accounts for some 60 percent of the objectionable sulfur dioxide admitted to the atmosphere annually. Techniques are therefore imperative to allow for the burning of high sulfur coal with minimum atmospheric pollution.

Experimentation with a fluidized bed of coal has proved by actual tests to be more than 90 percent effective in eliminating this sulfur dioxide from the products of combustion and thus the atmosphere levels of nitrogen oxide and particulate matter have also been reduced below the levels as provided by the Environmental Protection Agency (EPA).

The conceptual technology of fluidized bed material refinement was originated by the petroleum industry to produce high grade chemicals and gasoline. The Energy Research and Development Administration (ERDA) is presently working in joint concert with industry to get the implementation of fluidized bed combustion for power generation purposes. A non-technical description of the "fluidized" process is given as follows.

In a fluidized bed combustor, an upward stream of air suspends small particles of limestone (or dolomite) and coal (Fig. 1-21). The coal burns, generating heat in this fluid-like suspension. Heat is then transferred at high rates, even higher than conventionally realized by pulverized coal processes.

Essentially, limestone particles in the bed react with the released sulfur dioxide to form calcium sulfate. This inert substance ultimately is discharged with the ash, as opposed to releasing it to the atmosphere. If greater quantities of limestone are introduced, an even lower output of sulfur dioxide will be released to the atmosphere, which is the optimum end sought by this process. For example, when burning a high sulfur coal in a fluidized bed with a calcium to sulfur ratio of 2:1, the gases released to the atmosphere are well within the limits as set by the EPA for sulfur dioxide emission. Still lower sulfur dioxide values may be had by the increased adjustment of the calcium sulfur ratio.

The Argonne National Laboratory near Chicago is examining the behavior of mercury, lead, beryllium and flouride, all present in trace concentrations in the products of coal combustion. It seems these contaminates are also reduced in the fluidized

Fig. 1-21. Fluidized bed system (courtesy of Energy Research and Development Administration).

bed process of combustion, thus even enhancing the advantages of fluidized bed combustion even further (Fig. 1-22).

The fluidized bed process also transacts combustion at a much lower temperature than do conventional systems. The fluidized bed operates at an average combustion temperature of 1600 degrees Fahrenheit, as opposed to the average temperatures of 3000 degrees Fahrenheit. This reduction of the intensity of the combustion temperature causes a decrease in the amounts of nitrogen oxide released from the combustion process. The creation of this contaminate is chiefly affected by the temperature of combustion, the higher the temperature and the higher the incidence of nitrogen dioxide.

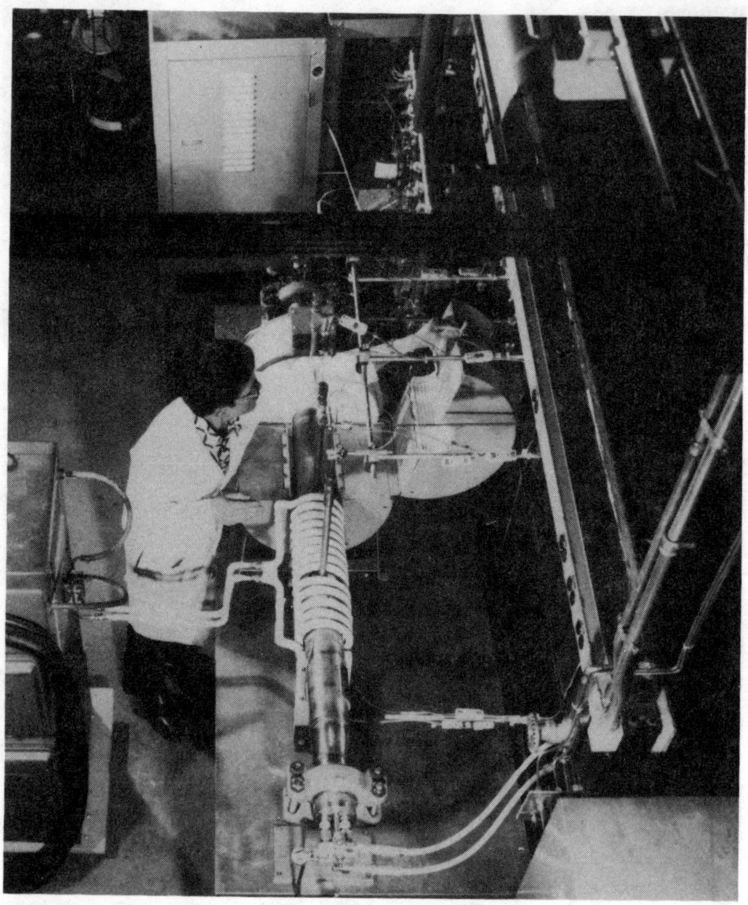

Fig. 1-22. Fluidized bed test (courtesy of United States Department of Energy).

UNDERGROUND COAL GASIFICATION

In this process coal in its naturally occurring solid form (in veins) is gasified deep in its deposits 500 to 3,000 feet below the surface thus increasing coal's recoverable portion. This form of gasified coal will patly serve as a substitute for natural gas and oil, for both power generation, industrial processes and chemical synthesis (Figs. 1-23 through 1-26).

Of the four trillion tons of coal, only one-half trillion tons may be economically mined by current techniques. The larger balance of 3.5 trillion tons of deposits are either too deep for extraction, in seams too thick or too thin for economic recovery, or so mixed with rock and entrained moisture to be profitably ex-

Fig. 1-23. High-Btu gasification pilot plant (courtesy of Energy Research and Development Administration).

tracted. At least one-third of this stated balance (1.2 trillion tons) could be recovered by the gasification process of extraction. The bulk of these deposits are in the Rocky Mountain states and in Alaska. These deposits are very deep in the earth and can only be had at great expense to the environment.

The environmental impact of recovery of these otherwise difficult deposits by the gasification process would be almost negligible. The conventional mining techniques cause a consequent "drop" in the earth's surface following the removal of coal deposits, whereas the removal by gasification only removes the relatively volatile percentages of the coal. The more solid and massive portions of the coal are left in place, resulting in a correspondingly less "void of removal" to be filled by surface drop".

The ERDA in its experimental plant in Laramie, Wyoming, has had encouraging results with the application of Gasification. By drilling vertical wells into the coal seam and by developing permeability by burning a connecting pathway between these drilled wells, access to the deposits results. This development of interior passageways by the burnt connecting paths as achieved by surface manipulation provides for the necessary introduction system. In this system air at low pressure and great volume may be injected to the deep interconnected tunnels and returned to the surface to yield a low Btu value gas which may be piped directly to power generating utilities or other utilizing systems.

Fig. 1-24. Another view of the high-Btu gasification pilot plant (courtesy of Energy Research and Development Administration).

The low Btu gas as offers only about 10 percent to 20 percent of the heating value of natural gas or coal gas as produced by its destructive distillation. As achieved by the first experiments of the ERDA at its Hanna, Wyoming, station, an average of 2 million cubic feet of gas was produced each day, for six months, (Figs. 1-27 and 1-28). The gas as recovered had an average heat value of 126 Btus (per cubic feet). The experimentation by the ERDA lasted a year and represented the longest single duration of gasification yet conducted, except by Russia, where three small electrical generating plants are operating on coal gasification products. Had the gas as produced from the Hanna project been used for the generation of electric power, about 1 million watts of electricity would have been generated, which is approximately

Fig. 1-25. Artist's concept of a low-Btu coal gasification plant (courtesy of United States Department of Energy).

enough to satisfy a town of 600 to 1000 people for the duration of the year's test.

Sandia and University of California Projects

The second test by ERDA which is at the time of this writing underway at the Sandia project in Albuquerque, New Mexico, is producing gas at the rate of some 4 million cubic feet each day

Fig. 1-26. A process development unit called Synthoil will convert coal into 30 barrels of low sulfur, low ash fuel oil (courtesy Energy and Development Administration).

Fig. 1-27. Underground coal gasification (courtesy of United States Department of Energy).

Fig. 1-28. The train is loaded with coal for a trip to midwestern utilities (courtesy of United States Department of Energy).

with a slightly richer heat value of approximately 150 Btu per cubic foot. This second test is one which has benefited by the results obtained empirically by the first test at Laramie. It is also being monitored by extensive instrumentation which will provide valuable data for future experimentation.

Another project of ERDA is being evaluated from the results of tests conducted at the Lawrence Livermore Laboratory by the University of California (Figs. 1-29 through 1-31). This experimentation is conducted in the following manner. An array of

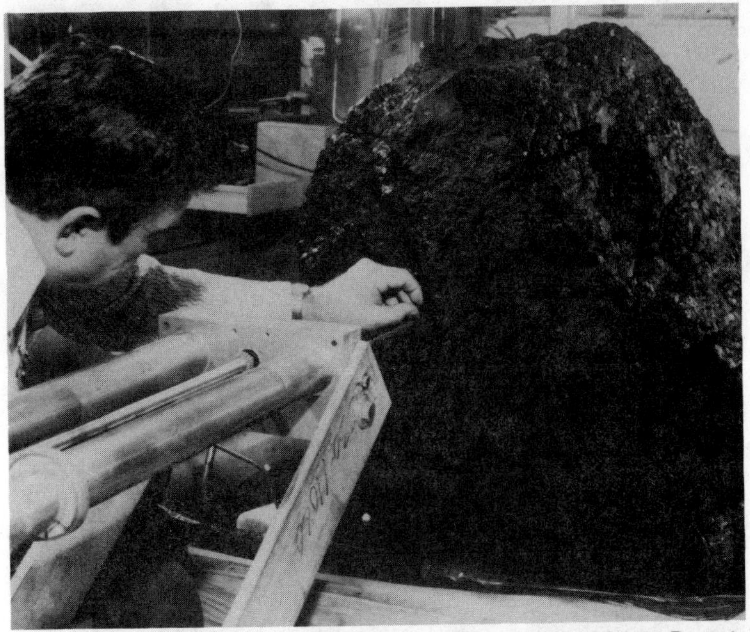

Fig. 1-29. Measuring coal permeability (courtesy of United States Department of Energy).

holes,each having a diameter of 24 inches are drilled into the earth and down into the coal seams, which are from 500 to 3,000 feet in depth. The holes are drilled in a diameter upon the surface of about 750 feet, in order to subtend an appropriate area of interior connection. Chemical explosives are then loaded down to the bottom of the holes and detonated, which will break up the coal seams into more permeable rubble, forming an underground bed. This bed will be injected with steam and oxygen from the surface and will cause the gasification of the coal in its veins. The gas so produced by this technique is expected to have an average value of from 300 to 400 Btus per cubic foot. This gas as released may be further enriched at the surface to provide a synthetic natural gas.

MERC Project

Collateral experimentation is being conducted by the ERDA at its Morgantown Energy Research Center (MERC) in West Virginia. The concept of running almost parallel boreholes to the much thinner coal seams is being tested to ascertain the feasibility of gasification for seams that would otherwise prohibit the economical extraction of such thin seams. This concept is applied as follows. A number of holes are bored vertically down to the

Fig. 1-30. Simulating underground coal gasification (courtesy of United States Department of Energy).

coal seams and then are bent to a direction which will actually parallel the run of the seam. These horizonally or seam plane holes are referred to as deviated wells and are drilled in the interior of the ground to lengths of several hundred feet in plane with the run of the seam. Using natural cracks in the coal seam and additional vertical wells if needed, the coal situated between such deviated wells are then gasified by the injection of air or air and steam to yield a low Btu gas for power generation. The deviated wells as bored expose large areas of coal and replace the greater number of holes as would be required by deep mining techniques, which would be unacceptable in populated eastern areas.

The MERC project people refer to their concept as the "longwall generator," due to its functionally similar aspects with longwall mining. Both produce a desirably low degree of surface drop to fill the void created by the removal of the coal as extracted underneath the surface. Thus, the gasification process of the extraction of coal serves the twin objectives of recovering otherwise uneconomically considered deposits and reducing the consequent surface disruption peculiar to other conventional methods of extraction. Extensive tests related to the gasification process are currently being conducted in respect to the pollution of both air and water used.

COMPOSITION OF PETROLEUM

Petroleum, which is the given name for Rock Oil, originated from organic matter deposited in marine sediments and is largely

Fig. 1-31. Coal gasification (courtesy of United States Department of Energy).

held in sedimentary rocks or pockets. There is still much unknown about the entire sedimentary cycle and origin of petroleum.

Petroleum is composed of many complicated compounds, mostly of carbon, hydrogen, and minor oxygen, nitrogen, and sulfur traces. The carbohydrate compounds appear in various series, each having intrinsic features. Some series members offer high gasoline content, while others have relatively little. Some members yield high lubricating oils, such as paraffin-based oils, which are light and of more fractionable value than asphalt-based oils. These are heavier, offer less gasoline and lubricant and are chiefly used as fuel oils.

The containing rocks of commercial oil pools are sands, sandstone, limestone, conglomerates and, rarely, fissured shale or igneous rocks. The enclosing sedimentary rocks are almost invariably marine in origin, although in rare cases fresh water sediments are the host rocks. Oil is found in rocks of all periods, from the earlies fossiliferous rocks (late Cambrian), almost 500 million years old, to the late geologic period, the Pliocene. Over all the world, beds of Tertiary age or those formed in the last 70 million years are the most prolific in yielding petroleum.

ORIGIN OF OIL AND GAS

By the overwhelming accumulation of evidence, petroleum is of organic origin from the lower forms of marine life. There still exist differences of opinion as to just what the organic material was and how it became converted from its plant and animal life forms into its present forms of petroleum and gas. The chief factors of the organic theory were deduced from the following evidences. Many oils are optically active, and only oils of organic derivation exhibit this characteristic. Nitrogenous compounds are constituents of petroleum, and in nature are confined to plant and animal life. Most oils contain chlorophyll porphyrins which are organic.

It is now generally agreed that organisms such as diatoms and algae that thrive abundantly near the surface of the sea are the most important sources of petroleum digestion. After the organic remains of these marine organisms precipitated to the bottom of the sea floor, bacteria commenced to convert this matter in its oxygen free environment to reduced portions of protein and hydrocarbons. The theory is also advanced that radioactivity had a contributory effect in this molecular rearrangement, as radioactive substances such as uranium have been closely found in and around geologic petroleum pockets. By laboratory experiments, it has been proven that radioactive emanations will convert some hydrocarbons to oily substances. During this formation and conversion, natural gas is also formed. This is predominately methane, which is also formed in peat swamps.

ACCUMULATION AND MIGRATION OF OIL

The transformation of organic matter does not in itself give rise to an oil pool. The oil itself would remain interlayered and sponged up in the sedimentary rocks of its birth were it not for certain forces or phenomena which, after acting upon the created petroleum, cause it to migrate into pools or pockets of great enough volume to make drilling for it worthwhile. The forces which cause migration are compaction of the oil entrained muds, buoyancy, capillarity, gravity, water currents, gas pressures, cementation and bacterial action.

Compaction is believed to be the chief force of migration, causing the oil and gas to move generally in a lateral and upward direction from its source rocks into carrier beds. Source muds and sediments may contain up to 80 percent water. The weight of sediments deposited on top of source muds gradually compacts them. The entrained fluids are squeezed out into places of less pressure, such as the pore spaces in sands. It is estimated that compaction removes upwards of 50 percent of the water when burial has reached 1,000 feet, and some 85 percent of water at depths of 4,000 feet. As burial of source muds progresses, there appears to be an almost continuous migration of the fluids into more porous beds.

Capillarity is a stimulus to migration. Where oil-wet shales are in contact with water-wet sands, the water will move by capillarity from the coarse sand pores into the fine pores of shale, displacing the oil into the adjacent sandstones.

Buoyancy adds its affects by virtue of the weight differences in oil, water and gas. The oil and gas naturally rise to the surface of any enclosure of which these three substances are a part.

Gravity functions where no water is present. The oil gravitates to the bottom limit of porous rock until impermeable layers are met.

Currents of underground water will flush water along with its self, redepositing the oil in entrapments where the oil will separate itself from the water stream upon entering appropriate chambers or receivers of rock holds and faults. Water can quite often be used to assist the release of oil from certain pockets during drilling operations.

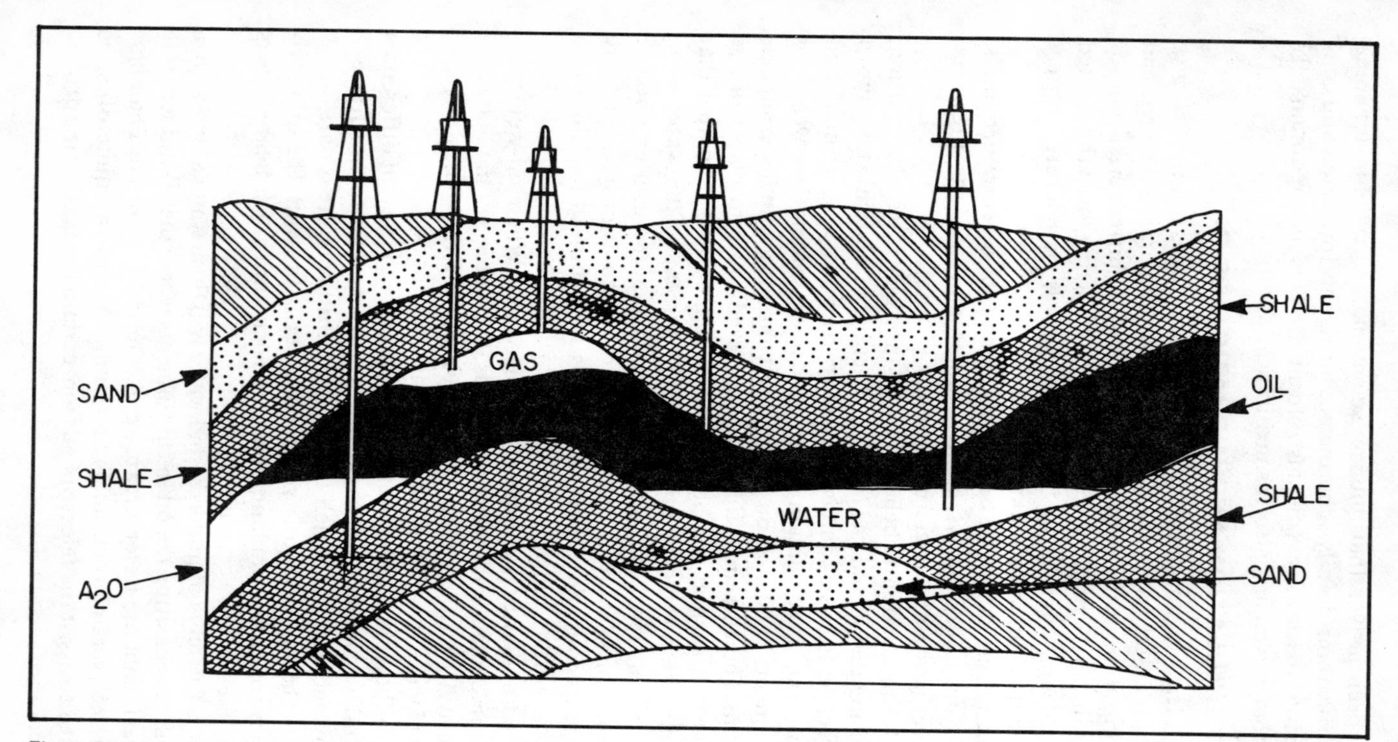

Fig. 1-32. Layers of material to be drilled through to reach oil vary considerably.

Gas pressure aids the movement of oil by combining with a given volume of oil, consequently making that given volume lighter. Also the movement of oil upward from its point of generation carries along with it a small but steady amount of oil.

Bacteria lend themselves to the movement of oil by creating gas pressure during their functions as decomposing agents. They create voids in the limestone by dissolution to provide porosity for the oil to seek, and the gas exerts its pressure to drive this oil into eventual pockets. Some bacteria are believed to have an affinity for certain solid rock, thus driving off the oil in displacement.

Ultimate accumulation is brought about by the eventual entrapment in rocks which, by nature, are both porous and permeable. The most suitable rocks for forming oil reservoirs are sands and sandstone. Cavernous limestone rocks often yield prolific flows of oil, as in some of the Mexican and Iranian fields.

A confining cap rock must be present to retain the accumulated oil; shale and clay are the most commonly found cap rocks. Good cap rocks form effective seals to underlying oil and gas for scores of millions of years, but poor permeable rock will permit the slow escape and loss of mobile hydrocarbons.

REMOVAL OF PETROLEUM

When one speaks of removing oil from its entrapments, we immediately think of the last "gusher" we saw on television. This gusher is only the result of the trapped gas above the oil itself. Although the oil may be under some pressure, it is never under enough to really pump itself out. After the cap rock is pierced by the drilling operation and the gas pressure has resided, the pocket must be pumped from that point on. This is done either by mechanical pumping or by pressurizing the pocket by letting water down into the confining pocket rock, thus literally "floating" the oil up out of the depths.

In Figs. 1-32 and 1-33 are sketched geologic "stratigraphs," representing the most common occurrences of oil fluids. We see in Fig. 1-32 that the layers of material to be drilled through vary considerably, even with wells located exactly adjacent to each other. Even under the best of today's sonar graphs and depth scanning electronics, the actual placement of the main reservoir is pretty much a hit or miss trick.

An examination of Fig. 1-33 will reveal the pocket-like appearances that oil occurs in. Oil invariably has a compressed pocket of gas on its surface.

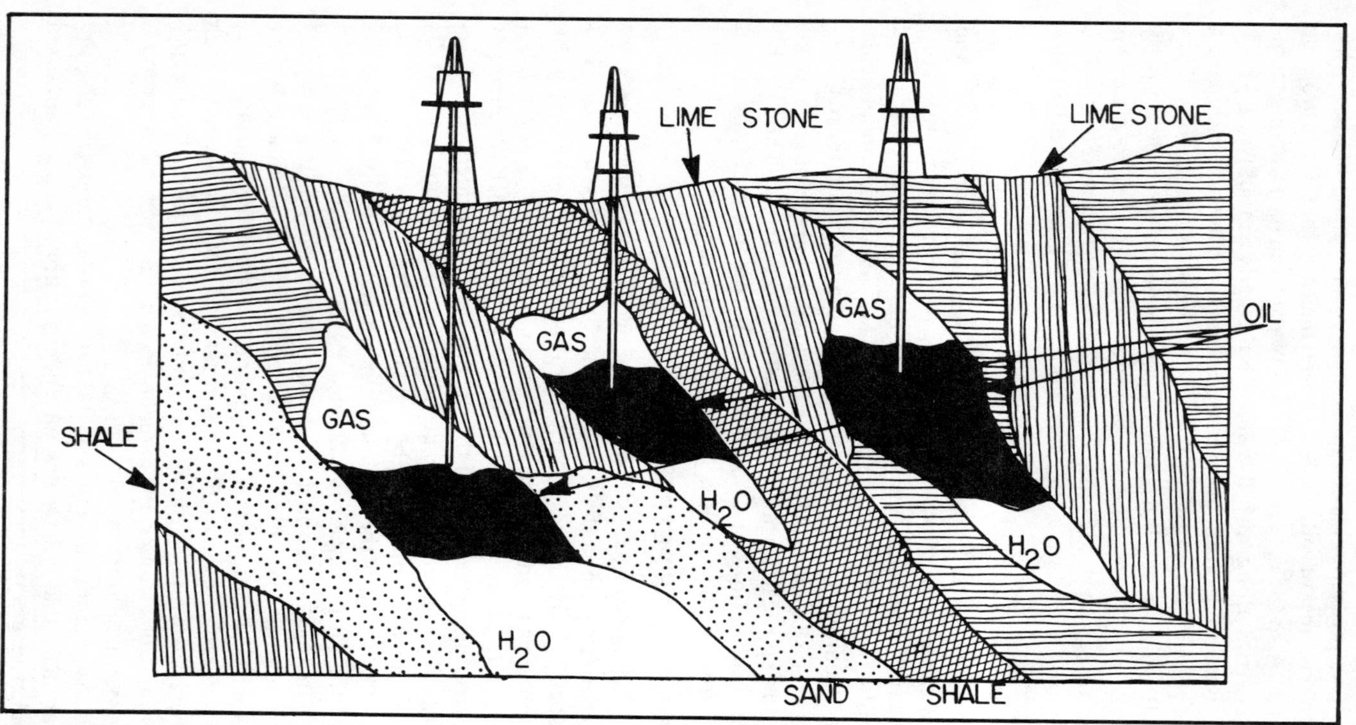

Fig 1-33. Oil occurs in pockets.

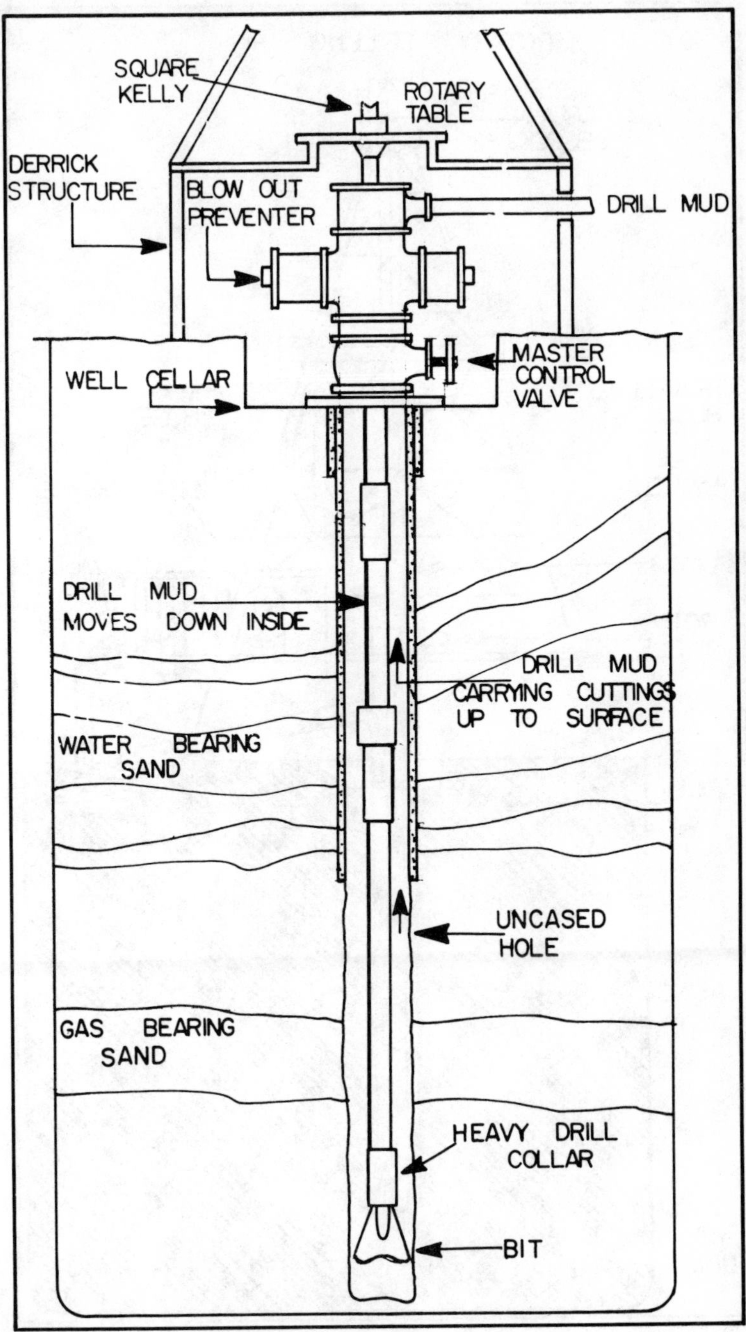

Fig. 1-34. The basic rotary drilling setup.

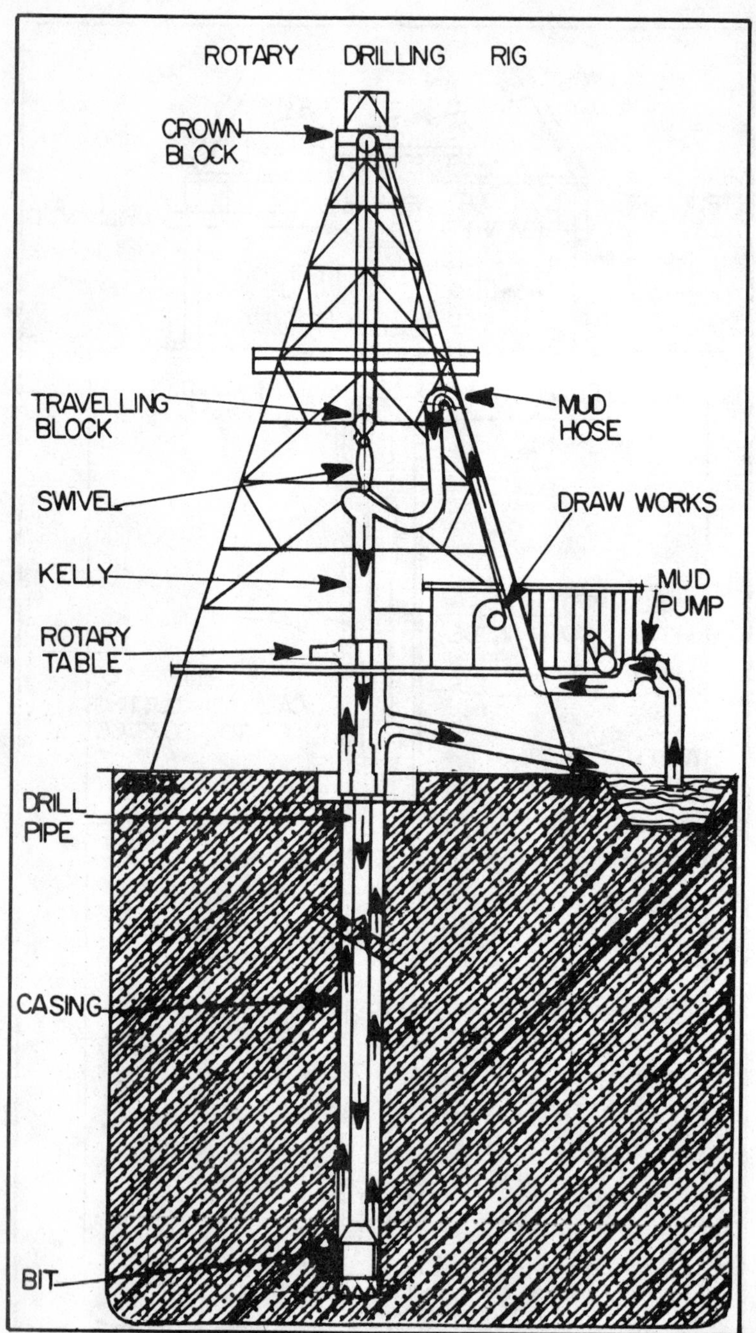

Fig. 1-35. Basic drilling rig.

Fig. 1-36. Offshore oil drilling (courtesy of United States Department of Energy).

A look at Fig. 1-34 will illustrate the basic rotary drilling set up. Although the rigs used are far more sophisticated, this illustration represents the fundamentals of rotary drilling (Figs. 1-35 through 1-37). Using this rotary method of drilling, holes as deep as 20,000 feet may be bored.

METHODS OF PETROLEUM EXTRACTION

Dating back to the first commercial oil well, which was sunk back in 1859 near Titusville, Pennsylvania, the technology of locating, drilling, pumping and refining oil has evolved to one of the most exact sciences as practiced by modern day man. Petroleum had in fact been released from the earth much earlier than 1859. The Persians used oil as far back as 6000 years ago. The Venetian traveler Marco Polo witnessed a fountain from which oil sprang in great abundance, inasmuch as 100 ship loads might be

Fig. 1-37. Strategic petroleum reserve (courtesy of United States Department of Energy).

taken from it at one time. The people of Baku, which is on the Caspian Sea, used this oil for burning.

Thus, as the early explorers of North America discovered, petroleum at one time seeped out of the ground naturally. It was quite often found floating on the surface of lakes and rivers.

In 1846 Dr. Abraham Gesner of Nova Scotia perfected a technique of obtaining oil from coal, which was named *kerosene.* A company was formed to manufacture it. This availability on a commercial basis of a substance which offered desirable burning for illumination and heating created an interest in liquid fuels. Attention was returned to the naturally occurring oil which was only years before considered an obstacle to be dealt with in pursuit of other substances.

One such place where oil was seen and noted in abundance was at Oil Creek, Pennsylvania (so named because of the oil found on the surface of this creek). In 1855 Professor Benjamin Silliman of Yale College, after conclusively proving the worth of petroleum as a substitute for coal oil, then proceeded to form a business group to determine whether petroleum could be profitably taken from the ground for commercial endeavors.

This question was answered when the business group engaged a retired railroad conductor by the name of Edwin L. Drake. Drake, with the contracted assistance of William A. Smith, an expert in salt well drilling, wasted no time in "bringing in" the first oil well in the United States during the summer of 1859.

PETROLEUM REFINEMENT

Having "brought home" a gusher or oil find is only the beginning of a long and technically detailed journey which the "raw black gold" will travel before being refined for industrial or commercial use. The oil is first drawn from its earthen hold by its own pressure caused by the gas head pressure above it, forcing itself up to the drill platform take-off. Or it is pressured out by admitting gas or water from above, traveling down the bore tube by separate lines.

The first stop for the newly released oil is the gas-oil separator, where the oil is separated from any entrained gas it may have (Fig. 1-38). This facilitates its transfer to the field storage tanks seen about drilling operations, known as the "tank farm." The field pipeline pumping station then drafts the oil from the field storage tanks and will transfer the crude to the loading docks of marine transport terminals to be shipped in oil tankers. Or it will pump the crude directly to the refinery. Even though the refinery may be many miles away, the pipelines are more than adequate to pipe it there. The United States has thousands of miles of piping to handle the distance from oil field to refinery, as there has been such a particular pipe line from Texas to New York since World War II.

Once arriving at the refinery site, the crude is stored in the refinery tanks. It is mildly heated and screen filtered before it is delivered to the main "still," where fractional distillation takes place. Figure 1-39 diagrams the basic construction, in theory, of the conventional fractionating tower. Figure 1-40 shows us the basic inside story of the catalytic cracking unit, which was only block diagrammed in Fig. 1-38.

FRACTIONATING TOWER

We observe in Fig. 1-38 that the crude oil, after being tempered and screened in the refinery storage yards, is now

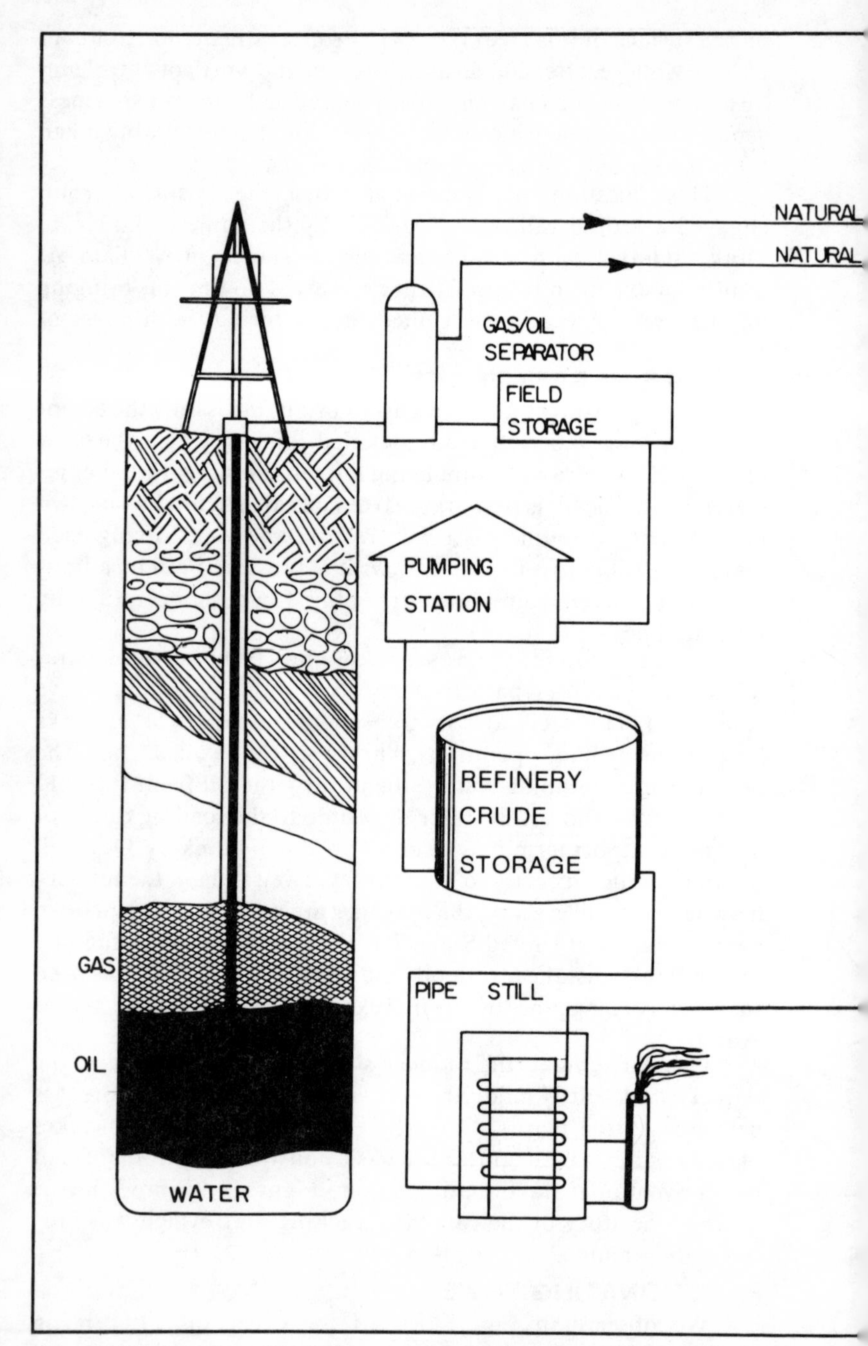

Fig. 1-38. A flow chart illustrating the refinement of crude petroleum.

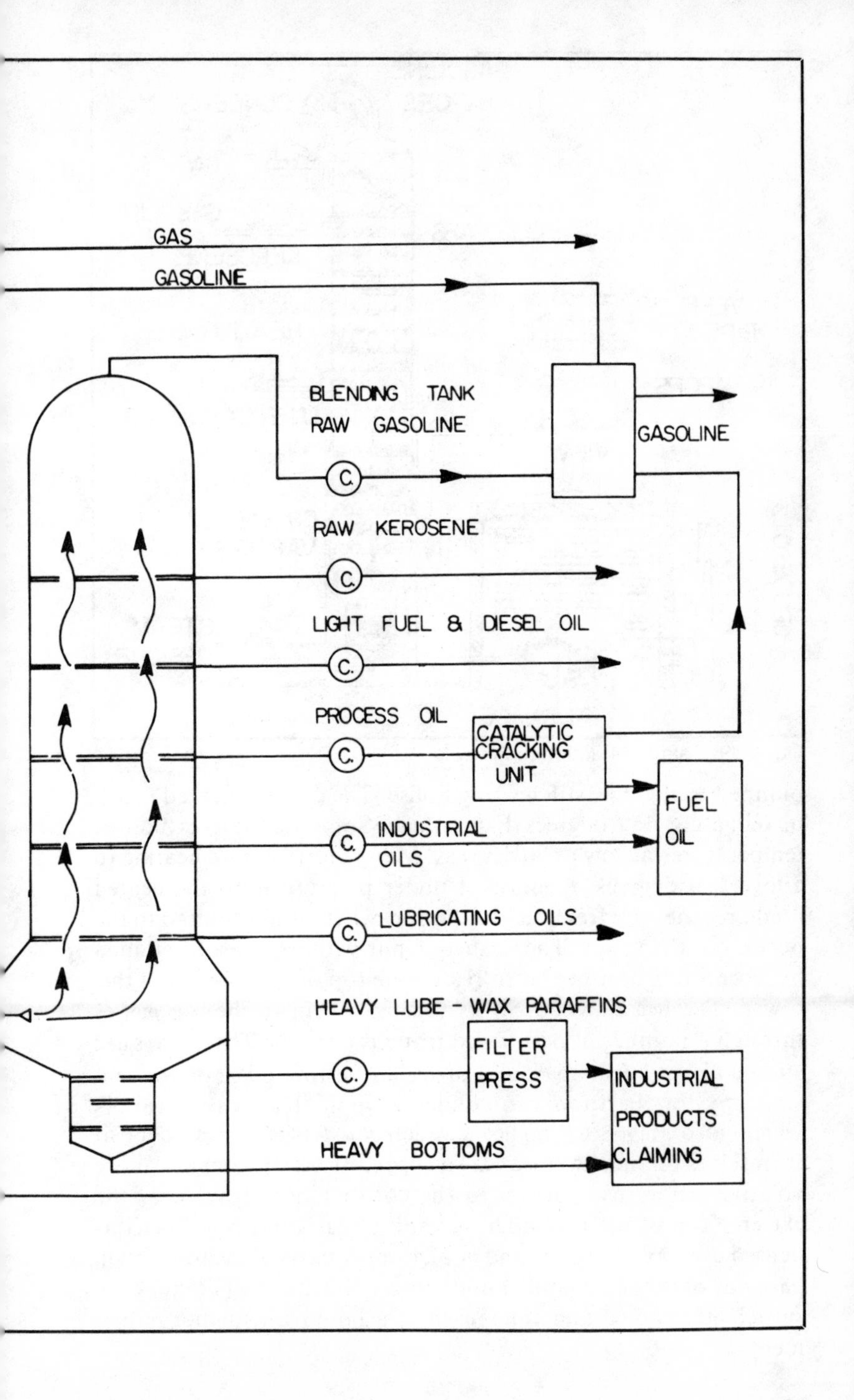
GAS
GASOLINE
BLENDING TANK
RAW GASOLINE
C.
GASOLINE
RAW KEROSENE
C.
LIGHT FUEL & DIESEL OIL
C.
PROCESS OIL
C.
CATALYTIC
CRACKING
UNIT
FUEL
OIL
C.
INDUSTRIAL
OILS
C.
LUBRICATING OILS
HEAVY LUBE - WAX PARAFFINS
FILTER
PRESS
C.
INDUSTRIAL
PRODUCTS
CLAIMING
HEAVY BOTTOMS

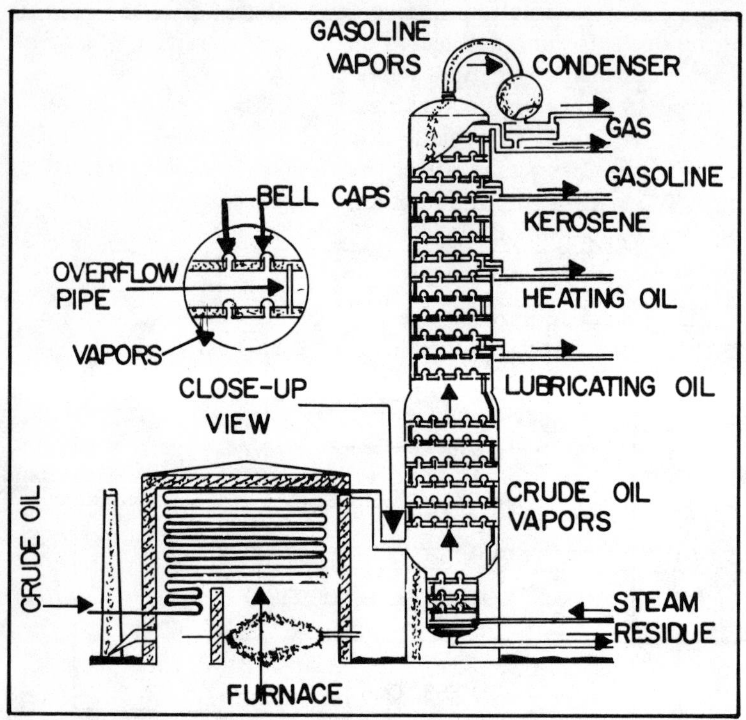

Fig. 1-39. Detail of fractionating tower.

pumped to the pipe still heating house. The crude is heated to approximately 650 degrees Fahrenheit. Some derivatives boil at temperatures as low as 70 degrees. Many more require heating to above 600 degrees, being kept under pressure until the heated crude reaches the fractionating tower, where it is admitted to the bottom of the tower. The flashed vapor with its multi-fractioned hydrocarbons now begins to rise to the top of the interior of this tower. The heated vapor in its vertical ascent to the top passes through a number of perforated troughs or trays. These trays are spaced about 2 feet apart and provide a resting place or condensing tray for the vapors to condense upon. The heavier vapors which also have the higher condensing temperatures deposit themselves on the lower trays. The progressively lighter and more volatile vapors pass on up to the cooler higher trays near the extreme top of the tower.Those vapors that still cannot be condensed even at the higher and cooler upper parts of the tower then pass out of the top chamber and into a condenser, where they are sufficiently cooled and condensed to a liquid for further refinement.

The heaviest fraction, known as residue, is collected at the bottom of the tower and released through a bottom valve for use as a heavy carbon-rich fuel, or as asphalt. The gas/oil tray is delivered to the catalytic cracking unit. The initially condensed oil is further refined by actually cracking the molecules to lighter ones. This process was creatively introduced to yield more aviation gasoline from the nominal amount previously realized by fractional distillation alone, which yielded only raw gasoline.

PROCESSED DERIVATIVES

From the continuous process fractionating method of refining crude petroleum and its many additional processes, we may expect to obtain many distiliates and processed derivatives. *Gasoline* is the most sought after product. Approximately 90 percent of the gasoline is consumed by automobiles and commercial vehicles. A part of the gasoline initially distilled is immediately piped to the blending operation where, along with parts of *kerosene*, a more volatile mixture is achieved for aviation and jet fuels. *Kerosene* is a lightly volatile combustible, of which approximately 5 percent is used for domestic fuels, farm equipment and as a part of aviation and jet fuel mixtures.

Diesel fuel is chiefly used for motive power fuels and for light domestic oil burners. Consisting of fuel oil classifications from grade #2 up to and including grade #6, *light* and *medium fuel oils* are the mainstays of industry. Residual and heavy fuel oils require special industrial burners. With the advent of special high-temperature and corrosion-resistant *lubricants,* engine designers were given greater latitudes in their selection of metals. They specified close tolerance fits of bearings for high speed and high-temperature super heat steam and gas turbines. Greases are essentially low-speed, heavy lubricants, resistant to high temperature dripping and providing long term uninterrupted applications.

Petroleum coke, a by-product of cracking, is used as a heavy fuel. Liquified petroleum gases have gained popularity for domestic uses and for many farm and agricultural operations. Petrochemicals constitute a great variety of organic and inorganic compounds: *ethylene, propylene, butylene, isobutylene, cyclohexane* and *phenol.* The inorganic compounds include *ammonia* and *hydrogen peroxide.* Petrochemicals also yield secondary petrochemicals such as *synthetic detergents, synthetic rubber, butadiene, styrene, nylon, orlon* and *dacron.* Included are many polyethylene plastics, disinfectants, antiseptics, shampoos, van-

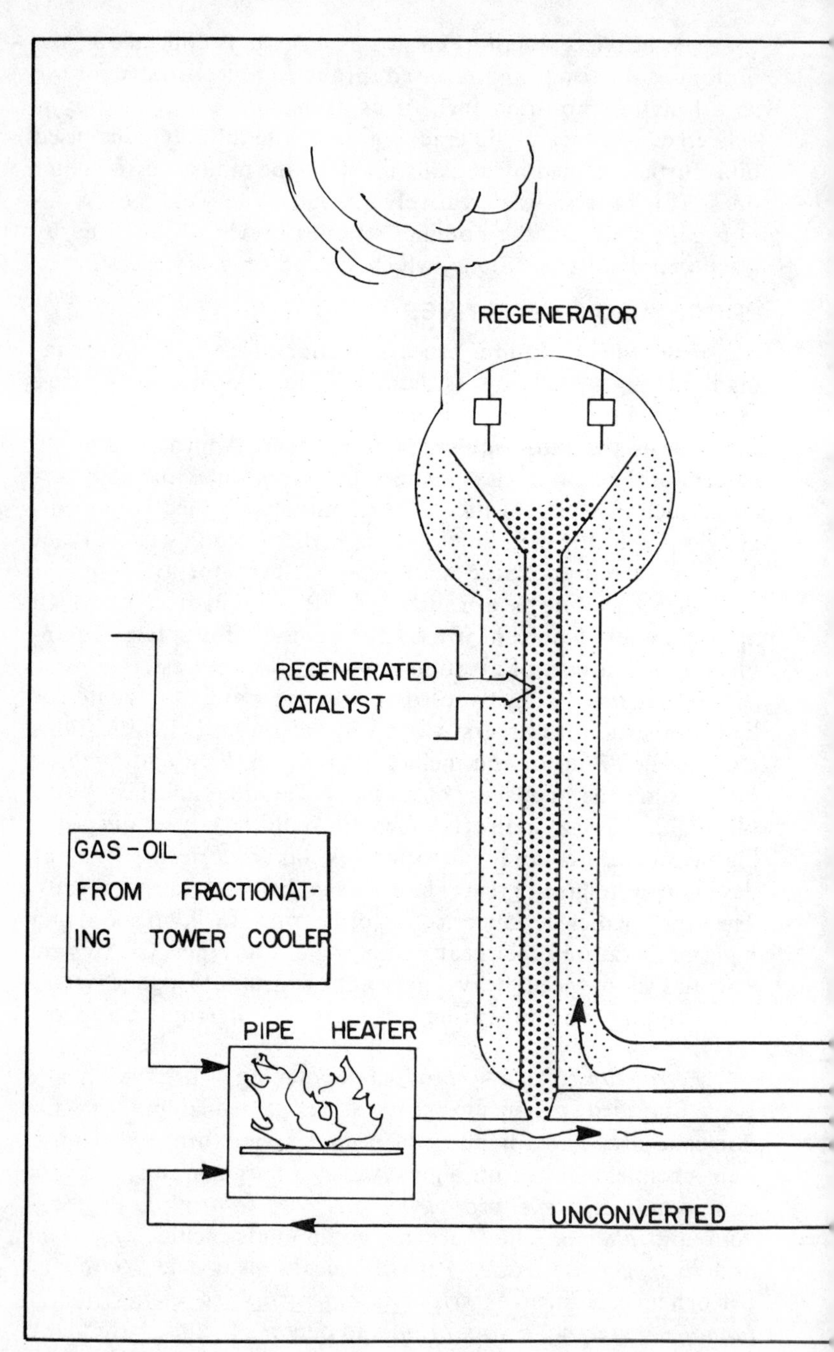

Fig. 1-40. Catalytic cracking unit.

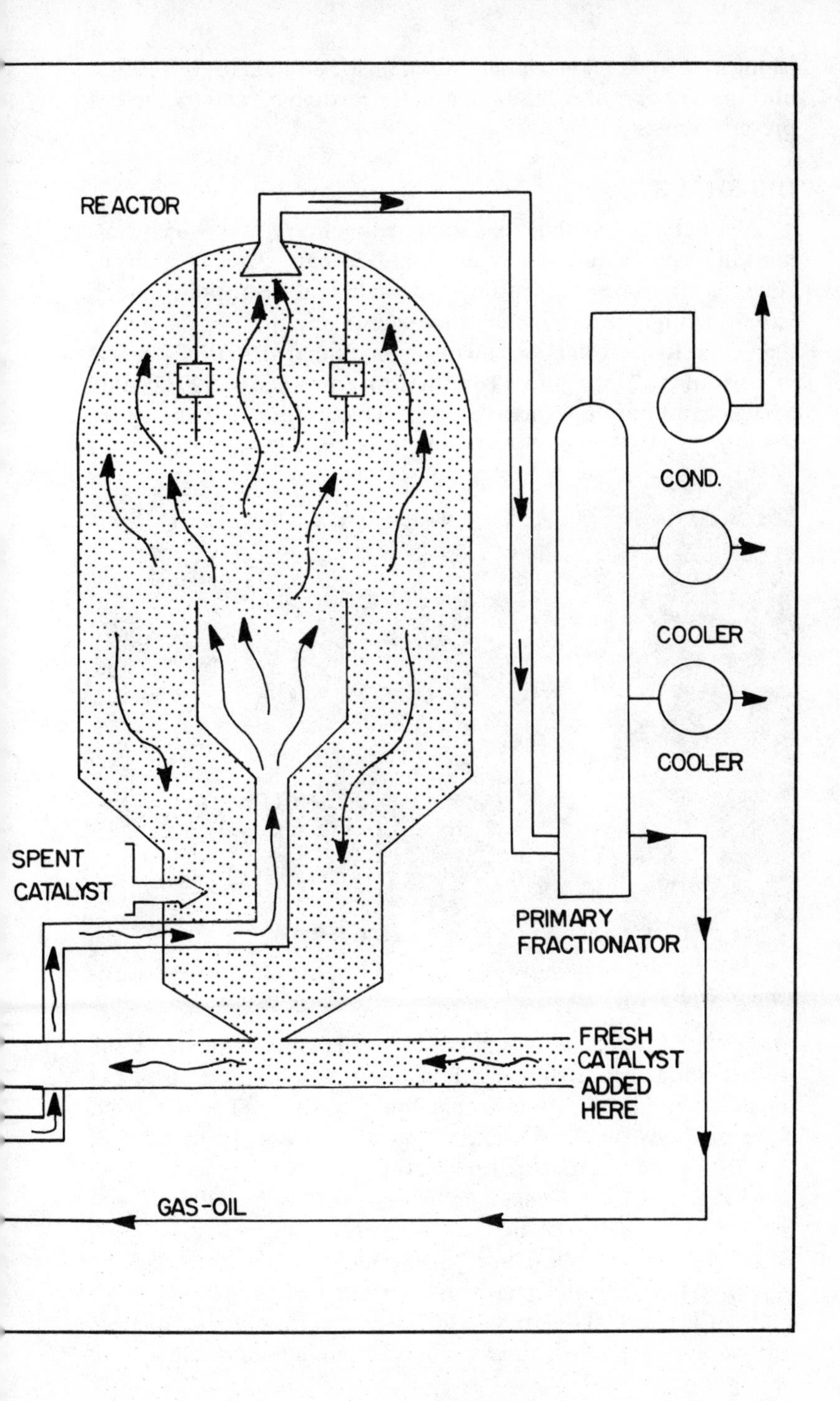

REACTOR
COND.
COOLER
COOLER
SPENT
CATALYST
PRIMARY
FRACTIONATOR
FRESH
CATALYST
ADDED
HERE
GAS-OIL

ishing and cold creams, hand lotion, lipstick and many cosmetics, not to mention the many synthetic medicines and treatment prescriptions.

OIL SHALE

Essentially, oil shale is a dark sedimentary rock which contains in its pores sufficient organic matter that will yield vapors of condensable oil upon heating. Oil shale had been pretty much left alone until now, in spite of the fact that in one deposit alone, the Green River formation, there are an estimated 960,000,000,000 barrels. In the Mahogony Ledge there are another 90,000,000,000 barrels, an enormous reserve when we consider that the entire world's production of oil from 1859 to 1957 amounted to

Fig. 1-41. In-situ processing of oil shale (courtesy of United States Department of Energy).

some 90,000,000,000 barrels. In fact, the best estimate is the United States' shale rock contains about 150 billion tons of extractable oil. It is so unfortunate that we haven't investigated more methods for the oils extraction when it is so badly needed.

However, much experimentation has been conducted to make the yield from oil shale more prolific (Figs. 1-41 and 1-42). In 1944, the Swedish Shale Oil Company applied what is known as the Ljundstrom "in-place" method of heating the shale by implanting electrical elements in the shale bed to raise the bed's temperature to 400 degrees celsius over a period of five months. The vapors were then drawn out and condensed, yielding the

Fig. 1-42. Gas combustion retort at an oil shale facility (courtesy of United States Department of Energy).

following per square meter of bed: 515 liters of gasoline, 160 liters of kerosene, 350 liters of heating oil, 80 liters of liquefied gas, 650 liters of sulfur-free gas, 350 kilograms of sulfur and 8 kilograms of ammonia. This Ljundstrom process was discontinued right after World War II when petroleum once again became available in Sweden, but it was productive and did work. The most recent attempt to utilize the virtually ignored oil shale potential is on behalf of the Standard Oil Company of the United States.

NATURAL GAS

In its unrefined state, natural gas is extremely rich in hydrocarbons and may be burned as collected. However, it is usually refined by condensing and further processing to remove some hydrocarbons and render the gas mechanically and chemically cleaner. As drawn from the deposits, it may be burned or further refined to yield liquefied petroleum gas (LPG) (Fig. 1-43). Natural gas generally offers from 1000 to 1200 Btus per cubic foot of available heat. When refined to butane and propane, it then yields between 2600 to 3200 Btus cubic foot as the LPG contains the richest hydrocarbons found in the petroleum gas. Basically, natural gas contains from 80 percent to 95 percent methane; the balance is made up of other hydrocarbons. Although there are a great many varieties of gases, only that gas actually occurring naturally in the petroleum pocket, either by itself or in contact

Fig. 1-43. This tanker transports natural gas in liquefied form (courtesy of United States Department of Energy).

with petroleum, will be recognized as natural gas. Other gases such as manufactured gas, water gas, producer gas, refinery gas, coal gas, etc., will be treated as primary fuels, but they are not actually resources in any naturally occurring form.

Natural gas may also be added to other by-product gases to enrich the total Btu content, or burned by itself. Generally, the gas as discovered upon breaking through the cap rock is stoped, piped to containing storage tanks, or left in place to be drawn from later. Natural underground pockets are also utilized for the containment of natural gas, until arrangements can be affected for its further distribution and use.

SELECTION OF PRIMARY FUELS

In assessing our natural resources from an engineering viewpoint, we're principally concerned with those commodities that may be either directly or indirectly converted into heat-producing matter or primary fuels. Thus, we view a forest as so many cords of dried Btus, mountainous mineral deposits as so many tons of calorie-yielding coal, and certain geological formations as offering potential petroleum and gas pockets for the extraction of their respective hydrocarbon-rich contents. As the artist and romanticist sees the beauty of nature, we as engineers and standard bearers for our technological society must appraise these natural resources for their chemical and calorific characteristics.

To date, the selection and utilization of our resources has not been indiscriminate. It has largely been based upon economics, as dictated by the proximity of a power plant's geography in respect to a forest, a coal-yielding area, or the plant's serviceability to railroads or highways.

Since the advent of large bulk transportation systems, newer and more economically superior means of coal mining techniques, more petroleum "finds," and faster operations for drilling out the crude oil and natural gases therein, we may base our selections for our primary fuels upon more advantageous considerations. These include: ecological effects relating to the products of combustion in respect to the environment, the competitive prices offered between coal, oil and gas suppliers, the reduction of manpower hours as made possible by the use of fuels such as oil, gas and pulverized coal, and the total efficiency offered by some fuels in respect to the possible mixture or additive aspects of the plant's by-product with a fuel.

Table 1-2. Comparative Evaluation of Fuels.

TYPE OF FUEL	BTUs/LB. or CU. FT. OR BY GALLON	PREPARATION FOR COMBUSTION	BURNING EQUIPMENT REQUIRED	PRODUCTS OF COMBUSTION
COAL				Ranging from
Anthracite	10,000 to 14,000/Lb.	Very minimal to	Moderate, to	very detrimental
Bituminous	11,400 to 14,600/Lb.	very extensive	extensive, from	sulfurous with
Sub-Bituminous	8,600 to 9.740/Lb.	Depending on	simple hand	heavy Carbon Monoxide
Lignite	7,300 to 8,000/Lb.	volume of Coal	fired plants to	oxide, to moderate
		required.	automatic stoker	smoke and pollutants.
			fired versions.	
OIL				
Grade #1	137,400/Gal.	None at all	Pot or Gun type	Depending upon the
Grade #2	139,600/Gal.	None at all	Pot or gun type	sulfur content of
Grade #3	141,800/Gal.	Room temp. req'd	Pot, Vapor, or Gun	oil, no objectional
Bunker-A, #4	145,100/Gal.	Room temp. req'd	Med or High Gun	products of combustion
Bunker-B, #5	148,800/Gal.	Must be preheated	High pressure Gun	or very little smoke
Bunker-C, #6	152,400/Gal.	Must be preheated	or Rotary Type	or visible soot, can
Residue	160,000 and up.	Extensive Prep.	Special Burners	contribute to pollution.
GAS	**BTU/CU. FT.**			
Natural Gas	950 to 1200	None at all	Conventional *	Virtually the cleanest
Butane (LPG)	3200	Vaporization	Conventional *	burning of all fuels,
Propane (LPG)	2550	Vaporization	Conventional *	Produces CO^2 and H^2O
Producer Gas	130 to 300	Adjusted *	Conventional *	with proper flame
Carb. Water Gas	500 to 600	Adjusted *	Conventional *	adjustment, no smoke
Coke Oven Gas	580 to 600	Adjusted *	Conventional *	or pollutants.

Note (*) Adjusted--Mixed or proportioned with other gases to provide optimum requirements.
Conventional--Can be efficiently burned with any commercial burner of any design.

In Table 1-2 is a comparative evaluation of the three primary fuels with which we will be dealing. Wood has been deliberately omitted because of its virtual disappearance from engineering use. Oil shale has not been included because the oils produced from them are treated as refined petroleum products and will be analyzed accordingly.

Foremost in our considerations for the selection of a fuel source will be the gross expense, which includes the transportation costs. Also important are the investment (initial and renewal) costs of any preparatory equipment required for the fuel's preparation before firing, the maintenance costs required for the fuel burning and preparation equipment, and a continued and uninterrupted supply of the fuel.

LIFE SUPPORT SYSTEMS

Lastly, but certainly with tantamount importance, is the role our selected fuel will assume in the very real life support system that we are all, as biological creatures, affected by. By life support system, I mean that total and interdependent "chain" or "cyclic system" by which we contribute, either as an asset or a liability, the by-products and end-products of our energy conversion techniques.

Figure 1-44 shows an ideal life support system showing the preferred phases or roles dictated by their respective functions. It will be noted in the ideal system that there exists no intermediary phase of contaminate reduction. This absence of the contaminate reducing phase is by virtue of the fact that we would hypothetically be using a fuel which yields no contaminating or polluting by-products.

CONTAMINATE REDUCTION

Also illustrated is a practical life support system (Fig. 1-45), whereby we have included the required intermediary phase of Contaminate Reduction. Contaminate reduction is achieved or brought about by combining the contaminates and pollutants with the required materials necessary for the eventual and ultimate return of these artificially created by-products to their natural slots in our earth's storage system. The material and matter used by this theoretical reduction phase comes from other natural phenomena capable of being destructively reduced or rendered compatible to the ultimate return of such contaminates to their neutral forms or compounds. These particular contaminates will correspondingly destructively associate themselves with

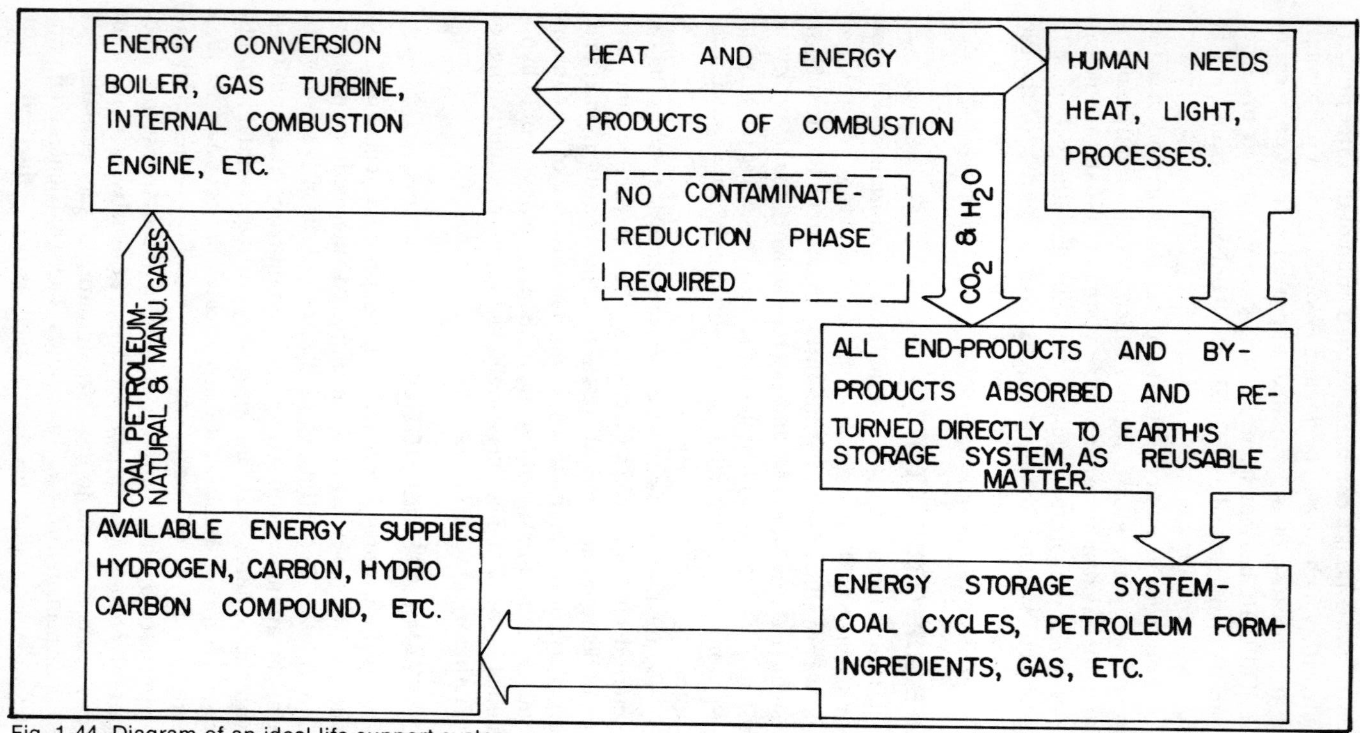

Fig. 1-44. Diagram of an ideal life support system.

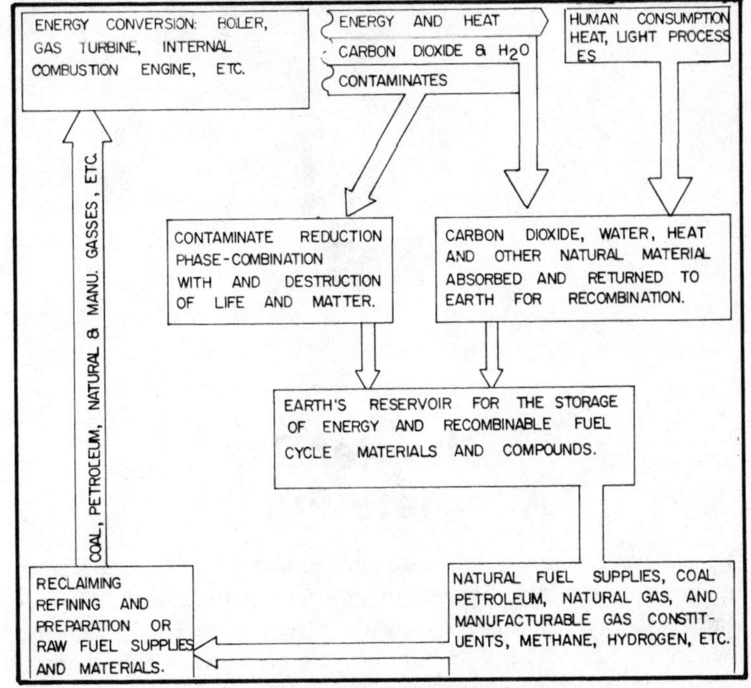

Fig. 1-45. Diagram of a practical life support system.

human and animal bio-functions through respiratory and digestive contacts, and contribute to the destruction of the organisms with which they combine with. Thus, the destructive reactions by the absorption of masses of the populace is one avenue of contaminate reduction. Other examples of the phase of contaminate reduction could be represented by the destructive rusting, erosion and corrosion of man-made metal structures.

Not all fuels have contaminate causing properties resulting from their inherent chemical makeup. Fuels composed exclusively of carbon or hydrocarbon elements only yield carbon dioxide and water vapor in the products of their combustion. All fuels to some extent or another contain some form or degree of contaminate and must be assumed as necessary evils to be dealt with accordingly. However, the manufactured fuels like hydrogen and methane gas are as near perfect combustibles as could be desired. Having been exposed then to the reality of what constitutes the contaminate reduction phase of the practical life support system, we would be wise to take whatever actions necessary to eliminate the contaminates or withdraw them immediately at the power plant.

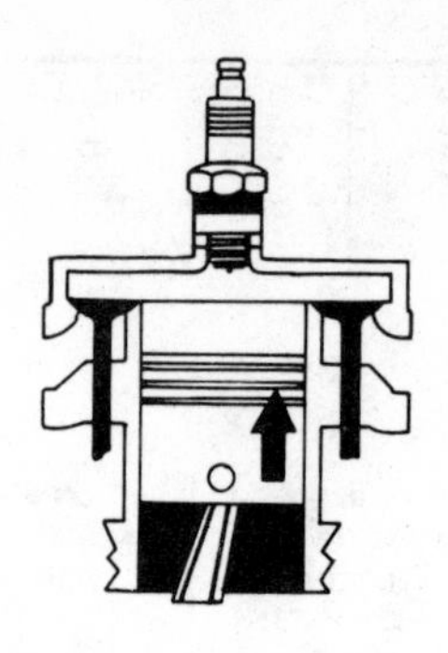

Chapter 2
Alternate Fuels

The *Bio-gas Generator,* although not new in the sense of a recent discovery, has received surprisingly little attention in the United States. However, it has realized much success in Europe and in countries where other fuel supplies are at more of a premium. Essentially, the bio-gas generator itself is a very uncomplicated arrangement of two main parts, the *digester*, where the bio-gas is generated, and a *receiver,* where the bio-gas is stored until use.

BIO-GAS AND ITS GENERATION

The term bio-gas describes the gas as evolved or liberated as a result of the chemical decomposition of certain decomposable or bio-degradable organic materials. The raw gas as collected generally contains from 50 percent to 85 percent methane. The rest is composed of carbon dioxide and hydrogen sulfide.

Methane has a heating value of approximately 1000 Btu/cubic foot. The hydrogen sulfide heating value is 845 Btu/cubic foot. The effective heating value of the bio-gas will depend directly upon the percentage of the methane and hydrogen sulfide content. The carbon dioxide provides no heat and does not enter into the combustion picture whatsoever.

What is most important about bio-gas is its availability. It can be generated by the decomposition of just about any organic matter imaginable. In most applications waste is used consisting of human feces, farm animal manures, grass clippings, sawdust, meat scraps, rags and straw, just to mention a few decomposable sources. The amount of methane liberated depends upon the organic richness of the material being digested.

In Fig. 2-1 is a working diagram of the bio-gas generator. It is comprised of only two parts, with interconnecting piping. The digester is the heart, or perhaps it should more correctly be called the "stomach" of the system. It is here where the "digestion" or gas-evolving decomposition takes place. The digestable products, whatever refuse or organic matter selected, are charged into the digester in a mixture, having enough water added to serve as a medium to affect decomposition. The digester must be absolutely tight, as the most important aspect of digestion is that it occurs in an oxygen-free atmosphere. The bacteria required for this particular type of decomposition are the *anaerobic* variety, meaning they must live in an oxygenless environment. Were oxygen permitted to find its way into the system, it would oxidize the bio-gas and stop the continued liberation of gas. Thus, we have an chemical operation which is termed anaerobic fermentation or digestion.

The fermentation process or the decomposition is best initiated and maintained in a temperature of between 80 to 95 degrees Fahrenheit. Below 60 degrees and above 140 degrees the bacteria become very inactive. This means that the digester must be housed or externally heated should the exposure to the environment predict temperatures above or below the temperatures given. However, the digester may be sufficiently heated by using a portion of its own gas to heat the enclosure of its occupancy.

An interior mixing device, such as a paddle, must be added to facilitate the occasional stirring or agitation required to promote continued and total decomposition. A motor-driven paddle is provided to stir the mixture. This stirring distributes the solid matter evenly throughout the digester, and prevents a surface scum from thickening on the surface of the mixture. The paddle drive should be run about 15 minutes per day. A time-actuated clock device will effectively assure this mixing.

The "slurry" is a solution of about 7 to 10 percent solid matter in water. It is charged to the digester through the filling hopper. The pipe of this hopper extends about two-thirds of the way down inside the digester. This pipe will prevent gas (which is in the upper part of the digester) from escaping out the filler valve upon subsequent chargings and help to prevent the entrance of undesired oxygen to the system. When it is determined that the last of

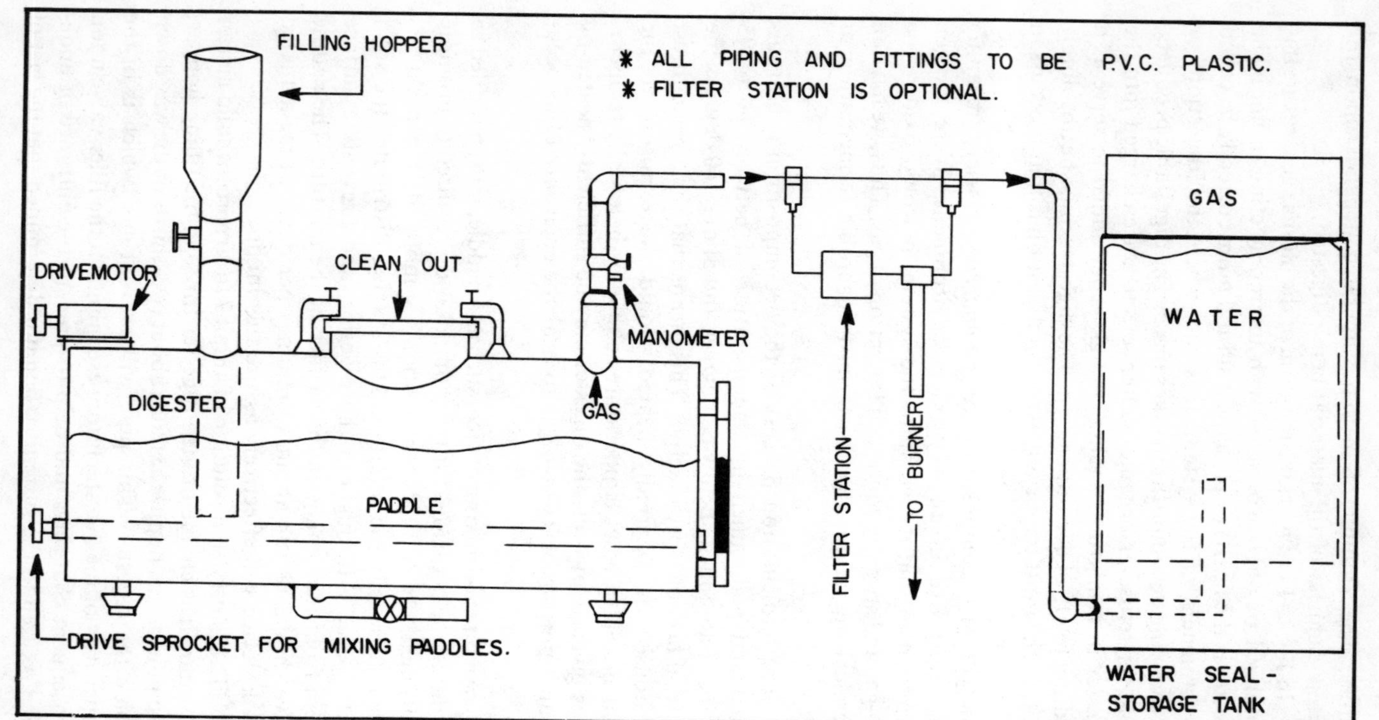

Fig. 2-1. Diagram of a bio-gas generator.

the carboniferous matter has been decomposed and the resulting gas liberated, the now depleted "slurry" may be dumped from the digester by way of the bottom valve, having first closed the gas discharge valve. The residue, which is high in nitrates and nitrogen compounds, is an ideal fertilizer for the soil. It may either be directly released to the soil or dried and used as a compost concentrate.

Filtering Station and Storage Receiver

In Fig. 2-1 is sketched a filtering station. The biogas may be filtered of its carbon dioxide and hydrogen sulfide by passing the gas through a bath of lime water, which will absorb the carbon dioxide and then through a filter containing ferric oxide which will remove the hydrogen sulfide. This is important if the fuel is to feed a gas engine or contact steel parts of any irreplaceable value. Filtering out the carbon dioxide is a favorable method of increasing the methane percentage of the gas per unit volume.

In experimenting with my own "rig", I have generated regularly from chicken manure, pig manure and dried grass an average of 2,000 cubic feet of bio-gas from approximately one ton of solid matter. Of the 2,000 cubic feet of biogas, approximately 60 percent was methane, yielding a calorific value of near 600 Btu/cubic foot of gas as collected, without filtering or carbon dioxide reduction.

The solution cannot become too acidic or too alkaline. Preferably the pH should be maintained between 6.8 to 7.5. No difficulty was had when using straight chicken manure, as it seems this material creates an almost perfect 7.5 pH.

The storage receiver is a PVC tank within a tank water-sealed arrangement. The pressure of the gas itself raises the top portion of the inverted inner tank to accommodate the incoming gas by further lifting action.

Digestion

The digester of my rig had 60 inches and was 144 inches in length. An old oil tank was used for this component. A cleanout opening was easily cut and a lid fitted. A very tight packing gland was fabricated for the outboard drive sprocket end of the mixing paddle, having three blades. A length of ½-inch pipe served as an axle. A ¼ horsepower motor was used with a clock timer to control mixing. All of the fittings were PVC plastic type. A sight glass is also of added value, although not imperative. A manometer was added to indicate pressure. The highest pressure obtained at any time on this particular system didn't exceed 0.5 psi.

The process of digestion, or the organic-methane producing breakdown, required an initial "startup" of 19 days. Bacterial activity was assisted in commencing by the addition of a "seeding" of already fermenting material. Otherwise a "start up" period of four to six weeks might have been required. The oxygen-bearing atmosphere was driven out of the system, and a desirable anaerobic environment was created by filling the system with carbon dioxide. This was accomplished by placing two pieces of dry ice in a can and connecting the can by hose to the digester manometer port. As the carbon dioxide was released by the warming dry ice, it entered the system, displacing the atmosphere inside. The carbon dioxide being heavier than air moved the air upward and out the gas discharge piping.

Life Support System

In the engineering profession, we're not required to consider the appearance or aesthetic values of our inventions or theories. If a device works and functions efficiently then it's satisfactory. If it functions as a coordinating life support system, utilizing available materials and producing energy and desirable by-products while contributing no objectionable pollutants to the environment, then we have an ideal system (Fig. 2-2).

The bio-gas generation system is one which could fulfill these requirements. The bio-gas generator, although not quite the utopia of science itself, does produce energy in the form of methane gas. It produces a desirable by-product of nitrogen-rich compost for plant fertilization and additional carbon dioxide which could be extracted for commercial purposes. If returned to the atmosphere, the carbon dioxide would re-enter the biological cycle of absorption by carboniferous green plants, which brings us back to the beginning of the next "cycle," utilizing the decomposition of these plants in our generator.

HYDROGEN PRODUCTION PLANT

Hydrogen is the most calorific, easiest to burn, cleanest burning and least objectionable of any of the primary fuels we have at our disposal. Hydrogen will ignite at a minimum temperature of 1,065 degrees Fahrenheit and will unite with oxygen to yield a heat release of 325 Btus per cubic foot, or 62,000 Btus per pound. Two moles of hydrogen uniting with one mole of oxygen form water vapor. Thus two volumes of hydrogen will unite with

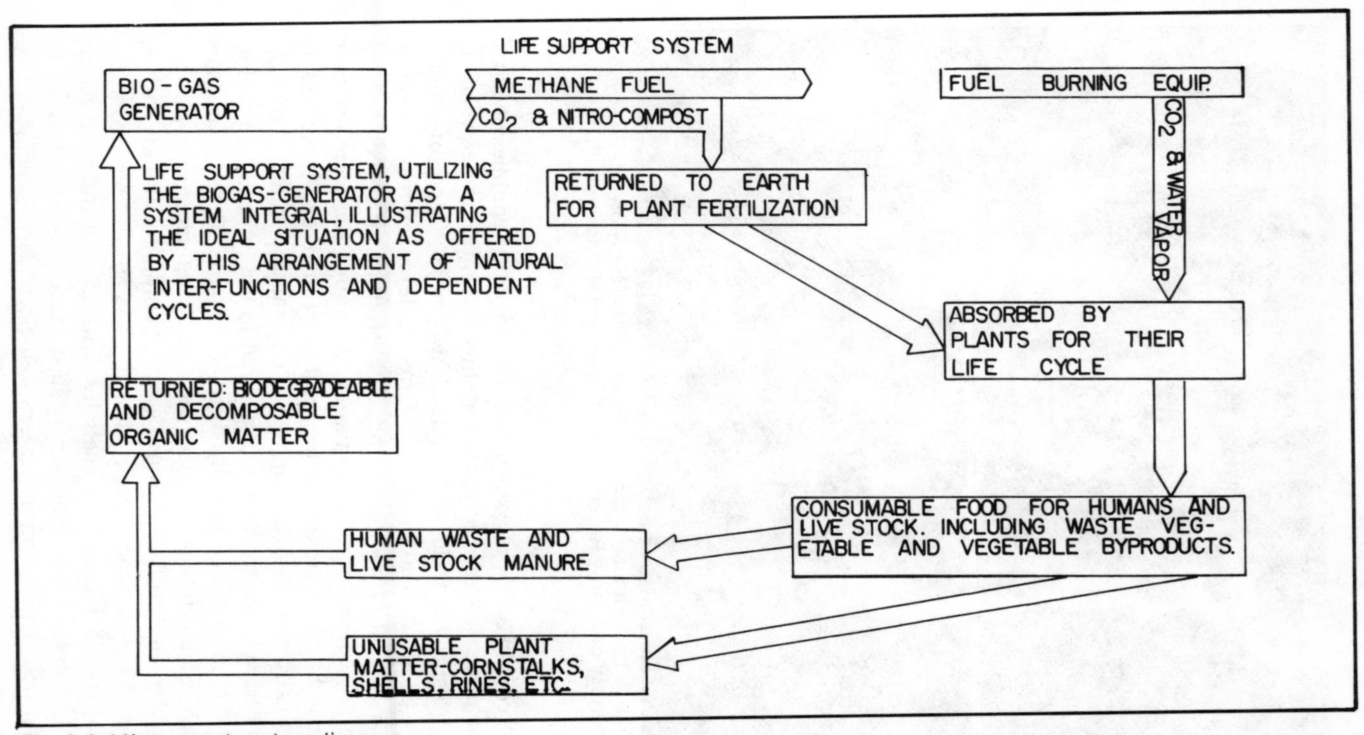

Fig. 2-2. Life support system diagram.

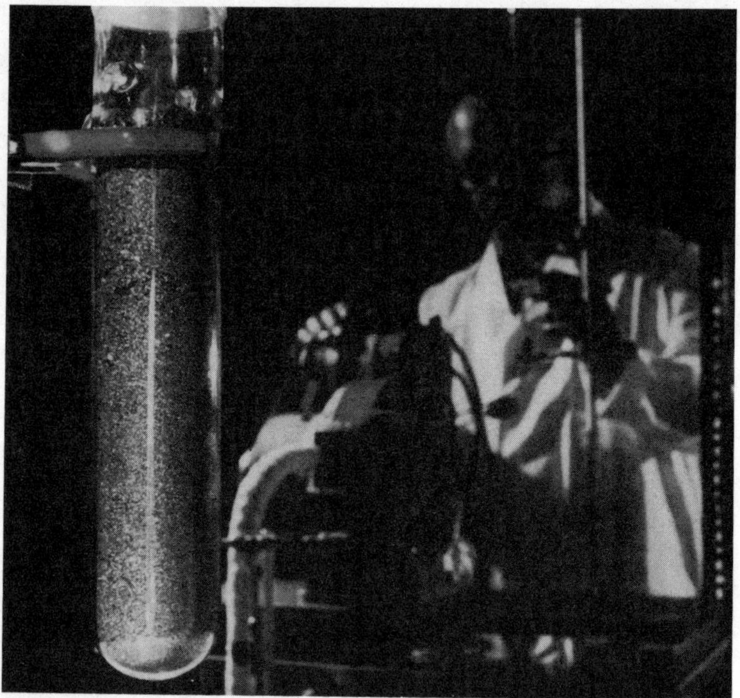

Fig. 2-3. The test tube in the foreground is filled with iron titanium hydride (courtesy of Energy Research and Development Administration).

one equal volume of oxygen to form two equal volumes of water. One mole weight of oxygen combining with two moles weight of hydrogen, will yield 44 pounds or two moles weight of water.

Of all the known fuels, hydrogen offers the greatest heat release—62,000 Btus per pound. The products of its combustion are absolutely harmless to nature. Although there are traces of uncombined hydrogen in the air, it must in all occasions be synthetically produced, as opposed to naturally collected. Hydrogen may be synthesized or manufactured in a number of ways (Fig. 2-3). It may be fractionally distilled in parts by the refinement of petroleum-based supplies, such as oil refinement. Hydrogen may be reduced from natural gas by breaking its bonds with its parent hydrocarbon molecules. It may be parted from the company of bituminous coals by destructive distillation of the coal to coke. Or hydrogen may, in addition to other means, be separated from common water in an electrolytic generator.

ELECTROLYTIC GENERATOR TECHNIQUE OF PRODUCTION

This technique was first discovered in the late 1700s. If a current of electricity was to be passed through a solution of water made slightly acidic (or electrolyte solution) by the addition of a hydrogen acid, hydrogen gas accumulated at the negative (-) terminal of the submerged electrode, called the cathode. At the same time, oxygen gas accumulated at the positively charged submerged electrode or the *anode* (+). The gases were collected above the electrodes that it was respectively liberated from, and was confirmed to be hydrogen and oxygen.

The slightly diluted H_2SO_4 acid had, as it always does when added to water, ionizes. The H_2 and the SO_4 radicals separate while they have the opportunity to do so in a water solution. The hydrogen, in order to replace the negative electrons for which SO_4 represented, then obtained this negative deficiency by hanging about the cathode or the negatively charged electrode. The hydrogen, then being a complete entity or molecule, had no place to go but up, right out of the solution. This also released oxygen gas. As soon as two oxygen atoms were left homeless by the exited hydrogen, they combined to form a complete molecule of free oxygen and proceeded to float up and through the surface of the electrolyte and be captured by the scientist who waited with open test tubes.

Technically, the electrolytic process of hydrogen production will be defined as the liberation of both hydrogen and its respective parts of oxygen, through ionic conduction, provided by a suitably mixed electrolyte by the passage of a direct current through the solution. Theoretically one hydrogen molecule should be released from the electrolyte for every two electrons picked up from the passing current. Since a molecule of hydrogen contains two atoms (being diatomic) for every two atoms, one complete hydrogen molecule will be liberated from solution.

Thus, hydrogen and oxygen were liberated and could be once again burned or chemically combined in an exothermic reaction and returned again to water. The water could be again broken down to its respective components of hydrogen and oxygen. The process is absolutely reversible, the direction depending on whether energy is applied or released from the gases in the form of combustion.

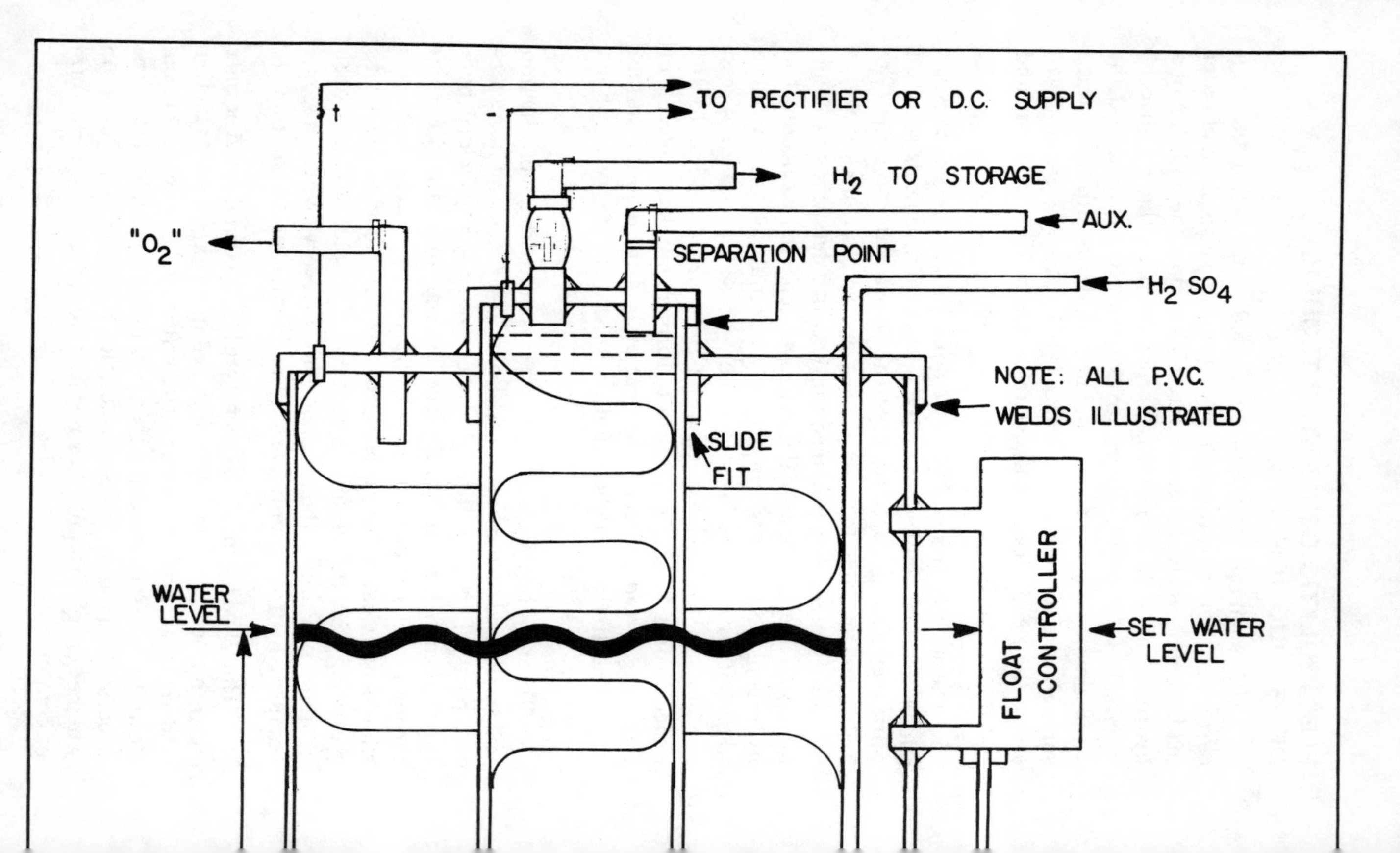

TO RECTIFIER OR D.C. SUPPLY
H_2 TO STORAGE
AUX.
"O_2"
SEPARATION POINT
H_2 SO_4
NOTE: ALL P.V.C. WELDS ILLUSTRATED
SLIDE FIT
WATER LEVEL
FLOAT CONTROLLER
SET WATER LEVEL

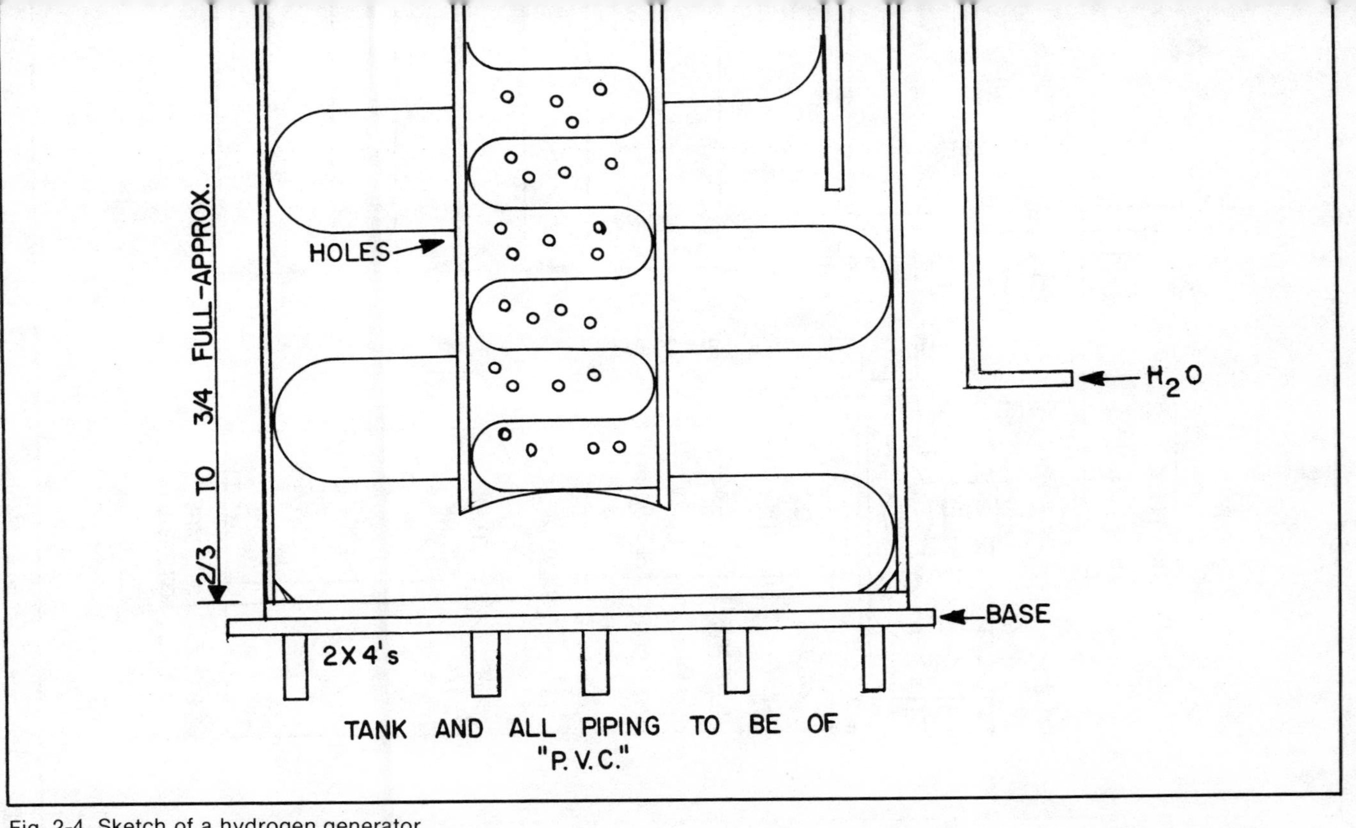

Fig. 2-4. Sketch of a hydrogen generator.

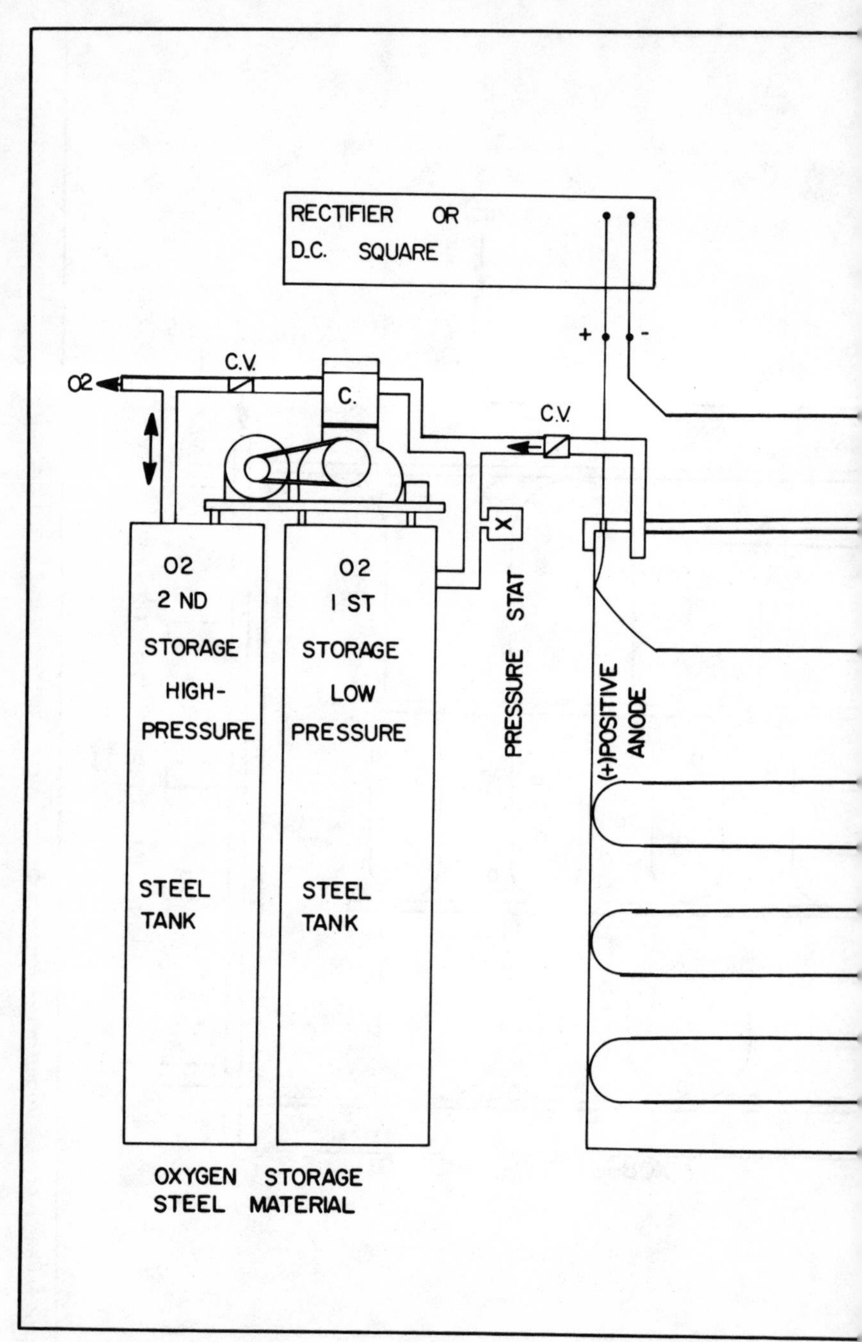

Fig. 2-5. Hydrogen generation plant.

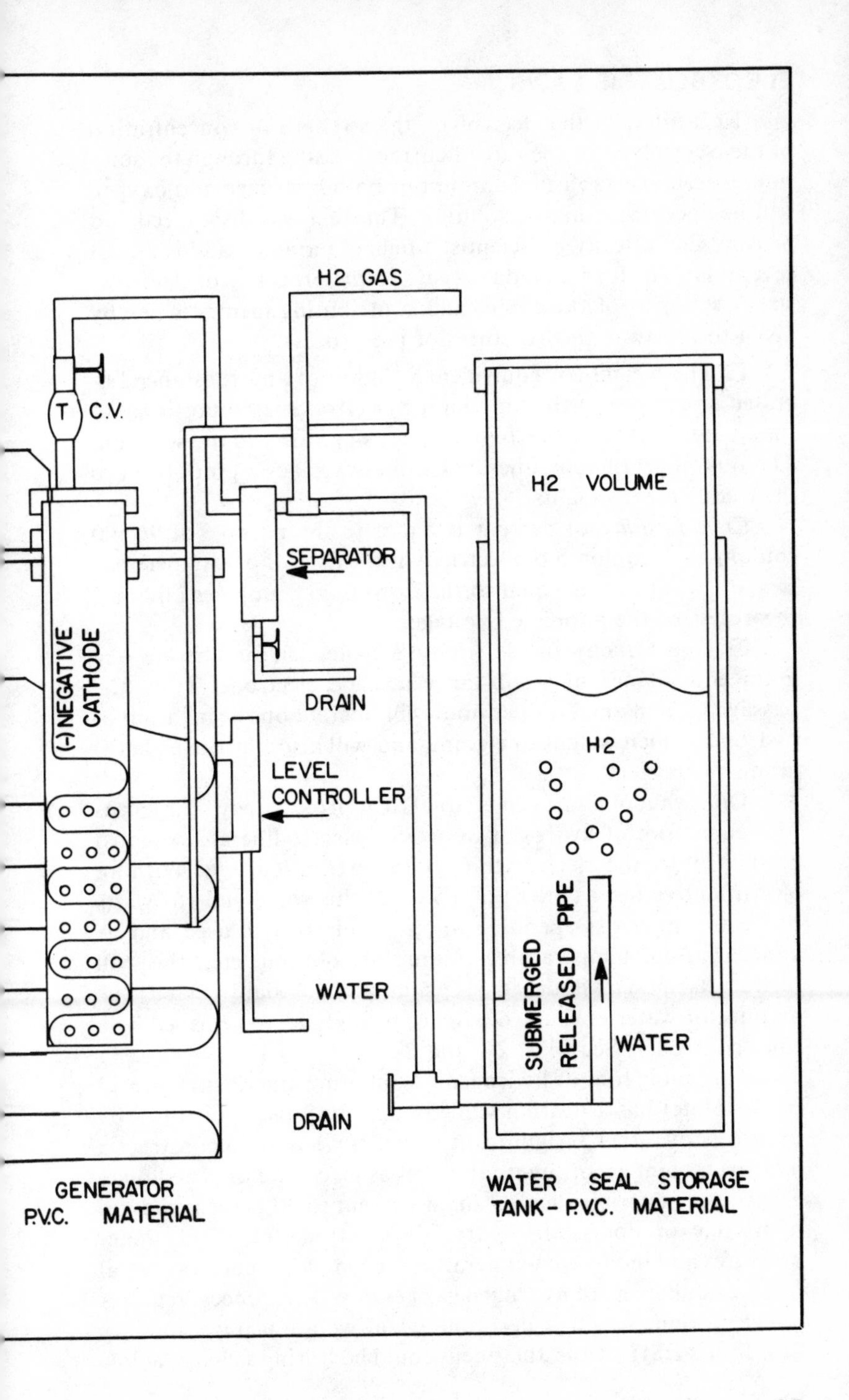
H2 GAS
T
C.V.
SEPARATOR
(-) NEGATIVE CATHODE
DRAIN
LEVEL CONTROLLER
WATER
DRAIN
H2 VOLUME
H2
SUBMERGED RELEASED PIPE
WATER
GENERATOR P.V.C. MATERIAL
WATER SEAL STORAGE TANK - P.V.C. MATERIAL

ELECTROLYSIS LAWS

Regardless of the electrolyte, the strength or concentration of the electrolyte, or the rate of current passing through the solution, a certain established amount of both hydrogen and oxygen will be liberated from the solution. This fact was discovered and proven scientifically by scientist Michael Faraday, and is one of several laws he formulated in regard to the processes of electrolysis. A synopsis of the extent and contributing factors is hereby given to illustrate the exactness of the process:

☐ The weight (or equivalent volume) of any substance liberated at any one of the functioning electrodes is proportional to the quantity of electrons (current) passing through the solution. The quantities thereby liberated will always be in proportion to their equivalent weights.

☐ A *faraday* of current is equal to the passage of 96,500 coulombs. A coulomb of electricity is the flow of one ampere per second. A faraday is equal to the flow of 26.8 amperes an hour, regardless of the impressed voltage.

☐ One faraday of electricity will deposit or liberate one gram equivalance of a substance at each electrode. Thus, the passage of 26.8 amperes per hour will liberate one gram atom of hydrogen which weighs one gram, and will also liberate eight (8) grams of oxygen.

One gram of hydrogen is equal to approximately four tenths of a cubic foot of hydrogen we would wish to liberate, we must send through the electrolyte 67.0 amperes of current. With an electromotive force (EMF) of 2.5 volts, this will equal 167 watts of electric power. Depending upon the electrolyte used and its concentration, temperature, volume of solution, etc., this will liberate itself on an hourly basis per amperes actually passed. The amount of water consumed will accordingly be replenished by a float controller. See Figs. 2-4 and 2-5.

This cubic foot of hydrogen gas (at atmospheric pressure-15 psi absolute) has a heating value of 325 Btu's per cubic foot. By itself it is of rather insignificant value. If we were to construct a hydrogen plant according to the sizes as given and used adequate electrolyte material, such as an 18 percent to 30 percent sodium hydroxide or potassium hydroxide solution, we could, under pressures and moderate temperatures, produce as much as several hundred cubic feet of hydrogen gas per day. This process requires direct current, which is desirable when we are using secondary cells for its EMF. Thus, these cells could be storing a charge at low points in the normal demand cycle of the generating stage and releasing the charge at other periods, thus creating hydrogen and oxygen gas.

Remember that for every pound of hydrogen generated, there is an additional eight pounds of oxygen. The sale of oxygen to such users as welding supply dealers or industrial buyers could more than compensate for any expenses incurred with the production of the hydrogen.

With a little imagination we could think of many prime sources for which to "move" a small dc generator. The voltage required through the solution, even considering "over voltage" would not exceed three volts. The amperage is the decisive factor. Naturally, the least electrolyte resistance present, the more current will be given to flow with the same EMF. Thus, the system is largely limited in only the amount of current the user wishes to send through the electrolyte. This maximum current value will be restrained by the ability of the conductors in terms of ampacity and the area of electrode surface. The actual size plant the reader may elect to construct will depend on the room he has to accommodate the gas storage tanks and the volume of hydrogen and oxygen he wishes to produce.

CATHODE-ANODE DISTANCE

No matter what electrolyte solution the reader intends on using, he should concentrate the materials to provide a resistance of not too much in excess of 0.5 ohm. This value of ohmic resistance will be controlled in large part by the particular solution concentration, the temperature of the electrolyte (the higher the temperature, the less the resistance), and the spacing of the electrodes from each other. This distance is a factor which requires explanation.

The distance between the *cathode* (-) in the center of the hydrogen generator and the *anode* (+), which is represented as the nickel wire wrapped about the inside of the units outer shell, is dictated inherently by the diameter of the generator itself. Thus, the larger the diameter, which means the more volume it can hold, the greater will be this distance or cathode-anode separation. We can overcome this problem, within reason, by increasing the height of the generator to make up for the reduction in diameter. However, this will result in a much higher generator. The builder then must compromise with this aspect ratio: diameter of shell increase of resistance due to distance between electrodes versus the gain in conductivity gained at the expense of making the unit that much higher.

The materials used in the generator and hydrogen storage units are commercially available (poly-vinyl chloride (PVC) plastic) to offer chemical inertness to the effects of acid and hydroxide. The piping is also PVC. The builder, with any skill at all, will have no trouble assembling this producer plant in one weekend.

SUGGESTED CONSTRUCTION DETAILS AND MATERIALS

The following materials may be increased in size, gauge or in any manner that will accommodate the demands or conditions of the builder. Refer to Figs. 2-4 and 2-5. Material for the generator and hydrogen storage tank are of PVC. The thickness of all parts should be not less than 3/16-inch. These parts will be subject to an internal pressure of up to 15 psi. The pipe wall thickness can be less, but ⅛-inch should be the minimum used. In Figs. 2-4 and 2-5 I have indicated junction points to be epoxy welded by using welding marks, as all welds will be of the "fillet type". Reinforcement may be used if thought necessary.

The cathode and anode consists of nickel steel wire to withstand corrosion. However, any chemically inert metal may be substituted providing it has relatively low ohmic resistance. Flat strip may also be employed; the main concern is to provide as great an exposed area as possible. For a generator having an inside diameter of 4 feet and a height of 6 feet (two-thirds of this depth to contain electrolyte), a length of wire 1200 feet in length would yield approximately 1000 turns of coils having a circumference of 12 feet. This inside large diameter wire serves as the anode (+). The cathode, which is wound inside of a submerged perforated tube approximately one foot in diameter, can be wrapped internally with 1500 feet of wire. This would provide 500 coils of 3-foot diameter. The inner cathode tube extends only to within about 6 inches of the bottom of the shell proper, and is perforated up to a point which will not be less than 1 foot below the water/electrolyte level at all times. Otherwise, oxygen would be mixed in the hydrogen gas collected.

The oxygen storage tanks may be of steel. The oxygen first enters the low pressure stage, whereupon a pressure switch will close a relay circuit when the pressure reaches its set point (up to 15 psi). This will activate the compressor and pump the oxygen to the high pressure side, until the low pressure side is returned to approximately atmospheric.

The hydrogen passes from the generator through a check valve and is then passed through an optional separator to separate any moisture. The gas storage tank is of the liftable top, water-sealed variety, such as the bio-gas system's tank. A feed system may be incorporated which could serve a bio-gas unit if a twin system were contemplated. A float control, similar to a boiler waterfeed system, is illustrated to provide makeup water as the water is depleted by its liberation.

The dc power supply can be had from a low-voltage high amperage rectifier. Or it may come directly from a dc source. such as a battery of secondary cells, or a dc generator, driven by a prime mover. The only factors are that the voltage should not exceed three volts, preferably 2.5 volts, and that the amperage is discretionary. It may be regulated by a rheostat, or from the transformer preceeding a rectifier, if ac is used for initial power. The connecting conductors will be gauged according to their anticipated current carrying loads; at least a #12 should be considered.

Over-pressure protection should be considered, in the event that the compressor should fail, such as a safety valve on the generator piping and on the high pressure side of the oxygen storage tank. Care should be exercised with maximum application, as hydrogen has a high ignition point. It will nevertheless explode violently when mixed with air, in the presence of even the smallest spark. Vent all lines carefully.

The concentrate, H_2SO_4, or hydroxide solution is admitted to the generator by a separate submerged line. Hydrometer readings should be taken regularly to assure optimum electrolyte density resistance characteristics. Electrical metering should be added for amperage and voltage checks.

Although this is but a very rough representation of the system, it provides the rudiments of operation and should "seed" considerations for the study of alternate energy. The advantages of hydrogen are well worth the expense of thought and experimentation.

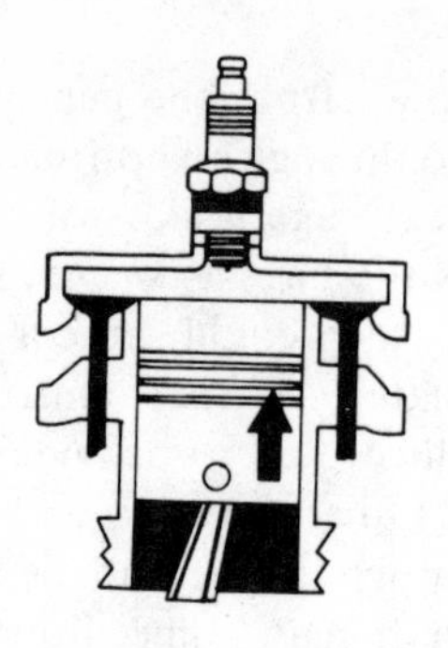

Chapter 3
Primary Fuels

The reader, after having acquainted himself with the origin, manner of geologic formation and availability of our primary fuels as discussed in the preceding chapters, should now be able to commence a qualitative and quantitative appraisal of the as delivered fuel commodity. It is expected that the reader can convert the greatest thermal output from the least raw input.

Regardless of the particular power plant or the fuel source assigned for use therein, a certain predetermined maximum value of efficiency can be realized from its "as designed" operation. Quite naturally, the engineer has no influence over design problems or the selection of fuels he is expected to fire. Nor is it within the engineer's responsibility to provide satisfactory results from a conversion boiler for which the fuel being fired is of improper character, such as burning oil in a converted coal boiler or gas in a converted oil or coal boiler. The engineer cannot be held accountable for inefficiency when the problem lies in the equipment itself. The only recourse the engineer has is to make the appropriate recommendations to his superiors that because of certain design limitations, only a given percentage of the theoretical may be had. An investigation should be considered in order to ascertain the economical aspects of investing the initial capital to purchase the proper equipment, which in turn will yield a more proper dividend in savings of otherwise wasted operating expenses.

KNOWING YOUR EQUIPMENT

Having arrived at the important fact that the engineer must contend with and make do with what he has, we're ready to undertake the investigative inquiry as to just what is the best efficiency that can be expected from the equipment. Obviously, there is no one equally applicable set of rules or instructions that could possibly cover all of the many varieties of power plant operations and elaborate on each, with such specific definition, as to enable an engineer to conduct an efficiency survey page for page, in a step by step manner, right out of the book. The countless idiosyncracies and peculiarities of each and every power plant would far exceed the scope of an entire library. Thus, this efficiency manual is by no means intended to substitute the equipment manufacturers' instructions and guides, but rather to serve as a supplement to the necessary data as provided by the best authority on the equipment itself—the manufacturer.

It is suggested, then, that a through knowledge of the equipment be gained by a study of all available manufacturers' data and publications before any attempt is made to make any alterations or modifications to the equipment. It may prove to be that just returning the equipment back to its factory specifications may result in a desirable effect in the equipment's performance. As in many such instances, I can remember that after being called for a service problem on an oil burner, the problem was remedied by simply referring to the instructions and data given by the manufacturers' guide and returning all adjustments back to their specified set points. Had the personnel only resorted to the guide, they could have remedied the situation themselves and saved my service charge.

I have routinely made it a point to study the manufacturers' specifications myself before I even open my tool box. I will usually pull out of the service manual from the approximately 300 or so manuals that I carry in my vehicle and enter the plant for an observation and discussion of the problem with the plant's attendant before I would even unload my service kit. Experience has taught me that a good many problems are corrected with nothing more than an understanding of the design functions of the equipment. A screw driver can put the linkages and control points back to their correct settings. A can of spray solvent cleans out the dirt accumulated as a result of covers being left off relay panels and circuit boxes.

PRE-SURVEY REQUIREMENT NUMBER ONE

Before proceeding further with the actual survey, the reader will make a collateral study of the designed functions and operating specifications of his equipment, as set forth by the manufacturer of the equipment. If such manuals or guides are unavailable, then either contact the equipment's manufacturer or your service organization. Many times "business assuring cooperation" will be extended to the customer in the form of technical data lent for the purpose of equipment maintenance.

To exemplify the imperative need for the technical familiarity that can be provided only from the manufacturer's data and instructions, let us consider a hypothetical situation. We determine from our survey that we have an excess of secondary air being admitted to our combustion process. How can we intelligently expect to correct this problem if we are lacking the manufacturers' recommendations for secondary air adjustment?

No matter how much air we admit to the combustion chamber, we cannot raise our carbon dioxide reading any higher than 10 percent. This problem could be indicative of any one of a number of related faults: a deficiency in refractory, poor turbulence caused by air diffusers being out of alignment, air leakage into the boiler setting or even improper atomization at the gun tip. The reader can visualize that without a thorough command of the equipment's ability and inherent limitations, we're at a loss to intelligently assign the problem to either operational or design. Many situations will be encountered whereby the fault may be in the equipment itself, and beyond correction by adjustment alone.

In this chapter on primary fuels will be found an exhaustive dissection of the individual fuel's qualitative and quantitative characteristics, properties and potential as a heat yielding combustible. A study of the fuel's chemistry will benefit us with the potential of the fuel in terms of what maximum heat releases can be achieved under ideal combustion techniques.

The chapter on combustion techniques will detail the preferred methods of preparing our fuels for combustion. It will outline the prerequisites necessary for perfect and complete combustion and give instructions for testing the products of combustion for further comparative purposes in gathering our information necessary for an efficiency survey.

The chapter on boilers, steam generators and steam system appurtenances is intended to provide the reader with a broad generalization of the equipment used in absorbing the heat as released by the combustion processes. It will direct the reader in acquiring the essential data collectable from these critical points of our system. Although these chapters are in fact generalizations, they are nevertheless applicable to any and all power plant systems of the fossilized fuel classification. Regardless of the individuality of the particular power plant, there exist certain basic collectable values and "read outs." When correctly applied by means of simple mathematical formulas, the efficiency or output/input ratios may be evidenced for any given set of factors and situational values.

This efficiency manual is thoughtfully structured to serve just such facility and purpose: in educating the reader as to what particular energy exchange or conversion process is being brought about: thermal, electrical or mechanical. Also considered are what indications or values are required to determine the effectiveness of the exchange or conversion process under study. The reader will learn the critical points or stages where these indications or readings must be accordingly taken from or measured at, and what instruments or meters are needed in the collection of measurements.

After having conducted our survey, the results, expressed in terms of percentages of efficiency, will in themselves dicate what improvements or attention are required and in exactly what area of the system. After the initial survey, it is expected that periodic surveys will be conducted at fixed intervals of such increments which will assure that future operational deviations from our established optimum percentages will be observed and corrected before the system may be permitted to increase its waste index. We will for practical purposes ascribe such waste index as unavoidable or unrecoverable losses due to design defficiencies, as opposed to operating deficiencies. We just aren't going to tolerate operational inefficiency.

COAL ANALYSIS

Coal is a solid combustible fuel. Its heating value ranges from a low of 6,300 Btu/lb. up to and including 15,670 Btu/lb. This depends upon its quantitative analysis in respect to its composition of carbon, which may comprise from 32 percent to 98 percent; its percentage of volatile matter, which may represent from 2 percent to 45 percent; and its sulfur content, ranging from 0.3 percent to a maximum of 5 percent.

Coal is a classification given to the solid fossilized form of carbon, volatile matter, sulfur, ash and other entrained impurities including moisture. Generally, the condition and quality will be dependent upon the geographic location of its extraction, the rank of its age, or the metamorphic time nature has had the chance to work upon coal and consolidate its properties. Thus,

Table 3-1. Coal Data.

		PROXIMATE ANALYSIS-PERCENT				ULTIMATE ANALYSIS-PERCENT					
		MOIS-TURE	VOLATILE MATTER	FIXED CARBON	ASH	S.	H.	C.	N.	O.	HEATING VAL. BTUs/Lb.
Anthracite	PENNA.	4.4	4.8	81.8	9.0	0.6	3.4	80.	1.0	6.2	13,130
Semianthracite	ARK.	2.8	11.9	75.2	10.1	2.2	3..7	78.	1.7	4.0	13.360
BIUMINOUS COAL:											
Low Volatile	MD.	2.3	19.6	65.8	12.3	3.1	4.5	75.	1.4	4.2	13,220
Medium Volatile	ALA.	3.1	23.4	63.6	9.9	0.8	4.9	77.	1.5	6.2	13,530
High Volatile-A	KY.	3.2	36.8	56.4	3.6	0.6	5.6	79.	1.6	9.2	14,090
High Volatile-B	OHIO	5.9	43.8	46.5	3.8	3.0	5.7	72.	1.3	14.	13,150
High Volatile-C	ILL.	14.8	33.3	39.9	12.	2.5	5.8	59.	1.0	20.	10,550
SEMIBITUMINOU:											
Rank-A	WASH.	13.9	34.2	41.0	11.	0.6	6.2	58.	1.4	23.	10,330
Rank-B	WYO.	22.2	32.2	40.3	4.3	0.5	6.9	54.	1.0	33.	9,610
Rank-C	COLO.	25.8	31.1	38.4	4.7	0.3	6.3	50.	0.6	38.	8,580
Lignite	N. DAK.	36.8	27.8	30.2	5.2	0.4	6.9	41.	0.7	46.	6,960

S - SULFUR H - HYDROGEN C - CARBON N - NITROGEN O - OXYGEN the term coal actually subtends the sedimentary carbonaceous substances from lignite, which exists as a soft crumbly coal of low heat value, up to the highest developed rank of meta-anthracite, which is a hard carbon-rich coal. This coal has next to no volatile matter, virtually no moisture and little if any ash or objectional impurities. Table 3-1 provides an analysis of coals by state, for the most representative ranks of the eleven most significant considerations.

It will be shown by an examination of Table 3-1 that the highest heating values are offered by the high volatile "A" bituminous and semi-anthracite coals. A reference to their analysis will explain this situation. Conversely, when we examine the lower values as provided by the sub-bituminous and lignite varieties, we will note a corresponding decrease in their fixed carbon and volatile matter and a marked increase in their moisture content. It can be deduced that the heating values to be expected from coal, per pound, will depend exclusively upon the percentages by weight of the combustible elements of carbon, hydrogen, and sulfur.

Dulong's Formula

Table 3-1 provides an approximate analysis, which is arrived at by physically subjecting the sample to distillation, buring and weighing. The ultimate analysis is achieved by the actual chemical investigation of the sample's constituent composition. From the ultimate analysis, we compute with reasonable accuracy the anticipated heating value of the coal as a combustible by applying Dulong's formula. This considers the actual and ultimate chemical analysis of the coal in given percentages of its total weight, multiplying each by its proven individual calorific value and yielding a figure of Btus per unit of weight, which in our calculations will be the pound.

$$\text{Btu/lb.} = 14{,}500 \times \textit{percent of C.} + 62{,}000 \times \text{percent of } (H_2 - O_2/8) + 4{,}050 \times \text{percent of S.}$$

Literally, this formula is read as follows. The heat value, in Btus per pound, is equal to 14,500 times the percentage (by weight) of carbon, plus 62,000 times the hydrogen (minus one-eighth of the oxygen present) plus 4,050 times the percentage of sulfur present in the coal. This is a pretty simple and realistic formula when you stop and think about it. There are but only three combustible constituents in coal, or for that matter any fossilized fuel—carbon, hydrogen and sulfur. Carbon has a pure heat value of 14,500 Btus/lb. Hydrogen yields 62,000 Btus/lb. and sulfur provides 4,050 Btus/lb.

Thus, by having an ultimate analysis of a combustible such as coal in this instance, and knowing the calorific values of carbon, hydrogen and sulfur, we may very simply, by the application of Dulong's formula, arrive at a near absolute expectation of the maximum heat release to be expected from our test sample. This is providing, of course, that no excess oxygen is supplied and all of the carbon, hydrogen and sulfur unite with their required number of oxygen atoms. This proper combination of elements and its related generation of heat during chemical combination is known as *combustion*, *exothermic* combustion to be exact. More on this concept will be covered in the chapter on combustion techniques. For now, let us limit our conclusion to the fact that there are but three principles which are involved in the combustion process. Each of these three elements has a fixed and predetermined heat release value, which will be realized only if the correct amount of oxygen is combined with all of the combustible elements available. The total heat given off for any given substance, such as coal, is governed by the collective and sum percentage expressed by factional parts of the sample's weight, usually in parts of a pound. The heat value of each is multiplied by the percentage of which it represents, to yield a total or effective total of Btus per pound of sample. For example, let us consider a piece of coal (weighing one pound). The chemist tells us by his submitted ultimate analysis that the coal is composed of the following:

sulfur-%	hydrogen-%	carbon-%	nitrogen-%	oxygen-%--	Btus ???
0.5	7.0	80.0	1.0	20.0	XX,XXX

Arranging our necessary data according to Dulong's formula, we have

$$\underset{\text{carbon}}{\underline{14{,}000 \times 80}} + \underset{\text{hydrogen} \quad \text{oxygen}}{\underline{62{,}000 \times (7.0 - 20/8)}} + \underset{\text{sulfur}}{\underline{4{,}050 \times 0.5}} \quad \text{which yields=}$$

$$\underset{\text{carbon \%}}{\underline{11{,}200 \text{ Btus}}} + \underset{\text{hydrogen \%}}{\underline{2{,}790 \text{ Btus}}} + \underset{\text{sulfur \%}}{\underline{202.5 \text{ Btus}}} \quad \text{equaling=} \quad \underset{\text{total/pound}}{\underline{14{,}192.5 \text{ Btus}}}$$

Thus, with basic math and the chemist's ultimate analysis, we have determined the maximum available heat release for coal, under perfect combustion, meaning proper molecular/atomic combination and provided oxygen.

Heat Potential

We have learned that coal, as with all other fossilized fuels we will be analyzing, is composed by percentages of carbon, hydrogen and sulfur, along with other minute elements which because of their relative insignificance in the combustion process will be ignored. These percentages are considered by fractional parts of the weight of the combustible being examined. We can further state, as has been illustrated by the preceding formula, that each representative element (or combustible value) contributes its own individual heat input to the total heat release of the combustible, whether it be coal, oil or gas.

In addition to the roles played by the combustible elements themselves, we must also consider the physical aspects of achieving this "as computed heat potential." As we have discussed, this heat release will only be realized if the theoretical amount of oxygen can be combined with the combustible. Any less than the required amount will cause starvation, lessening the amount of combustion and the amount of heat thereby released from this exothermic reaction. As all firemen and engineers know, incomplete combustion will result which produces smoke, a low carbon dioxide reading, a high carbon monoxide reading and a waste of fuel. Because oxygen or more appropriately the mixture of air required for satisfactory and complete combustion is equally applicable to not only coal but oil and gas alike, a thorough discussion of air required will be given in the chapter on combustion techniques.

Combustion Equipment

The most widely used combustion equipment used exclusively for coal includes the automatic *stoker*, which may feed the coal by means of an endless chain of grating. A screw system passes the coal from hopper to the combustion chamber. The method of hand firing has largely been replaced by the modern mechanical techniques. Since the 1920s, the pulverization of coal has received much attention and development, producing the modern day pulverized system. The pulverized system consists of a pulverizing mill, where the coal is finely pulverized; and a burner arrangement, where the super-fine coal dust is mixed with the proper quantity of air and introduced to the boiler's heat exchanging surfaces (Figs. 3-1 and 3-2).

Cyclone Furnace

It should be noted that in the pulverized system, the combustion chamber is still part of the burning apparatus. It can cool the elements of combustion below their optimum temperatures to retard or limit combustion if turbulence, time and ignition temperature are not adequate enough to promptly assure com-

Fig. 3-1. Coal pulverizer (courtesy of Babcock and Wilcox Company).

plete combustion at its earliest moment. This problem of utilizing the heat exchanging surfaces of the chamber to house the not quite completed combustible mixture has been very satisfactorly surmounted by the advent of the *cyclone furnace*, first appearing in power plants on an experimental basis as early as 1944 and proving itself to be very efficient (Fig. 3-3).

The cyclone furnace is an arrangement where the coal is first pulverized to a lesser fineness as required for pulverized systems (95 percent through a #4 mesh screen as opposed to the #200 mesh screen necessitated by the pulverizer). Thus, the cyclone can function with less critical preparation, which is a marked advantage. Due to the inherent design of the cyclone furnace, extraordinary air-coal admixture is accomplished. The coal and air are introduced to each other in a high speed whirling manner. Combustion is completed before the gases are admitted to the heat exchanging surfaces. Heat releases as high as 800,000 Btu/hr. and temperatures as high as 3,500 degrees Fahrenheit are realized by this system.

We could very honestly say that the cyclone furnace and the

Fig. 3-2. Detail of a coal pulverizer (courtesy of Babcock and Wilcox Company).

Babcock and Wilcox Company, who really invested the research and development necessary for its perfection, have provided the "pivotal hinge" by which we may revert back to the use of coal as a universal fuel for industry and commerce. Certainly no other single fuel has offered so much to our civilization and its technical development than has coal. With the cyclone furnace we may realize efficiency, cleanliness and economy almost paralleling that of its nearest competitor, oil. It is also a timely occurrence that we may now return to coal in large scale and efficient use. The United States has ample coal reserves to adequately last us until atomic energy has been perfected in all respects. Our access to petroleum and gas reserves are positively lessening and the demand is equally increasing.

OIL ANALYSIS

Oil, which is the processed result of petroleum, can be accordingly considered as a liquid form of coal. For all practical purposes, oil contains the same principal elements for combustion as does coal, namely carbon, hydrogen and sulfur. Each lends its

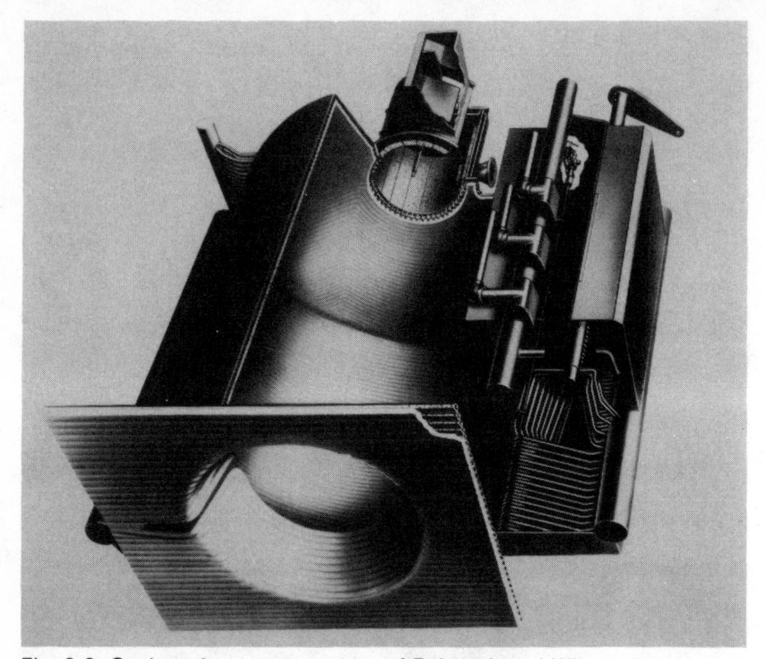

Fig. 3-3. Cyclone furnace (courtesy of Babcock and Wilcox Company).

individual share in the total sum of heat release upon completed combustion, per its percentage of weight.

Oil Grades

From the viewpoint of engineering, we will consider the six classifications of oil in respect to their relative characteristics as a fuel oil, paying primary consideration to each classification's worth as a combustible. This will be largely denoted by the oil's ability to produce heat or its calorific value. Attention will also be given to any special requirements dictated by the oil's properties in regard to preparation for combustion, such as preheating. With the heavier grades #5 and #6, we will observe that unless these oils are kept above a certain temperature, we cannot even pump them, let alone admit them to a combustion chamber for possible heat release.

By far, oil in any of its six main grades is a superior combustible, pound for pound, than is coal. First, oil has a higher Btu value per pound than an equal amount of coal. Oil may be burned with substantially less apparatus than coal. No ashes or by-products are left by the combustion of oil. Oil may be utilized with much less space than that necessarily allotted for coal. The storage of oil may be conveniently affected by underground storage tanks. Oil is much cleaner in respect to the dust and refuse encountered by coal. Generally, higher efficiencies may be realized by the more favorable combustion qualities of oil, as opposed to that of coal.

Ready for the bad news? Oil is a more expensive fuel, even on a Btu basis. Coal has more Btus per dollar purchased than does oil. Our oil reserves, although not nearing the danger point of exhaustion, nevertheless becoming alarmingly lowered, causing us to drill deeper and search farther for our supplies. A large portion of our supplies are being imported from foreign reserves. This fact by itself is one to be deferentially dealt with. With the advent of the cyclone furnace for which the Babcock and Wilcox Company has provided a great deal of contributing research, coal may now be burned with the desirable efficiency realized by oil burners. Needless to say, the initial investment of coal burning apparatus and the advanced pulverized coal system is far more expensive than may required by oil. We may very soon find ourselves forced to return to coal, if the situational pressure of oil does not drastically improve in terms of availability and financial affordability.

As depicted by Table 3-2, which analyzes grades #1, #2, #4, #5 and #6, the heating values in *Btus per pound* are highest in the lower number grades, which contain the greatest percentages of hydrogen, and decrease progressively toward the higher number fuels. It will also be noted that the higher numbered grades, from right to left, are heavier in respect to the weight per gallon and therefore contain more Btus per gallon. When we refer to the heavier grades as offering more heat per gallon, #6 grade has more Btus per gallon than #5, and #5 oil has more heat than #4, etc. Customarily, in speaking of fuel oils, we refer to the Btu/gallon and not by the pound, as in solid fuels.

Determining the Maximum Available Heat Release of Oil

As with any fuel sample, the heat release available can be determined by computations based on the fuel's chemical or ultimate analysis. Or a specified weight of the sample may be placed in a *bomb calorimeter*. With just the required amount of oxygen, the admixture is then ignited. Because the entire bomb calorimeter is submerged in water, the increased temperature of the water is then indicative of the heat released by the absorption of such heat. Table 3-2 gives us the "arrived at estimates" for all of

Table 3-2. Range of Analysis of Fuel Oils: #1 Through #6.

GRADE OF FUEL OIL:	NO. #1	NO. #2	NO. #3	NO. #5	NO.#6
WEIGHT-PERCENT					
Sulfur	0.1-0.5	0.5-1.0	0.2-2.0	0.5-3.0	0.7-3.5
Hydrogen	13.3-14.1	11.8-13.9	10.6-13.0	10.5-12.0	9.5-12.0
Carbon	85.9-86.7	86.1-88.2	86.5-88.2	86.5-88.9	87.-91.0
Nitrogen	Nil-0.1	Nil-0.1	XXX	XXX	XXX
Oxygen	XXX	XXX	XXX	XX	XXX
Ash	XXX	XXX	XXX	0-0.1	0.01-0.5
GRAVITY:					
Degrees A.P.I.	40-44	28-40	15.30	14.22	7.22
Specific........(High)	0.825-	0.887-	0.966-	0.972-	1.022-
(Low)	0.806	0.825	0.876	0.922	0.922
Pounds/Gallon (AVG.)	6.80 lb.	7.10 lb.	7.90 lb.	8.10 lb.	8.50 lb.
Pour Point, Deg. F.	0 to -50	0 to -40	-10 to +50	-10 to +80	+15 to +85
Flash Point, Deg. F.	90- 100	100- 110	110- 130	110- 140	140- 160
VISCOSITY:					
Centistrokes @100 d.f.	1.4-2.2	1.9-3.0	10.5-65.0	65- 200	260- 750
SUS @100 Deg. F.	XXX	32- 38	60- 300	XXX	XXX
SSF @122 Deg. F.	XXX	XXX	XXX	20- 40	45- 300
Water & Sediment percent	XXX	0- 0.1	tr. -1.0	0.05-1.0	0.05- 2.0
Estimated Btu Per Pound	19,750	19,500	18,900	18,300	17,800
Estimated Btu/Gallon	136,000	138,500	145,000	147,000	148,000-152,000
Required Pre-Heat, Deg. F.	XXX	XXX	135- 150	150- 190	200- 250
Relative Cost/Btu	118	100	76	60	51

the grades exclusive of #3 oil, which is disappearing from the engineering picture and thus will not be included.

Ideally, the heat of the oil sample should be determined by laboratory tests. If the degrees A.P.I. are known, a close estimate of the maximum heat content may be had by a formula I have used for some years, with very close results, which is from the U.S. Navy boiler labs:

$$\text{Btus per lb.} = 57.9 \times (\text{the degrees A.P.I. of the sample}) + 17{,}687$$

Another empirical formula to be considered is based upon the known carbon and hydrogen percentages (by weight):

$$\text{Btus per lb.} = 13{,}500 \times \%\text{C. plus } 61{,}000 \times \%\text{H}$$

Regardless of the manner of which the maximum available heat release is obtained, a near accurate value must be had in order to make any efficiency survey whatsoever. The heat content will, as is true for any fuel, be realized only with the optimum amount of oxygen admitted to the combustion process. This oxygen is provided naturally from the ambient atmosphere, which causes us to consider the actual cubic feet of air required. The percentage of oxygen will therefore remain a fixed, unalterable percentage of this gross amount. The aspects of combustion for oil will be dealt with more technically and inclusively in combustion techniques, as will the aspects of time, temperature and turbulence.

Oil is a combustible substance. When combined with the proper and ideal amounts of oxygen under optimum combustion techniques, it will release a quantity of heat per pound of oil, or gallon of oil burned, as per the analysis given in Table 3-2. Oil is essentially a liquid compound and an admixture of carbon, hydrogen sulfur and certain small amounts of impurities such as water, sediment, etc. The actual elements of combustible nature are the carbon, hydrogen and sulfur percentages. We know that these exist in fixed percentages, with each having their respective allotted heating values.

The more carbon per percentage of the oil, the heavier the oil will be and the more Btus will thereby be provided per gallon, as is illustrated by a study of Table 3-2. The heavy oils, #6, and #5, have the most Btus per gallon and progressively decreasing with the lower numbers. The relative cost per gallon also decreases with the higher heat values of the oil grades, the heaviest oils having the lowest relative costs.

You may be asking how this can be so. The more desirable the heat affordability of the fuel, the cheaper it becomes! The heavier fuels must be prepared with a correspondingly greater investment of time and expense in terms of fuel oil equipment, in addition to the more elaborate and expensive combustion equipment involved.

ANALYSIS OF GAS FUELS

The gaseous form of our fuel supplies is represented chiefly by natural gas, which is drawn directly from the ground, and by manufactured gas, as produced by either the coking of coal or by other destructive distillation processes. Like fuels of the solid and liquid varieties, gaseous fuels may be qualitatively and quantitatively analyzed in respect to their potential as heat releasable or calorific substances, for combustible purposes.

An examination of Table 3-3 will provide the reader with a comparative illustration of the differences in combustible content (by volume) for five selected samples of natural gas. It will be noted that the most influencing factor of the gases' calorific value is the content of carbon and hydrogen. Again, as with coal and oil, the total effective heat release may be computed by analyzing the gases' constituency of combustible elements. Multiply this by the percentage by which each individual gas is present by, and add these collectables to produce a maximum expectant heat release under optimum combustion conditions.

Gas fuels provide on an overall basis the most desirable properties of any fuel. Being in the gaseous or vaporized form, gas may very easily be brought into intimate contact with the oxygen from the air supply. Virtually no smoke or incomplete combustion should be had. And ideal efficiencies may be realized with a minimum of combustion equipment.

Because of the ease with which gaseous fuels may be burned, the cleanliness of their use, the efficiency of their combustion, and the absolute minimum of preparation and combustion equipment required for their burning, you may expect to pay a premium for this fuel. You're absolutely right! Gas is the most sought after of all fuels.

Tables 3-3 and 3-4 will acquaint the reader with the expected heat values and components of natural gas, and the individual heat values affordable by the constituent gases. Although very close calorific determinations can be arrived at by computations from chemical analysis of the fuel in question, the only absolute and error free analyses acheivable are had by subjecting the fuel sample to the bomb calorimeter test. Unfortunately, this test

Table 3-3. Data on Natural Gas Samples From Five States.

SOURCE OF SAMPLE: ANALYSIS NO.#	PENNA. 1	SO. CAL. 2	OHIO 3	LA. 4	OKLA. 5
ANALYSIS BY VOLUME (%)					
COMBUSTIBLE CONSTITUENTS:					
H^2 Hydrogen	XXXX	XXXX	1.82	XXXX	XXXX
CH^4 Methane	83.40	84.00	93.33	90.00	84.10
C^2H^4 Ethylene	XXXX	XXXX	0.25	XXXX	XXXX
C^2H^6 Ethane	15.80	14.80	XXXX	5.00	6.70
C0 Carbon Monixide	XXXX	XXXX	0.45	XXXX	XXXX
$C0^2$ Carbon Dioxide	XXXX	0.70	0.22	XXXX	0.80
N^2 Nitrogen	0.80	0.50	3.40	5.00	8.40
O^2 Oxygen	XXXX	XXXX	0.35	XXXX	XXXX
H^2S Hydrogen Sulfide	XXXX	XXXX	0.18	XXXX	XXXX
ULTIMATE PRECENTAGE BY WEIGHT:					
S Sulfur	XXX	XXXX	0.34	XXXX	XXXX
H^2 Hydrogen	23.53	23.30	23.20	22.68	20.85
C Carbon	75.25	74.72	69.12	69.26	64.84
N^2 Nitrogen	1.22	0.76	5.76	8.06	12.90
O^2 Oxygen	XXXX	1.22	1.58	XXXXX	1.41
SPECIFIC GRAVITY (REL. TO AIR)	0.636	0.636	0.567	0.600	0.630
Heating Valve, Btu/Cu. Ft.	1,129	1,116	964	1,002	974
Heating Value, Btu/Lb. weight	23,170	22,904	22,077	21,824	20,160

Table 3-4. Heating Values of Manufactured Gases, Average Btu Per Cubic Foot.

NAME OF GAS	SPECIFIC GR. (Air = 1.00)	Btus
Hydrogen	0.07	325
Carbon Monoxide	0.96	322
Methane	0.55	1,012
Ethane	1.05	1,792
Propane	1.56	2,590
Butane	2.07	3,370
Isobutane	2.07	3,363
Pentane	2.49	4,016
Isopentane	2.48	4,008
Neopentane	2.48	3,993
Hexane	2.97	4,762
Acetylene	0.91	1,499
Naphthalene	4.42	5,854
Methyl Alcohol	1.11	868
Ethyl Alcohol	1.59	1,600
Ammonia	0.59	441
Hydrogen Sulfide	1.19	647
Producer Gas (Lean Mixture)	0.85	170
Coke-oven Gas (Lean Mixture)	0.41	600
Carbureted Water Gas	0.66	534
Blast Furnace Gas (Avg.)	1.01	84.0
Reformed Gas, Refinery	0.49	530
Bio-Gas	0.62	600

necessitates an expensive investment of money and technical expertise which very few power plants are ready to meet. This book will avoid the actual technical discussions related to conducting such tests. As a rule, such tests are exclusively performed by qualified chemists in very extensively equipped labs.

Generally, an adequate figure of the fuel's maximum expected heat release may be reached by the application of the more or less empirical formulas as provided herein, in conjunction with the calorific tables are presented. These tabulated and correlated facts and figures have been exhaustively checked and cross-checked to insure their present day accuracy and usability.

Gaseous fuels, regardless of their origin or manufacture, may be both *qualitatively* (type or specific constituency) and *quantitatively* (amount or percentage of constituency) analyzed to determine the gases total or aggregate heating value. This analysis is undertaken by a study of the identified component parts of the fuel as composed in the sample under analysis. When the proportions (quantative percentages) of the sample are ascertained, it then becomes but a routine mathematical problem of multiplying each identified element by its predetermined calorific value, as related by the respective table. Deduct any non-combustible percentages in addition to any moisture or water vapor present.

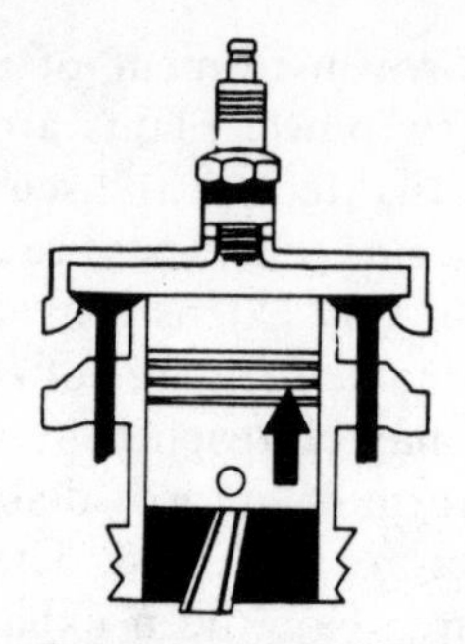

Chapter 4
Combustion Techniques

Up to this point in our study of primary fuels, we have limited our interests to the maximum or theoretical heat content of each respective fuel. The values as given in their respective analysis will only be realized by applying ideal combustion techniques to their chemical combination with oxygen in the combustion chamber—regardless of the particular fuel being fired or the combustion process being utilized. There are a number of variables which both individually and in concert will affect the completeness and efficiency of the heat, or that percentage of heat, from being effectively released to the heat absorbing surfaces of the boiler or heat exchanger and thus affecting the efficient heat release of the fuel as fired.

As related, this efficiency manual is by no intent provided to serve or function as a symposium on the techniques of combustion. Before a practical attempt can be launched by the reader to implement or adjust his equipment to increase its efficiency, then certainly a general idea of the ideal combustion requirements should be considered for educational benefit. The combustion techniques which will be discussed herein are essential to any and all fossilized fuel burning plants, irrespective of the fuel fired, the type of boiler used or the peculiar manner of pre-combustion preparation given to the fuel. Again, the subjects and areas presented for consideration are general and of universal application. Only after the reader has collaborated with his pertinent manufacturers' data and recommendations will he be able to apply any specific remedies or adjustments to his own peculiar combustion process. The reader will from this point on make reference to the primary fuel as fired in his plant, where particular references are made back to the fuel's composition or heat content.

COMBUSTION RUDIMENTS: TIME

As many aspects of scientific phenomena may be broken down into certain basic rudiments of explanation, so also may the science of combustion. The first rudiment of combustion is *time*. The combustible along with just the right proportion of oxygen (from the air) must remain in the combustion chamber long enough to completely and intimately combine chemically. Since all of the combustion aspects involving the main components of carbon, hydrogen and sulfur are of the exothermic or heat releasing type of reaction, the combustible mixture must therefore be delayed or accommodated long enough to give opportunity for the products of such combustion to release their full heat content, to assure their own propogation, and to liberate the greatest proportion of this available heat energy to the boiler. The time our combustion products remain in the combustion chamber and in the boiler itself is affected by the volume of combustible and air admitted to the boiler per unit of time. The admission of combustible and such air as required for its combustion is directly controlled by either the draft, as created by the chimney in natural draft situations, or by the admission rate as forced into the combustion chamber in plants.

COMBUSTION RUDIMENTS: TURBULENCE

The second rudiment in our scientific approach to combustion is manifested by the imperative requirement that the combustible and all components of the combustible thereof be brought into intimate contact with each other. The chemical reaction of oxidation is one which by nature of the reaction conducted is consummated with atomic proximity, meaning atom to atom and molecule to molecule distances. Our elements of combustion, such as carbon and hydrogen, certainly cannot be expected to combine with the oxygen provided if they are any further apart than an "electron's reach." Thus, by the action of *turbulence*, which also takes into account the atomization or reduction of combustible to its smallest possible size for further ease in exposing as much combustible to the provided oxygen, is meant the physical admixing or confrontation of all elements of the combustion process, which more often than not denotes agitation or turbulence.

COMBUSTION RUDIMENTS: TEMPERATURE

The third rudiment of combustion is the act of *temperature*. All combustible elements have a certain temperature at which

Table 4-1. Minimum Ignition Temperatures of Combustibles in Degrees Fahrenheit.

COMBUSTIBLE	TEMP.	COMBUSTIBLE	TEMP.
Acetylene	582	Isohexane	543
Allyl Alcohol	712	Kerosene	491
Ammonia	1204	Methane	1170
Anthracene	882	Methyl Alcohol	878
Benzene	1076	Methyl Chloride	1170
Benzyl Alcohol	802	Methyl Formate	456
Butane	826	Naphtha	450
Isobutane	1010	Naphthalene	1038
Carbon, Bituminous	765	Nitrobenzene	900
Carbon, Semibituminous	870	Nonane	545
Carbon, Anthracite	1110	Octane	446
Carbon Monoxide	1128	Paraffin	473
Carbon-Disulfide	248	Paraldehyde	460
Charcoal	650	Pentane	527
Di-ethylene Glycol	775	Petroleum Ether	624
Ethane	882	Phenol	1319
Ethylene	910	Propane	898
Ethyl Alcohol	738	Propyl Alcohol	822
Ethyl Benzene	1027	Isopropyl Alcohol	853
Ethyl Chloride	963	Propylene	856
Ethylene Glycol	775	Propylene Dichloride	1035
Gasoline (Regular)	536	Propyl Ether	372
Heptane	451	Propylene Glycol	790
Hexane	478	Sulfur	470
Hexyl Alcohol	572	Toluene	1026
Hydrogen	1065	Triethylene Glycol	700
Hydrogen Sulfide	558	Turpentine	464

they will most promptly combine with other elements of provisional character. This temperature is known as its *ignition temperature* or *kindling temperature.* Table 4-1 provides the ignition temperatures as required of 54 elements. After the elements have been ignited, this area of combustion must remain at the temperatures specified. This is known as flame propagation or the continuity of the cumbustion process. Generally, as long as the combustion chamber itself has adequate refractory to return a portion of the heat back to the flame area, propagation will continue on its own.

The temperature of the combustion process is affected by both the turbulence and the time the combustible components may be permitted to remain in contact. Turbulence exerts its influence by creating a through homogeneity of all of the participating elements of the combustion (carbon, hydrogen, sulfur and oxygen). Naturally, the more vigorously the agitation caused by the mechanically created turbulence becomes, the more throughly mixed will the participating elements be, preferably all within "an electron's reach" of each other to assure complete and perfect combination, which technically spells out perfect combustion. Thus, the entire picture of combustion may be represented by the three "Ts" time, temperature and turbulence. All three have a concerted and interrelated influence upon the final effect of combustion. If one aspect is to be deficient, than it will most definitely affect the other two, and in an adverse manner for sure.

For the purposes of evaluating the combustion process in terms of efficiency or from the viewpoint of how completely or perfectly the combustion is being carried out in the combustion chamber, we will first examine the temperature criterion.

We have arrived at certain known maximum heat releases which can be near accurately expected by the complete combustion of each and every fuel discussed, as specified in our preceding analysis of primary fuels. This, of course, is in most cases an academic and theoretical starting point. Since every investigation must start some place, then we will initiate our survey at the point of "thermal creation," which will for our purposes be the combustion chamber and the maximum temperatures which may be released there.

As specified, each and every fuel has a fixed maximum value, as per its quantitative and qualitative constituency. Referring to Table 3-1, Pennsylvanian anthracite coal yields a maximum of 13,130 Btus per pound fired.

When we review Table 3-2, which analyzes oil, we will find that grade #6, will yield a maximum of 17,800 Btus per pound. By turning to Table 303 we find that Oklahoma's natural gas will offer 20,160 Btu/lb. Since this is a gas, we will consider it as 974 Btus/cu. ft. Summarizing, every fuel has its own computable heating value if, and only if the theoretically necessary quantity of oxygen is provided for its combustion. We can in no way contain our computation to oxygen alone, since the atmosphere is composed of only 21 percent oxygen and approximately 79 percent nitrogen by volume, which is equal to approximately 23 percent oxygen and 77 percent nitrogen by weight. We will by necessity integrate the presence of the non-contributing nitrogen into our calculations. Therefore, when the term "air required" for combustion is given, it will be assumed that it is actually the oxygen in the air that is utilized, and proper corrections have been given for the useless and heat consuming nitrogen which also is unfortunately introduced into the combustion process.

THEORETICAL FORMULA FOR DETERMINING MAXIMUM COMBUSTION TEMPERATURE

Now I think we're ready to handle the first formula of combustion—the maximum temperature which may be reached if we are able to provide exactly the ideal 100 percent of air required for complete combustion. Our formula will be using only the 100 percent air required, as any excess air will only dilute the combustion process and serve to significantly lower the maximum temperature to be ideally achieved.

$$\text{theoretical flame temperature (Deg. F.)} = \frac{\text{(heat of combustion as taken from Analysis)} + \text{(sensible heat in fuel and sensible heat in air)}}{\text{(weight of products of combustion or volume of combustion)} \times \text{(mean value of specific heat)}}$$

Let us take for example 1 pound of Pennsylvanian anthracite coal, which we know to have a heating value of 13,130 Btus/lb. This pound of coal requires 15 pounds of air to be completely burned. I will also specify that the sensible heat in the fuel and in the air totals 50 Btus. Further, the mean specific heat of the products of combustion is 0.24. Having the necessary ingredients, which for the record have just been plucked out of thin air to facilitate our example, we will apply them to our formula:

$$\frac{13{,}130 \text{ Btus} + 50 \text{ Btus}}{1 \text{ lb.} + 15 \text{ lb} = 16 \times 0.24} = \frac{13{,}130 + 50}{16 \times 0.24} = \frac{13{,}180}{3.84} = 3{,}450 \text{ degrees F.}$$

Having applied the figures which are fair approximations, we arrive at the theoretical flame temperature of approximately 3,450 degrees Fahrenheit, which is very near what may be expected. Now that we've ascertained that a theoretical flame temperature can be achieved by applying certain existing factors and constants, let us learn of the origin and reasoning of these factors.

CHEMISTRY OF COMBUSTION CALCULATIONS

To many readers this discussion, which will be direct and of practical use, may be elementary. I'm sure combustion and the molecular principles by which it stands are basic to any engineer's education. Because this manual will deal in the abstracts and ideals, then a refresher in high school level chemistry will acquaint the reader with many terms and constants to be utilized hereafter.

The Molar Concept

When computing the weights and combining ratios of gases as related to combustion, it will be of great assist to consider them on the *molar* or *mole* basis. Simply, a mole is the *atomic or molecular weight* of a gas or substance expressed in pounds. The atomic weight of carbon is accepted as 12, or in its gaseous form there is one atom to the molecule, being therefore *monatomic*. A mole of carbon therefore weighs 12 pounds. Likewise, hydrogen is a *diatomic* molecule or mole having a mole weight of 2.0, since the atomic weight of hydrogen has been affixed as 1.0. Oxygen, whose atomic weight is given as 16, exists as a diatomic molecule. It would then have a mole weight of 32 lbs/mole. Thus, to determine the mole weight of any element, molecule or compound, all that is required is to multiply its atomic weight by the number of atoms it adjoins in a molecule and write pounds after the symbol.

Essentially, this concept of moles was introduced into engineering to facilitate the very same evaluation problems as we are doing in combustion calculations. Further, if we were to consider any number of gases—any oxygen, hydrogen, methane, carbon dioxide and sulfur dioxide having mole weights of 32 lbs., 2 lbs., 16 lbs., 44 lbs., and 64 lbs. respectively—all of these gases would occupy the same identical volume in cubic feet at the same pressure. Thus, it is evidenced by this rationale that all molecules of the respective gases must be of the same volumetric sizes. Only the corresponding weight for each is different. At 60 degrees Fahrenheit atmospheric oxygen has a mole weight of 32 and occupies 379 cu. ft. Hydrogen has the same volume, 379 cu. ft. The phenomena relating to gases and their respective weights, sizes, combining ratios and mole characteristics can be summed up by very specific laws of physics.

Law of Combining Weights

All substances, atoms and molecules combine in regard to very simple and definite weight relationships. These relationships are exactly proportional to their molecular weights per the constituents. For example, carbon, whose molecular weight is 12, combines with oxygen, which has a molecular weight of 32, to form carbon dioxide, which has a molecular weight of 44. So we may equate this in a molar formula to say that 12 pounds of carbon *plus* 32 pounds of oxygen will yield a mole of carbon dioxide weighing in at 44 pounds. Also, this molecule of carbon dioxide, because it behaves as a perfect gas, will occupy no more room than either a molecule of either carbon or oxygen.

One volume of oxygen *plus* one volume of carbon *equals* one volume of carbon dioxide. Although one might expect 1 + 1 to equal 2, in the case of combining molecules this is not so. When one molecule or atom combines chemically with another molecule or atom, a new molecule is formed which requires no more space or volume than the molecules required before its combination.

Avogadro's Law

Equal volumes of different gases at the same temperature and pressure contain the same number of molecules. If we were to have a number of containers, each having a volume of 379 cu. ft. and being under just atmospheric pressure and all the containers were at 60 degrees Fahrenheit, then each and every container holding whatever gas placed therein would have the same number of molecules. Only the weight of each container would differ. Since we're using the constant 379 cu. ft., then each container when weighed would identify itself by its individual mole weight of either 2 lbs., 16 lbs., 32 lbs. or 44 lbs. for the respective volumes of hydrogen, methane, oxygen, and carbon dioxide. See Figs. 4-1A through 4-1D.

Having a familiarity of the basic laws of physics, we'll move on to the next consideration—the actual exothermic reaction which our combustibles undergo in the combustion chamber. Exothermic, as you will recall from your high school chemistry days, is that reaction or combination of elements which give off heat during combination. An *endothermic* reaction absorbs or takes in heat during combustion. Since we're in the business of demanding the release of heat from our reactions, we will primarily be concerned with exothermic reactions.

Since our interest in chemical combinations are exclusive to those elements or molecules provided by our combustibles, which

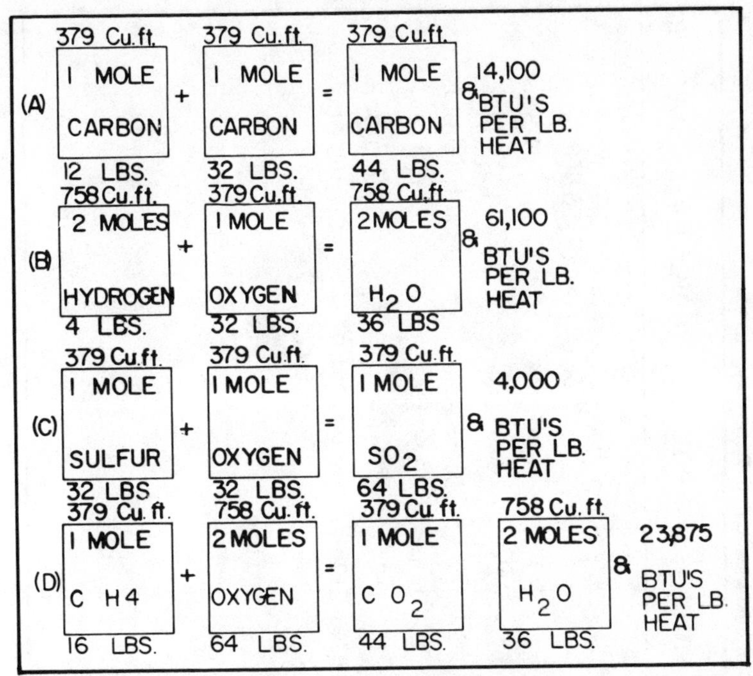

Fig. 4-1. Molar combination concept. (A) Carbon dioxide. (B) Water. (C) Sulfur dioxide. (D) Carbon dioxide and water.

in turn are provided by the fuel we're firing, we'll accordingly confine our study to the combination and heat release resulting from some elements and compounds. See Table 4-2.

Table 4-2 gives us the mole combinations for our combustibles using pure oxygen, which of course can only occur in a laboratory. As in the process of combustion, we will be utilizing air. To convert Table 4-2 for our practical purposes, we will add nitrogen to all of our computations; for every mole of oxygen used, 3.76 moles of nitrogen will also be introduced. Thus, we will add 3.76 moles of nitrogen to each side of our equation. Our actual reactions of combustion then become by the following examples:

$$CO \text{ TO } "CO^2" = 2CO+0^2+3.76N^2 = 2CO^2 + 3.76\,N^2$$

$$CH^4 \text{ TO } "CO^2+H^2O = CH^4+2O^2+(3.76)N^2 = CO^2+2H^2O+7.52N^2$$

$$"C" \text{ TO } "CO^2" = 2C+O^2+3.76N^2 = 2CO+3.76N^2$$

$$"H" \text{ TO } "H^2O = 2H^2+O^2+3.76N^2 = 2H^2O+3.76N^2$$

It will be noted that the nitrogen takes no part in the end product. The only effect nitrogen provides is to rob a substantial amount of the heat released. Thus, we will visualize that nitrogen is really a chemical bandit. It provides nothing whatsoever and

Table 4-2. Mole Combinations for Combustibles Using Pure Oxygen.

COMBUSTIBLE	REACTION	MOLES	POUNDS	HEAT RELEASED/LB.
Carbon To "C0"	$2C + 0^2 = 2C0$	2 + 1 = 2	24+32=56	3,960 Btus
Carbon To "$C0^2$"	$C + 0^2 = C0^2$	1 + 1=1	12+32=44	14,100 Btus
"C0" To "$C0^2$"	$2C0 + 0^2 = 2C0^2$	2 + 1=**2**	56+32=88	4,345 Btus
Hydrogen To H^20	$2H^2 + 0^2 = 2H^20$	2 + 1=2	4+32=36	61,100 Btus
Sulfur To $S0^2$	$S + 0^2 = S0^2$	1 + 1= 1	32+32=64	4,000 Btus
Methane	$CH^4+20^2=C0^2+2H^20$	1+2=1+2	16+64=80	23,875 Btus
Hydrogen Sulfide	$2H^2S+30^2=2S0^2+2H^20$	2+3=2+2	68+96=164	7,100 Btus

only serves to detract from the combustion process, which is one of the foremost reasons that we will be making efforts to limit the amount of excess air required in all applications. Even though an extra amount of oxygen may prove beneficial in securing completed combustion, for every 23 percent by weight of oxygen we admit to the combustion process, along with it comes 77 percent by weight of nitrogen. The nitrogen has to be heated along with the actual chemical products of the combustion.

To exemplify just how pernicious nitrogen is, for every 100 lbs. of air admitted we know we will have 23 lbs. of oxygen and 77 lbs. of nitrogen. Assume a specific heat for both oxygen and nitrogen as being 0.24. For every 23 lbs. of oxygen actually required for combustion, then 0.24 × 77, which equals 18.5 Btus, are taken unnecessarily by the nitrogen.

By referring back to our theoretical flame formula, we will further prove the heat deleting property of the presence of nitrogen. We should convince ourselves of the absolute deficiency realized by the admixture of excess air in the combustion process:

$$\text{flame temp degrees F.} = \frac{(\text{1 lb of carbon + 2.7 lb of } 0^2) \text{ (equaling 14,100 Btu's)} + \text{(sensible heat in mixture) (taken at 50 Btus)}}{\text{total weight of 3.7 lbs} \times \text{specific heat of 0.24}}$$

$$\text{flame temp.} = \frac{14{,}100 + 50}{3.7 \times 0.24} \text{ which equals } \frac{14{,}150}{0.888} \text{ equaling 15,000 degrees F.}$$

Therefore by using only the exact amount of oxygen, we can reach a temperature of approximately 15,000 degrees Fahrenheit. However, since we will be relying upon the ambient atmosphere to provide us with our oxygen, by only 23 percent by weight, the balance of 77 percent may be thought of as nitrogen (neglecting inert gases and water vapor). By including nitrogen into our formula by its respective percentage in the air, we now have the formula:

$$\text{flame temp. degrees F.} = \frac{\text{(1 lb of carbon+11.53 lb of air) (equaling same 14,100 Btu's)} + \text{(sensible heat in mixture) (taken as 50 Btus)}}{\text{total weight now of 13.53 lbs} \times \text{specific heat of 0.24}}$$

$$\text{flame temp. degrees F.} = \text{(1 lb of carbon+11.53 lb of air)} + \text{(sensible heat in mixture)}$$

$$\text{flame temp.} = \frac{14{,}100 + 50}{13.53 \times 0.24} \text{ which equals } \frac{14{,}150}{3.25} \text{ equaling now } \underline{\text{4,350 degrees F.}}$$

We have clearly proven that the normally proportioned nitrogen as mixed with the air will reduce a pure oxygen/combustible reaction from an achievable temperature of 15,000 degrees down to a temperature of 4,350 degrees. It will be quite obvious then that sometimes in an attempt to correct a high carbon monoxide problem by the increase of air, we may be raising our carbon dioxide by the decrease of the carbon monoxide. By so doing we may be more substantially robbing our initial flame temperature in the attempt.

ACTUAL AIR REQUIRED FOR COMBUSTION

By having prescribed the theoretical amounts of oxygen required and the necessary but heat detracting percentage of nitrogen also required, by virtue of their respective proportions in atmospheric Air, we may now knowledgeably compute the actual air required for combustion in our formulas:

air required=11.53 × % of carbon + 34.34 × % of (H^2-0^2/8) + 4.29 × % of sulfur

This formula tells us that the actual air required for the combustion of any fuel is equal to 11.53 pounds of air for every pound of carbon present in the combustible, plus 34.34 pounds of air for every pound of hydrogen present, minus one-eighth of the oxygen already in the fuel to produce water vapor, plus 4.29 pounds of air required for every pound of sulfur present in the fuel.

From Table 3-2, we find that #6 oil has the following precentages of combustible—carbon, 87 percent; hydrogen 9.5 percent, sulfur 3.5 percent; and no oxygen. Applying this data to our formula, we would then require:

11.53 × 87 percent + 34.34 × 9.5 percent ++ 4.29 × 3.5 percent

which necessitates 13.76 lbs of air per pound of #6 oil fired. Because we conventionally regulate our air by its volume, this 13.76 lbs of air is equivalent to 178.88 cu. ft. based on the factor that one pound of air occupies approximately 13 cu. ft. at 60 degrees Fahrenheit. When we more practically consider that our oil, in this case #6, contains 8.5 lbs/gallon, we will multiply 8.5 × 179.0 and obtain the desired 1,513 cu. ft. of air required for the combustion of one gallon of #6 oil.

We have to this point learned, by our investigations of primary fuels, the theoretical oxygen required and the theoretical air actually required, that we may, for any fuel considered, ascertain the theoretically ideal proportions of air required for the combustion of any given fuel In addition, we have proven that we can compute the flame temperature both by formulations taken by the fuel's maximum heat release and the amount of air/combustible involved in the reaction. All of these computations have involved only the theoretical amounts of oxidizing agent required.

There is little else that the engineer himself can either change or modify. He certainly cannot change the size of the combustion chamber, the design of the burner, the degree of turbulence of the air introduced to the combustion chamber, the amount or present condition of the refractory of the combustion chamber, or its effectiveness in assisting the propagation of the necessary refracted heat. There are, however, certain operational variables which the engineer may use to some degree of correction. The area of combustion efficiency by which the engineer may apply his skill and knowledgeable expertise to the greatest extent is in the modulation of the combustion air, in an effort to secure the maximum heat release from the combustible fired with the very least air admitted in the process.

I feel we should have now an adequate background in the analysis of the combustion process to examine the products of the combustion processes for which we have comprehensively covered thus far. The following recapitulation will be given to consolidate the technical essence of our collective study. The purpose of our investigation into the combustion process as given herein is to enable us to conduct a combustion analysis, consisting of a chemical evaluation of the products of combustion as done in our boiler; and an evaluation of the thermal efficiency of the boiler and combustion equipment, requiring thereby both a chemical and thermal survey for which we are now able to perform.

As you will recall, the component percentages of air are represented by 23 percent and 77 percent by weight of oxygen and nitrogen respectively. In the analysis of the products of combustion, we will be comparing and computing in terms of volume. Our percentages of oxygen will be taken as 21 percent by volume, and the balance will be allotted to nitrogen, which is 79 percent of the remaining volume. Although air also contains very minute quantities of inert gases and water vapor, for our calculations air will be assumed to be composed of only 21 percent oxygen and 79 percent nitrogen.

Referring to Figs. 4-2A, we will observe the theoretical products of combustion first for one pound of carbon, which when combined with the required 11.53 lbs. of air will produce 12.53 lbs. of flue gas, having a theoretical analysis of 21 percent carbon dioxide and 79 percent nitrogen. A more practical, but still theoretical, combustion process is diagrammed in Figs. 4-2B whereby one pound of #6 fuel oil having the composition as shown unites proportionally with the oxygen required producing its respective percentages of carbondioxide, water and sulfur dioxide. This analysis is given as theoretical because the process

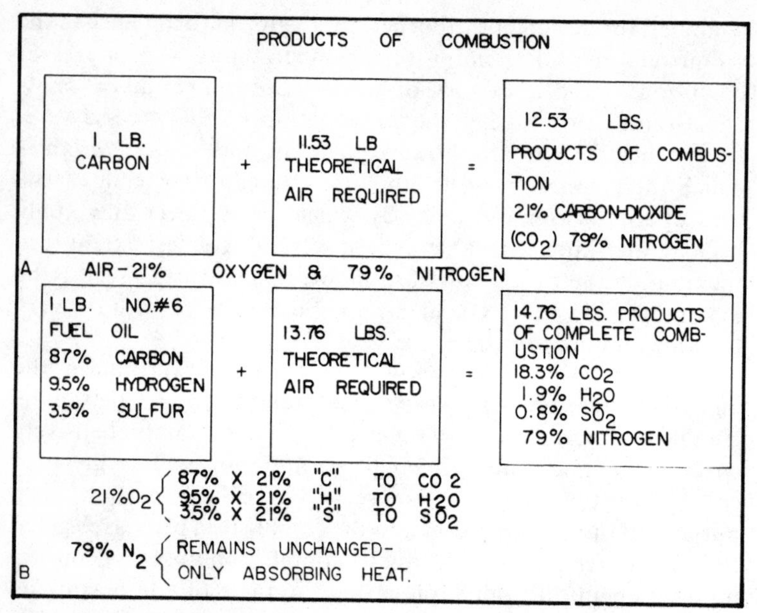

Fig. 4-2. Products of combustion. (A) One pound of carbon. (B) No. 6 fuel oil.

is utilizing just the exact 100 percent of air required. No reduction of the initial volume resulting from the inevitable condensation of water vapor is considered in this approximation. A more realistic approximation of the maximum carbon dioxide percentages are set forth in Table 4-3 along with the attending formulas for arriving at practical maximum carbon dioxide values.

I cannot overemphasize the importance of orsat analysis of the products of combustion. It is only by the intelligent interpretation of the orsat indications of percentages of carbon dioxide, carbon monoxide and oxygen, along with the inherent temperature differences as taken at the most intense combustion point and the exiting stack gases, that any efficiency evaluation may be had.

FACTORS AFFECTING THE ACTUAL COMBUSTION PROCESS

As repeated references have been made in regards to all of our calculations utilizing only 100 percent air required, we will now advance our study to the practical processes of combustion. We must understand why our processes will yield far from the ideal analysis as observed under theoretical circumstances.

Table 4-3. Maximun Carbon Dioxide Ranges Attainable for Combustibles Listed.

TYPE FUEL-COMBUSTIBLE	PERCENT CO^2 BY VOLUME
Fuel Oils	14.25% to 16.35%
Natural Gas Fuels	11.60% to 12.70%
Anthracite Coal	19.30% to 20.00%
Bituminous Coal	17.70% to 19.30%
Charcoal	18.60%
Lignite, Coke	19.20%
Pure Carbon	21.00%
Tar and Pitch	17.50% to 18.40%
Coke Oven Gas, (6% "C0")	9.23% to 10.60%
Black Liquor, Without Saltcake	18.60%
Blast Furnace Gas	24.60% to 25.30%
Wood, Average-Dry	20.10% to 20.50%
Tan Bark	20.10%
Bagasse	20.90%

$21.0 - 0^2$

$$\text{(A) Maximum } C0^2 = \frac{21.0\ C0^2}{21.0 - 0^2}$$

$$\text{(B) Percent Excess Air} = 100 \times \frac{0^2 - C0/^2}{0.264\ N^2 - (0^2 - C0/2)}$$

$$\text{(C) Maximum } Co^2 = \frac{100\ (C0^2 + C0)}{100 - 4.78\ 0^2 + 1.89\ C0}$$

$$\text{(D) Maximum } C0^2 = \frac{31.3\ C + 11.5\ S}{1.504\ C + 3.55\ H^2 + 0.56\ S + 0.13\ N^2 - 0.45\ 0^2}$$

First, due to the impossibility of bringing all of the combustible elements into such intimate contact with the limited oxygen in the introduced air as admitted to the combustion chamber, in the short time as necessitated in most situations, a much greater volume of air is required to avail the process with the percentage of oxygen required. This excess may be anywhere from 300 percent, as required in stoker or grate-fired coal plants; to 125 percent to 200 percent, as required by the typical oil-fired plant; to a low of 110 percent to 150 percent for the best maintained gaseous fuel plants. Thus, it is essentially the imperfection in the design of the combustion equipment and lack of through and absolute admixing of the primary and secondary air delivered to the area of combustion that circumstantially warrants the excess as must be supplied. This has a devastatingly reducing effect upon the carbon dioxide content as sampled from the products of combustion, along with a high oxygen content and lowered flame temperature. The increased volume of gases must consequently be brought up to flame temperature along with the "contributing" elements of the combustion process.

As can be appreciated by an evaluation of the various fuel compositions, in addition to carbon, there are also percentages of hydrogen and sulfur present in all fuels. These components also require their prorational amounts of the 21 percent by volume of oxygen admitted for total combustion, which has a taxing effect on the carbon dioxide percentage of the aggregative composition as sampled. That portion of oxygen used by the hydrogen will not reveal itself even as a substituting volume in the orsat reading. The water, or the water vapor thus formed, will condense in either the sampling tube or the cooling jacketed burette. It will create somewhat of a void in respect to the volume as originally represented by itself. The effect of sulfur oxidizing itself to sulfur dioxide will not be revealed in the standard orsat test; however, an orsat may be ordered with a special sulfur dioxide analyzer provided, as is featured by the Hays model 621 A. The additional sampling of the sulfur dioxide portions in the flue gas will provide a more definitive picture of the total combustion completeness, but a more through a comprehensive standard of completeness will be realized by experimentation upon the reader's own plant. What may be a relatively ideal orsat indication for one plant firing even the same delivered fuel, due to the variances in combustion chamber and process techniques, might prove unattainable in another plant. Table 4-3 offers more realistic ranges of carbon dioxide percentages as realized in actual practice.

INTERPRETATION OF THE PRODUCTS OF COMBUSTION

Much comment has thus far been given to carbon dioxide, vapor, carbon monoxide and free oxygen as sampled from the products of the combustion process. These products are in addition to the "riding nitrogen," which will be ignored in our discussions. It's just there and not much can be done either about it or to it. Its excess is indicated exclusively by the prorational amounts of free oxygen as indicated by the flue gas sample.

Carbon Dioxide

Theoretically, we desire to achieve the highest carbon dioxide percentage as possible. The maximum of course is 21 percent; however, as is corraborated by Table 4-3, this high value will probably only be realized in the combustion of pure carbon. All other fuels fall far short of this value. Percentages on the average of from 14 percent to 16 percent will be realized in oil-fired plants operating under very optimum conditions. I would be very happy if I found carbon dioxide percentages above 12 percent. These values can be achieved, and we must attempt to do so in our plants.

The percentage of carbon dioxide in the flue gas sample is essentially an index as to the excess or deficency of air (oxygen) and an indication of how well the combustion process is performing. Carbon dioxide indicates the air situation by providing a direct readout of what percent of the "captured volume" of gas sample (which is usually 100 cubic centimeters) actually has participated in the combustion process and has accordingly reacted to form carbon dioxide. We could very well reason then that were we to have a 21 percent carbon dioxide reading, then certainly all of the air has been used to contact and unite with the carbon and oxidize it to carbon dioxide. However, as has been related, when firing fuels with hydrogen or hydrogen compounds, some of the oxygen will be lost in forming water vapor. Were we to burn pure hydrogen, then we would find no carbon dioxide reading at all. This is why the maximum carbon dioxide in each fuel and each plant must be evaluated by empirical independence.

Generally speaking, a low carbon dioxide reading will evince an unwanted excess of air, with an accompanying high oxygen reading and no carbon monoxide. Conversely, a high carbon dioxide reading is satisfactory, providing there is no carbon monoxide present. A high percentage of carbon dioxide with indications of carbon monoxide tells us that we have arrived at our preferable carbon dioxide percentage at the cost of not enough air to thoroughly unite all of the carbonaceous elements in the combustible. As an example, let us consider the combustion of just one pound of carbon. This would require 11.53 pounds of air. If only the 11.53 pounds of air were admitted and complete combustion resulted, we could expect to have a reading of 21 percent carbon dioxide in our orsat. However, if we were to provide 23 lbs. of air to this same pound of Carbon, which is twice the amount of air required, or 200 percent, then we could only expect to have a maximum of 10.5 percent carbon dioxide in our sample. Accordingly, if we were to only admit one-half of the required 11.53 lbs. of air to this one pound of carbon, which would be approximately 5.75 lbs.' then the carbon dioxide would probably read about 21 percent, but with a considerable carbon monoxide level. Certainly all of the available oxygen would unite with the carbon. There would be a great amount of carbon left starved for oxygen. This would be evidenced by the great carbon monoxide level. In addition, we would probably be promptly informed by the air pollution authorities that a citation was being sent in the mail for producing too much smoke, as much unburned carbon would also result.

Carbon Monoxide

This gas results from the incomplete combustion or reaction of carbon to oxygen. This can be caused by an inadequate supply of air, or it may be due to poor combustion techniques. It has been my experience that this problem is more often a result of improper combustion of the fuel, poor combustion chamber condition, deteriorated refractory, poor burner operation, improper atomizing of oil or a host of contributing factors. The usual manner by which the carbon monoxide is made to disappear from the flue gases is to open the air delivery ports to maximum open position, and to cut back on the oil so far as to increase the excess air to about 300 percent to 500 percent. The matter is assumed under control, regardless of the fact that the efficiency has decreased to about 20 percent in the process.

The only remedy for carbon monozide, after assuring yourself that the fire is receiving at least the required amount of air, is to immediately investigate your particular combustion equipoment.

You wouldn't admit cold water to your boiler to bring down the pressure, would you? Of course not, and it's just as bad to remedy a high carbon monoxide reading by throwing open the air shutters. Again, carbon monoxide indicates incomplete combustion. By taking an orsat sample, we will determine if it is from an insufficient supply of air or from a fault in the combustion process—which means either the design, manner of operation, or the condition of the firing equipment. Were we to have a reasonable carbon dioxide reading and just a trace of oxygen, then this would clearly indicate that it's a process problem. The process itself is receiving adequate air. On the other hand, were we to find an exceptionally high percentage of carbon dioxide, no free oxygen and smoke, then our first recourse would be to increase the air supply slightly and see if that doesn't repair the situation. But from this writer's experience, high carbon monoxide percentages have invariably been caused by combustion equipment faults and improper firing techniques rather than from insufficient air.

Free Oxygen

Free oxygen is a direct indication of excess air. We want absolutely no oxygen to appear in our flue sample. If we are to maximize our combustion, then it must be done with only the minimum amount of air (oxygen) permissible, with no carbon monoxide or smoke and the highest carbon dioxide percentage achievable. The oxygen reading will be helpful in only two respects. First, if there is any question of the possibility of not enough air in the process, as questioned by a high carbon dioxide percentage, then the least presence of oxygen in the sample will rule out the insufficient air probability. The air/fuel ratio should then be adjusted until the oxygen disappears from subsequent samplings. Second, the oxygen reading will indicate the presence of air leakage into the boiler flue passages from leaks in the casing. This, of course, will not be encountered with positive draft boilers, but can be a literal pain with negative draft furnaces. It is very difficult sometimes to determine whether the excess air is coming from the combustion chamber or in through the side of the boiler casing.

Only separate sampling analysis will pinpoint the actual area where the free oxygen is gaining entrance between the burner and the breeching. Generally, free oxygen can be at first assumed to indicate an excess of combustion air, usually secondary Air to the combustion chamber. Only by orchestrating the readings of carbon dioxide and carbon monoxide will the matter be conclusively resolved.

Table 4-4 is given to provide the engineer with a handy reference guide to correlate the effects of excess air, carbon dioxide readings and exiting flue gas temperatures. This table is surprisingly valid, as it represents a proven approximation of the components of the combustion process which work in concert to affect the efficiency or deficiency in the total combustion process. To provide the engineer with a further approximation of actual operating ranges of excess air, I have provided, with the permission of the Babcock and Wilcox Company, Table 4-5 which sets forth the usual amounts of excess air, as found in practice for the combustion processes and fuels related therein.

DRAFT IN THE COMBUSTION PROCESS

Because of the misunderstanding and misconstrued significance of the effects of *draft* and its possible effects on combustion efficiency, a thorough definition of draft and its fundamentals will be needed. Although draft can and does affect all three factors of combustion—time, temperature and turbulence—by virtue of automatic control draft often lies dormantly at the root of many problems.

Draft is the flow of the atmosphere (air) caused by the heated products of combustion which, going up the chimney, are being continuously replaced by the colder and heavier primary and secondary air coming from the boiler room and admitted to the boiler for further combustion. This continuous flow or draft as it was first called, behaves according to all "displacement phenomena," meaning that it may be described and measured by its actual flow in terms of volume (as so many cubic feet or cubic meters per minute).

We know that when air (or any gas or fluid) becomes heated, it naturally tends to become less dense. This is another way of saying that a given volume will increase in size at a proportionate and predictable rate or curve in respect to its final absolute temperature. Let's say that we have a container with a volume of 100 cubic feet and a temperature of 60 degrees Fahrenheit, which is equal to 460+60=520 degrees. If we were to add enough heat to this 100 cubic feet of gas to double its absolute temperature to 520 × 2 = 1040 degrees (absolute), or 1040 - 460 = 580 degrees Fahrenheit, we would have also doubled the volume of the gas to a new volume of 200 cu. ft. The number of molecules in the container remain the same.

Table 4-4. Carbon Dioxide Stack Temperature Efficiency Guide.

EXITING FLUE GAS TEMPERATURE DEGREES F								
EXCESS AIR %	"CO^2%"	300	400	500	600	700	800	900
		RESULTING LOSS IN PERCENT OF COMBUSTION EFFICIENCY						
00.0%	15.0%	10.7	12.7	14.8	16.8	18.8	20.8	22.8
02.5%	14.5%	10.9	12.9	15.0	17.1	19.2	21.2	23.3
0.50%	14.0%	11.0	13.1	15.3	17.4	19.5	21.6	23.8
07.5%	13.5%	11.1	13.4	15.6	17.7	20.0	22.0	24.3
10.0%	13.0%	11.3	13.5	15.8	18.1	20.5	22.5	24.9
15.0%	12.5%	11.5	13.8	16.2	18.4	20.7	23.1	25.5
20.0%	12.0%	11.6	14.0	16.5	18.8	21.4	23.7	26.2
30.0%	11.5%	11.8	14.4	16.8	19.3	22.0	24.3	26.9
40.0%	11.0%	12.1	14.7	17.3	19.8	22.6	25.1	27.8
50.0%	10.5%	12.4	156.0	17.8	25.0	23.3	25.8	28.8
60.0%	10.0%	12.6	15.4	18.3	21.2	24.0	26.8	29.7
70.0%	9.5%	12.9	15.7	18.8	21.8	24.8	27.8	30.8
82.0%	9.0%	13.3	16.3	19.4	22.6	25.8	28.8	32.0
97.5%	8.5%	13.6	16.8	20.1	23.5	26.8	30.0	33.5
112.0%	8.0%	14.0	17.5	20.9	24.5	28.0	31.5	35.0
130.0%	7.5%	14.5	18.3	21.8	25.5	29.3	33.0	36.8
148.0%	7.0%	15.1	18.9	22.9	26.8	30.8	34.8	38.8
172.5%	6.5%	15.7	19.8	24.0	28.2	32.3	36.7	41.0
197.0%	6.0%	16.5	20.8	25.5	29.8	34.3	39.0	43.4
231.5%	5.5%	17.3	22.2	27.0	32.0	36.7	41.5	46.5
266.0%	5.0%	18.3	23.6	29.0	34.3	38.6	45.0	50.2
318.0%	4.5%	19.5	25.5	31.4	37.3	43.2	49.0	54.8
370.0%	4.0%	21.1	27.6	34.2	40.7	47.4	53.7	60.5
406.0%	3.5%	22.8	29.5	37.0	43.1	51.0	57.8	66.1
EXCESS AIR-%	"CO^2"	300	400	500	600	700	800	900
EXITING FLUE GAS TEMPERATURE DEGREES F.								

Table 4-5. Usual Amounts of Excess Air for Combustion Processes (courtesy of Babcock and Wilcox Company).

FUEL USED	COMBUSTION PROCESS-EQUIPMENT	EXCESS AIR-% BY WEIGHT
Pulverized Coal	Completely Water-Cooled Furnace for Slag-Tap or Dry Ash Removal	15% to 20%
" " "	Partially water-cooled furnace for dry ash removal	15% to 40%
Crushed Coal	Cyclone Furnace-Pressure or Suction Type	10% to 15%
Standard Coal	Spreader Stoker	30% to 60%
" " "	Water-Cooled Vibrating Grate Stoker	30% to 60%
" " "	Chain and Traveling Grate Stokers	15% to 50%
" " "	Underfeed Type Stokers	20% to 50%
Fuel Oils, #1-#6	Oil Burners, Register Type	5% to 10%
" " "	Multi-fuel Burners and "Flat Flame"	10% to 20%
Natural, and	Register Type Burners	5% to 10%
Coke Oven/Refinery Gases	Multi-fuel Burners	7% to 12%
Blast Furnace Gas	Inter-tube nozzle type burners	15% to 18%
WOOD, and Bark	Dutch Oven (10%-23% through grates) and Hofft Type Furnaces	20% to 25%
Bagasse	All Furnaces	25% to 35%
Black Liquor	Recovery Furnaces for Kraft and Soda-Pulping processes	5% to 7%

Notice that we are dealing with absolute temperatures and not the everyday Fahrenheit temperatures. The absolute temperature may be had by simply adding 460 degrees to the reading of any Fahrenheit thermometer. As a refersher to the reader, absolute zero, which is technically 459.69 degrees below zero Fahrenheit, is that temperature where there is absolutely no heat at all in a substance. Consequently, all molecular activity ceases. This is equivalent to minus 273 degrees Centigrade. Thus, the freezing point of water would be 460 + 32 degrees = 492 degrees absolute. The boiling point of water at standard conditions would then be 460 + 212 degrees = 672 degrees absolute.

Back to draft. Any given volume of a gas will expand or contract directly and proportionately as its absolute temperature. If we increase the absolute temperature of a volume of gas twice, then its dimensions will increase twice and obviously its density will decrease by one-half. If we lower the absolute temperature of a given quantity of gas by one-half, then we will be decreasing its original volume by one-half. We will likewise be doubling its density, or mass per unit volume. We can equate this idea by the following formula:

absolute temperature increase = volume increase + density decrease
absolute temperature decrease = volume decrease + density increase

When we admit air from the boiler room, typically at 60 degrees, or 60 + 460 = 520 degrees absolute, to the combustion process in the boiler, we raise its temperature to the typical combustion temperature of 2500 degrees, or 460 + 2500 = 2960 degrees absolute. We have at the same time increased the volume of the original gases by 520/2960, or approximately 5.7 times. We have simultaneously decreased the density of this air volume by 5.7 times.

Of course, this temperature of 2500 degrees Fahrenheit will be substantially reduced by the absorption of heat by the Boiler before it is released to the atmosphere. Or it should be at any rate, so we will next consider the temperature of the volume of gas as it enters the smoke stack, having been cooled down from 2500 degrees Fahrenheit to perhaps 400 degrees Fahrenheit.

Assuming that we were burning a pound of typical combustible in our boiler, which would require a typical 13 lbs. of air for its combustion we will then be considering a total of 14 lbs. of combustion products. Our problem now is reduced to the increased temperature of the 14 lbs. of spend products of combustion, which are now heading up the stack, at an absolute temperature of 860 degrees. The temperature of these exiting gases are, as we said, 400 degrees Fahrenheit, or more importantly 460 + 400 = 860 degrees absolute. We have then, from the heat added by combustion, increased the temperature (absolute) of this 13 lbs. of admitted air from 520 degrees absolute to 860 degrees absolute. We could say that we have elevated its absolute temperature by 520/860, or almost 1-1/6 times its original absolute temperature. We could, by the related functions as discussed, expect to have the volume increased by 1-1/6 times, which is exactly what has happened. Because this initial volume of gas has had its volume increased by 1-1/6 times, its density, or specific weight in lbs. per cu. ft. has been equally and oppositely reduced. Where the original 13 lbs. of gas had occupied the reduced volume of 13 × 13.5 = 175.5 cu. ft. at boiler from room temperatures, the gases as they are proceeding to the stack at their elevated temperature of 860 degrees are now expanded 1 1/6 times in volume, to an increased final volume of 1.6 × 175.5 = 280.8 cu. ft. in the stack.

Thus, our stack gases at 860 degrees absolute occupy 280.8 cu. ft. at a specific density of approximately 0.046 lb. per cu. ft. Originally, this same volume of 13 lbs. occupied (in its unheated state) only 175.5 cu. ft. and weighed at this volume 0.075 lb. per cu. ft. at its cooler and consequently more dense state. To recapitulate this most important comparison between the gases at their two points of temperature considerations, "T-1," or the original temperature of 520 degrees absolute, and the final temperature, "T-2," of the stack gases at 860 degrees absolute, have a volume and density difference. A difference of volume is increased to 280.8 - 175.5 = 105.3 cu. ft. above original volume, and a density difference is 0.075 - 0.046 = 0.029 lb. per cu. ft. in weight from its original density at boiler room temperatures.

Density Difference

We have just revealed the all important principle of draft—the difference in densities of two gases, or fluids, before and after becoming heated from original to final temperatures. In the particular case of air, and its form as the products of combustion. The evidenced differences are in the density of air at boiler room temperatures and the correspondingly lesser density resulting from its elevated temperature at the stack. We have clearly shown that the differences in densities of our hypothetical boiler with associated temperatures will produce, by related temperature differences, a change in densities of the traveling and participating gases. The boiler thereby acts as a heat pump displacing the hot gases into the stack and to the atmosphere and admitting colder replacement air.

The figure of density difference of our example—0.029 lb. per cu. ft. between the air supplied and the stack gases, is really typical of what may be expected in actual operation. As the temperatures of the stack fluctuate, and if air preheaters are installed or if cold air is admitted to the boiler for combustion, this figure will change to accommodate all of the variables as mentioned. For now, we will consider a difference in air densities of 0.029 lb./cu. ft.

A "container" was spoken of earlier to contain a certain volume of gas. We may very appropriately consider our stack as a container, only not having a lid on it. Taking a typical stack of 50 feet in height, this stack-container will then contain or confine a column of heated gas having a height of 50 feet. The gas is at the approximate temperature of the stack gases, or 860 degrees absolute. This 50-foot column of gas contained in our stack will have a specific density of 0.046 lb./cu. ft., or a difference of 0.029 lb./cu. ft. from that of an equal column of atmospheric air the same height, outside of the stack, and awaiting to enter the boiler. The heavier atmospheric air, at the given 60 degrees as mentioned, will have a specific density of 0.075 lbs./cu. ft., which creates our pressure or density difference of the derived 0.029 lb./cu. ft.

A column of combustion gas 50 feet high at 860 degrees absolute as contained in our stack will exert a pressure of 50 times 0.046 = 2.3 lbs. per square foot at the very lowest point of the stack. An equal column of atmospheric air 50 feet in height at 520 degrees absolute will exert a pressure of 50 times 0.075 = 3.75 lbs. per square foot, a difference in pressure as shown of 3.75 - 2.30 = 1.45 lbs./sq. ft. Since we wish to express this difference in terms of lbs./sq. in. we divide the 1.45 lb./sq. ft. by 144 and arrive at 1.45/144 = 0.010 psi.

Water Draft Indicators and U Tubes

A pressure of only 0.010, or one-hundredth of a pound per square inch, would indeed be very difficult to measure by any common instrument. We know that a barometric pressure of water at atmospheric pressure—14.696 psi absolute—will have a height of 33.9 feet. Thus, it would be to our advantage to use the equivalent height of water for the measurement of such a minute difference in pressure. In fact, water is the medium by which slight pressure differences are accurately metered. A pound per square inch of pressure may be dramatically indicated by its equivalence in inches or feet of water, one psi equaling 2.307 feet. By using a *manometer*, or water tube, very minute variations in pressure may be amplified by the equivalent height of water it will draw up an attached tube, when the other open end has the weight of the atmosphere pressing upon its surface.

Were we to attach a water draft indicator to our stack in question, the 0.010 psi difference in pressure would be sufficient to raise a tube of water 0.010 × 2.307 feet, or a distance of approximately 0.277 inches, clearly observed by a draft indicator. The construction of water level, or "U tubes" is to arrange the connected end of the "U" tube in such a manner that it will be on an incline to the vertical open atmospheric end. The drawn water will travel further due to its angularity with the open end. An inclined draft tube would cause the water to travel perhaps as much as 3 or 4 inches to equal the 0.277 inches vertically, as

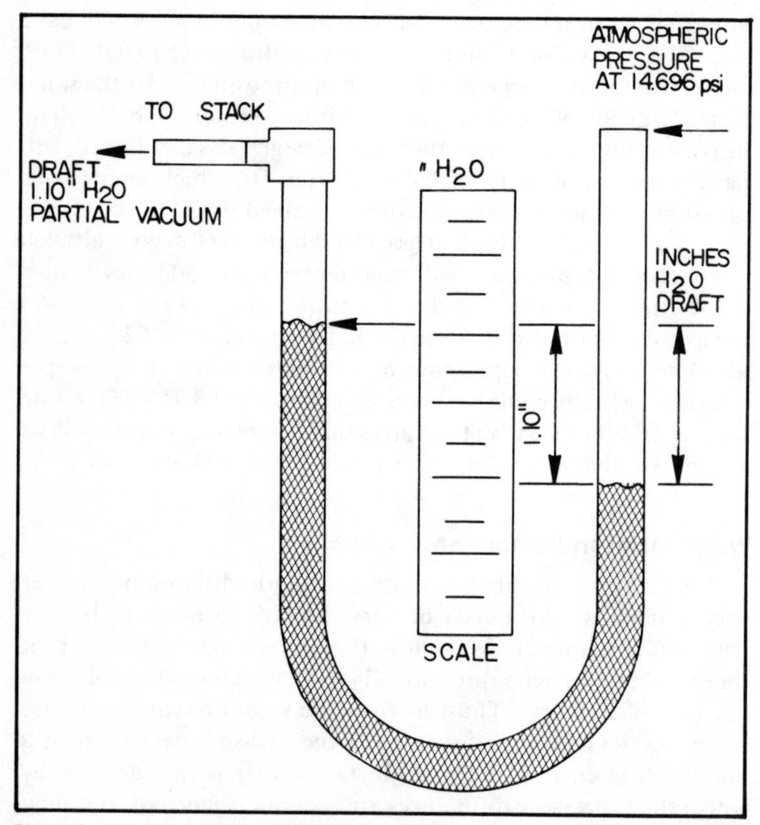

Fig. 4-3. "U" tube draft indicator.

would be required were it constructed as just a simple "U" tube device. See Figs. 4-3 and 4-4.

The creation and maintenance of draft for the purposes of combustion efficiency can best be explained by each respective aspect of combustion. We will then return to the "three Ts" of combustion—from turbulence and temperature. Primarily, a sufficient enough draft or "swept furnace displacement" should be maintained at all times. It should be adequate enough to assure that all of the products of combustion will be removed from the boiler-combustion chamber as promptly as they have been created by combustion. There should be enough time for these heat-bearing gases to have sufficiently and optimumly transferred their heat to the boiler surfaces, no sooner and no later.

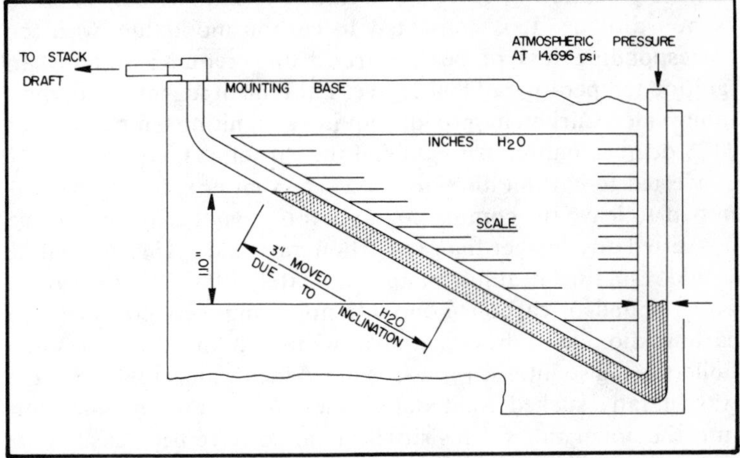

Fig. 4-4. Inclined draft indicator.

Theoretically, a balanced draft is sought to be achieved whereby the pressure inside of the boiler-combustion chamber is equal to the atmospheric pressure outside of the boiler. However, this delicate balance would be most difficult to maintain with any degree of consistency. So a slight negative pressure (less than atmospheric) is then needed to assure that the combustion gases will not tend to leak out of the boiler and into the boiler room. The gases tend to have an inward propensity for the atmosphere to very minutely percolate through the boiler setting and thus somewhat cool it but in not so great amounts as to unnecessarily dilute the combustion chamber with large cooling drafts or currents. Thus, we would have chosen between the lesser of two evils, the possiblity of outward leakage or the possibility of inward leakage, by attempting to keep on the negative side of an otherwise preferred balanced combustion chamber pressure situation.

Time

Draft may cause the components of the combustion process to leave the intended area of combustion earlier than desired by too strong a draft. With too great a draft, the proper time required for combustion will thus be shortened. The combustible and oxygen from the air supply will thus be separated before complete combustion may be accomplished. Particles of combustible will be pulled out of the refractory area and cooled below their ignition temperature upon the relatively cool walls of the boiler parts. A sizable portion of carbon will not be terminally converted to carbon dioxide. It is converted to carbon monoxide, with the corresponding loss of heat. Carbon only requires roughly an ignition temperature of 800 degrees Fahrenheit to ignite to carbon monoxide.. Carbon monoxide requires an ignition temperature of 1128 degrees Fahrenheit. Thus, if the carbon is not immediately converted to carbon dioxide, the carbon monoxide may be permitted to leave the combustion area too prematurely and not be converted any further than to carbon monoxide. This, of course, would result in a heat loss of approximately 1,000 Btus of heat for every pound of carbon monoxide not completely converted to carbon dioxide. I have actually witnessed the draft in some boilers being so intense (upwards of 0.5 inch water) that the flame was literally sucked right out of the combustion chamber and into the boiler tubes. This situation could have been easily corrected by merely reducing the prevalent draft intensity by opening the draft regulator slightly. The draft must not be too mild either, as a draft under 0.05 may cause the flame and the products of combustion to remain in the combustion chamber too long, retarding the flame propagation. Maintain just enough draft to assure the adequate and timely displacement of all products of combustion by the required rate of firing. Consult your manufacturers' recommendations.

Turbulence

The flame pattern and its designed purpose for thorough homogeneity of the fuel-air mixture prior to and during firing is affected to an appreciable degree upon optimum combustion chamber pressures, usually at about 0.02 inches water. The primary and secondary air shutters and delivery modulation vanes and the blower output in general are based upon certain expected back pressures or negative pressures in the combustion chamber. With an excessively high chamber pressure, the volume of air will tend to become choked by the "stuffed" combustion products. The air supply will then be lessened in proportion to that needed for complete combustion. If too negative of a pressure is encountered, there will consequently be a greater than computed pressure difference across the air control devices including the blower, which will deliver more of a supply of air than what is required. The effects of excess air delivery are chiefly responsible for low carbon dioxide, high oxygen and a reduction of flame temperature. If the excess supply of air delivered is admitted in such a manner as to abort or interfere with the flame propagation, then the presence of carbon monoxide will be evidenced, caused by the cooling effect of the intermediate stage transition of carbon to carbon monoxide and then to carbon dioxide. It will be deduced then that before an effective and meaningful flue gas analysis of the products of combustion is done the correct specified draft must be ascertained.

Temperature

The variations of draft from that of the ideal, both in an excessively higher as well as lower level, will affect the desired temperature of Combustion by several distinct routes. An excessively negative draft will cause cool atmospheric air to enter the boiler and combustion chamber through the porous walls of the setting and through all leaks and flaws in the boiler setting construction. This excessive inward leakage will cool the gases, reducing the maximum attainable heat release of the fuel fired and thus wasting fuel. The boiler is forced to higher than required positions of firing to compensate for the cooler products of combustion. The admittance of cooler atmospheric drafts of air may cool the carbon monoxide molecules and inhibit their progressive transition and exothermic reaction to the desired carbon dioxide, sacrificing approximately 1,000 Btus per lb. of combustible fired. The boiler becomes forced to higher rates of firing to accommodate this loss. With the higher rates of firing will be witnessed even higher intensities of negative draft to cause even more excess air to be unwantedly admitted.

With an inward draft, having certain random distributions other than that of through the flame itself, the air so admitted could seek an enveloping path around the flame and between the flame and refractory. The refractory would be cooled, reducing its radiant or reverberatory character and effectively defeating the purpose of the refractory, which is to return a portion of the heat back to the flame constituency to promote rapid combustion.

We have considered the fundamentals of draft, the effects of draft and the preferred manner of dealing with draft. Establish only enough draft or "furnace volume sweep" to facilitate the adequate removal of the products of combustion, while staying on the slightly negative side of the barometric. Generally, I have found that a draft of approximately 0.02 to 0.08 inches of water in the combustion chamber will provide a sufficient displacement for most liquid and gaseous fuels. Coal plant operation with natural draft alone will require considerably greater draft intensities, ranging anywhere from 0.05 to 0.5 inches of water depending upon the thickness of the fuel bed, temperature of the flame (average), height of stack, percentage of firing rating and other related variables.

Induced and Forced Draft

The rudiments as examined thus far relate to the creation of what is known as natural draft, obviously because it is created naturally. We may also establish an artificial draft by either induced or forced techniques. *Induced draft* is mechanical draft created by a fan or blower in either the breeching between the boiler and stack or in the base of the stack. Many years ago in some plants, a steam jet was placed in the stack. The gases were aspirated up the chimney. This stack jet technique was also utilized in railroad locomotives where the discharged steam from the cylinders was sent to the stack jet, thus inducing the necessary draft for combustion air to enter the firebox.

Forced draft is that technique almost universally employed today in all power plants for either coal, oil or gas-fired boilers. The forced draft is literally blown into the boiler wind box by a properly designed blower fan. Regulation is accomplished by choking the air suction to the blower, discharging air from the blower by air registers, regulating the exit damper position connecting the boiler to the stack breeching or all of these in concerted coordination. It must be recognized that regardless of how tall the stack of the plant may be, the stack will still exert a draft effect upon the internal volume of the boiler and combustion chamber when heated gas begins to flow through the stack. This negative pressure will add to that positive pressure as created mechanically by either forced or induced draft.

Draft Regulator

In plants using natural draft for both primary and secondary air delivery, the intensity of the draft will be determined by the height of the stack. The higher the stack, the greater the height of expanded and less dense gas it can contain, as balanced by the more dense and cooler atmospheric air contained about the stack in an equal height. The draft then may be controlled immediately by admitting cool air to the stack at its base to "kill the draft". Or the ashpit and fire doors may be positioned to choke the air admitted at the demand of the draft intensity. However, when the draft is controlled entirely by the air admission method, a partial vacuum or negative pressure will still exist in the boiler proper. The draft is still there, pulling on all areas of the combustion chamber in an attempt to equalize itself. The obvious answer to this problem is the use of a draft regulator, which is simply a hinged and adjustably weighted flap placed in the boiler stack breeching (usually). When once adjusted, it will tend to keep the draft intensity in check.

The draft regulator works by simply having a swinging flap or door, opening only inwardly, with a balance weight of an ounce or so placed at its selected gravity point. As the draft increases in its negative or inward intensity, the door or flap that is positioned between the negative draft and the positive atmospheric pressure

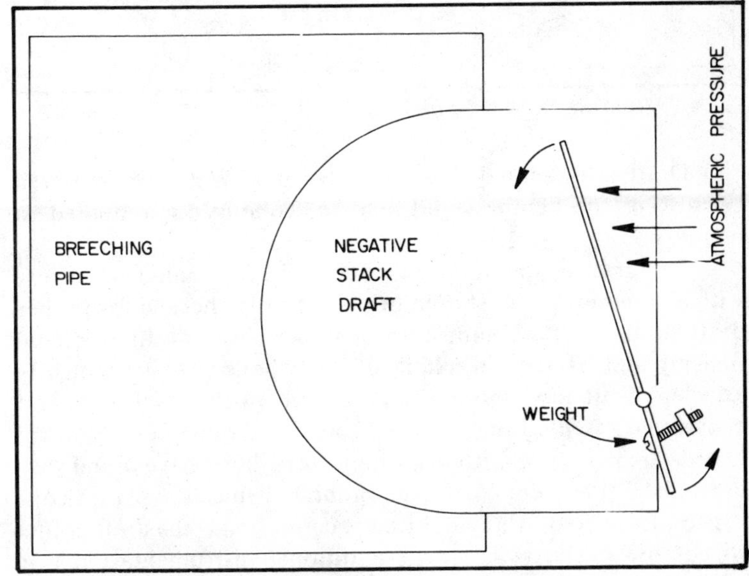

Fig. 4-5. Side view of a draft regulator.

will then be pressed inward to equalize this pressure difference. As the door is being pushed inward, the atmosphere is rushing into the breeching and up the stack. This will dilute the lighter hot gases in the stack and cause this now diluted admixture of gases to be more closely the weight of the atmosphere, thus "killing the draft." When the draft is either above its desired intensity or corrected by outside dilution from the regulator having been opened, the weight which has been accurately set for just this purpose will tend to pull the door shut, thus keeping out the atmosphere or restoring the draft. See Figs. 4-5 and 4-6.

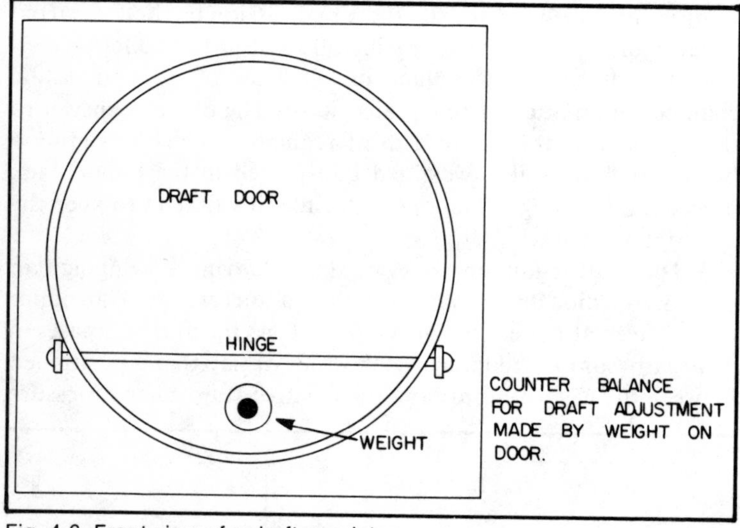

Fig. 4-6. Front view of a draft regulator.

Draft is measured in inches of water. Water was selected because of the lightness and relative economy as compared to mercury.

The draft of the boiler, or each and every boiler in battery with a common stack, should be periodically checked for proper draft, both in the combustion chamber (over the fire in coal boilers) and at the breeching of each boiler. A permanently attached draft tube should be connected to the stack base and considered essential to the boiler room log. Most boiler control boards are equipped with draft indicators, but I have found very few indeed that were correct or calibrated since the plant's conversion from coal. Many old timers could "read" the draft points so comprehensively that little additional instrumentation was required for combustion.

CARBON MONOXIDE RELATED EFFICIENCY LOSS AND DETERMINATION

The complete and perfect combustion of carbon compound fuels result terminally in the formation of carbon dioxide, water vapor and oxides of any other elements of the fuel source, such as sulfur dioxide, sulfur trioxide, nitrous oxides, etc. The actually incurred products of combustion will of course be indicative of the original qualitative analysis of the fuels' chemical constituency. Each representative element will be present in the sampling burette as obtained at the stack connection.

The combustion of carbon, hydrogen and sulfur, (excluding nitrogen) are of the exothermic reaction, which is heat releasing, as opposed to the endothermic type, which will absorb heat from the "field of action." The amount of heat, expressed either in Btus per pound of combustible burned or in "K" calories per kilogram of combustible fired, will of course yield a fixed and identifiable maximum quantity of heat units.

Carbon may be oxidized during this exothermic reaction in one of two possible states or final compounds. The ideal oxidation of carbon will yield carbon dioxide. Its optional resultant may appear in the form of carbon monoxide. Carbon dioxide will yield in its final combination, or oxidation end product, a total of 14,540 Btus per pound of carbon burned (or oxidized). Carbon monoxide, which is formed usually from the inadvertent lack of oxygen in the combustion area, will yield but a maximum heat release of only 4,380 Btus per pound. Thus, we may readily deduce that there is a difference of 14,540 - 4,380 = 10,160 Btus lost when carbon is oxidized.

Another view of this difference in oxidation heat release is to consider carbon monoxide as an aborted attempt to reach its naturally intended perfection to the state of carbon dioxide—at the expense (paid for by the power plant fuel bill and passed on to the public) of a difference of approximately two-thirds of its possible or potential heat release. It needs then no further economic analysis to reinforce the claim that carbon monoxide is a bandit of fuel and one that we cannot afford to tolerate, especially at todays exorbitant fuel costs. Carbon monoxide may be and will be present in the products of any imperfect or incomplete combustion of any fuel—solid, gaseous or liquid. If the combustible contains carbon, then it may escape the combustion process without transformation to its complete oxidation to the ideal carbon dioxide "end product." This refugee from the combustion process can readily escape and carry off with its almost two-thirds of the intended maximum available heat in its incompleted exit from the boiler furnace.

Correcting Incomplete Combustion

The correction of incomplete combustion can be affected by any first class Engineer by referring to basic combustion science. For a typical boiler installation, even the presence of only 2 percent of carbon monoxide in the flue gas sample will mean an average loss (calculated from average #6 fuel oil) of 1,185 Btus lost for every pound of fuel fired. We could consider it to be approximately equal to from 5 to 8 percent heat loss for every pound or gallon fired. If we had a hole in our fuel storage tanks that leaked out a steady flow of fuel at this continuous rate, you can bet your job that you'd be sealing up that leak in a hurry. You would lose your job if the plant management were to learn of such an uncorrected waste of precious fuel.

But how about the same loss that's going up the stack 24 hours a day for every operational day? Doesn't that rate equal attention? Not only is it a loss from the standpoint of operational efficiency, but it's an objectional and overt contribution to the pollution of our atmosphere. Let's contain that carbon monoxide where it belongs—in the combustion chamber—and keep it there until we can convince it to convert to the heat releasing end product of carbon dioxide.

Just how much heat are you losing from your present combustion process? Assuming of course that you are a first class engineer, you will have at your disposal an orsat test kit and the experience in employing the same in collecting and analyzing a flue gas sample. Now that you've analyzed your latest gas sample, we'll consider, for the specific purpose of this investigation of possible carbon monoxide presence, the measured percentage of carbon monoxide collected in the sample taken. This will be the difference of the carbon monoxide subtracted from the total sum of the carbon dioxide and oxygen readings.

Having at hand the percentage of carbon monoxide expressed as a percentage, we'll insert this percentage in the following formula as carbon monoxide. Heat lost in Btus per carbon consumed in the fuel as fired (obtained from a chemical analysis of the fuel) will then equal:

$$\text{Heat loss (Btus/lb.) "H"} = \frac{C \times 10{,}160 \times \text{"C0"}}{CO^2 \times \text{"CO"}}$$

Whereas CO_2 = percent carbon dioxide   C = percent carbon in fuel consumed.

10,160 = heat lost by difference in incomplete combustion of carbon from carbon monoxide (CO) to carbon dioxide (CO)

Let us consider the following illustrative example, which is about as typical as will be encountered in the average oil-fired steam plant, with fairly attentive personnel in charge.

flue gas analysis: CO_2-10 percent; CO-2 percent; carbon in fuel consumed-70 percent

applied to our formula: Btus lost from carbon monoxide =

$$.7 \times \frac{10,160 \times 2}{10 + 2} = .7 \times \frac{20,320}{12} = .7 \times 1,693 =$$

a grand total of 1,185 Btus lost from 2 percent of carbon monoxide

It will be noticed that it doesn't take very much carbon monoxide in the fuel sample (products of combustion) to represent a very sizable loss in the consumed fuel. The higher the percentage of carbon in the fuel, the higher will this same loss (in percent carbon monoxide) ultimately reveal itself.

Heat Loss Formula

We may further apply our deductions to express the percentage of fuel or carbon lost by the presence of carbon monoxide We find the following formula available for representing the determined heat loss in Btus per pound of carbon consumed to the weight of fuel fired. We may of course substitute or rearrange the formula to accommodate-either weight in pounds, cu. ft., or gallons of fuel consumed, just as long as we maintain the units throughout the entire formulation. Here is the formula for the determination of loss in percent of fuel as fired, based upon the percentage of carbon content as identified by analysis:

$$\text{percent heat loss - percent (resulting from CO)} = \frac{\text{Btus/lb. lost by carbon monoxide (CO)}}{\text{max. Btu value of fuel (*)}}$$

(*) max. Btus determined by bomb calorimeter or other suitable test.

Btus lost from 2 percent CO from above example = 1,185 Btus

heat value of fuel/lb. (#6 L/S oil) = 19,000 Btus

$$\text{percent heat loss} = \frac{1,185 \text{ Btus}}{19,000 \text{ Btus}} = 0.06 \text{ percent heat lost from 2 percent CO in sample.}$$

Carbon Monoxide is the union of one atom of carbon and one atom of oxygen. This combination is only a partial or half consummation of the ideally desired carbon dioxide molecule which could and would result, providing the atom of carbon could be brought into intimate contact or proximity with the necessary two atoms of oxygen. When carbon is only partially oxidized to carbon monoxide, it will only liberate 4,380 Btus of heat per pound of carbon (thus incompletely oxidized). Carbon dioxide on the other hand will yield, after being completely oxidized, a maximum of 14,540 Btus of heat for the same amount (1 lb.) of carbon consumed in the combustion process. We therefore have, as a result of the incomplete combustion of carbon to its potential carbon dioxide compound, a substantial loss of 10,160 Btus of heat per every pound of carbon consumed and incompletely rendered to carbon dioxide. This loss, or more importantly the quantitative percentage of its appearance in the flue gas sample (drawn from the products of combustion), may by the application of the formulas provided herein be expressed as a near accurate account of the actual Btus lost due to the quantitative presence of the detected carbon monoxide.

SUMMARY OF IMPORTANT POINTS

We may, with a reasonable degree of accuracy, predict by chemical or ultimate analysis the "expected maximum heat release" of a fuel, either by relating to its theoretical analysis by laboratory report or by having a test specimen submitted to a bomb calorimeter test. These results will provide us with a "starting point figure" of what expectable heat in Btus per pound or by the cubic foot of combustible we may realize under complete and perfect combustion, under ideal conditions.

By determining the Qualitative and Quantitative composition of the actual combustible matter in the fuel, we may determine the amount of oxygen required for its complete combustion to its respective oxides. Referring to the formulas given for the amount of air required for the combustible matter, we may arrive at a theoretical air value needed for our particular fuel.

After having conducted an orsat gas analysis of the products of combustion from the exiting flue gases, we may then be in a position to adjust the actual air admitted to the combustion process. This orsat analysis will tell us how effective our combustion process is in respect to uniting the components of combustion with respect to the air/fuel ratio applied to achieve combustion. The respective readings of carbon dioxide, carbon monoxide and oxygen will each indicate, by their relationships with the combustion flame temperatures, an overall presentation of the efficiency of the combustion process. This collected data when applied to the tables provided will indicate the point of present efficiency. The data will point the engineer in the proper direction for increasing the efficiency of the concerted operation by indicating the critical area of needed attention.

After having arrived at the maximum heat release obtainable for the fuel under study, the maximum flame temperature may then be calculated by the formula provided herein. The maximum flame temperature will be a deducing factor in determining the efficiency of the combustion process and a theoretical "starting point" for our computation in respect to the efficiency of the boiler as a heat absorbing entity. The initial flame temperature is an integral factor in the very simple input/output relationship which will indicate the thermal efficiency of our boiler. The maximum flame temperature, as we have learned in the preceding evaluations, is affected directly by both the achieved completeness of combustion and the amount or percent of excess air admitted.

We have admitted the absoluteness and theoretical fabric of our discussions on the computationally arrived at figures of combustion values, and have given appropriate attention to the practical as opposed to the theoretical in our study of excess air required. This subject will have to be dealt with by the engineer in respect to the operating peculiarities of his own equipment as mentioned. Certain basic parameters and guiding hints may certainly be discovered by reviewing the facts and figures thus far presented. The actual excess air required for any individual plant can really best be determined by empirical trials in conjunction with the manufacturer's data and recommendations.

Any combustible fuel may be calorifically analyzed to ascertain its maximum available heat energy, which can be released under ideal conditions. The actual percentage of the heat so released from the fuel will depend upon the effectiveness of the combustion process, which takes into account the completeness of combustion and the amount of excess air supplied above that which is actually required.

To determine the thermal efficiency of the boiler, it is necessary to determine the initial flame temperature by either a pyrometer (or any other physical means) or by near theoretical calculations. The intensity or temperature of the combustion process will be directly affected by the amount of excess air and the completeness of combustion.

The optimum situation of combustion is to exact the most heat from the fuel fired. This can, in most instances, be secured by maintaining the highest percentage of carbon dioxide attainable in the flue sample with absolutely no carbon dioxide nor free oxygen present in the flue gas sample. This will most generally also assure that the flame temperature, or combustion intensity, is at the most achievable level.

There exists, by virtue of each and every type, variation and condition of fuel burning plant, a certain inherently achievable maximum efficiency which can be attained under the operational

Table 4-6. Fuel Combustion Survey: Part A.

Name of Company: ____________________

Location-Mailing Address: ____________________

Survey Requested By: ____________________ Title: ____________________

Date of Survey: ____________________ Time Required: ____________ Material: ____________

Survey Conducted By: ____________________ Reason: ____________________

PART " A " **FUEL - COMBUSTION SURVEY** **TYPE FUEL** ____________________

$$\frac{\text{BTUs} \quad \text{Lb / CuFt.}}{\text{Calorific Value by Analysis}} \quad \frac{\text{BTUs} \quad \text{Lb / CuFt.}}{\text{Actual Heat Released per Unit - by Combustion}} \quad \frac{\text{BTUs / Combustion}}{\text{BTUs / Analysis}} = \frac{\_\_\_\_\_\_\_\_\_\_ \%}{\text{Efficiency}}$$

1) Calorific Formula: Btus/Lb. = 14,500x%c + 62,000x%(H^2-0^2/8)+4,050x%s. (by weight)

2) Btus Released by Combustion

$$:= \frac{\text{Measured Actual Flame Temperature of Combustion (degrees F.)}}{\text{Weight of Combustible (+) Weight of Air Supplied (X) Avg. Sp. Heat}}$$

3) Theoretical Flame Temp. (degrees F.)

$$= \frac{\text{Calorific Value of Fuel (+) Sensible Heat in Air (+) Fuel}}{\text{Weight of Provided Air/Fuel Mixture (X) Avg. Sp. Heat}}$$

4) Short Method of Temperature Eff.

$$: = \frac{\text{Actual Measured Flame Temp. by Pyrometer}}{\text{Theoretical Flame Temperature by Formula}} = \_\_\_\_\_\_\_\_\_\_ \%$$

5) Theoretical Air Required : = (In Lbs.) = 11.53 x%C. (+) 34.34 x% (H^2-0^2/8) + 4.29 x%S. (Dry Air)

6) Percent Excess Air Supplied

$$= \frac{\text{Theoretically Required Amount of Air}}{\text{Actually Delivered Amount of Air}} \text{ (X) } 100 = \_\_\_\_\_\_\_\_\_\_ \%$$

characteristics and procedures as mechancially dictated by this plant in respect to its mechancial ability to convert the energy of the fuel it fires whether or not this "built in" ceiling of maximum efficiency is 50, 60, 80, or 99.99 percent the inherent percentage is nevertheless present as a product of the plant's related characteristics as a fuel utilizing entity. It is therefore the responsibility of the engineer to determine just exactly what that maximum ceiling for efficiency is and, of equal importance, to achieve it.

SURVEY FORM

No degree of efficiency whatsoever may be attained without an intelligently and logistically conducted survey to ascertain an evaluation of the plant's status in respect to operating efficiency. This data, when applied to the engineering formulas as provided herein, will direct the engineer in regard to the efficiency of the plant at present and what specifically must be done to upgrade the percentage of efficiency to a more optimum value.

In Table 4-6 is the thermal-fuel survey. This form requires the collection of pertinent information: heating value of fuel, calculated air required and the related formulas required for the meaningful translation of the applied data as collected. In this form will be found the pertinent formulas given for each evaluation or computation. I have included the formulation required in each section and sub-section to facilitate the computations, which will save the engineer the otherwise time-consuming cross references necessary from the manual.

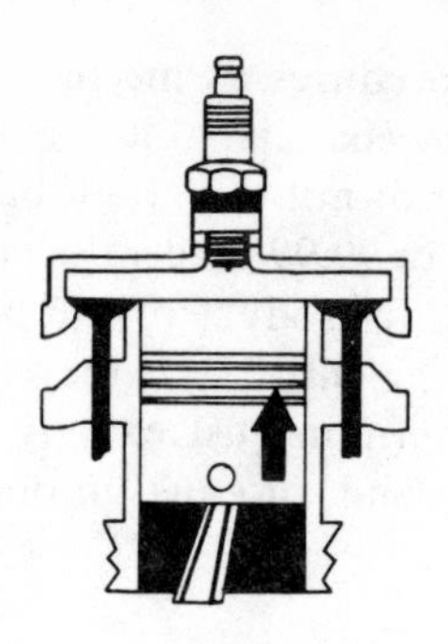

Chapter 5
Boilers and Steam Generators

By design, the purpose of a boiler or steam generator is to absorb the heat released in the combustion chamber and transfer this heat to the medium or substance contained within the boiler or generator. There are several prominent perspectives we may view when we discuss the functions, capacities and characteristics of a boiler or steam generator. The function and thermal characteristics of a boiler make it absorb the greatest amount of heat applied, known as *input* per amount of heat actually released in the combustion chamber by the fuel so supplied, known as *heat release.* The end product of this exchange of heat energy from the fuel to the boilers heat absorbing surface area will result in the certain measurable quantity of steam (or heated water) generated by the heat input. This generated steam is known as and hereafter to be referred to as output.

It's interesting to note the manner by which heat transfer is accomplished, in respect to the boiler being of the fire-tube, water-tube or any other novel varieties (Figs. 5-1 through 5-4). The types of water circulation for each variety are natural or forced circulation. The heat-bearing flue gases (or actual products of combustion) are made to travel across and through the heat exchanging surfaces by either natural draft (as created by "chimney effect") or the more modern system of forced draft, which may be provided by mechanical or steam jet effects.

The manner or modulation of steam output, as affected by combustion, is controlled chiefly by the circumstances of end-point control as related to the "set-pressure" at which the system is expected to reach and maintain, also known as header pressure.

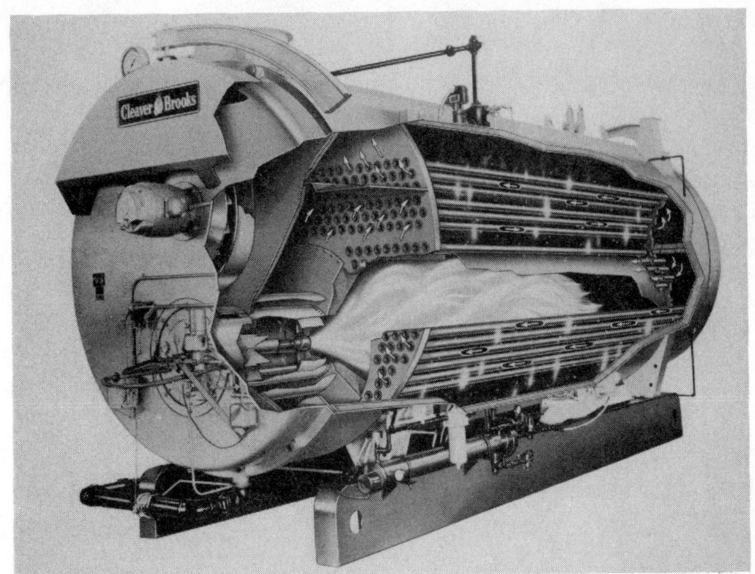

Fig. 5-1. Cutaway of a fire-tube boiler (courtesy of Cleaver-Brooks Co).

This aspect is regulated by the degree of instrumentation, or automatic controlling devices incorporated in the boiler/combustion equipment. The degree, or extent of control may be entirely automatic, semi-automatic or manual. Invariably, the combustion process will be actuated and controlled by the set header

Fig. 5-2. Delta water-tube boiler (courtesy of Cleaver-Brooks Company).

Fig.5-3. A "CN" burner (courtesy of Cleaver-Brooks Company).

pressure expected to be maintained, regardless of the demand conditions incurred.

The boiler or generator may be of the coal, oil or gaseous fuel fired variety. Figures 5-5 through 5-20 show some boilers and burners. The boiler may utilize a combination of these fuels, either separately or simultaneously, as "availability/cost" ratios might contemporaneously favor. It has been by practice the custom to utilize as much gas as is available and then returning to oil when the permissible gas allotment has been consumed in dual-fuel plants. However, in the larger utility plants where coal is the principal and only fuel, this "prime recourse" system is not so enjoyed.

Fig. 5-4. Commercial water-tube boler (courtesy of Cleaver-Brooks Company).

DISCUSSION OF VARIABLES

It must be emphatically underscored that there are only a given number of variables which are under the control or command of the engineer. These *variables*, as opposed to the fixed characteristics or functions of the plant, then will be the topics of our discussion.

The engineer, of course, has no ability to change or modify the boiler he has been assigned to operate. If this boiler be of the fire-tube design, then the fire-tube design it shall be and nothing is going to change that. If the boiler is one of the saturated steam type, then there is nothing the engineer can do about this problem either. He must just persevere with saturated steam until it is

Fig. 5-5. Multi-fuel burners on power station boilers (courtesy of Babcock and Wilcox Company).

deemed economically advantageous by the plant's management to install a superheater bank. This offers less condensation problems and longer turbine blade life.

Fig. 5-6. Multi-fuel oil-gas burners installed in a combustion chamber (courtesy of Babcock and Wilcox Company).

Fig. 5-7. High pressure oil burners (courtesy of Babcock and Wilcox Co.).

If the plant is exclusively set up for the firing of #6 oil, then there is absolutely nothing the engineer can do to enjoy the benefits of firing pulverized coal. The engineer's only hope is that soon his company will make the large expenditure for a cyclone furnace.

The practical concept that I'm expounding is the engineer should be held accountable only for those variables not inherently "fixed" by virtue of actual construction or functional features of the equipment. Conversely, the engineer does have certain responsive variables at his disposal. The operational and accordingly controllable functions as follows:

☐ The air/fuel ratio, as provided by "superimposed trim" of the system's combustion regulation, as incorporated in all instrumentation, automatic or otherwise. It affects ideal combustion characteristics and combustion efficiency.

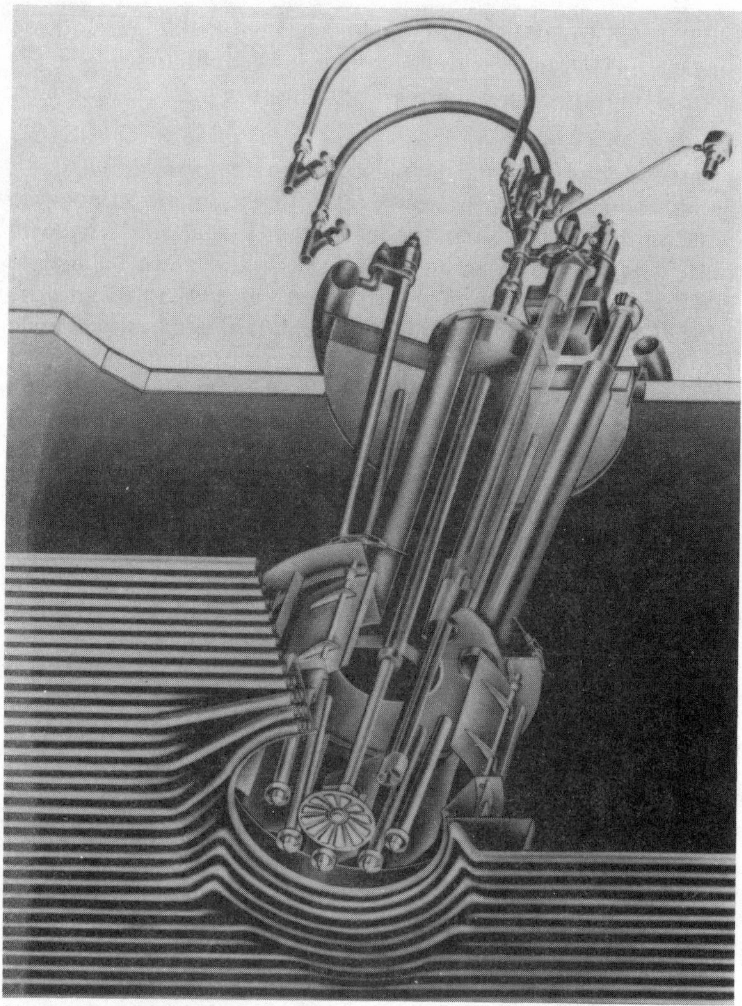

Fig. 5-8. Steam atomizing oil burner (courtesy of Babcock and Wilcox Co.).

☐ The draft, or damper regulation, may be adjusted to retain the flue gas heat within the boiler proper for greatest and exhaustive heat transfer and thermal efficiency.

☐ Fuel oil temperature as required preparatory to admission to the burner.

☐ Number of boiler units to be selectively put "on the line" to meet the load requirements, which in itself will dictate total thermal efficiency.

☐ The proper feed water treatment to control the deposition of thermally insulating (and dangerous) scale and salt compounds.

This will impede the rate of heat transfer to the water and thus adversely affect the thermal efficiency. Also in this consideration are the effects soot and ash have on the thermal efficiency of the unit. The presence of these products of combustion also perform an insulating effect upon the heat transfer rate and reduce the thermal efficiency.

□ The water level and its desirable admission per rate of evapora-

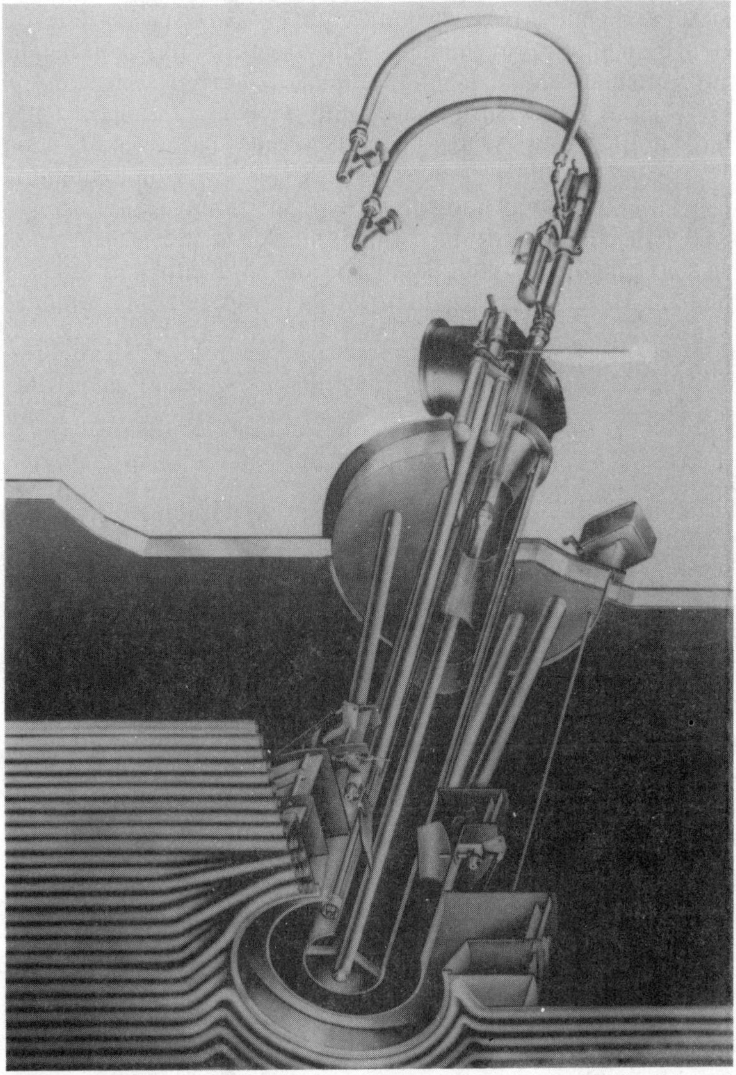

Fig. 5-9. Steam or air atomizing oil burner (courtesy of Babcock and Wilcox Company).

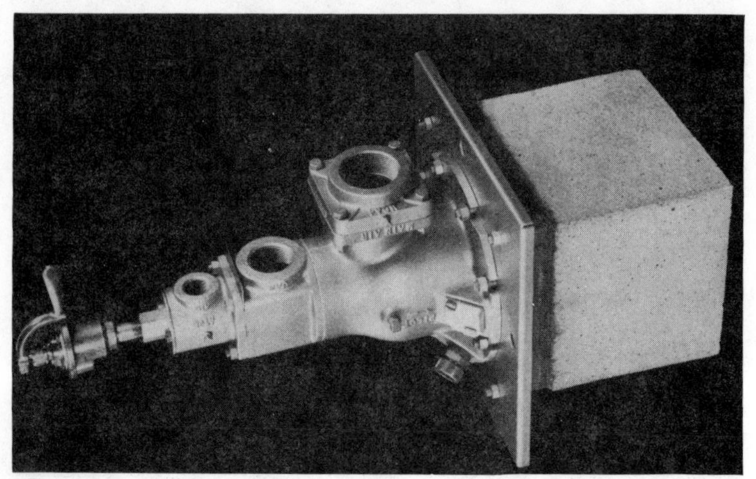

Fig. 5-10. A dual-fuel burner (courtesy of North American Manufacturing Company).

tion is influential to efficiency and under the control of the engineer. By improper or unresponsive level control, an abnormally high level may be constantly maintained which would provide a high degree of moisture in the steam and reduce its total evaporative heat content per cubic foot generated.

☐ The amount of boiler blowdown as required to reduce concentrates is related to feed water treatment. The misadjustment of this variable can account easily for up to 10 percent of thermal loss. Most continuous blowdown is, as the term describes, continuous, and can have an unbelievable accumulative effect over a 24-hour period.

☐ The proper and required insulation of the boiler, piping and all boiler system appurtenances can easily account for another 2.5 percent to 15 percent of heat loss. The steam and heated water leaks have proven to be of equal loss.

☐ Improper adjustment of the combustion modulation control system is often responsible, for the direct waste of heat. This modulation control is of great value in maintaining a constant heat input and load matching response. Nothing is so unavoiably wasteful of heat and so detrimental to the destruction and deterioration of refractory and fatiguing of boiler metal than the "on-off" cycles of load balancing as practiced in many of the power plants that this inspector has had the dubious distinction of visiting. A boiler, when once put on the line, should be kept on the line. The reduction of the steam generated should be controlled by anticipation, as provided by a properly and sensitively adjusted

Fig. 5-11. A packaged automatic dual-fuel burner (courtesy of North American Manufacturing Company).

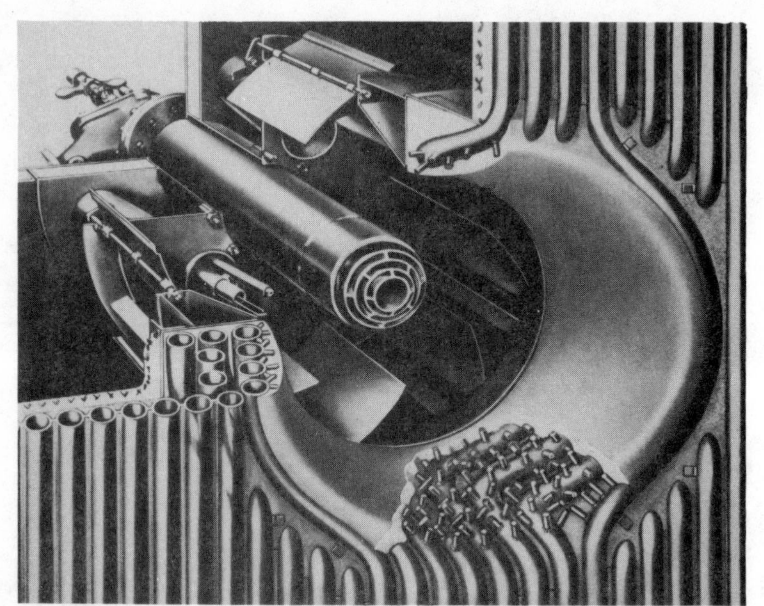

Fig. 5-12. A dual-fuel burner (courtesy of Babcock and Wilcox Company).

Fig. 5-13. An oil to gas converter packaged system (courtesy of North American Manufacturing Company).

Fig. 5-14. View of a Faber burner (courtesy of E. Keeler Company).

combustion modulator, and not by retiring the combustion process and having to bring the boiler once again up to operating temperature. The expansion and contraction brought about by the thermal changes of the "on-off" cycle are directly responsible

Fig. 5-15. A Faber "WB" burner unit (courtesy of E. Keeler Company).

Fig. 5-16. Another view of a Faber burner (courtesy of E. Keeler Company).

for the loss of more boilers and the waste of more fuel than any other single evil associated with power plant efficiency and safety. □ The comprehensive knowledge and investigative data acquired of the plant's inherent and relative values of efficiency is important, of course. The application of this data and intelligence in the form of co-related charts, operating statistics and proven optimum parameters can be used to obtain in a lucid and predictable index of the plant's ability and limitations to efficiently serve as an energy converting mechanism.

The reader will gain much valuable insight and instructional suggestions by a thorough study of the boiler manufacturer's operating manual or by the acquisition of a modern textbook on steam boiler operation. While this manual by virtue of its intent and inclusiveness will provide the reader with the rudiments and essentials of efficiency and the manner in which such evaluation and testing is to be carried out, a provisional study of preferential engineering practice will enlighten the reader with sound and applicable techniques and theory imperative to the reduction of efficiency data to actual operational performance. The topics of

Fig. 5-17. A Faber fuel pumping unit (courtesy of E. Keeler Company).

Fig. 5-18. Another Faber operation (courtesy of E. Keeler Company).

combustion and firing rate modulation will be given preference, as these two integrals are of chief concern in respect to their critical effects upon the overall picture of the plant's efficiency.

COMBUSTION CONTROL

Whether the system in question is full automatic—as the Foxboro, Hagan, Bailey, Hays, Leeds and Northrup, Taylor, Republic, Moore, Fisher are—or is totally manual, the ratio of air and fuel can be fine tuned. The commonest excuse this inspector can recall hearing is, "But the system is fully automatic, and there's nothing I can do about the combustion or its admixture of Air/Fuel proportions." Every system is provided at the factory with appropriate air/fuel adjustments to make the necessary corrections as mentioned. All that is required is for the engineer to

Fig. 5-19. This Faber unit is in an Ohio hospital (courtesy of E. Keeler Company).

Fig. 5-20. Some boilers (courtesy of E. Keeler Company).

sit down with the manufacturer's instructions and acquaint himself with these provided adjustments.

When a familiarity is had of the necessary adjustments, the engineer should then acquire an orsat kit. With the assistance of his fireman or assigned maintenance mechanic, he will proceed to run the boiler through its firing swing from low fire up to high fire. He takes a sufficient number of flue gas samples and flue gas temperatures at all firing positions. The data is transferred onto a

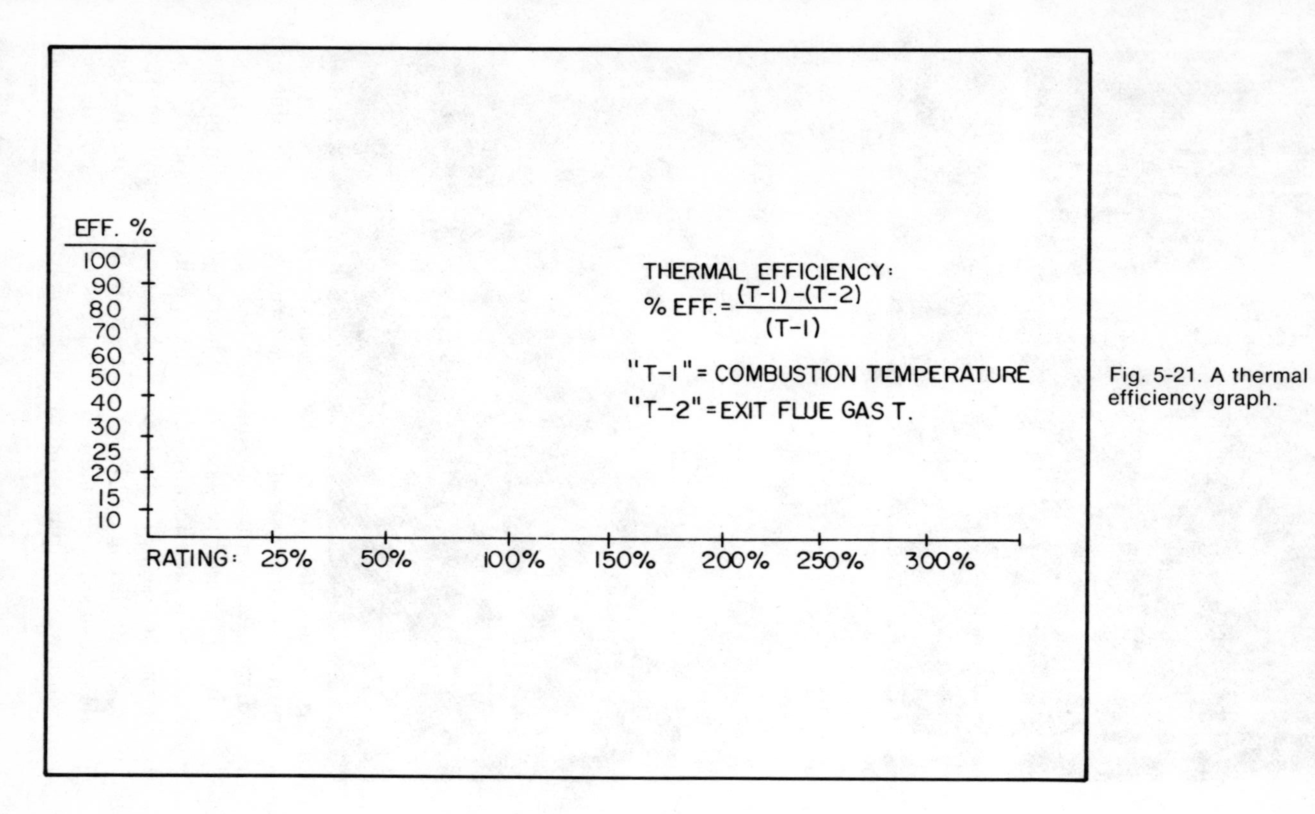

Fig. 5-21. A thermal efficiency graph.

form as suggested by Figs. 5-21 through 5-23. Once having this information collected, the engineer may then make empirical adjustments of air/fuel ratio, draft intensity (or "casing" pressure) combustion temperature, exiting gas temperature and steaming rates for each firing position.

These results will, in themselves, evidence the efficiency of the boiler and combustion process for each point as tested and display the presence of maladjustments and/or poor design by graphical representation. Thus, if a boiler would prove to be 82.4 percent thermally efficient at low fire and only 78.5 percent efficient at full fire, then it obviously indicates a lack of heat transfer which may indicate improper air shutter linkage and fuel starvation at maximum swing. Or it could suggest that the boiler is being forced beyond its "rating." There may be the very real problem of excess draft caused by an improperly designed damper arrangement. Regardless of the many possible theories, it remains that such a test will viably prove the presence of problems. When investigated by the learned engineer, it will almost dictate its own remedy.

It is of importance to remind the reader that even though the boiler manufacturer may advertize that the boiler may be fired at 200 percent to 300 percent of the boiler's rating, (given either in boiler horsepower or lbs. steam/hr.) with no adverse effects to the safety of the boiler, you will, by forcing the boiler above its rating, be putting a far greater amount of combustible through the boiler. The boiler has less chance of extracting the heat out of this greater amount of combustion, which is to say that by exceeding the "rating," from which and only which the efficiency guarantee is stipulated at, you will be decreasing the efficiency. This combustion rate/efficiency ratio is one which must be given consideration when the load demand is one of fluctuating character. It may, by proven tests and trials, be more economical to "cut-in" an another boiler, rather than to experience the drop in combustion efficiency which would result by having just one boiler operated at 50 percent efficiency or even less. This, again, will be a matter of the pecularity of the engineer's own equipment and the boiler's operating rating/efficiency, which will be clearly indicated by the recorded data and graphs. There exists, to be determined, by the engineer, a certain firing rate/efficiency ratio which will answer the question, when raised by the load demand, whether or not the boiler should be forced beyond its rating. Or by economy should another boiler be "cut-in?"

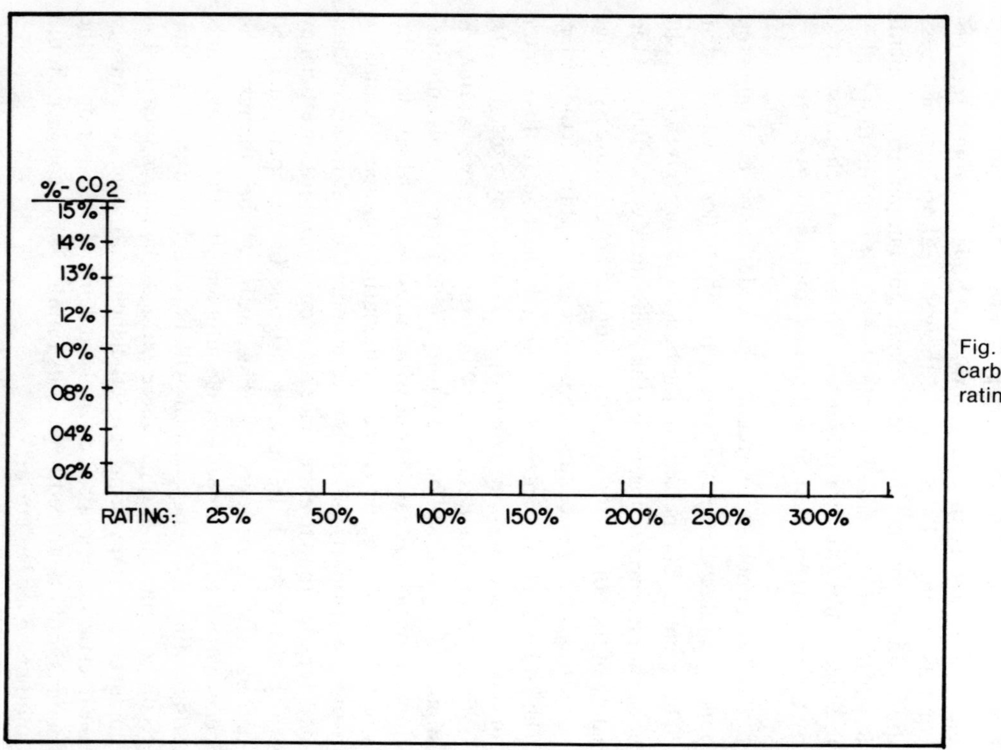

Fig. 5-22. Graph of the percentage of carbon dioxide taken by orsat at each rating position.

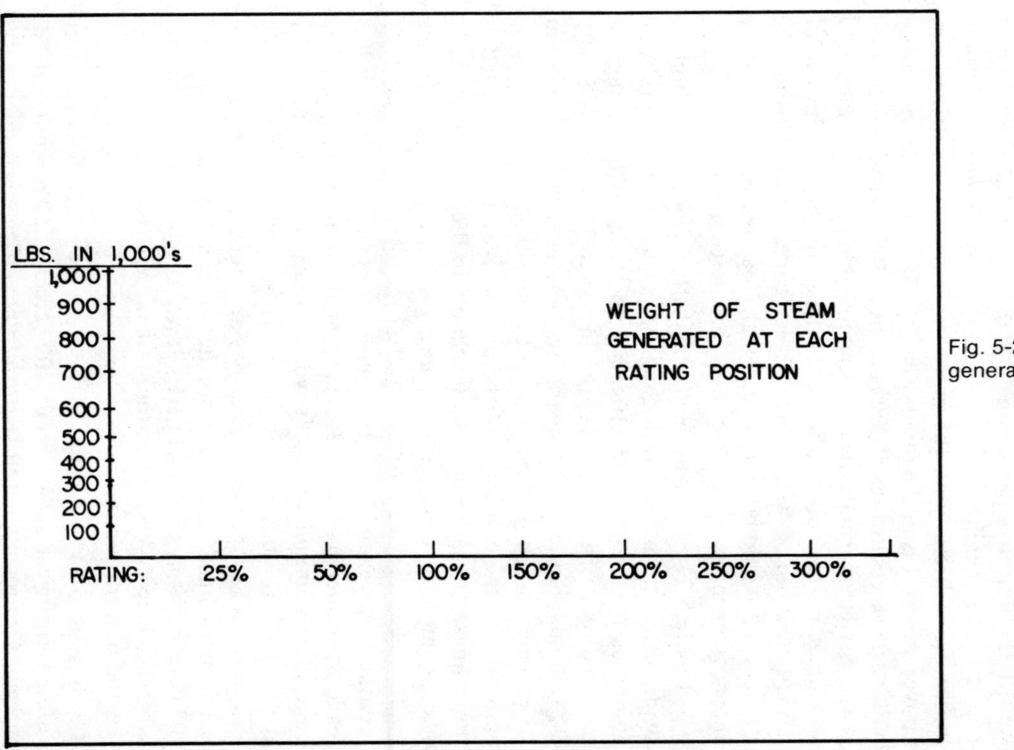

Fig. 5-23. Graph of the weight of steam generated at each rating position.

COMBUSTION/FIRING RATE EFFICIENCY

As discussed previously, every boiler has a definite "rating" as established by the manufacturer. This may be expressed in boiler horsepower, pounds of steam/hr. at a fixed pressure or in its Btus output per hour. Regardless of the manner of expressing this maximum efficient output, the fact remains unchanged that the manufacturer is held responsible only for efficiency as certified to at the rating stipulated. At 100 percent rating the boiler will be capable of absorbing from the stipulated volume of flue gases a certain amount of the heat therein. It only stands to reason that if we fire the unit at a percentage less than the rating, we will not realize the total ability of the boiler's efficiency. Even more importantly, we will enjoy all the more less efficiency at firing rates above or in excess of the rated capacity. As in the case with many boilers, at excess ratings of 200 percent to 300 percent, such a loss is witnessed in thermal efficiency that it would prove economically advantageous to "cut-in" another boiler to the header. Instead of having one boiler work at twice its capacity and losing the efficiency, it would be wiser to divide the load between two boilers and have each work at close to its 100 percent rating.

Figure 5-21 compares the thermal efficiency over the boiler's swing from 25 percent rating up to 300 percent. Figure 5-22 compares the combustion efficiency in terms of percentage of carbon dioxide from 25 percent rating up to 300 percent . Figure 5-23 compares the actual steam/hr. generated per rating, from 25 percent to 300 percent. It is suggested that these graphs be employed in the collection of the data as required and tested at the ratings, or maximums, as reachable in the engineer's plant. These comparative evaluations then should be incorporated in the plant's log and reviewed from time to time to assure their updated accuracy. An example of their use has been illustrated. When testing a boiler having a 250 percent maximum rating, at 150 boiler HP nominal rating, it will be observed that this boiler achieves maximum values of both thermal efficiency, combustion efficiency and evaporative efficiency at 100 percent rating. There is a decrease in all values as the boiler is forced beyond its nominal 100 percent rating (150 boiler HP). It will also be observed that the same efficiencies are also reduced at ratings substantially below the 100 percent point.

In Figs. 5-24 through 5-26, we have tested the boiler at four different firing rates: 75 percent, 100 percent, 150 percent and 250 percent. Our results tell us that optimum efficiency occurs at 100

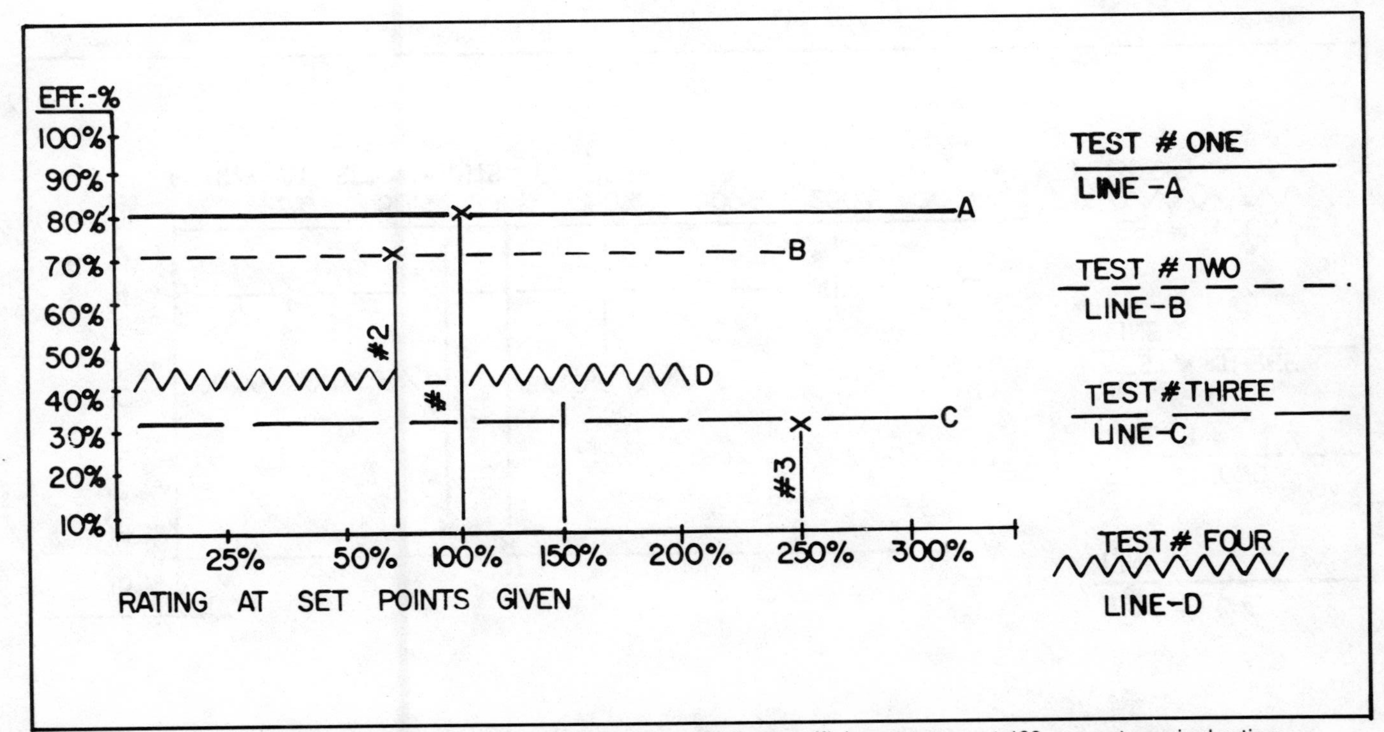

Fig. 5-24. The boiler has been tested at four different firing rates. Optimum efficiency occurs at 100 percent nominal rating.

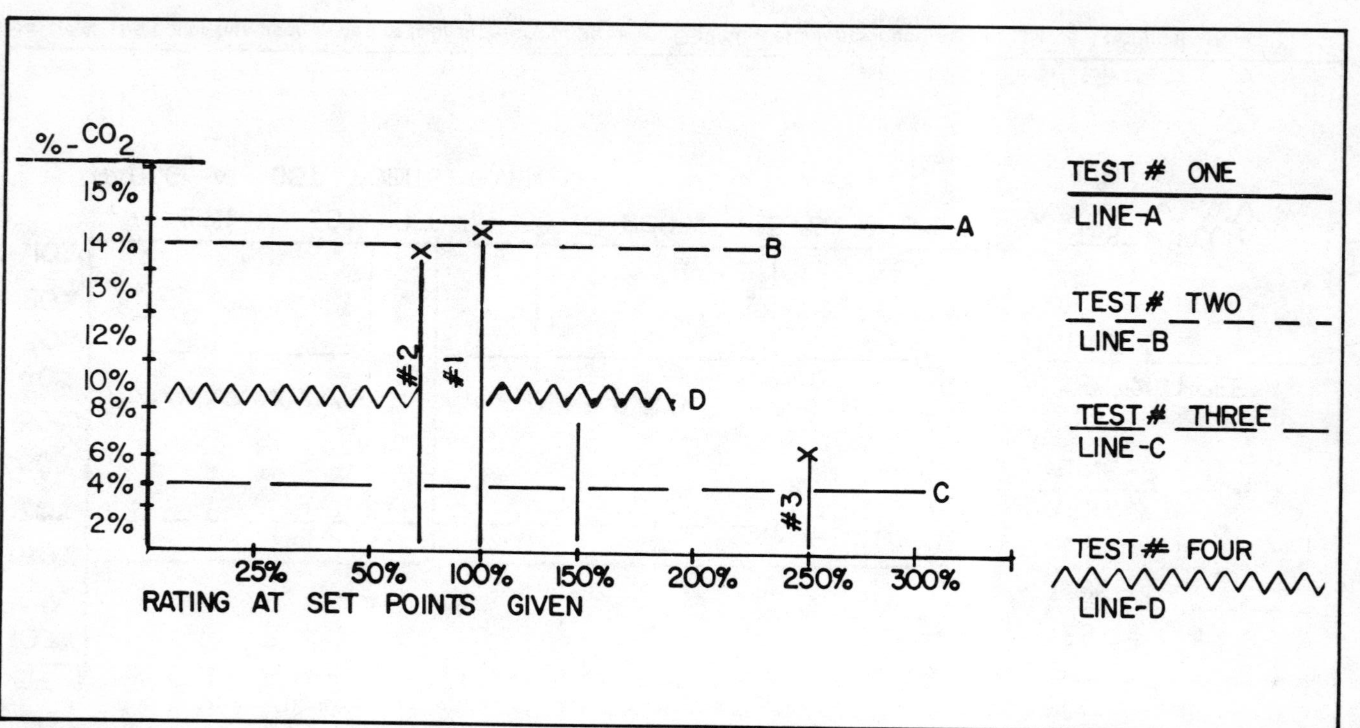

Fig. 5-25. Carbon dioxide percentage graph.

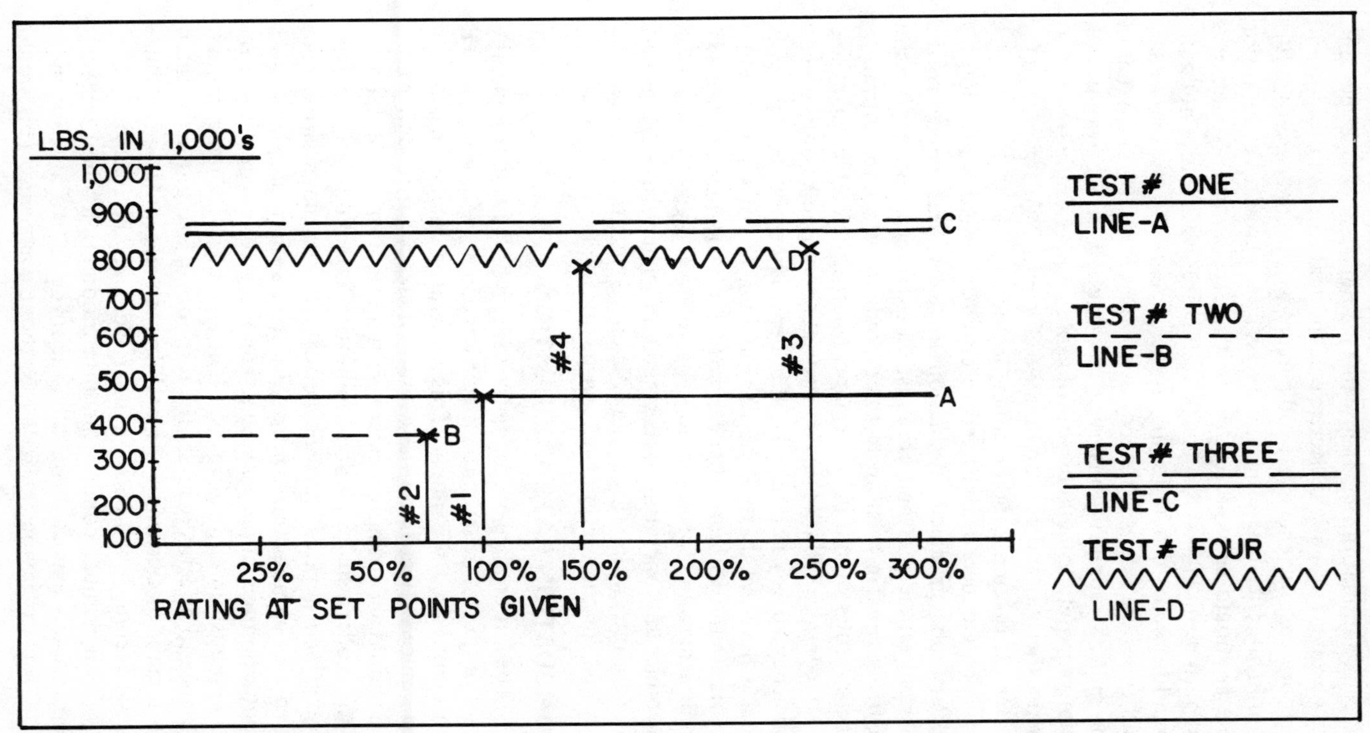

Fig. 5-26. Steam evaporation graph.

percent nominal rating. Below this rating, such as at 75 percent, we observe but a slight decrease in thermal, carbon dioxide and evaporative efficiencies. At 150 percent we observe a sharp reduction in all efficiencies and, relatively, not as great an evaporation rate as should be expected. As at 100 percent rating, we observe an output of 400,000 lbs./hr. At 150 percent we realize not quite an additional 400,000 lbs/hr., at the expense of over two-and-a-half times the loss of efficiency at 100 percent. At 250 percent of the boiler's rating, we observe that although we are evaporating approximately 850,000 lbs/hr. of steam, we are doing so at a loss of the thermal and carbon dioxide efficiencies. We would be far better off putting another boiler on the line.

The engineer is expected to conduct similar tests upon his boilers, one graphical evaluation for each, and determine the thermal efficiencies and combustion efficiencies for each point of firing rate, in relation to the manufacturer's nominal rating. These results should be a permanent part of the boiler plant log. It is recommended that this boiler trial be conducted only after an efficiency survey is done of the boiler and combustion process. This will assure that the readings as obtained at selected rating points are the optimum that can be had, in that they are the maximum achievable for and at any given percentage of rating, with the boiler unit operating at its greatest efficiency. Specific instructions for integrating this rating efficiency evaluation into the data required for the survey itself will be given later.

COMBUSTION RATE MODULATION

The importance of having the exact input of heat to equal the exact output of heat, as required by the load, will be considered and emphasized separately. This imperative integral has a very pronounced effect upon the overall efficiency of the boiler unit, and should accordingly receive separate attention. In the majority of the modern steam plants, combustion rate modulation is purposely incorporated into the network of instrumentation provided for combustion control. In semi-automatic plants, this modulation technique may be to some degree independent from the combustion control per se. It is still of influencing enough presence to effectively affect the response in firing rate to match the demand as created by the load. In manual plants, the modulation is accomplished by the human element, the fireman.

With manual combustion rate modulation, the firemen is appropriately expected to match the firing rate of his equipment to the steam leaving the boiler, without overfiring. This would lift the safety valves and waste steam. The header pressure should not drop so much as to have to open up the firing to such a point as to exceed the rating by several hundred percent, and thus cause a drastic loss in efficiency.

This combustion rate modulation is accomplished in automatic systems by any number of electro-mechanical, pneumatic or electro-pneumatic devices as the manufacturer may design into his particular control system. One point in common to all systems, regardless of their functioning character or principles of operation is that their sensitivity or response to load changes may be accordingly misadjusted. It is a prerequisite to the meaningful survey to ascertain that this function of combustion rate modulation, which is in itself so critical and influencing, be accurately and sensitively proportioned to the operational requirements and load changeability response of the plant.

The prime objective of combustion rate modulation is twofold. The first is to regulate the quantity of combustion in Btus as necessitated to generate the steam taken by the load, in such a manner as to maintain the least drop in header pressure. The second is to regulate or proportion the actual heat input to the actual heat output. Thus, if the load is normally demanding 250,000 lbs./hr., then it is expected of the modulator system commanding the combustion process (oil burner, coal feeding grate, gas controller, etc.) to only deliver that 250,000 lbs./hr. of equated heat input. By providing less than the commanded 250,000 lbs./hr., pressure would be reduced to the point where it could only be compensated by an abnormally high combustion rate. This would be wasteful in respect to the excess in firing rate brought about by this "catchup" manner of compensation of the load's demand and satisfaction by generated replacement. The combustion rate modulator can be practically described as a device which should proportionally match the steam being consumed by the load to the heat release as provided by the combustion process in the boiler unit.

One of the two points of foremost concern in respect to the entire picture of plant efficiency is the combustion control, which as controls or proportions the air/fuel ratios at all points of rating or steaming capacities. The control appropriately meters the exact percentages of air and fuel as required and confirmed by actual orsat tests of the products of combustion. The combustion rate modulator, which is a built-in system, may consist of anything from a simple balancing-potentiometer coil motor

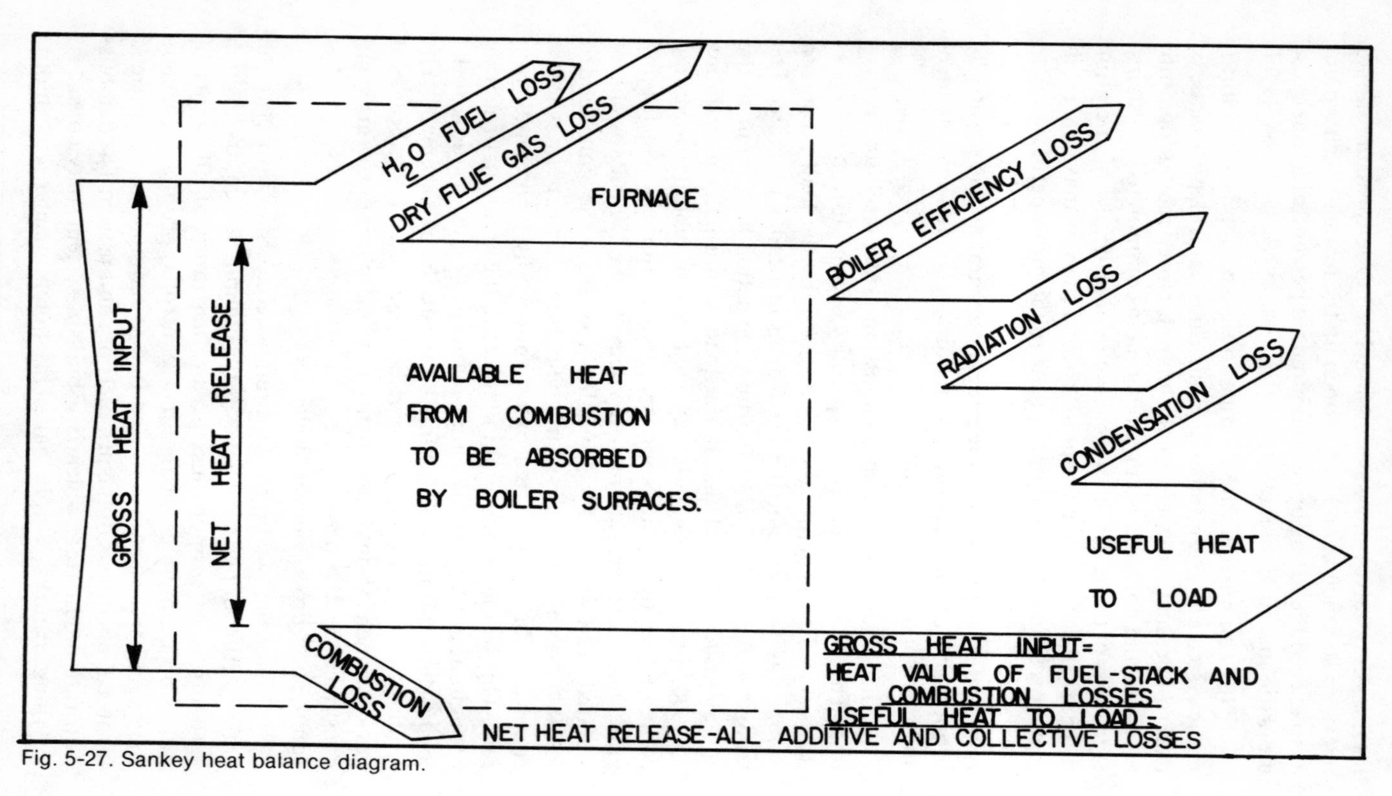

Fig. 5-27. Sankey heat balance diagram.

arrangement up to an integrated "input-output signal anticipation and response" network.

The engineer will, after having assured himself that the combustion control and combustion rate modulation system are in functioning within designed parameters of adjustment and response, proceed to collect the specified data and information as is required by the fuel-combustion and boiler-heat exchanger survey sections. Part A of the survey is included later in the chapter. The total survey, which academically is of thermal effects, will then provide the engineer with both vital operating efficiency intelligence and the preliminary data necessary to conduct a conclusive/efficiency/rating study.

Although efficiency to some extent or another may be had by every boiler, regardless of type or age, several variations will offer what appears to be a distinct advantage in respect to burning oil and gas. Representative of package firetube boilers is the Cleaver-Brooks unit. The reputation of this particular firetube package is known throughout the industry. Next is a discussion on boiler selection by the Cleaver-Brooks Company.

ENERGY CONSERVATION BEGINS IN THE BOILER ROOM!

Soaring fuel costs, high maintenance costs, and questionable fuel availability are now the chief factors in boiler replacement! Many of the old reasons for boiler replacement still apply. Plant expansion requires more steam. Summer versus winter loads indicate it's more practical to have two boilers instead of one. Other reasons are smoke abatement or emission regulations, downgrading of operating pressure on old equipment by the boiler inspector; erratic steam pressures, etc.

Today, however, high operating and maintenance costs are cutting into profits. This is the most important reason for replacing old, inefficient boilers with the most modern, most efficient packaged boilers available.

In addition, questionable fuel availability and allocations are forcing management to burn the fuel which they have (or can get) in the most efficient manner. Again, packaged boilers are the answer.

In the good old days (about seven years ago) No. 6 was 6 cents gal. Now 35 cents to 40 cents per gallon is not uncommon. This rapid rise in oil and gas costs, and all energy costs, is forcing profit conscious managers and owners to focus on boiler efficiency and replacement. See Tables 5-1 through 5-3.

Table 5-1. No. 2 Oil Fuel Burning Rates at Various Efficiencies (courtesy of Cleaver-Brooks Company).

Boiler Size BHp	Fuel-To-Steam Efficiency																			
	60.0	62.5	65.0	67.5	70.0	72.5	75	76	77	78	79	80	81	82	83	84	85	86	87	88
100	40.0	38.5	37.0	35.5	34.0	33.0	32.0	31.5	31.0	30.5	30.5	30.0	29.5	29.0	29.0	28.5	28.0	28.0	27.5	27.0
125	50.0	48.0	46.0	44.5	42.5	41.0	40.0	39.5	39.0	38.0	38.0	37.5	37.0	36.5	36.0	35.5	35.0	35.0	34.5	34.0
150	60.0	57.5	55.0	53.0	51.0	49.5	48.0	47.0	46.5	46.0	45.5	45.0	44.5	43.5	43.0	42.5	42.0	41.5	41.0	41.0
200	79.5	76.5	73.5	71.0	68.5	66.0	64.0	63.0	62.0	61.5	60.5	60.0	59.0	58.5	57.5	57.0	56.5	55.5	55.0	54.5
250	99.5	95.5	92.0	88.5	85.5	82.5	79.5	78.5	77.5	76.5	75.5	74.5	74.0	73.0	72.0	71.0	70.5	69.5	68.5	68.0
300	119.5	115.0	110.5	106.5	102.5	99.0	95.5	94.5	93.0	92.0	91.0	89.5	88.5	87.5	86.5	85.5	84.5	83.5	82.5	81.5
350	139.5	134.0	129.0	124.0	119.5	115.5	111.5	110.0	108.5	107.5	106.0	104.5	103.5	102.0	101.0	99.5	98.5	97.5	96.0	95.0
400	159.5	153.0	147.0	141.5	136.5	132.0	127.5	126.0	124.0	122.5	121.0	119.5	118.0	116.5	115.0	114.0	112.5	111.0	110.0	108.5
500	199.5	191.5	184.0	177.0	171.0	165.0	159.5	157.5	155.5	153.5	151.5	149.5	147.5	146.0	144.0	142.5	140.5	139.0	137.5	136.0
600	239.0	229.5	220.5	212.5	205.0	198.0	191.5	189.0	186.5	184.0	181.5	179.5	177.0	175.0	173.0	171.0	169.0	167.0	165.0	163.0
700	279.0	268.0	257.5	248.0	239.0	231.0	223.0	220.0	217.5	214.5	212.0	209.0	206.5	204.0	201.5	199.0	197.0	194.5	192.5	190.0
800	319.0	306.0	294.5	283.5	273.5	264.0	255.0	251.5	248.5	245.0	242.0	239.0	236.0	233.5	230.5	227.5	225.0	222.5	220.0	217.5

Table 5-2. No. 6 Oil Fuel Burning Rates at Various Efficiencies (courtesy of Cleaver-Brooks Company).

Boiler Size BHp	Fuel-To-Steam Efficiency																			
	60.0	62.5	65.0	67.5	70.0	72.5	75	76	77	78	79	80	81	82	83	84	85	86	87	88
100	37.0	35.5	34.5	33.0	32.0	31.0	30.0	29.5	29.0	28.5	28.5	28.0	27.5	27.0	27.0	26.5	26.0	26.0	25.5	25.5
125	46.5	44.5	43.0	41.5	40.0	38.5	37.0	36.5	36.0	36.0	35.5	35.0	34.5	34.0	33.5	33.2	33.0	32.5	32.0	31.5
150	56.0	53.5	51.5	49.5	48.0	46.0	44.5	44.0	43.5	43.0	42.5	42.0	41.5	41.0	40.5	40.0	39.5	39.0	38.5	38.0
200	74.5	71.5	68.5	66.0	64.0	61.5	59.5	58.5	58.0	57.0	56.5	56.0	55.0	54.5	54.0	53.0	52.5	52.0	51.5	50.5
250	93.0	89.5	86.0	82.5	79.5	77.0	74.5	73.5	72.5	71.5	70.5	69.5	69.0	68.0	67.0	66.5	65.5	65.0	64.0	63.5
300	111.5	107.0	103.0	99.0	95.5	92.5	89.5	88.0	87.0	86.0	85.0	83.5	82.5	81.5	80.5	79.5	79.0	78.0	77.0	76.0
350	130.0	125.0	120.0	115.5	111.5	107.5	104.0	103.0	101.5	100.0	99.0	97.5	96.5	95.5	94.0	93.0	92.0	91.0	90.0	89.0
400	149.0	143.0	137.5	132.5	127.5	123.0	119.0	117.5	116.0	114.5	113.0	111.5	110.0	109.0	107.5	106.5	105.0	104.0	102.5	101.5
500	186.0	178.5	171.5	165.5	159.5	154.0	149.0	147.0	145.0	143.0	141.0	139.5	138.0	136.0	134.5	133.0	131.5	130.0	128.5	127.0
600	223.0	214.0	206.0	198.5	191.5	184.5	178.5	176.0	174.0	171.5	169.5	167.5	165.5	163.5	161.5	159.5	157.5	155.5	154.0	152.0
700	260.5	250.0	240.5	231.5	223.0	215.5	208.5	205.5	203.0	200.5	198.0	195.5	193.0	190.5	188.0	186.0	184.0	181.5	179.5	177.5
800	297.5	285.5	274.5	264.5	255.0	246.5	238.0	235.0	232.0	229.0	226.0	223.0	220.5	217.5	215.0	212.5	210.0	207.5	205.0	203.0

Table 5-3. Natural Gas Fuel Burning Rates (Therms/Hr.) at Various Efficiencies (courtesy of Cleaver-Brooks Company).

Boiler Size BHp	Fuel-To-Steam Efficiency																
	60.0	62.5	65.0	67.5	70.0	72.5	75	76	77	78	79	80	81	82	83	84	85
100	55.8	53.6	51.5	49.6	47.9	46.2	44.7	44.1	43.5	43.0	42.4	41.9	41.4	40.9	40.4	39.9	39.4
125	69.8	67.0	64.4	62.0	59.8	57.7	55.8	55.1	54.4	53.7	53.0	52.3	51.7	51.1	50.4	49.8	49.3
150	83.7	80.4	77.3	74.4	71.8	69.3	67.0	66.1	65.2	64.4	63.6	62.8	62.0	61.3	60.5	59.8	59.1
200	111.6	107.2	102.0	99.2	95.7	92.4	89.3	88.1	87.0	85.9	84.8	83.7	82.7	81.8	80.7	79.7	78.8
250	139.5	133.9	128.8	124.0	119.6	115.5	111.6	110.1	108.7	107.3	106.0	104.6	103.4	102.1	100.9	99.7	98.5
300	167.4	160.7	154.5	148.8	143.5	138.6	133.9	132.2	130.5	128.8	127.2	125.5	124.0	122.5	121.0	119.6	118.2
350	195.3	187.5	180.3	173.6	167.4	161.6	156.2	154.2	152.2	150.2	148.3	146.5	144.7	142.9	141.2	139.5	137.9
400	223.2	214.3	206.0	198.4	191.3	184.7	178.6	176.2	173.9	171.7	169.5	167.5	165.3	163.3	161.4	159.4	157.6
500	279.0	267.8	257.5	248.0	239.1	230.9	223.2	220.3	217.4	214.6	211.9	209.3	206.7	204.2	201.7	199.3	197.0
600	334.8	321.4	309.0	297.6	287.0	277.1	267.8	264.3	260.9	257.5	254.3	251.0	248.0	245.0	242.0	239.1	236.3
700	390.6	374.9	360.5	347.2	334.8	323.2	312.5	308.3	304.3	300.4	296.6	293.0	289.3	285.8	282.3	279.0	275.7
800	446.4	428.5	412.0	396.8	382.6	369.4	357.1	352.4	347.8	343.4	339.0	335.0	330.6	326.6	322.7	318.8	315.1

It boils down to this; if you can amortize the cost of a new packaged boiler in 1-2 years, due to greatly reduced annual fuel costs (due to much higher boiler efficiency), then these savings turn into increased profits—every year after amortization. In addition, you're burning your fuel in the most efficient manner possible! Actually, for the same output of steam, you burn less fuel with the more efficient packaged boiler.

An example of this current profit-oriented thinking is the Worchester Textile Company of Centerdale, Rhode Island. They compete with low labor cost foreign manufacturers in the woolen and polyester fabric market.

Two of their four boilers were 30 years old and were originally designed for coal, but later converted to rotary cup oil operation. Erratic performance and high maintenance costs were "standard procedure."

Mr. Chester Reynolds, Chief Engineer, determined that the two oil units were operating at about 60 percent efficiency. the two "newer two-pass units" were somewhat better, but not reliable.

The local Cleaver-Brooks representative helped study the problem and recommended one new 800 Hp CB boiler, with guaranteed 84 percent fuel-to-steam efficiency, guaranteed steady performance and substantial savings in fuel oil. A new boiler room was built and all new accessories were installed.

In the first six months of operation, the new boiler saved a predicted average of $8,500/month fuel cost, and the complete installation was amortized in about 10 months. On an annual basis, savings (future profits) will be $100,000 per year. Fuel oil conservation will be over 300,000 gallons of No. 6 oil per year.

Thus, one new, extremely efficient packaged CB firetube boiler replaced four older, inefficient units. The effect on the company's profits is very obvious!

☐ Study your maintenance costs carefully! Annual pressure vessel, burner, or refractory problems are very expensive. Consider a new packaged boiler!

☐ Conversion of an old boiler designed for coal firing can become a costly lesson in "throwing good money after bad." Frequently, expected efficiencies will not occur on a conversion job due to the inherent inefficient design of the old unit. Consider a new packaged boiler!

☐ Since a new boiler is "automatic" in operation, labor savings are a definite potential. Constant operator attendance is not required.

☐ Run your own stack temperature CO_2, O_2 test on your existing equipment. Or call in a Cleaver-Brooks sales engineer to help with your test. He has the calibrated equipment.

☐ Study and use the methods to measure the fuel-to-steam efficiency of your existing equipment. Then compare tangible savings which can be derived from the purchase of a new CB tube boiler.

☐ The same comparison method can be used to evaluate savings obtainable by selection of a new CB boiler rather than new Brand S or Brand K equipment. See Table 5-4.

THINGS YOU SHOULD KNOW ABOUT FIRE-TUBE BOILERS!

Packaged fire-tube boilers are different in design, and they do have inherent differences in fuel-to-steam efficiency, which is your true gauge of operating fuel costs. Some units remove heat from the hot combustion gases in two passes through the boiler, some in three passes, and a few in four passes.

If you were to call a heat exchanger manufacturer and ask for the heat exchange or output of a two-pass and a four-pass exchanger (each with the same heating surface, same inlet volume to be heated, and same heating medium) guess what he would tell you. He would state that the four-pass exchanger would produce a greater output. The point is that Cleaver-Brooks' four-pass design provides increased velocity, increased heat transfer, and greater efficiency from the same heating surface. A boiler is a heat exchanger!

Some fire-tube units are designed to transmit the heat from the hot combustion gases through 3 sq. ft. of fireside heating surface per boiler horsepower, some through 4 sq. ft./BHp, and a few through 5 sq. ft./BHp. For optimum boiler life, 5 sq. ft./HP should be demanded as the basis of design.

Some so-called "packaged" fire-tube boilers are no more than an agreement between a boiler manufacturer and a burner manufacturer. You can't expect the same performance from this type of unit that you would get from a true packaged unit where the boiler manufacturer designs, builds, and tests his own burners on his own boilers.

Study the air-fuel ration control systems. Does the unit have an accurate-positive method of metering air and fuel over the operating range or just a couple rods and linkages for low and high fire?

Table 5-4. Guaranteed Fuel-To-Steam Efficiencies on Model CB Fire-Tube Boilers at 50 Percent and 100 Percent of Rating (courtesy of Cleaver-Brooks Company).

BOILER SIZE BHp	* Gas Fuel				* No. 2 Oil				* No. 5 and No. 6 Oils			
	15# Design 10 PSIG Operation		150# Design 125 PSIG Operation		15# Design 10 PSIG Operation		150# Design 125 PSIG Operation		15# Design 10 PSIG Operation		150# Design 125 PSIG Operation	
	% of Load		% of Load		% of Load		% of Load		% of Load		% of Load	
	50%	100%	50%	100%	50%	100%	50%	100%	50%	100%	50%	100%
100	83.5	83.5	80.5	81.0	86.5	86.5	83.5	84.0	87.0	87.0	84.0	84.5
125	82.0	82.0	79.0	79.5	85.0	85.5	82.0	83.0	85.5	86.0	82.5	83.5
150	83.0	83.0	80.0	80.5	86.0	86.5	83.0	84.0	86.5	87.0	83.5	84.5
200	83.5	83.5	80.5	81.5	87.0	87.0	84.0	84.5	87.5	87.5	84.5	85.0
250	82.0	82.0	79.5	80.0	85.5	85.5	83.0	83.0	86.0	86.0	83.5	83.5
300	82.5	82.5	80.0	80.5	86.0	86.0	83.5	83.5	86.5	86.5	84.0	84.0
350	83.0	83.0	80.5	81.0	86.5	86.5	84.0	84.5	87.0	87.0	84.5	85.0
400	83.0	83.0	80.0	80.5	86.0	86.5	83.5	84.0	86.5	87.0	84.0	84.5
500	84.0	84.0	81.0	81.5	87.0	87.5	84.0	85.0	87.5	88.0	84.5	85.5
600	84.0	84.5	81.0	82.0	87.5	87.5	84.5	85.5	88.0	88.0	85.0	86.0
700	84.5	84.5	81.5	82.0	87.5	88.0	84.5	85.5	88.0	88.5	85.0	86.0
800	84.5	84.5	81.5	82.0	88.0	88.0	85.0	85.5	88.5	88.5	85.5	86.0

* Fuel Heating Values: No. 2 Oil – 140,000 BTU/Gal.
Natural Gas – 1,000 BTU/Cu. Ft. No. 5 & 6 Oil – 150,000 BTU/Gal.

NOTE: CB FTSE includes Radiation and Convection boiler heat losses to the boiler room.

The boiler that is designed for ease of maintenance will be maintained! Does the pressure vessel have large front and rear access doors and can they be easily opened for inspection of tube sheets, tube attachment, and refractory? (Access on wetback designs is limited and can be extremely difficult.) It is an easy function to pull the oil burner for nozzle inspection and maintenance?

There are major differences in packaged boiler designs which must be studied and evaluated. Let's look at the over-riding decision element—the true efficiency of this machine called a boiler!

DIFFERENT TYPES OF EFFICIENCY

Combustion efficiency is the effectiveness of the burner only and relates to its ability to completely burn the fuel. The boiler or heat exchanger has little bearing on combustion efficiency. A well designed burner will operate with as little as 15 to 20 percent excess air, while converting all combustibles in the fuel to useful energy.

Thermal Efficiency is the effectiveness of the heat transfer in a heat exchanger. It does not take into account boiler radiation and convection losses. (e.g. from boiler shell, water column, rear door, etc.).

Radiation and convection losses will vary with boiler size, operating presure, and from one manufacturer to another! CB boilers probably have the lowest losses of any units on the market due to shell insulation and rear head design. These losses are considered constant in Btu/hr., but become a larger percent loss as firing rate decreases.

Boiler efficiency is a general term. It can mean thermal efficiency or fuel-to-steam efficiency. It should be defined by the manufacturer before being used in any economic comparison.

Fuel-to-steam efficiency is the ratio of BTu output divided by Btu input and is the correct figure to use when determining fuel costs. It includes radiation and convection losses to the boiler room. See Table 5-5.

This direct indicator of fuel costs is made up of two components. The first component is the stack temperature or temperature of the exhaust gases (dry and water vapor) leaving the boiler. The function of a good boiler or heat exchanger design is to remove as much heat as possible from the hot gas leaving the burner. Thus, the lower the stack temperature, the better job

Table 5-5. Model CB Boilers: Percentage of Radiation and Convection Losses (courtesy of Cleaver-Brooks Company).

	100-350 BHp		400-800 BHp	
Firing Rate (% of Load)	OP. Pressure = 10 PSIG	OP. Pressure = 125 PSIG	OP.Pressure = 10 PSIG	OP. Pressure = 125 PSIG
25%	3.8%	5.1%	3.2%	4.4%
50%	1.9%	2.6%	1.6%	2.2%
75%	1.3%	1.7%	1.1%	1.5%
100%	1.0%	1.4%	0.8%	1.1%

*Add to stack loss and subtract result from 100% to obtain Fuel-to-Steam Ef-Ficiency. **NOTE:** These losses can run considerably higher on competitive equipment.

the boiler or heat exchanger is doing, and the less heat loss up the stack.

The second component is the effectiveness of the burner. High and constant CO^2 readings (with no CO and little O^2) throughout the boiler operating range (25-100 percent of rating) prove that you have a good burner and a good air-fuel ratio control system.

The combination of these two components, plus use of recognized charts for the specific fuel, results in a stack loss in percent. This stack loss is a measure of the amount of heat carried away by the dry flue gases (unused heat) and the moisture loss. You then add radiation and convection losses to the stack loss, subtract from 100 percent, and obtain the fuel-to-steam efficiency.

Stack Temperature — CO^2 test method to determine fuel-to- steam efficiency

This is a practical, accepted method of predicting or job site checking fuel-to-steam efficiency. Equipment Required includes flue gas analyzer to check CO^2, O^2, and CO, stack temperature, room thermometer, charts for heat loss up the stack for various fuels, and a correction chart for addition of radiation and convection losses.

The boiler has operated for several hours at a constant rate and stack temperature has leveled off. Measure CO^2, O^2, and CO in flue gas. Subtract room temperature from stack temperature, to obtain net loss in temperature up the stack.

Refer to Tables 5-6 through 5-8 for proper fuel, and obtain heat loss up the stack. Add proper percent for radiation and convection losses. Subtract from 100 percent to get approximate fuel-to-steam efficiency.

Table 5-6. Percentage of Stack Loss for No. 2 Oil (courtesy of Cleaver-Brooks Company).

DIFFERENCE BETWEEN FLUE GAS AND ROOM TERMPERATURES IN DEGREES FAHRENHEIT

%	200	220	240	260	280	300	320	340	360	380	400	420	440	460	480	500	520	540	560	580	600	620	640	660	680	700	750	800	850	900	950	1000
	24.1	25.8	27.7	29.3	31.3	33.9	34.8	36.4	38.2	40.0	42.9	44.8	45.5	47.0	49.0	50.8	52.4	54.3	56.0	57.9	59.6	61.5	63.5	65.0	66.8	68.8						
3.5	21.7	23.1	24.8	26.2	27.8	29.2	31.7	32.5	33.9	35.3	36.9	38.5	40.0	41.7	43.1	44.8	46.1	47.8	49.4	50.9	52.2	53.9	55.7	57.0	58.3	60.0	63.8	67.8				
4.0	19.9	21.2	22.5	24.9	25.2	26.5	27.9	29.2	31.7	32.0	33.3	35.8	36.0	37.3	38.7	38.7	40.0	41.4	42.9	44.1	45.5	46.9	48.1	49.8	50.9	52.1	53.8	57.0	60.2	63.9	67.1	
4.5	18.4	19.7	20.8	22.0	23.2	24.4	25.6	26.9	28.0	29.3	30.4	31.8	32.9	34.2	35.6	36.7	37.8	39.0	40.1	41.2	42.5	43.8	45.0	46.3	47.4	48.8	51.8	54.6	57.8	60.9	63.9	66.9
5.0	17.2	18.5	19.5	20.7	21.7	22.7	23.8	24.9	26.0	27.1	28.2	29.4	30.3	31.5	32.7	33.8	34.9	35.9	36.8	38.0	39.2	40.1	41.7	42.4	43.7	44.7	47.4	50.1	52.9	55.8	58.3	61.2
5.5	16.3	17.4	18.4	19.4	20.4	21.3	22.3	23.3	24.3	25.4	26.3	27.3	28.4	29.4	30.6	31.4	32.4	33.6	34.5	35.3	36.4	37.4	38.4	39.	40.3	41.7	44.0	46.5	49.0	51.8	54.1	56.5
6.0	15.6	16.5	17.4	18.3	19.3	20.4	21.2	22.0	23.0	23.9	24.9	25.8	26.8	27.7	28.6	29.5	30.4	31.4	32.3	33.1	34.2	35.0	36.0	36.9	37.9	38.9	41.0	43.5	45.8	48.0	50.3	52.8
6.5	14.9	15.7	16.7	17.5	18.4	19.3	20.1	20.9	21.8	22.7	23.6	24.5	25.3	26.1	27.0	27.8	28.8	29.6	30.6	31.3	32.3	33.0	34.1	34.8	35.7	36.5	38.7	40.8	42.9	45.1	47.5	49.7
7.0	14.4	15.3	16.0	16.8	17.8	18.4	19.3	20.1	20.9	21.7	22.4	23.2	24.1	24.9	27.5	26.5	27.3	28.1	28.9	29.8	30.5	31.4	32.3	33.0	33.8	34.6	36.5	38.6	40.5	42.7	44.7	46.6
7.5	13.9	14.6	15.4	16.2	16.9	17.7	18.5	19.2	20.1	20.7	21.3	22.2	23.0	23.8	24.5	25.2	26.0	26.8	27.5	28.2	29.0	29.8	30.6	31.3	32.2	32.9	34.8	36.5	38.5	40.3	42.3	44.2
8.0	13.5	14.3	14.9	15.7	16.3	17.1	17.7	18.5	19.3	20.0	20.7	21.4	22.1	22.8	23.5	24.2	25.0	25.7	26.3	27.0	27.8	28.5	29.2	30.0	30.8	31.5	33.2	35.0	36.8	38.5	40.2	42.1
8.5	13.2	13.8	14.5	15.2	15.8	16.5	17.3	17.8	18.6	19.3	20.0	20.6	21.3	21.9	22.6	23.3	23.9	24.5	25.3	25.9	26.7	27.3	28.0	28.8	29.4	30.1	31.8	33.5	35.2	36.9	38.7	40.2
9.0	12.8	13.4	14.1	14.7	15.1	16.0	16.7	17.3	17.9	18.6	19.3	20.0	20.6	21.2	21.8	22.4	23.1	23.8	24.4	25.0	25.7	26.3	27.0	27.7	28.3	28.9	30.5	32.1	33.8	33.5	37.0	38.5
9.5	12.5	13.2	13.7	14.3	14.9	15.7	16.3	16.8	17.4	18.1	18.6	19.3	19.9	20.5	21.1	21.7	22.4	22.9	23.5	24.1	24.8	25.4	26.0	26.7	27.2	27.9	29.4	31.0	32.5	34.0	35.5	37.2
10	12.3	12.8	13.4	14.0	14.6	15.2	15.7	16.3	16.9	17.5	18.1	18.7	19.3	20.0	20.5	21.0	21.6	22.2	22.8	23.4	24.0	24.6	25.1	25.8	26.3	27.0	28.3	29.9	31.4	32.9	34.4	35.7
11	11.8	12.4	12.8	13.8	13.9	14.5	15.0	15.5	16.2	16.7	17.2	17.8	18.3	18.7	19.4	20.0	20.5	20.9	21.5	22.0	22.6	23.1	23.7	24.2	24.8	25.3	26.7	28.0	29.4	31.8	32.1	33.5
12	11.4	11.8	12.5	12.9	13.4	13.9	14.4	14.9	15.4	15.9	16.4	16.9	17.4	17.9	18.4	18.9	19.5	20.0	20.5	20.9	21.4	22.9	22.4	22.9	23.5	24.0	25.2	26.5	27.8	29.0	30.2	31.7
13	11.2	11.6	12.1	12.5	12.9	13.4	13.9	14.3	14.7	15.3	15.8	16.3	16.7	17.2	17.7	18.1	18.6	19.1	19.6	20.1	20.5	21.1	21.3	21.8	22.3	22.8	24.0	25.2	26.3	27.5	28.8	30.0
14		11.3	11.8	12.2	12.6	13.0	13.4	13.8	14.3	14.8	15.3	15.6	16.2	16.5	16.9	17.4	17.8	18.3	18.7	19.2	19.7	20.2	20.6	21.0	21.4	21.8	22.9	24.1	25.2	26.2	27.4	28.6
15			11.4	11.7	12.4	12.6	13.1	13.5	13.8	14.3	14.8	15.3	15.6	15.9	16.4	16.7	17.3	17.7	18.1	18.4	18.9	19.4	19.8	20.3	20.6	21.0	22.0	23.1	24.2	25.2	26.2	27.3

Table 5-7. Percentage of Stack Loss for No. 6 Oil (courtesy of Cleaver-Brooks Company).

%	DIFFERENCE BETWEEN FLUE GAS AND ROOM TEMPERATURES IN DEGREES FAHRENHEIT																														
CO_2	200	220	240	260	280	300	320	340	360	380	400	420	460	440	480	500	520	540	560	580	600	620	640	660	680	700	750	800	850	900	950	1000
3.0	24.5	26.5	28.5	30.2	32.2	34.5	36.5	38.2	40.4	42.2	44.4	46.8	48.2	50.0	52.3	54.3	56.3	58.2	60.3	62.0	64.1	66.2	68.1	70.1								
3.5	21.8	23.4	25.2	26.8	28.6	30.4	32.1	33.8	35.5	37.4	39.0	40.6	42.2	44.0	45.6	47.5	49.2	51.0	52.8	54.0	56.0	57.8	59.9	61.1	63.0	64.9	69.0					
4.0	19.8	21.2	22.8	24.2	25.7	27.3	28.8	30.2	31.6	32.5	34.8	36.3	37.8	39.4	40.8	42.2	43.8	45.1	46.9	48.2	49.8	51.2	52.9	54.2	56.0	57.8	61.1	65.0	68.9			
4.5	18.2	19.4	20.8	22.2	23.5	24.8	26.2	27.4	28.8	30.4	31.5	33.0	33.2	35.4	37.0	38.1	39.4	41.0	42.2	43.5	45.0	46.3	47.9	49.0	50.1	51.9	55.0	58.2	61.8	65.1	68.5	
5.0	16.8	18.0	19.3	20.4	21.7	22.8	23.2	25.3	26.6	27.8	29.0	30.3	31.4	32.6	33.8	35.3	36.2	37.5	38.8	39.8	41.0	42.3	43.8	44.9	46.1	47.5	50.1	53.6	56.3	59.8	62.3	65.8
5.5	15.8	16.8	18.0	19.2	20.3	21.3	22.5	23.5	24.6	25.8	26.9	28.0	29.2	30.2	31.4	32.5	33.5	34.7	35.8	37.0	37.9	39.2	40.1	41.3	42.3	43.8	46.1	49.1	52.0	54.7	57.8	60.1
6.0	14.8	15.8	16.9	18.0	19.0	20.0	21.1	22.0	23.1	24.2	25.2	26.3	27.3	28.2	29.3	30.3	31.3	32.3	33.5	34.3	35.3	36.5	37.5	38.3	39.7	40.5	43.0	45.8	48.2	50.9	53.5	56.0
6.5	14.3	15.2	16.1	17.1	18.0	18.9	19.9	20.8	21.8	22.8	23.7	24.6	25.5	26.5	27.5	28.5	29.4	30.4	31.4	32.3	33.4	34.3	35.1	36.1	37.1	38.0	40.2	42.8	45.1	47.6	49.9	52.1
7.0	13.5	14.4	15.3	16.2	17.1	17.9	18.8	19.7	20.6	21.5	22.4	23.3	24.2	25.0	25.8	26.8	27.7	28.6	29.0	30.2	31.2	32.2	33.0	33.9	34.9	35.8	37.9	40.1	42.1	44.4	46.8	49.0
7.5	13.0	13.8	14.6	15.2	16.3	17.3	18.0	18.8	19.7	20.5	21.4	22.2	22.9	23.7	24.6	25.4	26.3	27.2	29.4	30.5	31.2	32.1	33.0	34.9	35.9	37.9	40.0	42.0	44.1	46.1		
8.0	12.5	13.3	14.1	14.8	15.7	16.4	17.3	18.0	18.8	19.6	20.4	21.2	21.9	22.7	23.5	24.2	25.0	25.8	26.6	27.4	28.2	29.0	29.9	30.6	31.5	32.1	34.1	36.0	38.0	40.0	41.9	43.9
8.5	12.2	12.8	13.6	14.4	15.1	15.7	16.6	17.3	18.0	18.7	19.6	20.3	21.0	21.6	22.5	23.3	23.9	24.7	25.5	26.2	26.8	27.6	28.2	29.1	29.9	30.8	32.6	34.2	36.2	38.0	39.9	41.8
9.0	11.7	12.4	13.2	12.8	14.6	15.3	15.9	16.6	17.4	18.1	18.8	19.5	20.2	20.8	21.6	22.3	22.9	23.7	24.4	25.0	25.7	26.5	27.1	27.9	28.7	29.4	30.1	32.9	34.6	36.3	38.0	39.9
9.5	11.4	12.1	12.7	13.4	14.1	14.7	15.4	16.0	16.7	17.5	18.1	18.7	19.4	20.0	20.7	21.4	22.1	22.8	23.5	24.0	24.7	25.4	26.1	26.8	27.5	28.1	29.8	31.2	33.2	34.9	36.4	38.1
10	11.2	11.7	12.3	13.0	13.7	14.4	14.8	15.5	16.2	16.8	17.5	18.2	19.7	19.4	20.0	20.6	21.3	21.9	22.6	23.2	23.8	24.5	25.1	25.8	26.4	27.0	28.7	30.1	31.8	33.5	35.0	36.7
11	10.6	11.3	11.8	12.4	12.9	13.5	14.2	14.7	15.3	15.8	16.5	17.0	17.6	18.2	18.8	19.4	20.0	20.6	21.2	21.7	22.3	22.9	23.5	24.1	24.8	25.2	26.8	28.1	29.8	31.2	32.5	34.1
12.	10.2	10.7	11.3	11.7	12.3	12.8	13.4	13.8	14.5	15.1	15.6	16.2	16.7	17.2	17.8	18.8	19.4	19.9	20.4	21.0	21.6	22.1	22.7	23.1	23.8	25.8	26.4	27.9	29.1	30.5	31.9	
13		10.3	10.8	11.3	11.8	12.3	12.8	13.3	13.8	14.4	14.8	15.4	15.8	16.3	16.8	17.3	17.9	18.4	18.9	19.3	19.8	20.4	20.9	21.4	21.9	22.4	23.8	24.9	26.2	27.5	28.9	30.0
14		9.8	10.4	10.8	11.4	11.8	12.3	12.8	13.3	13.7	14.3	14.7	15.2	15.6	16.2	16.6	17.1	17.5	18.0	18.5	18.8	19.4	19.9	20.4	20.9	21.2	22.5	23.7	24.9	26.1	27.2	28.5
15			10.2	10.6	11.0	11.4	11.8	12.4	12,7	13.7	13.7	14.2	14.6	15.0	15.4	15.8	16.4	16.8	17.3	17.7	18.2	18.6	19.0	19.5	19.9	20.3	21.5	22.6	23.8	24.9	25.9	27.1
16				10.3	10.7	11.1	11.5	11.8	12.3	12.8	13.3	13.7	14.0	14.4	14.8	15.3	15.7	16.2	16.6	16.9	17.4	17.9	18.2	18.8	19.1	19.5	20.6	21.6	22.7	23.8	24.8	25.9

Table 5-8. Percentage of Stack Loss for Natural Gas (courtesy of Cleaver-Brooks Company).

DIFFERENCE BETWEEN FLUE GAS AND ROOM TEMPERATURES IN DEGREES FAHRENHEIT

% CO_2	200	220	240	260	280	300	320	340	360	380	400	420	440	460	480	500	520	540	560	580	600	620	640	660	680	700	750	800	850	900	950	1000
3.0	23.1	24.4	25.9	27.2	28.6	30.0	31.3	32.8	34.1	35.8	36.9	38.2	39.8	41.0	42.2	43.8	45.0	46.3	47.8	49.0	50.0											
3.5	21.2	22.5	23.8	24.9	26.1	27.2	28.4	29.6	30.9	32.0	33.2	34.4	35.8	36.8	37.9	39.2	40.3	41.6	42.8	43.8	45.0	46.2	47.7	48.3	49.8							
4.0	19.9	20.9	22.0	23.1	24.1	25.1	26.2	27.2	28.3	29.4	30.4	31.8	32.5	33.8	34.8	35.8	36.8	37.8	38.8	39.9	40.9	42.1	43.0	44.1	45.2	46.2	48.8					
4.5	18.9	19.9	20.9	21.8	22.7	23.6	24.5	25.5	26.4	27.3	28.3	29.3	30.2	31.2	32.2	33.0	34.0	34.9	35.9	36.8	37.8	38.6	39.8	40.4	41.5	42.6	44.8	47.2	49.8			
5.0	18.0	18.9	19.8	20.6	21.4	22.2	23.1	24.0	24.9	25.8	26.8	27.5	28.3	29.1	30.1	30.9	31.8	32.5	33.6	34.3	35.7	36.2	36.9	37.8	38.8	39.7	41.8	43.8	46.0	48.2		
5.5	17.4	18.1	18.9	19.8	20.5	21.2	22.1	22.9	23.8	24.5	25.2	26.2	26.9	27.8	28.5	29.2	30.0	30.8	31.8	32.3	33.2	34.1	34.9	35.8	36.3	37.3	39.2	41.0	43.0	45.3	47.2	49.0
6.0	16.8	17.4	18.2	18.9	19.6	20.4	21.1	21.8	22.7	23.3	24.1	24.9	25.5	26.2	27.0	27.8	28.4	29.2	30.0	30.8	31.5	32.2	32.9	33.8	34.3	35.2	36.8	38.8	40.4	42.5	44.3	46.2
6.5	16.3	16.9	17.6	18.4	19.0	19.8	20.4	21.1	21.8	22.4	23.2	23.8	24.5	25.2	25.9	26.5	27.2	27.9	28.7	29.2	30.0	30.9	31.4	32.1	32.8	33.5	34.6	36.8	38.4	40.3	42.0	43.8
7.0	15.8	16.5	17.1	17.8	18.4	19.1	19.8	20.4	21.0	21.8	22.3	22.9	23.6	24.2	24.9	25.5	26.2	26.8	27.4	28.0	28.8	29.4	30.0	30.8	31.2	32.0	33.8	35.3	36.8	38.3	40.0	41.8
7.5	15.5	16.1	16.7	17.2	17.9	18.5	19.1	19.8	20.3	20.9	21.5	22.2	22.8	23.3	24.0	24.6	25.2	25.8	26.4	26.9	27.7	28.2	28.8	29.4	30.1	30.8	32.2	33.8	35.2	36.8	38.3	39.9
8.0	15.2	15.7	16.3	16.9	17.4	18.0	18.6	19.2	19.8	20.3	20.9	21.5	22.1	22.8	23.2	23.8	24.4	25.0	25.5	26.0	27.2	27.2	27.8	28.4	29.0	29.5	31.0	32.4	33.8	35.4	36.8	38.2
8.5	14.9	15.4	15.9	16.5	17.1	17.6	18.2	18.7	19.3	19.8	20.4	20.9	21.4	22.0	22.5	23.1	23.7	24.2	24.8	25.3	25.8	26.4	26.9	27.4	28.1	28.6	29.9	31.3	32.8	34.2	35.4	36.8
9.0	14.6	15.2	15.7	16.2	16.6	17.2	17.8	18.3	18.8	19.3	19.9	20.4	20.9	21.4	21.9	22.5	23.0	23.5	24.1	24.5	25.2	25.8	26.2	26.7	27.2	27.8	29.0	30.3	31.8	33.0	34.3	35.7
9.5	14.6	14.9	15.4	15.9	16.4	16.9	17.4	17.9	18.4	18.9	19.5	19.9	20.4	20.9	21.4	21.9	22.4	22.9	23.4	23.8	24.4	24.9	25.4	25.9	26.4	26.9	28.2	29.4	30.8	32.0	33.3	34.5
10	14.2	14.6	15.2	15.6	16.1	16.6	17.1	17.5	18.1	18.5	19.0	19.5	20.0	20.4	20.8	21.4	21.8	22.4	22.8	23.3	23.8	24.2	24.8	25.2	25.8	26.2	27.4	28.6	29.8	31.2	32.2	33.4
11		14.4	14.7	15.2	15.6	16.1	16.5	16.9	17.4	17.8	18.4	18.8	19.3	19.6	20.2	20.5	20.9	21.4	21.9	22.3	22.8	23.2	23.7	24.2	24.6	25.0	26.2	27.2	28.3	29.5	30.8	31.8
12			14.4	14.8	15.2	15.6	16.1	16.5	16.9	17.3	17.8	18.2	18.6	19.0	19.4	19.8	20.2	20.6	21.1	21.4	21.9	22.3	22.8	23.2	23.6	24.0	25.1	26.1	27.2	28.3	29.2	30.3

Example No. 1

Assume 15 lb. design, 100 HP CB boiler, fired on gas at 100 percent of rating. You see stack temperature of 340° F. and room temperature is 80° F. (340-80 - 260° F. Net). You measure CO^2 of 10 percent with no CO. At 260° and 10 percent CO^2, you get stack loss of 15.6 percent going up the stack.

Add 1.0 percent for radiation and convection losses—15.6 percent plus 1.0 percent = 16.6 percent.

100 - 16.6 = 83.4 percent fuel-to-steam efficiency.

Example No. 2

Assume 150 lb. design, 400 HP CB boiler, fired on No. 6 oil at 100 percent of rating. You see stack temperature of 450° F., and room temperature is 70° F. (450 - 70 - 380° F. Net.). You measure CO^2 of 13 percent with no CO. At 380° F. and 13 percent CO^2, you get stack loss of 14.4% going up the stack.

Add 1.1 percent for "radiation and convection losses".

14.4% plus 1.1% = 15.5%
100 - 15.5 = 84.5% Fuel-to-Steam Efficiency

CONCLUSIONS

If you are concerned about your present fuel costs, you can obtain a job-site stack temperature - CO^2, O^2 test on your present equipment to see "where you are" regarding efficiency. Call your Cleaver-Brooks representative. He has the calibrated equipment, and the charge will be minimal. If you are adding a new boiler or are building a new plant, compare and evaluate guaranteed fuel-to-steam efficiency. If you select a boiler because of a lower first cost—and this unit is 3 percent, 4 percent or more percent less efficient than the higher price boiler—you will pay the difference many times over during the life of that boiler! The percent increase in fuel costs is greater than the nominal percent decrease in fuel-to-steam efficiency! A 3 percent drop in efficiency increases fuel costs 3.8 percent (85 percent versus 82 percent). A 5 percent drop in efficiency increases fuel costs 6.3 percent (85 percent versus 80 percent). A 7.5 percent drop in efficiency increases fuel costs 9.7 percent (85 percent versus 77.5 percent). A 10 percent drop in efficiency increases fuel costs 13.5 percent (85 percent versus 75 percent). How is this proven? A 200 Hp unit has an output of 6,700,000 Btu/hr. Assume No. 6 oil @ 150,000 Btu/gal.

		Boiler "X"	Boiler "CB"
Fuel-to-Steam Efficiency	=	80%	85%
Output + Effic. = Input	=	8,375,000 Btu/Hr.	7,882,000 Btu/Hr.
Input -- gal/Hr.	=	55.8	52.5
55.8/52.5	=	6.3% Increase in fuel usage	
	=	6.3% Increase in fuel costs due to 5% decrease in fuel-to-steam efficiency.	

Courtesy of Cleaver-Brooks Company

CONDUCTING THE SURVEY

Reference will be made throughout the following instructions to the survey form part A (Table 5-9). The first half of this form appeared in Chapter 4. Essentially, we will be taking temperatures with the use of pyrometers and high temperature thermometers and applying our deductions to the form itself.

The carbon dioxide, carbon monoxide and oxygen readings will be collected at the flue gas breeching, at a point selected close to the boiler, assuring that no intervening air leakage will adversely affect the orsat's accuracy. The sampling point for combustion may accordingly be taken at any point after the combustion chamber. Combustion should be absolutely completed by the travel point where the sample will be extracted.

The temperatures to be collected will be directly influenced by the exact point in which the sensing element will be placed. The initial combustion temperature, T-1, must be taken at the hottest area of the fire or the most intense point of the combustion process. This will be our thermal "starting point." The temperature, T-2, will accordingly be taken immediately following the last and furthermost part of actual heat absorbing surface. If a boiler is equipped with an economizer or an air preheater, then the temperature shall be taken immediately following that last point of actual heat transfer. Otherwise, we will not be giving credit to the boiler as constructed, for heat absorbed or not absorbed.

With a simple boiler, having no secondary transfer surfaces, the temperature and flue gas analysis may be taken from the very same orifice, or port, after the fact that no external air leakage has been adequately assured. The entrance of any air would substantially interfere with both temperature and flue gas readings. The temperature should be very skeptically observed to ascertain that it is the actual temperature of the products of combustion and not the radiation of some part of the transfer surface. Be sure the sensing element of the thermometer or pyrometer is not placed in an air pocket or in the stream of an air current. Several small holes may advantageously be placed at points at and around the sampling point. Readings should then be taken until the engineer is convinced that the readings so taken are factual and representative of only that medium he is testing.

The following steps will be given for the collection of data. The progressive steps are suggested, as the engineer may wish to augment this procedure to accommodate his routine inspection.

Suitable pyrometers and thermometers will be secured. The pyrometer used should have a temperature 25 percent above the highest expected temperature of the combustion process to be sensed. Thus, if a flame temperature of 2,600 degrees F. is to be metered, the Engineer should select a pyrometer with a maximum point above 3,000 degrees to assure upper mid-scale reading. A high temperature liquid-filled thermometer may be used to sense the exiting flue gases. As in the majority of cases, this temperature will be below 800 degrees F., with averages more often found at 400-600 degrees. Do not use a liquid filled thermometer to sense the combustion temperatures. A pyrometer or thermocouple must be used for these intense areas.

The respective chemicals for the sampling of the products of combustion must be reasonably fresh and unsaturated. If this is the first time that orsat testing has been conducted in the plant for some time, chemicals viz—potassium hydroxide, pyrogallic acid and cuprous chloride—which are used for the absorption of carbon dioxide oxygen and carbon monoxide respectively. These chemicals will weaken themselves with age and decomposition unless they have been kept in sealed bottles away from sunlight.

A fuel calorific analysis will be obtained from a chemical testing laboratory. One may be done by conducting a calorimeter test by the fuel-bomb method. The available heat per pound of combustible will be noted in the spaced provided. In addition, the calorific value of the fuel by theoretical computation will also be noted, along with the theoretical flame temperature by computation. The actual amounts of air required for combustion may at this time be calculated also.

The actual air delivered may be determined by an "air-flow" meter, in conjunction with the area as provided for delivery through the air registers and positioning of the air shutters. This data will be applied to the air actually required to obtain the air efficiency.

Table 5-9. Fuel Combustion Survey and Boiler-Heat Exchanger Surveys.

Name of Company ____________________

Location-Mailing Address ____________________

Survey Requested By ____________________ Title __________

Date of Survey ____________________ Time Required __________ Material __________

Survey Conducted By ____________________ Reason __________

PART -"A" **FUEL- COMBUSTION SURVEY** **TYPE FUEL:** ____________________

$\frac{\text{Btus} \quad \text{Lb./CuFt.}}{\text{Calorific Value from Analysis}}$ $\frac{\text{Btus} \quad \text{Lb./CuFt.}}{\text{Actual Heat Released per unit-by Combustion}}$ $\frac{\text{Btus By Combustion}}{\text{Btus / Analysis}}$ = ____________________ % Efficiency

Calorific Formula: Btus/Lb = 14,500 x%C. + 62,00 x% (H^2-0^2/8) + 4,050 x%S. (by weight)

Btus Released by Combustion = $\frac{\text{Measured Actual Flame Temperature of Combustion (degrees F.)}}{\text{Weight of Combustible (+) Weight of Air Supplied (X) Ave. Sp. Heat}}$

Theoretical Flame Temp. (degrees F.) : = $\frac{\text{Calorific Value of Fuel (+) Sensible Heat in Air (+) Fuel}}{\text{Weight of Provided Air/Fuel Mixture (X) Avg. Sp. Heat}}$

Short Method of Temperature Eff. $\frac{\text{Actual Measure Flame Temp. by Pyrometer}}{\textit{Theoretical Flame Temperature by formula}}$ = ____________________ % Efficiency

Theoretical Air Required = (In LBS.) = 11.53 x%C. (+) 34.34 x% (H^2 -02/8) + 4.29 x% S. (Dry Air)

Percent Excess Air Supplied : = $\frac{\text{Theoretically Required amount of air}}{\text{Actually Deliverd Amount of Air}}$ (X) 100 = ____________________ % Excess%

Percent Excess Air by Orsat Analysis := $100 (X) \dfrac{O^2 - CO/2}{0.264 \, x\% \, N^2 \, (+) \, 1.89 \, x\% \, CO.}$ ________________ % Excess%

Maximum Percent CO^2 Attainable w/Fuel = $\dfrac{100 \times (CO^2 + CO)}{100 - 4.78 \, x\% \, O^2 \, (+) \, 1.89 \, x\% \, CO}$ ________________ % Max. Attainable

---

PART-"A" **BOILER-HEAT EXCHANGER EFFICIENCY** **TYPE UNIT**

Temp."T-2" ________________ (F.) Exiting - Final Flue Gas Temperature/Therometer

Temp. "T-1" ________________ (F.) Actual Combustion Process Temperature/Pyrometer

$\dfrac{T1 - T2}{T1}$ = ________ % Thermal Eff.

Theoretical Efficiency of Combustion/Boiler Combination = $\dfrac{\text{Actual Heat "Q" in Steam(X)Lbs. Generated(X)Time}}{\text{Calorific Value of Fuel(X)Weight of Fuel(X) Time}}$

Practical: Combustion/Boiler Efficiency Formula $\dfrac{\text{Lbs. Steam/Hr.(X) "Q"(X)F.E.}}{\text{Lbx. of Fuel Fired(X)BTU's/Lb.}}$ (X) 100= ________________ Operating Eff.

Percent Rating of Boiler Achieved by Operation $\dfrac{\text{Lbs. Steam Generated(X)Q(X)FE/ 34.5(T)}}{\text{= Manufacturers Rated Blr. H.P.}}$ ________________ % = Rated Blr. H.P.

Enthalpy of Steam/Lb.("Q") As obtained from TABLE S-1 =Heat of Evap.(+) Heat in water above 32 d.F.@Pressure

Factor of Evaporation (FE.) = (Q-t+32)/ 970.4 = $\dfrac{\text{Heat in Steam-Feed Water Heat @ 32}}{\text{Heat read. @ 212 d.F.-Unit Evap.}}$

Boiler H.P.=0.03(X)Lbs. Steam Hr. Generated(X)FE.OR B.H.P.="Q"(X)FE.per Hr. 34

The temperatures of the exiting flue gases after the last effective transferring point may now be applied to the temperature of the combustion process. The basic formula $\frac{("T\text{-}1")-("T\text{-}2")}{("T\text{-}1")}$ will indicate the boiler/combustion process efficiency.

The orsat tests are taken, The data thus obtained are applied to the formulas as given. The related excess air and combustion efficiency will be thus revealed.

Reference must be made to a steam table to determine the evaporative efficiency, boiler horsepower and operating efficiency. See Table 5-10, which is an abbreviated one.

The measurement of fuel fired may be had directly by weight in coal-fired plants, or multiplied by the gallons of fuel fired in oil plants. The weight per gallon of each fuel will be found in the chapter on primary fuels, or direct measurement can be made with a hydrometer. The weight of gas fuels is also given in cubic feet pound of fuel in the primary fuels chapter.

The data collected by the survey may then be incorporated in the efficiency-rating graphs as illustrated in Figs. 5-21 through 5-23. Readouts of operating efficiency for each and every percent of rating may then be had to guide the engineer in his load demand problems as necessitated when the operational efficiency of one boiler unit may be questioned by the economy of multiplc unit operation.

The timely tests and surveys at periodic intervals will provide a positive indication of the plant's maintenance of efficiency. The decline or regression of thermal efficiencies or combustion efficiencies will bear direct relation upon the heat exchangeability of the boiler and the effectiveness of the combustion process, all other factors remaining equal.

SHORT TEST METHOD

A comparatively non-technical method of establishing boiler unit efficiency may be had by installing a fuel oil meter immediately at the burner and a water meter on the feed water line. Apply the required conversion constants for pounds/gallon of oil and pounds/gallon water. A periodic check of the "unit" efficiency can be reduced by directly dividing the amount in lbs./steam generated (as metered by the feed water required for its replacement) by the amount of oil consumed in lbs./hr. Although this method lacks the conclusiveness as would be obtained by the more scientifically conducted and supported tests as mentioned earlier, it would nevertheless provide the engineer with a conveniently con-

Table 5-10. Properties of Saturated Steam and Water.

"Q" = ENTHALPY OF STEAM; BTUs/LB.							
TEMP. d.F.	PRESS. (a) (#PSIA)	WATER - BTU/LB.	STEAM-Q BTU/LB.	TEMP. d.F.	PRESSURE (#PSIA)	WATER - BTU/LB.	STEAM-Q BTU/LB.
32	0.885	0.000	1075.5	270	41.856	238.95	1170.6
35	0.0999	3.000	1076.8	280	49.200	249.17	1173.8
40	0.1216	8.030	1079.0	290	57.550	259.40	1176.8
45	0.1474	13.04	1081.2	300	67.005	269.70	1179.7
50	0.1779	18.05	1083.4	310	77.67	280.0	1182.5
60	0.2561	28.06	1087.7	320	89.64	290.4	1185.2
70	0.3629	38.05	1092.1	340	118.00	311.3	1190.1
80	0.5068	48.04	1096.4	360	153.00	332.3	1194.4
90	0.6981	58.02	1100.8	380	195.73	353.6	1198.0
100	0.9492	68.00	1105.1	400	247.26	375.1	1201.0
110	1.2750	77.98	1109.3	420	308.78	396.9	1203.1
120	1.6927	87.97	1113.6	440	381.54	419.0	1204.4
130	2.2230	97.96	1117.8	460	466.90	441.5	1204.8
140	2.8892	107.95	1122.0	480	566.20	464.5	1204.1
150	3.7180	117.95	1126.1	500	680.91	487.9	1202.2
160	4.7410	127.96	1130.2	520	812.52	512.0	1199.0
170	5.9930	137.97	1134.2	540	962.83	536.8	1194.3
180	7.5110	148.00	1138.2	560	1133.4	562.4	1187.7
190	9.3400	158.04	1142.1	580	1326.2	589.1	1179.0
200	11.526	168.09	1146.0	600	1543.2	617.1	1167.7
210	14.123	178.15	1149.7	620	1786.9	646.9	1153.2
212	14.696	180.17	1150.5	640	2059.9	679.1	1133.7
220	17.186	188.23	1153.4	660	2365.7	714.9	1107.0
230	20.779	198.33	1157.1	680	2708.6	758.5	1068.5
240	24.968	208.45	1160.6	700	3094.3	822.4	995.2
250	29.825	218.59	1164.0	705.5	3208.2	906.0	906.0
260	35.427	228.76	1167.4	(Critical Pressure of Steam)			

trived figure of the unit's relative "fuel consumption/steam output" index. This formula is as follows:

unit efficiency = steam generated (by meter) fuel admitted to burner (by meter)

CONSTRUCTING A "HEAT BALANCE" BY THE SANKEY DIAGRAM

A heat balance is a collective observation whereby all heat (in (Btus) is accounted for. This evaluation may best be initiated by reference to the simply constructed Sankey diagram (Fig. 5-2). The gross input is first considered as the available calorific value of the fuel "as fired." The next consideration is made subsequent to determining the heat loss from the moisture in the products of combustion (from the hydrogen in the fuel), the dry flue gas losses, resulting from incomplete combustion and hydrocarbons in the flue gas, and the computed combustion inefficiency loss from the combustion process or fuel burning equipment.

After arriving at the net heat release to the boiler surfaces, we may now consider the subsequent loses as represented exterior to the boiler/furnace proper. These losses chiefly consist of boiler efficiency loss radiation losses, and lastly the condensation losses, which may be from insulation problems, excessive distance of steam travel to the load or possibly by mechanical problems resulting from malfunctioning traps.

In essence we may, by having conclusive figures of all other factors of loss, deduce an unknown quantity. The entire sum of all losses will of course equal the gross heat input; thus, we have an interchangeable and reversible mathematical problem. Any factor will be the difference of all of the other members or components subtracted from the gross input. Likewise, the gross input will equal all of the combined and collective losses, including that quantity supplied to the load:

Gross input = glue gas losses + combustion losses
+ all external losses + heat to load

Useful heat to load = all external boiler losses + all internal boiler losses

Flue gas losses = gross input - combustion losses + external boiler losses + useful heat as delivered to load.

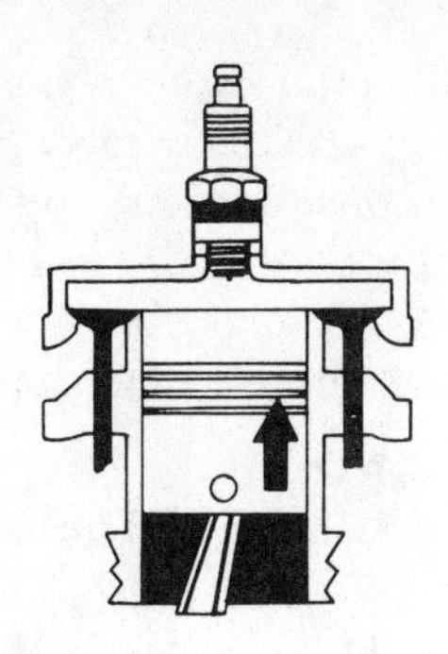

Chapter 6
Electrical Power

It is certainly no abstract expression or academically advanced theory that the total electrical power consumed by a plant is represented by the basic formula: *power delivered=useful power consumed + power wasted by resistance + power wasted by power factor losses.* Rarely may the losses and factors attributing to reduced efficiency and waste be so clearly and accurately analyzed for computative purposes as they can be in electrical engineering situations. It is such a frequently demonstrated phenomenon that not even the insignificant lone electron is able to leave the premises of deduction without having to give account for its whereabouts, in terms of either work done or energy wasted. All of our calculations and considerations will be conducted in accordance with and reference to the international units of electrical measurements. These measurements have been scientifically proven to be accurate to the very last electron. By utilizing these units of functional exactness, our figures will be equally assured of phenomenal precision.

In addition to the exactness enjoyed in all aspects of electrical engineering, the study of this physical branch of science is one of a profound and technically detailed structure. In our efforts to consider the power efficiency aspects we must adhere to the provisional parameters of efficiency as set forth by the intended scope of this manual and limit our interests to the appropriate metering, collection of required data, and the application of such data to the formulas related to the determination of efficiency.

Although it would certainly be to the reader's advantage to have at least a familiarity with electrical engineering, I will confine the technical considerations to meters and their respective circuit adaptation for data collection. There will be a brief commentary of each power circuit under analysis, with a short description of each circuit and a representative schematic.

DIRECT CURRENT AND ALTERNATING CURRENT SYSTEMS

When electricity generation first became a reality to be enjoyed by our populace, the prime question presented to authorities was, "Should the electricity be of the direct current (dc) or the alternating current (ac) mode of generation and distribution?" The question was resolved by the arguments and facts advanced by George Westinghouse, in favor of the alternating current mode of generation. This mode as such could best be utilized by recipients far from the source of generation. There would be the least reduction of circuit voltage as would be otherwise consumed by the anticipated "line drop" inherently associated by distribution copper losses, absorbed by the lengthly distribution lines required for delivery of the current demanded.

Although this acceptance was granted over the protests and rebuttal of Thomas Alva Edison, who favored the direct current mode for safety reasons, the ac mode was adopted in spite of the rationale of high voltage dangers to be encountered with ac distribution techniques, which have historically justified its adoption by the absence of the "latent dangers of high transmission potentials" and the advantages of economy offered by the reduced "I^2R" losses. Additionally, the universal endorsement of the induction motor alone is evidence of the advantages gained by the sanction of the national system of alternating current, as is the standard of this nation's electrical power network.

There are however, great and unique advantages to the use of direct current in industrial applications such as direct motor speed and torque regulation as offered by dc machinery and the use of dc in electrochemical processes. Many future energy conversion systems will, by virtue of the expected storage of electrical energy, be entirely of the dc circuitry. The direct use and interexchange processes of direct current can be effected more efficiently without the use of otherwise rectified ac supplies.

The losses incurred by conductor resistances, which are represented by the "I^2R" relationship and expressed mathematically in terms of wattage or equivalent heat loss, are inherent to both dc and ac systems. These are commonly referred to as resistance or non-spurious load losses. They can only be lessened by either the increase in the size of the conductors or by the reduction in the conducted current or *ampacity*. Every electrical device, motor, transformer and length of conductor is affected by the property of resistance to electron flow. The electrical device will electrically conform to the basic related functions of Ohm's Law, which deals exclusively with the values and interdependency of the circuit's *voltage* (or EMF), *current* (or *amperage*), and the circuit's inherent *resistance* (or opposition of electron movement). The representations of these functions are succinctly expressed in the formula: voltage (EMF) equals the current (amps) multiplied by the resistance of the circuit (in ohms). This formula may be transposed to identify the value of any of its members when the remaining two functions are known.

In dealing with ac circuits and their related power distribution systems, we have the additional (and frequently objectionable) characteristic of the reactive effects of spurious resistance, created by the ac mode's peculiar property of alternating its impressed voltages throughout its sinusoidal continuity of generation. The dc mode of course lacks this oscillation of output and resulting values of power and polarity. The output of the dc generator is commutated to provide for an even and constant value of power, with no polarity reversals. The ac generator, having no commutator, can only provide the rotationally effected and "as impressed" voltages corresponding to the variation of the magnetic flux densities, and subsequent alternating polarities, as experienced and brought out of the generator by the collecting armature.

It is particularly this alternation of power and polarity that attributes to the spurious resistance as mentioned. Acting through a purely resistive load, the reaction to this alternation of values is of no evidenced consequence. When this same circuit has in its electrical path any manner of *capacitance* or *inductance,* the spurious resistance will then be witnessed. More often than not, the reactive resistance will be far greater than the pure effective resistance, as would be experienced by a direct current of the same value of amperage and driving electromotive force. It is this peculiar characteristic of alternating current that impedes the effective work actually performed by the movement of its current through the circuit components of inductive and capacitive devices--the behavior of reactance.

This reactance, be it either of the inductive or capacitive variety, will impede the current so moved though the circuit by the voltage impressed upon it. It is directly proportional to the frequency of the circuit's alternating surges of current in instances of induction, and in direct but inverted proportions to the fluctuations of power in instances of capacitance. In inductance devices, such as motors, relays, transformers and reactors, the actual inductive reactance, which may be measured in ohms will be directly proportional to its inductance as computed in henrys or millihenrys. The frequency will remain, for all practical considerations, in power generation at a fixed 60 cycles per second. An inductive device is in fact frequency-sensitive. The capacative reactance of a capacitive circuit component will decrease in reactance as the capacitance of the capacitor increases. The capacitor is also a frequency-sensitive component. Because of the standard set frequency of power remaining at 60 hertz (Hz), the actual variable needed to be considered is the capacitor's ability to store a charge of electricity. This measurement of capacity is given in units called farads. The capacitive reactance of the capacitor actually decreases as its capacity increases.

Following is the mathematical relationship of inductive reactance and capacitive reactance, and related factors affecting the formulation of each as represented. It is only necessary to ascertain the amount of reactive loss in terms of ohms-reactance and ascribe this loss to either inductance or capacitance. Apply a correspondingly equal amount of the opposite reactance to the circuit to effect its correction.

CIRCUIT IMPEDANCE

The impedance of any ac circuit is mathematically represented by the square root of its resistance squared, plus the square of its net reactance. The net value of reactance is the difference between the two independent reactances of either inductance or capacitance. The smaller value is subtracted from the larger of the two and formulated, with Z expressed in ohms. R is the circuit's resistance to a dc measurement by an ohm meter. Xc is capacitive reactance and X_L is inductive reactance.

$$Z = \sqrt{R^2 + (X_C \text{ or } X_L)^2}$$

Let us assume a circuit having many induction motors being fed by an ac supply of 60 Hz. After having metered the resistance (with the power off), we find the ohmic resistance to be 12 ohms. We find by appropriate measurement that this circuit contains 18 ohms of inductive reactance and only 2 ohms of capacitive reactance. Naturally this will, by inductive reactance, be classed as an inductive loss. We must then, by the instructions of our formula, deduct the 2 ohms of capacitance reactance from the 18 ohms of inductive reactance, leaving a net reactance of 16 ohms. This 16 ohms will then be inserted as required and the formula now will appear as follows:

$$"Z" = \sqrt{12^2+(18-2=16^2)} \text{ equaling- } Z = \sqrt{12^2+16^2} = \sqrt{144+256} = \sqrt{400} = 20 \text{ ohms}$$

The actual reactances may be computed by the following formulas, or they may be measured by instruments such as the inductive bridge and capacitive bridge. The theoretical formulas for determining by component identification the respective values of inductive reactance and capacative reactance are:

	Formula for Inductive Reactance	**Formula for Capacitive Reactance**
	$Xc = \frac{1}{2 \text{ PI } fxC} = \frac{1}{6.28x60xC}$	$X^1 = 2 \text{ fl} = 6.28\times(f)\times(1)$
where:	Xc is reactance in ohms	X^1 is reactance in ohms
	2 constant = 6.28	2 constant = 6.28
	f = frequency in Hz	f = frequency in Hz
	C = capacitance in farads	l = inductance in henries

The impedance of an ac circuit, by virtue of its law of squares, may be represented geometrically by the right triangle method. The impedance will be construed as the hypotenuse of a right triangle. The sides represent the ohmic resistance and reactive or non-active value, also expressed in ohms. By applying the quantitative values of impedance, reactance and resistance to a squared graph, the resulting hypotenuse may be measured off directly with near accuracy. All circuits having all three properties of resistance, capacitive reactance and inductive reactance are classified as R_L circuits. Thus, a circuit having only resistance and capacitive reactance is classed as an Rc circuit. One in which only resistance and inductance are found is an R_L circuit. All circuits have resistance. To some minute detectable degree, all ac circuits contain both inductive and capacitive reactance. Many circuits, because of the insignificance as contributed by one reactance, will be classified by the reactance of its counter effect.

Even a transmission line, as straight and unaffected as it might be assumed, has in very long stretches a capacitive effect. Unless the minor contribution of the opposite reaction is of substantial influence, we generally consider an ac circuit as being either inductive or capacitive in nature.

In returning to our basic interrelationship as captioned by Ohm's Law where the EMF equals the current times the resistance, we may consider the ohmic or pure resistance to be tantamount to the Impedance. The combined effects of reactance yield an effect of impedance upon our ac circuit just as its equivalent pure resistance will impart in a dc circuit.

REVIEW OF CONCEPTS

In short, we have learned that the electrical power delivered to a device, circuit or motor will be accounted for by the useful or active power actually converted to equivalent work, plus the inactive power which is represented by resistance and reactance losses. In purely resistive circuits, the impedance is merely the ohmic resistance. That is found by measuring the resistance with an ohmmeter. In ac circuits, this value of impedance is affected not only by the actual conductor losses, but also by the concerted properties of reactance, as instituted by the circuit's inductive and capacitive characteristics. The impedance is a trigonometric sum of the independent effects of resistance and net reactance. The basic interrelated Ohm's Law of dc circuits then must accordingly be modified to accommodate the resistance to represent impedance in dealing with ac circuits. Ohm's Law for ac circuits then will be modified.

Both the inductive and capacitive reactance of any circuit may be computed or their equivalent values of inductive and capacitive reactance (X^1 X_c respectively) may be measured by instrumentation. Formulas for all computations and factors are included later. All values of voltage and amperage as dealt with in discussions of ac circuits are the effective values, unless otherwise specified. The effective values evidence an equivalent heating effect in Btus per amount of current passing through a heating element. One watt either of ac or dc will produce 3.41 Btus of equivalent heat. A descriptive presentation of the maximum, peak to peak, effective and average values of the ac sinusoidal wave is given in Fig. 6-1. A compendium of formulas and conversion factors is given in Table 6-1, which will serve to both refresh and acquaint the reader with the formulas we will be utilizing in our future investigations of efficiency considerations. Table 6-1 also contains formulas of related electrical engineering importance which might add to the academic facility of the reader's command in investigating the engineering aspects of his equipment.

I have prepared the ac sinusoidal functions and the effects both capacitance and inductance exert in their interrelated relationship with the power of a circuit and the circuit's *power factor* (PF). Figures 6-1 through 6-3 illustrate the functions of all

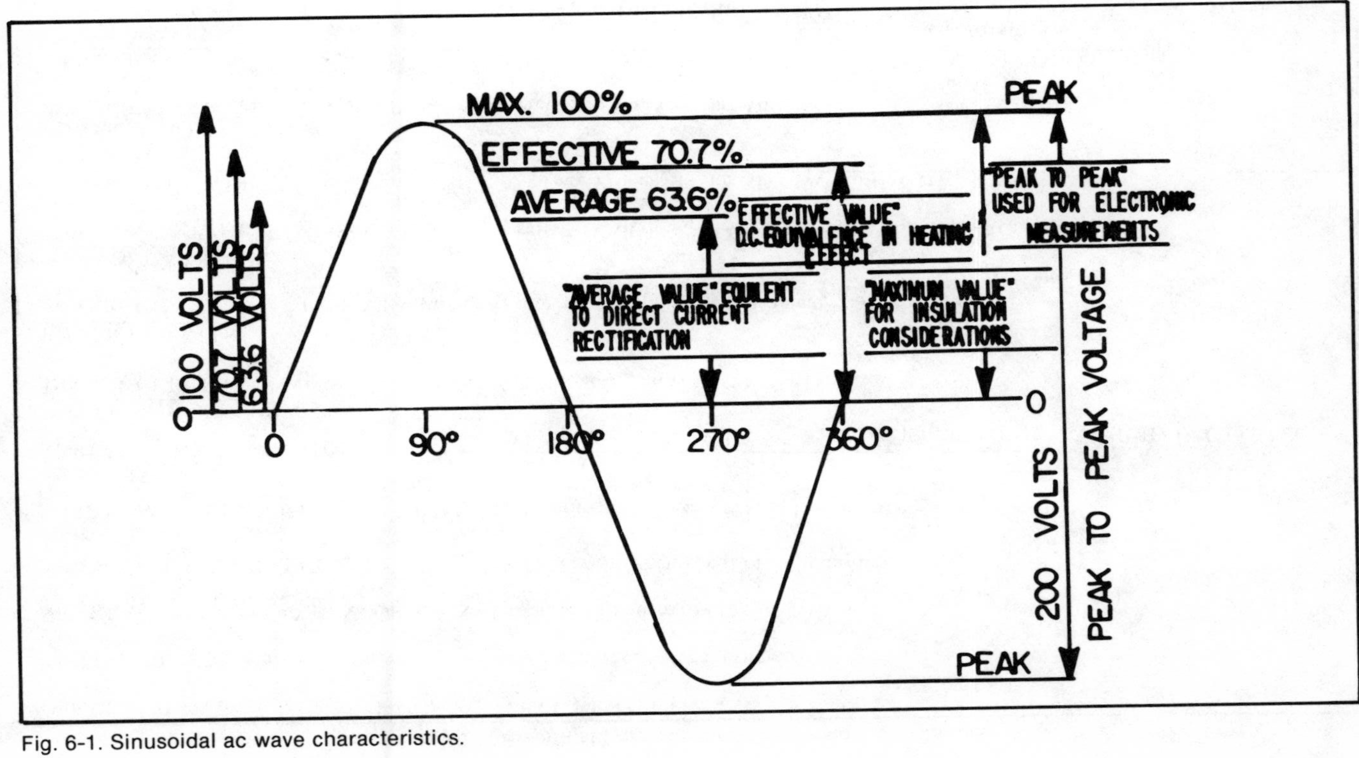

Fig. 6-1. Sinusoidal ac wave characteristics.

Table 6-1. Electrical Formulas and Conversion Constants.

Ohms Law(D.C.) E=IR; I=E/R; R=E/I Ohms Law (A.C.) E=IZ; I=E/Z; Z=E/I

IMPEDANCE "Z"= $R^2 + X^2$ Where: R=Ohmic Resistance, X=Reactance, (Net)

NET REACTANCE "X" = $X_l - X_c$ OR $X_c - X_l$ The subtrahend being the larger value.

AC CIRCUIT ADMITTANCE Z = $(R)^2 + (X)^2$ This is the reciprocal of impedance

1 1 1 1 1

X_l-REACTANCE, INDUCTIVE X_l =2 PI FL = 6.28(x)F(x)L; F=frequency, L=Inductance

X_c-REACTANCE, CAPACITIVE X_c = $\frac{1}{2\ PI\ F\ C} = \frac{1}{6.28(x)F(x)C}$;F=Hz C=Capacitance

POWER IN WATTS
DIRECT CURRENT "P" = E(x)I OR $P=I^2(x)R$ OR P=1/746 H.P. OR P=1 joule/sec.

POWER IN WATTS
AC CIRCUIT "P"=E(x)I(x)Power Factor OR "P"= $(\text{True Power})^2+(\text{Reactive Power})^2$

"VA" VOLT AMPERES
APPARENT POWER VA=in WATTS= $(\text{watts})^2+(\text{VARs})^2$; Where VARs=Volt-Amps-React.

POWER FACTOR
AC ONLY "PF" = $\frac{\text{True Power (as measured w/Wattmeter)}}{\text{Apparent Power (Volts (x) Amperes)}}$ PF= $\frac{\text{"R"}}{\text{"Z"}}$

POWER FACTOR
POWER FACTOR "PF" = cos o of Phase angle between Voltage and Amperage

POWER FACTOR
(3 PHASE) "PF" = For THREE PHASE SYSTEMS = $\frac{\text{True Wattage (Meter)}}{\text{Volts(x)Amps(x)1.73}}$ (1.73 = 3)

POWER FACTOR "PF" =
(2 Ph./4 wire) For Two PHASE SYSTEMS Using Four Wire Cuit. $\frac{\text{True Wattage (Meter)}}{\text{Volts(x)Amps(x) 2}}$

POWER FACTOR "PF" = (2 Ph./3 wire)

For Two PHASE SYSTEMS Using Three Wire Cuit. = True Wattage (Meter) / Volts(x)Amps(x)1.41 (1.41 = 2)

POWER FOR BALANCED THREE PHASE SYSTEM (Either Delta or Wye) = 1.732 (x) E (x) I (x) cos. o

ELECTRICAL HORSE POWER OUTPUT - for dc MOTORS

"H.P." = VOLS (x) AMPS (x) Motor Eff. / 746 = V.(x)A.(x)E. / 746

ELECTRICAL HORSE POWER OUTPUT - for ac MOTORS

"H.P." = VOLTS (x) AMPS (x) "PF" x Eff. / 746 FOR SINGLE PHASE

ELECTRICAL HORSE POWER POLY PHASE HORSE POWER

H.P. = VOLTS (x)AMPS(x)"PF"x Eff. (X) (#) / 746

(#)=1.73 for 3 Phase; (#)=2 for Two Phase/4 wire;(#)=1.41 for Two Phase/3 wire

ELECTRIC MOTOR EFFICIENCY DIRECT CURRENT MOTORS

% Eff. = H.P.(x) 746 / VOLTS x AMPS = H.P.(x) 746 / WATTS

ELECTRIC MOTOR EFFICIENCY FOR SINGLE PHASE MOTORS

% Eff. = H.P. (x) 746 / VOLTS(x)AMPS(x)P.F. = H.P. x 746 / V.xA.xP.F.

ELECTRIC MOTOR EFFICIENCY FOR POLY PHASE MOTORS

% Eff. = HORSE POWER (x) 746 / VOLTS(x)AMPS(x)P.F.(x)#-Phase constant

(#)=1.73 for 3 Ph.; (#)=2 for 2 Ph.4 wire; (#)=1.41 for 2 Ph. 3 wire circuits

(Table 6-1 continued)

PRIME MOVER HORSE POWER REQUIRED FOR GENERATOR

$$\text{H.P. Reqd./K.W. Output} = \frac{\text{KVA (or 1000 v.a.) x P.F. x (\#)}}{\text{Eff. of Generator (x) .746}}$$

(#)=1 for single phase; (#)=1.73 for 3 Ph. 3 wire; (#)=2 for 2 Ph. 4 wire (#)=1.41 for 2 Ph. 3 wire circuits. Note: For DC Generators, the "PF" and (#)=1.00

TRANSFORMER EFFICIENCY:

$$\text{\% Eff.} = \frac{\text{OUTPUT}}{\text{INPUT}} \text{ (X) } 100 = \frac{\text{OUTPUT}}{\text{output + losses}} \text{ (X) } 100$$

Where INPUT = OUTPUT = Total Core and Copper Losses + Heat Losses

$$\frac{\text{REACTANCE OF TRANSFORMERS}}{\text{EXPRESSED IN \% RATED VOLTS}} \quad \frac{\text{REACTIVE VOLTAGE DROP AT FULL LOAD}}{\text{RATED VOLTAGE}} \text{ (X) } 100$$

$$\frac{\text{PERCENT RESISTANCE}}{\text{EXPRESSED - \% K.W.}} : \text{\% K.W.} = \frac{\text{COPPER LOSS IN K.W.}}{\text{RATED OUTPUT IN K.W.}}$$

$$\text{PERCENT IMPEDANCE; \%} = \sqrt{(\text{percent resistance})^2 + (\text{percent reactance})^2}$$

$$\text{Voltage Regulation, \%} = \frac{\text{(No load Sec. Voltage)-(Full Load Sec. Voltage)}}{\text{Full Load Secondary Voltage}} \text{ (X) } 100$$

CONVERSION CONSTANTS & FACTORS

ONE AMPERE = ONE COULOMB; OHM's LAW-ONE AMPERE = ONE VOLT/ONE OHM RESISTANCE

ONE WATT = ONE AMPERE(X)ONE VOLT; ONE WATT = one joule per sec. ONE WATT = 1/746 H.P.

ONE WATT = ONE AMPERE(X)ONE VOLT with PF of 1.00 for A.C. CIrcuits.

ONE WATT = 3.41 B.T.U.'s; ONE WATT = 0.738 foot pounds per sec. ONE WATT = 10^7 erg/sec

ONE KILOWATT (1 KW) = 1,000 Watts; 1 KW = 1 VA (D.C.); 1 KW = 1 VA w/PF @ 1.0 (A.C.)

ONE HORSE POWER = 746 Watts, or 0.746 KW; ONE HORSE POWER = 33,000 Ft.Lbs./minute

ONE "HENRY" OF INDUCTANCE = The C.E.M.F. resulting from an INDUCTOR WHEN AN AMPERAGE of one AMP is flowing for one sec. at the impressed E.M.F. of ONE VOLT.

ONE "FARAD" OF CAPACITANCE = The storage of ONE COULOMB of electricity across the plates of a CAPACITOR, when a VOLTAGE OF ONE VOLT is impressed across it.

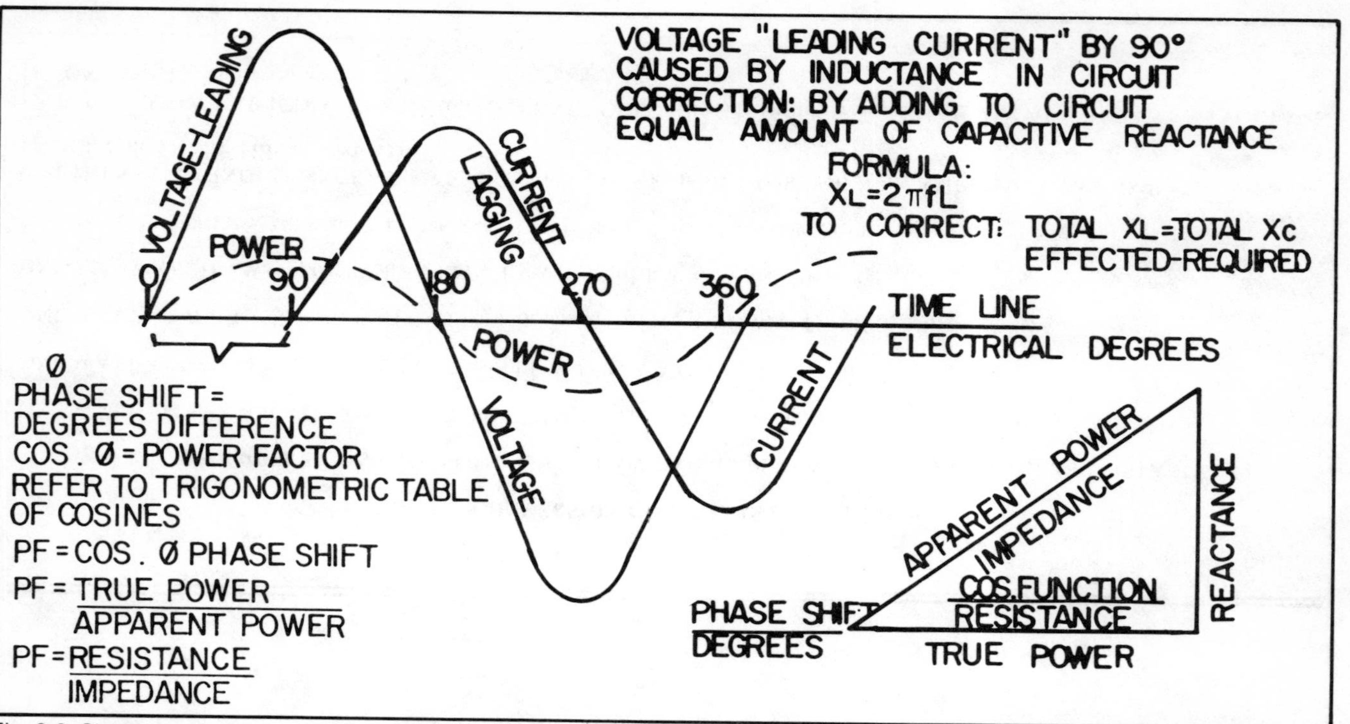

Fig. 6-2. Showing effects of inductive reaction, "X_L."

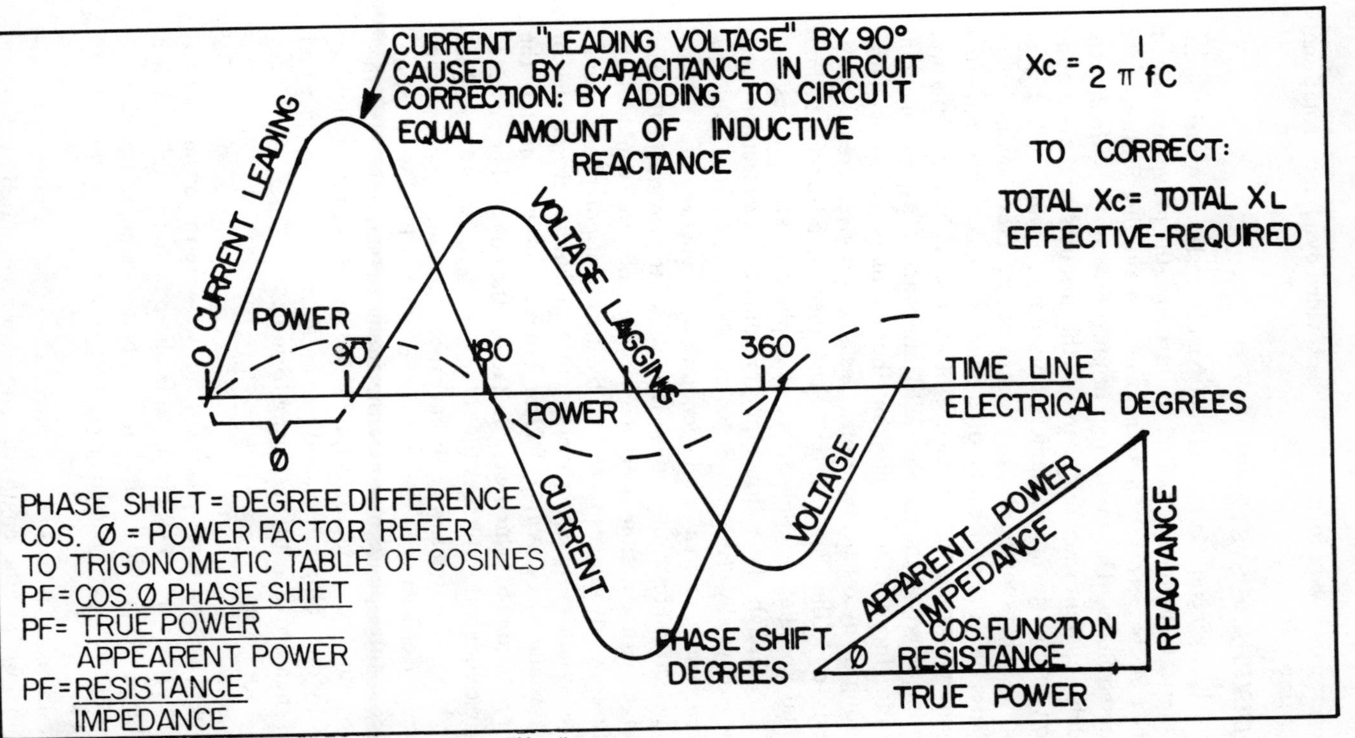

Fig. 6-3. Showing effects of capacitive reactance, "X_C."

generated ac wave forms as will be provided by a typical rotational generator.

POLYPHASE SYSTEMS

It is generally conceded that Nikola Tesla, an Austrian-born American engineer, was first to introduce the polyphase system of ac electric generation, transformation and distribution in the United States. The advantages of polyphase systems are immediately apparent when one examines the formulas for the computation of equivalent mechanical horsepower from an electric ac motor, when comparing the contributing factors against single phase, or more noticeably, direct current. In contrast with other founding fathers of the science of electricity, Nikola Tesla was both a genius of the pure as well as the practical aspects of engineering. In addition to his contribution of polyphase systems, among many other inventions, Tesla is also given credit for the Tesla coil, the Tesla turbine, and theories advanced in high voltage oscillation applications as applied to radio physics.

Although the rudimentary groundwork of electricity was chronologically laid years before Tela's affiliation with the science in such areas of dc generation, magnetic motor physics and electrical resistance, it was left to Tesla to provide the theories and experimentation necessary to convince the world of the efficiency advantages to be had by the use of polyphase ac electrical systems. The phase factor (corresponding to the number of phases and wires, as 1.41, 1.73 or 2.0) is a multiplier in all of the computations by which we will "factor" the power. The power of course equals the voltages times the amperage. See Table 6-2. In single phase systems, we only have the voltage times the amperage times the efficiency of the motor to deal with. If we were to consider a three-phase motor (either delta or star), we may then enjoy the "phase factor" of 1.73. We will accordingly multiply this by our volt-amperes and power factor-motor efficiency and will then arrive at the following formula:

$$\text{motor HP (3-Phase)} = \frac{\text{Volts} (\times) \text{Amps} (\times) \text{P.F.} (\times) \text{Eff.} (\times) 1.73}{746} = \frac{\text{V.A.} \times \text{PF} \times \text{eff.} \times 1.73}{746}$$

If, in place of a three-phase system, we were to consider a two-phase-four wire (2∅-4) circuit, our phase factor would then be a full two. We would realize an even better value, as follows:

$$\text{motor HP (2-0,4 wire)} = \frac{\text{Volts} (\times) \text{Amps} (\times) \text{P.F.} (\times) \text{Eff.} (\times) 2}{746} = \frac{\text{V.A.} \times \text{PF} \times \text{Eff.} \times 2}{746}$$

In using a two-phase, three wire (2∅-3) arrangement, although we save the lengths of conductor necessitated by the elim-

Table 6-2. Power Factors for Phase Angles.

PHASE ANGLE DEGREES-ϕ (COSINE)	POWER FACTOR PF=R/Z-%	PHASE ANGLE DEGREES-ϕ (COSINE)	POWER FACTOR PF=R/Z-%	PHASE ANGLE DEGREES-ϕ (COSINE)	POWER FACTOR PF=R/Z-%
0	1.000	30	0.8660	60	0.5000
1	0.9998	31	.8572	61	.4848
2	.9994	32	.8481	62	.4695
3	.9986	33	.8387	63	.4540
4	.9976	34	.8290	64	.4384
5	.9962	35	.8192	65	.4226
6	.9945	36	.8090	66	.4067
7	.9926	37	.7986	67	.3907
8	.9903	38	.7880	68	.3746
9	.9877	39	.7772	69	.3584
10	.9848	40	.7660	70	.3420
11	.9816	41	.7547	71	.3256
12	.9782	42	.7431	72	.3090
13	.9744	43	.7314	73	.2924
14	.9703	44	.7193	74	.2756
15	.9659	45	.7071	75	.2588
16	.9613	46	.6947	76	.2419
17	.9563	47	.6820	77	.2250
18	.9511	48	.6691	78	.2079
19	.9455	49	.6561	79	.1908
20	.9397	50	.6428	80	.1737
21	.9336	51	.6293	81	.1564
22	.9272	52	.6157	82	.1392
23	.9205	53	.6018	83	.1219
24	.9136	54	.5878	84	.1045
25	.9063	55	.5736	85	.0872
26	.8988	56	.5592	86	.0698
27	.8910	57	.5446	87	.0523
28	.8830	58	.5299	88	.0349
29	.8746	59	.5150	89	.0175
				90	0.0000

ination of the fourth wire in the above circuit, we lose the full strength of the phase factor. Instead of enjoying the multiplier of two, we now only have a multiplier of 1.41. This again is for two phase, three wire, as opposed to the two-phase, four wire system having the inherent factor of a full two.

TWO PHASE VERSUS THREE PHASE SYSTEMS

Why would we want to utilize any other system other than one which would give us the greatest output of power for the same given amount of power (volts times amperes)? Since we can have a phase factor as high as two in two-phase, four wire systems, why would we want to even consider using a system which would offer us less? Economics is the reason. We will save on millions of miles of copper, on labor involved in running the conductors, and on the costs of the larger sized equipment necessary to accommodate the use of two-phase, four wire systems. The United States had to resolve which of the two systems it would utilize, and one of the largest considerations was of economic interests and the savings offered by having a factor of two over that of a more economic system offering a close second place factor of 1.73. The three-phase system was decided to be preferential, not withstanding the minor power loss.

References will be made later to such schematics like Figs. 6-4 through 6-7 where we observe the wiring and related computational characteristics of the delta and star (or wye) systems of three-phase circuit arrangements and electro-geometric configurations. In contrast to the single-phase system, the polyphase system provides a more continual and less articulated supply of power to the field coils and armature of the motor receiving it.

Actually, the two or three phases do not occur at the very same instant. Otherwise, the phase factor for three-phase systems would be three and the two-phase systems factors would be as four and two, for four-wire and three-wire systems respectively. The phases of the three wire system are spaced or separated 120 electrical degrees apart, *not necessarily* 120 physical degrees. When the "3ϕ" circuit has more than just three field coils, its physical separations about the stator will naturally have to accommodate the number of field coils. In a six pole machine, the phases are still 120 electrical degrees apart, but now only 60 physical (or geometrical) degrees apart. The same theoretical idea is applied to two-phase systems, which are 90 electrical degrees apart. Due to the number of field poles they may be 90, 45, or even 27.5 phys-

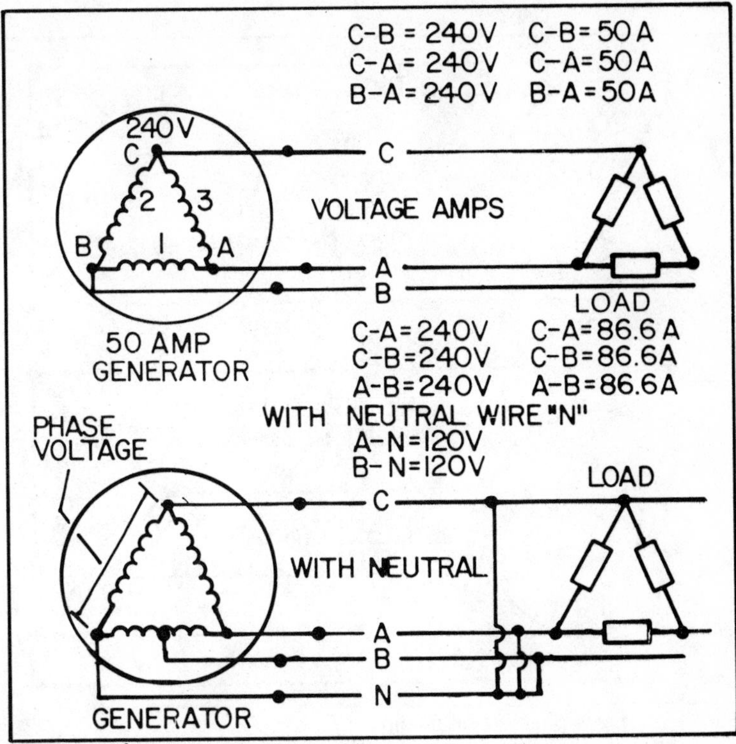

Fig. 6-4. 3 Ø delta phase relationship.

ical degrees apart, spaced upon the internal circumference of the motor's stator.

The current flowing through a three-phase system of both the delta and star circuitry can be summarized in this way. When the current or voltage in any one of the three phases is at its maximum value, the same values of the other two phases will be exactly half of the value of the first and will be flowing in the opposite direction of the first. Thus, if we measure the voltage in "phase-A" and find it to be 240 volts (+), then the other two phases, "B" and "C", will be found to be of only 120 volts each, with a relative (-) polarity. This property of three-phase systems is where the figure or phase factor number of 1.73 is derived from. It is therefore the square root of the number three-phase factor = 3.

The two-phase system is essentially the concerted magnetic effects of two separate and independent single-phase circuits, each being 90 degrees electrically separated, and physically separated in degrees by the quotient of the electrical degrees (90) divided by the number of pairs of field poles constituting the entire field of

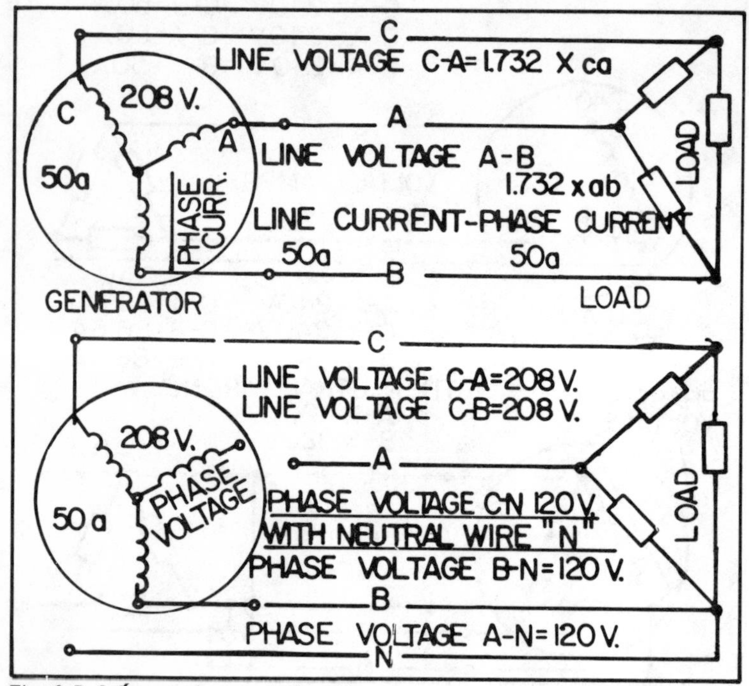

Fig. 6-5. 3 Ø wye phase relationship.

magnetomotive force (MMF). In two-phase, four wire systems, the resultant effect is likened to two separately connected circuits of single phase character, being 90 electrical degrees apart, as the power supply circuitry is actually composed of two separate unconnected circuits. In two-phase, three wire systems, in order to save on the use of the fourth line, which would provide a total division of circuitry, the two circuits are electrically connected. Both phases share a common line; the power is thus not quite the product of their individual values. In fact their effective value is the square root of the number of the combined phases—phase factor= 2=X0 which equals 1.41. We must sacrifice the value of two in the three wire systems in economizing to eliminate the fourth feeder line, and lose (2.0 - 1.41) 0.59 of the four wire value.

The higher the phase factor value, the higher the resulting horsepower of the motor. It can then be both visualized and computationally proven that in order of desirability, from an efficiency standpoint, two-phase, four wire with the phase factor of two is to be preferred over the lesser factor of 1.73 provided by three-phase systems. Three-phase systems are to be preferred over those of two-phase three wire, having only a factor of 1.41. Last

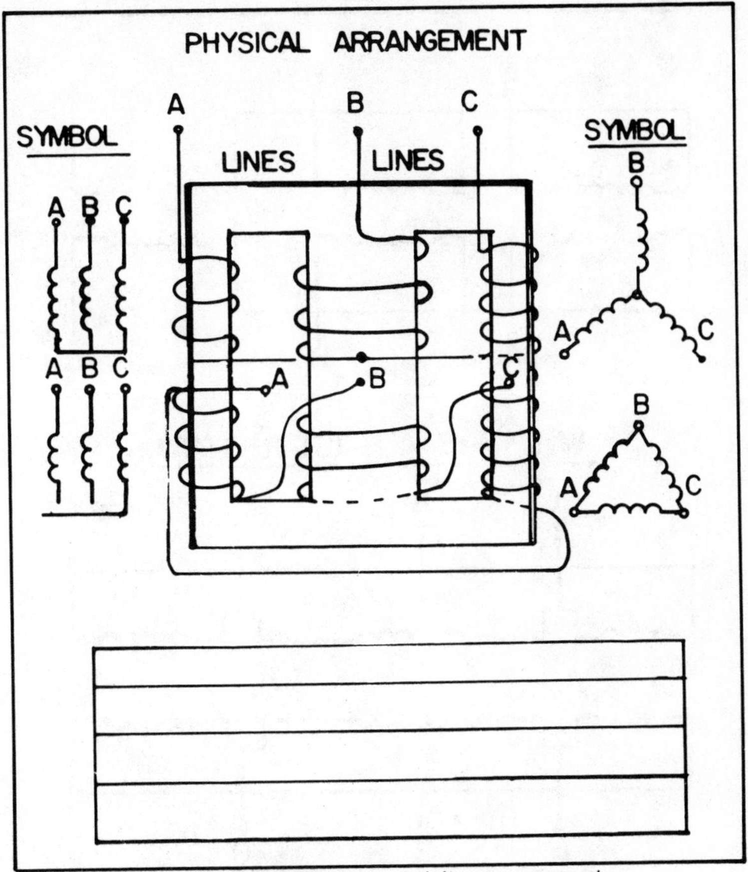

Fig. 6-6. Transformer geometry for wye-delta arrangement.

are single-phase systems for the heavy application of power, as it is generally conceded that single-phase motors are very costly in electrical energy consumed for applications above 1 horsepower.

It will be reasoned that the benefits offered by the polyphase mode of alternating current are, first, the greater efficiency realized by the larger polyphase motors and generators as compared to the single-phase varieties. Second is the greater available package horsepower that may be enjoyed by the two or three-phase motors. Third is the greater horsepower per amperage delivered, offering the very decisive advantage of smaller and fewer conductors, which signifies less "I^2R" losses per horsepower. In light of the "rotating field" of the polyphase system, the necessity of a starting circuit or phase splitting device is eliminated.

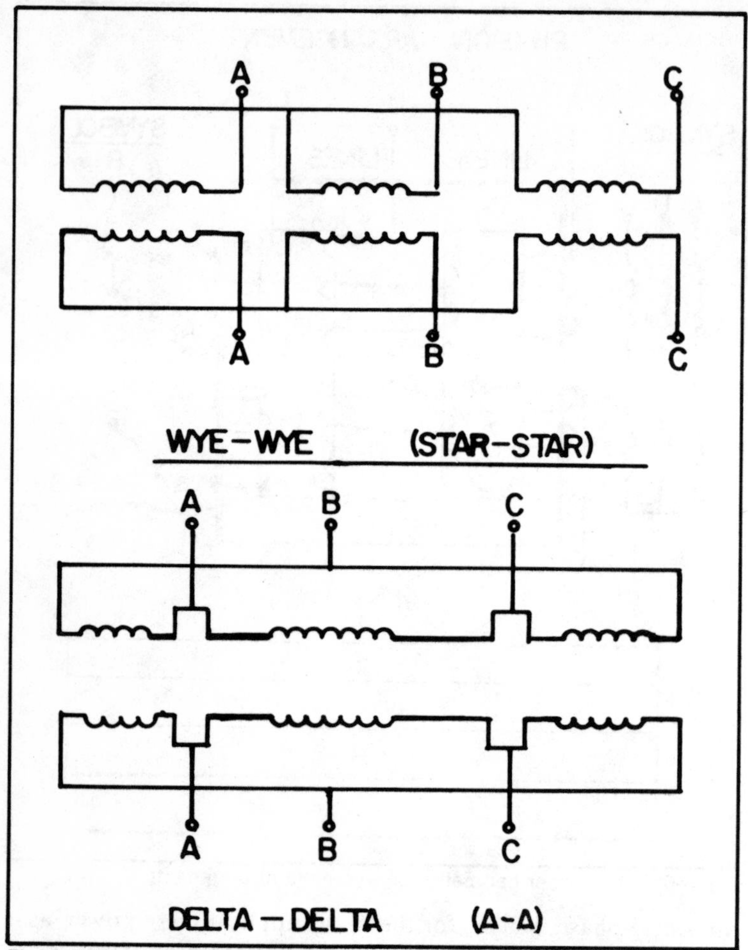

Fig. 6-7. Transformer geometry for wye-wye and delta-delta.

Parenthetically, one might rationalize that the more phases, the greater the benefits. Unfortunately this is not the exact situation. Although six-phase systems are in use for special applications, there is little to be effectively returned for any further "phase compiling" beyond the conventional three phases, in considering the additional expenses of such a multi-phase system.

CONVERSION TRANSFORMERS: TECHNICAL CHARACTERISTICS

Three-phase to two-phase (3-Ø /2-Ø) conversion is accomplished by the use of a "scott connection," which is an electrogeometric variation involving the three-phase supplied to the primary and the two-phase drawn from the secondary side of the configuration. Chiefly because of the economic advantages offered by generating and transmitting three-phase electric current, the three-phase system has become almost universal in our nation. There are, however, certain advantages with the adoption and use of two-phase circuits. Principally, the advantage is the higher phase factor of two inherent to the computations for two wire systems, as opposed to the phase factor of only 1.73 as offered by the conventional three-phase lines.

Many older businesses and buildings still have in service the earlier distributed two-phase four wire circuitry, and thus must be supplied at the meter side of the service with the required two-phase four wire supply. The manner most frequently employed for this electrical transition or conversion is to use a transformer so tapped and connected as to offer this change. The "scott," also referred to as the "T" connection—will generally be loaded to only 58 percent of the rated capacity of the primary side of the transformer.

The "scott" connection consists of an arrangement of two transformers. One transformer will have a "turns ratio" of 10:1, and the other will have a ratio of $\frac{1}{2}\sqrt{3}$:1; that is an .866.1 ratio (5 × 1.73:1). This arrangement may also be accomplished by having two separate transformers, with one having a 10:1 ratio and the other a 9:1 ratio. The 10:1 transformer is the main transformer and the other is the teaser. In the former example, the 8.66:1 teaser transformer is, as with the 9:1 (if substituted), connected by taps to the three star connected transformers to so proportion the secondary as to receive 57.7 percent of the primary's full rated voltage.

INSTRUMENT TRANSFORMERS: POTENTIAL AND CURRENT TYPES

When measurements of current and voltage are required in circuits above 5 amps and 600 volts, the generally used ammeters and voltmeters are usually not capable of withstanding these higher values. If the regular voltmeter was to be subjected to high voltages in excess of 600 volts and upwards to 23,000 volts, no value of resistance would be sufficient to prevent either a burnout of the meter windings or possible electrocution of the technician. Similar safeguards must also be adhered to in measuring high current values.

By shutting down the line power and installing the properly designed capacity current and/or potential transformer, the highest possible values of current and voltage may then safely be measured with the ordinary instrumentation. Depending upon the turns ratio, the current and voltage at the secondary of each transformer may be measured at a proportionate percentage of the line values. If a current transformer is used with a 100:1 turns ratio, a 5 ampere ammeter may be used to measure the 500 ampere line current. This 500 ampere current will then cause full 5 ampere meter deflection. If a 10 ampere meter movement or a 5 ampere shunt were used, then a half-scale deflection would result. Attention must be given for any and all multiplier and/or dividing factors to obtain a true reading.

A potential transformer is equally applied. We may use a standard 115-volt full deflection meter by connecting it to the secondary of a 20:1 stepdown potential transformer. With a line voltage of 2300 volts, a turns ratio of 20 to 1 would result in a secondary voltage of 115 volts. Each division of the meter scale would then indicate 20 volts. The wattmeter is also connected to high power lines through the use of current and potential transformers, either individually or in a combined transformer.

AVOIDABLE VERSUS UNAVOIDABLE ENERGY LOSSES

Electrical machinery, notwithstanding the relative simplicity of complex mechanical parts, will be exposed to the very same losses resulting from design limitations, imperfections and operational losses resulting from poor maintenance, improper control techniques, forcing of the machinery beyond its rated capacity, and a multitude of other efficiency reducing factors. Essentially, an electrical device is one of the most efficient entities for the conversion of energy invented, by virtue of its inherent simplicity and absence of dependent counter-accessories. Theoretically, a transformer which has no moving parts whatsoever would at first be assumed to be absolutely free of any losses. This rationale is nearly correct. A transformer has no actual physical losses as such. Because of design limitations in that the secondary windings can not actually be placed in the exact same place as the primary windings, a certain negligible air space between the two windings results from this mechanical separation. Consequently, a very small leakage flux will follow, causing slight inductive reaction. The reaction thus caused will provide for a reactive voltage drop when the transformer is operating at full load. This reactive voltage drop is included in part B of the electrical survey.

Are the efficiency reducing factors of design or operational origins? We certainly cannot change any losses resulting from a transformer's internal mechanical separation, no matter how serious the reactive voltage drop may become. If the problem is affected by improper cooling of the heat dissipating medium, then we can effectively adminster correction.

The majority of transformers have an efficiency above 95%, and many of the large utility company units work closer to 97 percent efficiency. In outlining the services we can actually perform for our Transformers, look at the transformer manufacturer's manual of maintenance in respect to changing the oil and conducting the periodic chemical tests for breakdown signs. Inspect transformer bushings to eliminate any foreign matter which would cause problems between terminals.

MOTORS AND GENERATORS

Motors and generators, by virtue of their moving and thereby wearing parts, do require certain operational attentions and periodic adjustments (Figs. 6-8 through 6-10). Chiefly, both motors and generators have actual mechanical losses of less than 10 percent, caused primarily by bearing and heat factors. Accordingly, the manufacturers' instructions and recommendations relating to the maintenance of these points will be followed. On dc and ac motors where brushes are employed for the collection of current or electrical control and slip rings (ac) and commutators (dc) are needed, skilled adjustment and maintenance is required which quite often may require factory service personnel.

Fig. 6-8. View of a large generator (courtesy Westinghouse Electric Corporation).

Fig. 6-9. This generator is being prepared for testing (courtesy Westinghouse Electric Corporation).

The electrical losses which motors and generators are subject to are due in large part to resistance (ohmic) and reactance (X^1), in ac machinery this is expressed by impedance. In addition to the losses experienced by impedance, we also have the problem of power factor to contend with, which plays a very large role in gobbling up the otherwise acheivable maximum efficiency of the motor or generator.

Fig. 6-10. Large electrical generators line an aisle of Westinghouse's manufacturing facility (courtesy Westinghouse Electric Corporation).

Power Factor and its effect on all electrical machinery, including transformers, will be evidenced by an examination of the related formulas as provided in Table 6-1, and in the electrical survey itself. It will be promptly recognized by the penalizing position the percent power factor occupies in all formulas in its dividend capacity, which naturally has a decreasing effect on the ultimate quotient of the formula for all horsepower computations. Power factor affects all electrical computations in an equally adverse manner. Of course, this consideration is unnecessary in direct current systems, as there can be no possibility for phase shift as contributed by the inductive and capacitive reactances. In Figs. 6-1 through 6-3 will be found ac graphs. Figures 6-2 and 6-3 will render a very graphic and illustrative explanation of the effects of reactance in shifting the phase of the voltage and current relationships, and how the shifted phase affects the power by adding a factor effect.

Power factor correction may be equally brought about by either adding enough capacitance into the circuit by the capacitance offered by a static capacitor, or by the addition of synchronous induction motors to the common line of power feed throughout the plant. The effect of a synchronous induction motor can, in addition to affecting its own operation without the effect of inductive reaction, serve to counteract the inductive effects of other non-synchronous induction motors by increasing the applied excitation voltage even higher to produce a leading power factor. A synchronous induction motor may be just placed on the line with no load attached whatsoever to affect just such capacitance, in which case the synchronous induction motor is referred to as a synchronous condenser. Whether the power factor of your plant is corrected by the static capacitor method or by the application of synchronous induction motors, the costs of these correction techniques will invariably pay for themselves and provide more current carrying capacity for the existing transformers, bus bars, conductors and circuit breakers. This is above and in addition to the savings directly enjoyed by the reduced power factor charge, which is an adjusted surcharge placed upon the users of electrical power who have an excessive and abnormal power factor problem.

Before the engineer commences the survey of his electrical equipment, it is suggested that he refresh himself with a detailed and comprehensive study of power factor and its correction. Much loss will be represented by the inductive effects of motors. The correction of this condition will no doubt need to be remedied as soon as possible. Unless a thorough command of the desirable and most advantageous manner of power factor improvement is attained, then certain errors in the selection of remedial equipment may inadvertently be made.

POWER FACTOR LOAD BALANCING

The advantages of balancing the circuit loads in terms of both resistance and reactance cannot be overemphasized. The impedance and consequently the current demands must be equally distributed between all three phases or however many phases there are at the plant transformer. By balancing the power factor as close as feasible across all of the supply circuits, it will enable the motors themselves to operate at a higher horsepower output. For unbalanced loads, the motors will progressively decrease in their outputs, while demanding even more current because of their true power demand. This results in a snowballing effect of eventual over-current conditions.

Don't be needlessly circumspect when connecting resistive lighting loads to inductive supply lines. The increase in the impedance caused by the added resistance will cause an increase in the power factor, enabling the existing motors to operate at a greater output. The power factor losses of the reactive current are only a relative function between the resistance and the reactance. With a fixed reactive load, either inductive or capacitive the additional resistance, even though it will increase the total circuit impedance, will prorationally decrease the angle of their relative angularity and decrease the power factor.

Make every attempt to keep the power factor balanced by equally dividing the inductive (or capacitive) load between all phases or load legs of the supply system. Fluorescent lighting may definitely be connected to inductive lines. If the voltage drop is a troublesome problem, then an autotransformer may be connected to the supply to facilitate a slightly higher voltage at little added expense. The added expense of an autotransformer would be more than offset by the gains brought about by the added savings resulting from reduced "I^2R" losses, as accomplished by the reduction in the reactive power consumption. Next is a treatise on power factor by the Sprague Electric Company.

WHAT IS POWER FACTOR?

Power factor is simply a name given to the ratio of *actual* power being used in a circuit, expressed in watts or kilowatts (KW), to the power which is *apparently* being drawn from the line, expressed in volt-amperes or kilovolt-amperes. The power factor ratio is of great importance in ac circuits, as this manual will show, although it has no significance in purely dc circuits.

To find the power required by an electrical load device, it is common practice in single-phase circuits to multiply the load current by the voltage applied to its input terminals. This product gives the apparent power only, and more elaborate measurements are needed to find the actual or useful power. A wattmeter will show the actual power, which can never exceed the apparent power but often is less.

For load devices containing only resistance, such as soldering irons and ovens, the actual and the apparent ac power are the same and the power factor is 100 percent. But many devices, such as ac motors and transformers, have also a property known as inductance—or "iron effects"—and these devices consume less actual or useful power than shown by the products of their operating currents and voltages.

Most of the latter devices can be identified by their construction, which usually employs coils of electrical wire wound in various ways around cores made of iron. The inductance of the coils causes the inequality between true and apparent power.

"HIGH" AND "LOW" POWER FACTOR

The power factor measurement expresses the ratio of actual to apparent power. (By "actual power" we mean working, or real, or true power used to produce heat or work. Apparent power is the product of volts times amperes, and may or may not be more than the actual power). When the two values are equal, their ratio is 1:1—or 1.0, or 100 percent. This is the highest power factor (unity) that can be obtained.

But if, for example, the actual power consumption is 400 watts in contrast with an apparent power demand of 1,000 watts, the ratio is 400:1,000 or 0.4, or 40 percent. In the latter example, the power factor is said to be "low" because 40 percent is low with respect to 100 percent. An electric motor with an actual-to-apparent power ration of 90 percent may be said to have a reasonably "high" power factor.

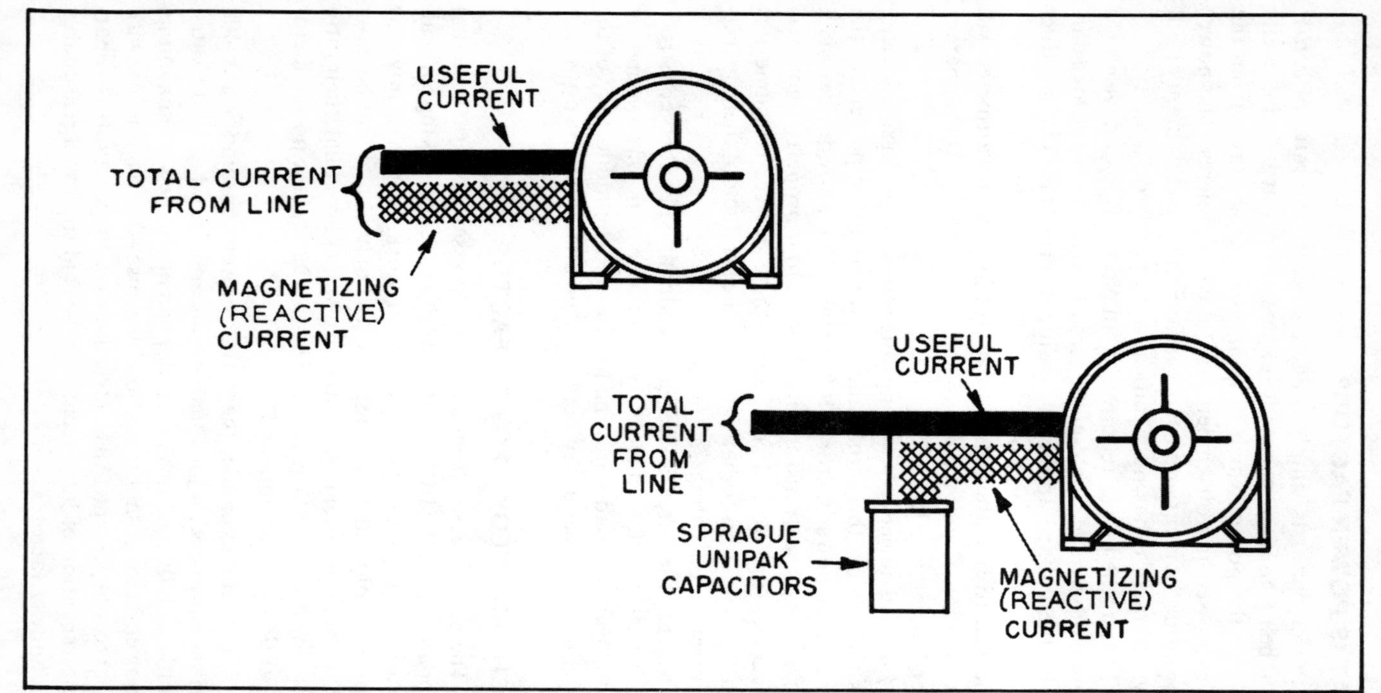

Fig. 6-11. An induction motor operating under partially loaded conditions without power factor correction. Also, the result of installing a capacitor near the same motor to supply the magnetizing current required to operate it (courtesy of Sprague Electric Company).

Having a low power factor means poor electrical efficiency, which is always costly—because the actual power consumption is less than the volt-ampere product. It is volt-amperes which the electric circuit sees, but it is watts that the load uses. The voltage in the system remains nearly constant. Therefore, if the volt-ampere value exceeds the watts consumption, extra current is being drawn through the power lines, which causes the volt-amperes to rise above the watts. Let's look at several reasons why low power factor is costly.

☐ Low power factor increases the power company's cost of supplying actual power, because more current must be transmitted and *this higher cost is directly billed to the industrial consumer* by means of power factor clauses in the rate schedule.

☐ Low power factor also causes overloaded generators, transformers, and distribution lines within the industrial plant; voltage drops and power losses are greater than they should be. All of this represents waste, and needless wear and tear on industrial equipment.

☐ Finally, low power factor reduces the load handling capability of the plant electrical system.

When the volt-ampere product (KVA) exceeds the actual power (KW), a power component known as *reactive power* (KVAR) is present since the operating current consists of two parts. One part of the current results in useful work; the second, known as the *reactive current,* does not. Reactive current is drawn by inductive load devices to develop the magnetic fields required for their operation. When corrective steps are not taken, this reactive current must be supplied by the power company's lines or the plant generator in addition to the current that contributes to useful work, thus causing the electrical supply system to work harder than really necessary (See Fig. 6-11). Purely resistive load devices, such as incandescent lighting, which draw no reactive current, do not create this kind of a problem.

Reactive power is the product of the reactive current and operating voltage. The greater the reactive current, in proportion to the useful current, the greater the reactive power and the lower the power factor.

Though reactive current is actually part of the total current indicated in an ammeter reading, reactive power does not register on a kilowatt-hour meter. However, the plant electrical lines must carry this extra current and power companies take it into account in their industrial billings.

The reactive current taken from the power company's lines—in addition to the current that contributes to useful work—necessitates the use of larger generators, transformers, bus bars, wires, and other distribution devices than otherwise would be needed. This, of course, increases the power company's costs for capital equipment. Hence, industrial consumers are often billed at adjusted rates which are increased in accordance with the plant's power factor figure.

In the plant, low power factor causes excessive voltage drops and power losses because the plant line conductors and distribution devices are often too small to carry both the reactive and working currents. If the plant wiring system is large enough for the existing load, it can be made to support additional loads by taking steps to eliminate the reactive current through power factor improvement.

The low power factor problem can be solved by adding corrective equipment to the plant circuit. There are a number of devices available which are designed and built for power factor correction purposes, including synchronous motors, synchronous condensers, and power factor correction capacitors.

Synchronous Motors and Condensers

Synchronous motors correct for power factor losses to some extent; however, they do this efficiently only when underloaded. When overloaded, they take reactive power from the line.

Synchronous condensers are generally used in the heavy-duty installations of power companies and work well, but they are large and very expensive. These devices are essentially synchronous motors, operated over-excited or without loads.

Power Factor Correction Capacitors

For general use by industrial plants, the most practical and economical power factor correction device is the capacitor (Fig. 6-12). Power companies use capacitors at power stations where an elaborate and expensive synchronous condenser installation is not justified.

Low power factor is caused by the inductive effects of "iron and coil" devices or loads, such as motors and transformers. The mechanics of these effects will be discussed later. For the present, let us say that the inductive part of a circuit can be balanced by the addition of capacitance. The capacitor current is used to furnish the magnetizing current required by the load and can supply all or part of that requirement.

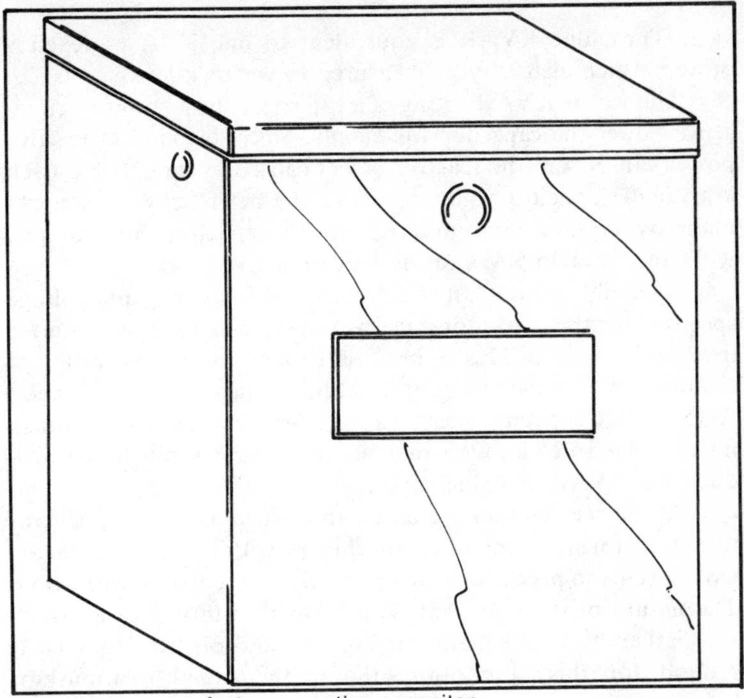
Fig. 6-12. A power factor correction capacitor.

Capacitors improve the power factor because the effects of capacitance are exactly opposite those of inductance. Adding capacitors to an inductive circuit essentially cancels the effect of the circuit inductance. The net amount of reactive power is thus reduced, and the power factor is consequently increased.

Some of the advantages offered by capacitors over other types of power factor correction devices are:

☐ They are substantially lower is cost.

☐ They can be readily moved from one point to another in an electrical distribution network, as required.

☐ They can be installed economically in a decentralized manner and require virtually no maintenance.

☐ The correction capability of a capacitor bank can easily be increased.

KVAR AND CAPACITOR RATINGS

To indicate their power factor correction capabilities, capacitors are rated in vars or kilovars. One var is equivalent to one volt-ampere of reactive power, and one kilovar (KVAR) equals 1,000 vars. Thus, one KVAR is equivalent to one KVA of reactive power, which also may be measured in var or kilovar units.

The var or KVAR rating of a capacitor shows how much reactive power the capacitor will supply. Since this kind of reactive power cancels out the reactive power caused by inductance, each kilovar of capacitor value decreases the net reactive power demand by the same amount. A 15 KVAR capacitor, for example, will cancel out 15 KVA of inductive reactive power.

It should be noted, however, that the frequency and voltage specified for the capacitor must be the same as the power source used in the plant. This is because capacitors furnish different amounts of reactive power at different frequencies and voltages. A 15 KVAR capacitor designed for 60-cycle circuits will furnish much less than 15 kilovars in a 50- or 25-cycle and will also produce less KVAR at reduced voltages.

All devices containing inductance, such as motors, generators, transformers, and other machinery with coils, require reactive currents to produce the magnetic fields needed for operation. The nature of these currents is described in this section, which shows them to be the main cause of low power factor. As a background for this description, the mathematical relationships between apparent power, actual power, reactive power, and power factor are also explained.

APPARENT POWER AND ACTUAL POWER

Apparent power is easily defined as the voltage applied to a circuit multiplied by the current which is then drawn. This is measured in volt-amperes and includes any reactive power that may be required by the load.

$$\text{Apparent power (in volt-amperes)} = \text{volts} \times \text{amperes.}$$

The actual power in watts consumed by an electrical load is the product of the load current, the applied voltage, and a third factor—the cosine of the phase angle, Θ. That is, power (in watts) = volts × amperes × cosine Θ.

The cosine of the phase angle accounts for reactive power. It appears in the equation because either inductance or capacitance causes a time difference between the peak of the ac voltage applied to the load and the peak of the ac current drawn by the load. Figure 6-13 illustrates a time lapse for a purely inductive circuit.

In inductive circuits, the voltage peak occurs first, and the current is said to "lag". In capacitive circuits, the current peak occurs first and the current is said to "lead." Either the lag or the lead is measured in degrees proportional to the amount of source generator rotation occurring during the time lapse. This number of degrees is the phase angle, Θ, and thus, in Figure 6-13, Θ is a lagging angle of 90°. Since most plant machinery is inductive in nature, the plant engineer is chiefly concerned with lagging currents.

In purely resistive circuits with no inductance or capacitance, the current and voltage peaks occur simultaneously, and are said to be "in phase". Here the angle Θ is always equal to O.

In the circuits containing resistance together with inductance, the lag angle is always less than 90°. How much less depends upon the relative amounts of each, or, more precisely, the $\frac{\text{inductance}}{\text{resistance}}$ ratio. The greater the inductance with respect to a given resistance, the greater the lag angle Θ. From a practical standpoint, however, this angle never reaches 90°, because some resistance is present in every circuit.

The fact that greater inductance produces a greater lag is mathematically reflected by the cosine figure. In trigonometry, the cosine of any angle from 0° to 90° lies between the values of 1 and 0 respectively. When $\Theta = 0°$ (as in a purely resistive circuit), cosine $\Theta = 1$, giving:

$$\text{actual power (in watts)} = \text{volts} \times \text{amperes} \times 1,$$

in which case the *actual power* and *apparent power* are *equal.* When $\Theta = 90°$ (as in a purely inductive —or purely capacitive— circuit), cosine Θ = O, and actual power (in watts) = volts × amperes × 0, which equals 0. (However, 90° is only a theoretical limit as explained above).

For a practical example, let $\Theta = 30°$. From trigonometric tables, cosine 30° = 0.86603; hence actual power (in watts) = volts × amperes × 0.86603, a typical case where actual power is much greater than 0, but still considerably less than the product of volts × amperes. The difference is due to reactive power, from the reactive currents described later in this section.

It logically follows that the addition of more motors—that is, more inductance—to an industrial plant circuit lowers the plant power factor. This is because:

$$\text{power factor} = \frac{\text{actual power}}{\text{apparent power}}$$

$$= \frac{\text{volts} \times \text{amperes} \times \text{cosine } \Theta}{\text{volts} \times \text{amperes}}$$

$$= \text{cosine } \Theta$$

As the angle of lag is increased by adding more inductance, the fraction represented by cosine Θ becomes smaller, giving a lower figure for the power factor.

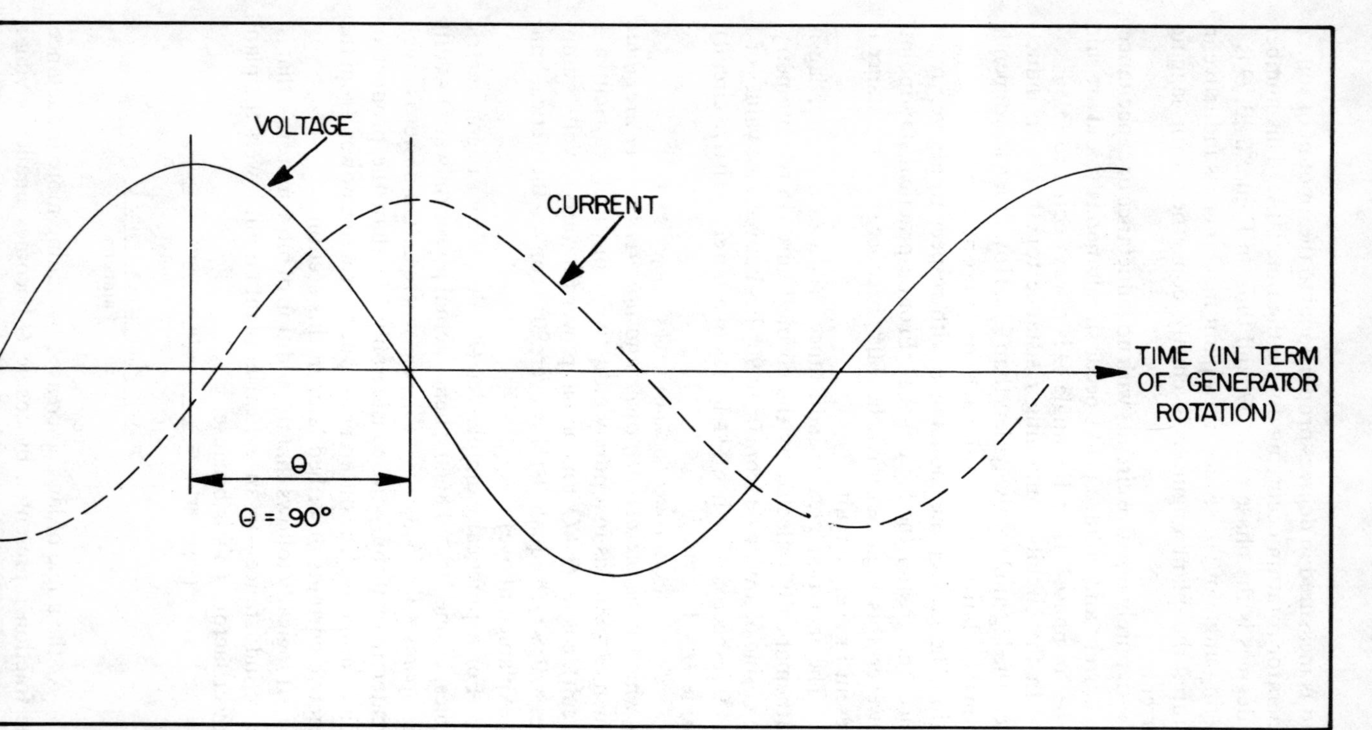

Fig. 6-13. Time lapse for a purely inductive circuit.

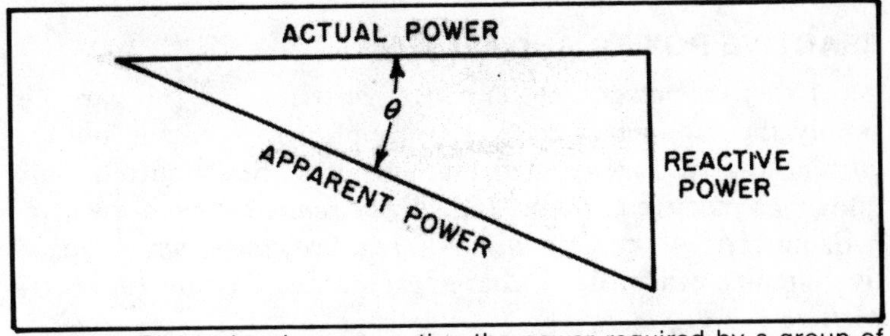

Fig. 6-14. Right triangle representing the power required by a group of induction motors (courtesy of Sprague Electric Company).

Consider the right triangle (Figure 6-14) representing the power required by a group of induction motors.

In this figure, the reactive power is rather small, and it is easily seen that the leg of the triangle representing actual power is nearly as long as the leg representing apparent power; thus, the ratio of actual power to apparent power (cosine Θ) approaches unity. Notice that in this case, the angle Θ is small (along with the reactive power leg of the triangle). In Figure 6-15, the number of motors in the original group was increased.

Now, the angle Θ has increased, and though both the reactive leg and the actual power leg have increased, the apparent power leg has become *relatively* longer. Thus, the ratio of actual power to apparent power has decreased. Since this ratio is equal to cosine Θ — that is, the power factor—the cause of low power factor is more readily seen. As the inductive load increases, the reactive power requirement increases, and the ratio of actual-to-apparent power—the power factor—decreases, causing the undesirable effects described throughout this manual.

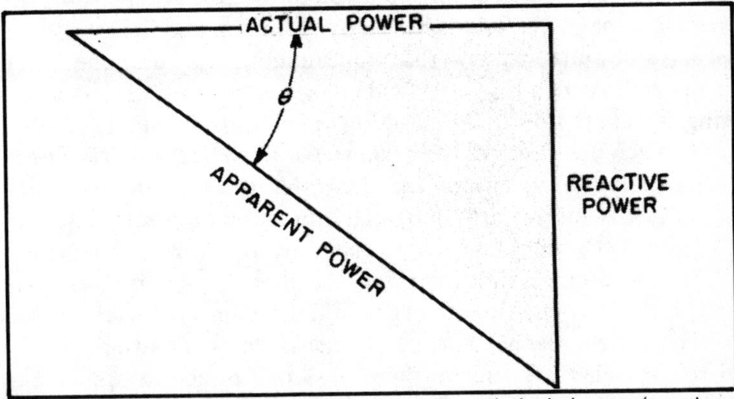

Fig. 6-15. The apparent power leg has become relatively longer (courtesy of Sprague Electic Company).

REACTIVE POWER AND CURRENT

Reactive power, measured in volt-amperes (or vars), is simply the product of the voltage applied to a circuit and the current drawn for magnetizing purposes. Such currents are known as reactive currents. Inductive circuits, which draw magnetizing currents, contain *inductive reactive power,* while capacitive circuits, which draw charging currents, contain *capacitive reactive* power.

Inductive reactive current produces a lagging phase angle, because this kind of current flows *after* the circuit voltage rises to full value. Capacitive reactive current produces a leading phase angle because this kind of current flows *before* full voltage is reached.

To understand the reactive currents associated with inductance, consider the iron-cored coil or electromagnet with dc applied as in Figure 6-16.

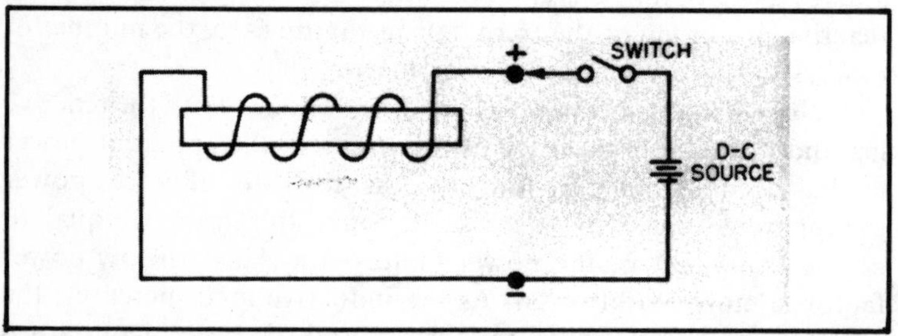

Fig. 6-16. Iron-cored coil or electromagnet with dc applied (courtesy of Sprague Electric Company).

The instant current flows through the coil upon closing the switch, a magnetic field is built up. This magnetic field reaches a peak value and remains at that value indefinitely—until the switch is opened. At this time, the field collapses, causing a brief collapsing current to flow in the direction opposite to the supply current.

When the same device is connected to an ac source (Figure 6-17), the alternating current causes a different situation. As in all ac circuits, the supply current changes direction at regular intervals—typically 60 times per second, the most common line frequency. When the circuit is first energized, the alternating current flows in one direction, creating a magnetic field, then falls back to zero when the field reaches a peak. As when opening the switch (Figure 6-16), a collapsing current then flows in a direction opposite to the incoming supply current. Each time the supply current changes direction, the field collapses; and each time the field

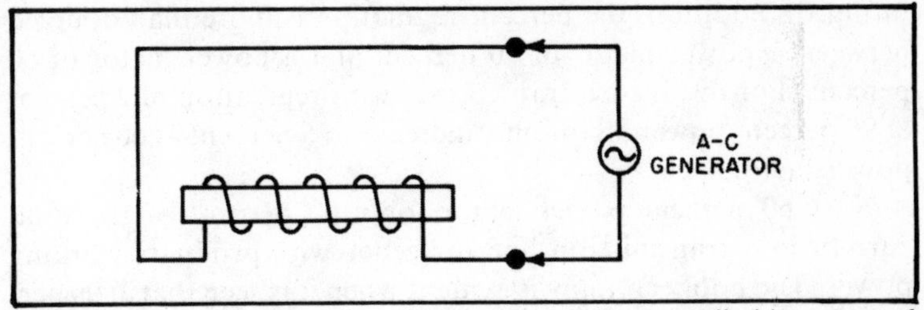

Fig. 6-17. An iron-cored coil or electromagnet with ac applied (courtesy of Sprague Electric Company).

collapses, there is a current flow in a direction opposing the supply current.

The collapsing current is known as the reactive current which, together with the source voltage, makes up the reactive power. Reactive power, however, is released and returned to the power source each time the field collapses, and so does not appear on the kilowatt-hour meter.

The situation is similar to the mechanical action of a spring compressed and then released. To compress the spring, energy must be used. Once the spring is clamped down, no more energy is required to keep it in place. When the clamp is released, the original energy that was applied to the spring, and then stored by the spring, is released. No useful work is done, and only "reactive" power is used. Similarly, the current in an electrical coil flows heavily until the field is set up, then drops to zero (or is stored), and finally releases its energy when its direction is reversed.

THE EFFECTS OF POOR POWER FACTOR

The cost and inefficiency of poor power factor were generally described earlier. More specific details of the effects of low power factor on generators, transmission lines, transformers, and feeders are given in this section.

Generators

Generators are normally rated in KVA. Therefore, when the reactive current needed by induction devices must be supplied by a generator, the generator's useful output is greatly reduced. A reduction in power factor from unity to 80 percent reduces kilowatt output by as much as 27 percent.

Transformers

Transformers are also rated in KVA similar to generator conditions as described above. At a power factor of 60 percent, the available kilowatt power is about 60 percent of its nameplate rating. In addition, the percent regulation is more than doubled between a power factor of 90 percent and a power factor of 60 percent. For instance, a transformer with regulation of 2 percent at 90 percent power factor may increase to 5 percent at 60 percent power factor.

At 60 percent power factor, only 60 percent of the total current in a transmission line or feeder wire produces working power. The poor economy is evident when it is seen that at 90 percent power factor, 90 percent of the current is useful, and at 100 percent power factor, it is all useful.

HOW CAPACITORS CORRECT FOR LOW POWER FACTOR

Capacitors correct for low power factor because the leading current fount in a capacitive circuit is opposite to the lagging current in an inductive circuit. When the two circuits are combined into one, the effects of capacitance tend to cancel the effects of inductance.

A properly chosen capacitor gives perfect cancelation, but too little or too much capacitance should be avoided. Too little will not give enough lag correction, while too much may produce a leading phase angle. This leading angle causes the same undesired effects as an equivalent uncorrected angle of lag.

Of the three major power factor correction devices previously mentioned, only synchronous motors and power factor correction capacitors are generally suitable for industrial plants. Synchronous condensers have very specialized application and are usually very expensive. Therefore, this discussion will involve only synchronous motors and static capacitors.

SYNCHRONOUS MOTORS

Synchronous motors are sometimes used in place of induction motors because of the synchronous motor's ability to maintain a high power factor. These motors can do many of the jobs commonly done with induction motors, and, if run underloaded, need no power-factor correction. When its load is quite light, or when no load is applied the synchronous motor can compensate for a low power factor caused by other equipments in the same distribution system. Induction motors, on the other hand, always tend to lower the power factor, and therefore are generally used in conjunction with power capacitors.

Usually, the cost of a synchronous motor installation is very high compared with the cost of the equivalent induction motors with corrective capacitors. Synchronous motors entail the added cost of controls and dc exciters, including the generator necessary for dc excitation. By and large, the maintenance problem is considerable.

POWER CAPACITORS

Capacitors are economical, both to install and maintain. They have no moving parts that wear out or endanger plant employees. No complicated motor-starting or adjusting procedures are needed.

Capacitors also afford a highly flexible method of power factor correction, since they can be installed almost anywhere, in any quantity. Capacitors are available to fit any and all motors, and can be placed at the points on a loaded line where they are most needed. No special foundations are required, since they have no moving parts and do not vibrate.

Capacitor losses are negligible, and if left on the line when motors are shut off, their power consumption is insignificant. They can be obtained for indoor or outdoor installations, and at any voltage level.

While individual capacitor units are guaranteed for one year, the actual life of such a capacitor is from 10 to 20 years. The length of life depends, of course, on operating conditions, such as the surrounding temperature and operating voltage.

How Capacitors Work

To understand how capacitors improve the power factor, it is necessary to go back to Figure 6-13 and the explanation of current lag in an inductive circuit. In that example, the current and voltage peaks are displaced by the theoretically maximum angle of 90°.

When ac is applied to a circuit having capacitance, an electrostatic field—rather than a magnetic field—follows the same cycle of building up and falling down as the field of the inductive circuit. In this case, the capacitor current reaches a peak value before the voltage across the capacitor reaches its peak, resulting in a leading current. Therefore, when an inductance and a capacitance are connected in parallel, current will flow back and forth between the inductor and the capacitor. If the currents were equal and no losses occurred in the circuit, no current would be required from the power source.

In actual practice, the generator or power supply must furnish current for circuit resistance and other losses, as well as for any differences between capacitor and inductor currents that might occur.

All of this means that with a properly chosen capacitor, no kilovar current flows back and forth between an inductive machine—for example, an induction motor—and the source of power, but only between the capacitor and the motor. The power transmission system is relieved of unnecessary drain on its capabilities, if the capacitor is located close to the motor. No matter at what location the capacitor is connected, the benefits always are gained from the point of installation back to the power source.

Placement of Power Capacitors

There are two methods of capacitor power factor correction. The first is group or bank installation in which many capacitors are connected to the line at some central point such as a switchboard or distribution panel. Generally, this method only serves to reduce the utility-company penalty charges.

The second (but most effective) method is the installation of individual capacitors directly at the source of poor power factor, that is, close to the motors. This has all the benefits of group installation plus the advantages of released system capacity, improved voltage levels and reduced power losses.

INDIVIDUAL CAPACITOR INSTALLATION

The installation of individual capacitors provides the greatest benefit in plants where wiring is being overloaded by induction motors. This is more concern in older plants, but also applies to new installations where growth has led to the addition of more motors. Power factor correction can, in such cases, mean the difference between an entirely new wiring job or the simple addition of the new equipment to the old lines.

Connecting Directly at Motors

The most effective correction is obtained when the individual capacitors are connected directly to the terminals of the motors, transformers, and other inductive machinery. Reactive current causes losses either between the induction machine and the correction device, or, when the correction device is lacking, between the machine and power source. Thus, the closer the capacitor to the machine, the lower the losses, and the greater the benefits.

Another advantage in placing capacitors directly at motor terminals is that both motor and capacitor can be switched on and

Table 6-3. Suggested Maximum Capacitor Rating When Motor and Capacitor are Switched as a Unit (courtesy of Sprague Electric Company).

Induction Motor HORSE-POWER Rating	NOMINAL MOTOR SPEED IN RPM 3600 Capacitor Rating KVAR	3600 Line Current %	2800 Capacitor Rating KVAR	2800 Line Current %	1200 Capacitor Rating KVAR	1200 Line Current %	900 Capacitor Rating KVAR	900 Line Current %	720 Capacitor Rating KVAR	720 Line Current %	600 Capacitor Rating KVAR	600 Line Current %
3	1.5	14	1.5	15	1.5	20	2	27	2.5	35	3.5	41
5	2	12	2	13	2	17	3	25	4	32	4.5	37
7½	2.5	11	2.5	12	3	15	4	22	5.5	30	6	34
10	3	10	3	11	3.5	14	5	21	6.5	27	7.5	31
15	4	9	4	10	5	13	6.5	18	8	23	9.5	27
20	5	9	5	10	6.5	12	7.5	16	9	21	12	25
25	6	9	6	10	7.5	11	9	15	11	20	14	23
30	7	8	7	9	9	11	10	14	12	18	16	22
40	9	8	8	9	11	10	12	13	15	16	20	20
50	12	8	11	9	13	10	15	12	19	15	24	19
60	14	8	14	8	15	10	18	11	22	15	27	19
75	17	8	16	8	18	10	21	10	26	14	32.5	18
100	22	8	21	8	25	9	27	10	32.5	13	40	17
125	27	8	26	8	30	9	32.5	10	40	13	47.5	16
150	32.5	8	30	8	35	9	37.5	10	47.5	12	52.5	15
200	40	8	37.5	8	42.5	9	47.5	10	60	12	65	14
250	50	8	45	7	52.5	8	57.5	9	70	11	77.5	13
300	57.5	8	52.5	7	60	8	65	9	80	11	87.5	12
350	65	8	60	7	67.5	8	75	9	87.5	10	95	11
400	70	8	65	6	75	8	85	9	95	10	105	11
450	75	8	67.5	6	80	8	92.5		100	9	110	11
500	77.5	8	72.5	6	82.5	8	97.5	9	107.5	9	115	10

* For use with 3-phase, 60 cycle NEMA Classification B Motors to raise full load power to approximately 95%.

off together. This arrangement ensures that the motor cannot operate without the correction of the capacitor, and that the capacitor is used only when needed. Another saving is gained with this method because, unlike the group or bank installation method, extra switches are not needed to switch the capacitors in and out of the line. The capacitor is switched with the motor control.

No complicated engineering studies are needed when capacitors and induction motors are connected together and operated as a unit. The size of the capacitor needed for each motor may be determined very simply from Table 6-3 by looking up the motor speed and horsepower and reading off the required capacitor value in KVAR. The table shows capacitor ratings in accordance with AIEE recommendations; these values will improve motor power factor to about 95 percent.

Flexibility of arrangement is another advantage of the individual capacitor installation setup. When plant layout is changed or machinery relocated, both the motor and its capacitor can be moved with a minimum of inconvenience.

Summary of Advantages

To summarize, the advantages gained with individual capacitor installations as follows:

☐ Voltage drops to individual motors are reduced, decreasing damaging heat due to excessive currents.

☐ Capacitors are switched in and out of service as they are needed. In this way, power factor is adjusted to overall load requirements, and better voltage regulation is obtained.

☐ The capacitor values required for individual motors are easily chosen from tables, thus reducing engineering problems.

☐ Individually corrected motors and other machines have a great flexibility in case of plant alterations and rearrangements. No matter where the motor is located, proper correction is assured.

The chief disadvantage of individual capacitor correction is that the smaller capacitor units have a higher price per kilovar than larger units. When many small motors on the same circuit need to be corrected it may be more economical to correct them in groups, using 10 KVAR to 15 KVAR rated capacitors connected to the feeder lines. A common practice is to correct motors larger than 10 HP with individual capacitors, and to correct smaller motors in groups.

GROUP OR BANK INSTALLATION METHOD

When the chief reason for power factor correction is to obtain lower electric bills or to reduce the current in primary feeders from a main generator or transformer bank, then the group installation method is more economical. In this case, large banks or racks of capacitors are installed at the main switchboard or at the generator. The overall plant power factor is then increased, but it should be remembered that no benefits to the plant distribution system are obtained on the load side of the capacitor installation.

Two other advantages of bank installation are that large banks have a lower price per kilovar than individual units, and are less expensive to install. The cost of the switches or circuit breakers, needed for such units, however, may partially offset the lower cost of this type of installation.

Switching

All or part of the bank can be switched on or off, either manually or automatically, depending on load requirements. In this way, only as much correction as is required for the load at a particular time need be switched in.

Summary of Advantages of Group or Bank Installation

Summarizing the advantages of the bank installation method bank installation improves overall plant power factor, effecting a reduction in power bills. Banks of capacitors have a lower cost per kilovar than individual capacitors. The cost of bank installation wiring is lower per KVAR than the cost of wiring individual capacitors to their associated motors.

Combinations of Installation Methods

In a large number of cases the most desirable results will be obtained from an installation of individual capacitors on larger motors and banks on main feeders or switchboards. To determine the correction needed, first establish the total kilovar requirement for the plant as discussed below. Then select individual motor correction from Table 6-3. As each larger motor is corrected, deduct its KVAR from the total; then the balance required can be installed as a single bank at the service entrance or subdivided at critical points around the plant. Remember again that the optimum correction is realized when capacitors are placed as near the source of low power factor as possible.

Avoiding Overcorrection

Remember that capacitors draw reactive currents opposing the direction of reactive current from inductive devices. Thus, avoiding overcorrection is important, because having too much leading kilovar current from capacitors is as undesirable as having too much lagging kilovar current from inductive machines.

Computing the Proper Correction

As with individual motor correction, it is easy to select the proper amount of correction needed by any particular system through the use of tables. Table 6-4 gives the amount of correction needed when the average plant power factor and the kilowatt load of the plant are known.

As a demonstration of the use of Table 6-4, consider, for example, a plant with an average power factor of 76 percent and an average power consumption of 400 KW. The power company supplying the plant includes a power factor penalty clause in its billings that increases the size of the bill when power factor is less than 85 percent. To avoid this expense, it is necessary to increase the plant power factor to 85 percent. To find the number of kilovars needed to obtain the higher power factor, proceed as follows. Under the first left-hand column marked "original power factor", look up "76 percent".

Follow the row of figures to the right of the "76 percent" value until you reach the one under the number "85" in the columns headed "corrected power factor". This figure is 0.235.

Multiply the number 0.235 by the kilowatt load of 400 KW. The product is the number of kilovars needed for correction. That is, 0.235 × 400 KW = 94 KVAR. Standard kilovar and voltage ratings may be obtained from Sprague's power capacitor catalog.

REDUCED POWER COSTS

An electric power company must provide the necessary power to all of its consumers. This includes their needs for reactive power, which does not register on the company's kilowatt-hour meters. To supply these needs, a power company must spend extra money to furnish higher rated generators, power lines, transformers, and other equipment on which the company receives no direct return. Lighter equipment, sufficient for the true power shown on kilowatt-hour meters, would be overloaded by the additional currents drawn for the reactive power. Since small conductors present greater resistance than large conductors, power losses and voltage drops in the company's distribution system would be excessive.

Consequently, most power companies make up for the added expense by including power factor penalty clauses in their industrial billings. The cost of reactive power or, as it is commonly known, *kilovar consumption,* may amount to a significant part of the monthly bill.

The consumer who supplies his own kilovars by installing capacitors can enjoy a substantial saving. To examine the gains possible from the installation of static capacitors, consider the case of a typical industrial plant with a power factor of 68 percent. The billing rates of the power company supplying this plant include a kilowatt demand charge of $1.67 per kilowatt, plus energy charges of:

2.00 ¢ per kilowatt-hour—first 10,000 kilowatt-hours
1.50 ¢ per kilowatt-hour—nest 10,000 kilowatt-hours
1.25 ¢ per kilowatt-hour—next 30,000 kilowatt-hours
1.00 ¢ per kilowatt hour—next 150,000 kilowatt-hours

Energy charges are based on the total number of kilowatt-hours of true power that must be supplied by the power company over a certain period. In addition, a power factor penalty clause requires that the billing be adjusted by the formula:

$$\text{Billing demand} = \frac{\text{maximum demand} \times 0.85}{\text{measured power factor}}$$

Electric bills show that during an average month this plant has a maximum demand of 350 kilowatts (KW) in addition to energy charges for 86,000 kilowatt-hours (KWH). The plant power factor is 68 percent.

The monthly billing is itemized as follows:

$$\text{Billing demand} = \frac{350 \times 0.85}{0.68} = 438$$

at $1.67 per KW . $ 732.00
Energy charges per above schedule $1,085.00
Total billing before power-factor correction....... $1,817.00

Now consider what happens to the monthly bills when the plant power factor is increased to 85 percent. The new monthly billing is as follows:

$$\text{Billing demand} = \frac{350 \times 0.85}{0.85} = 350 \text{ KW}$$

at $1.67 per KW . $ 585.00
Energy charges per above schedule $1,075.72
Total billing after power-factor correction $1,660.72

Thus, a montly electric bill reduction of $156.28 (that is, $1,660.72 instead of $1,817.00) has been gained by improving the plant power factor.

The capacitor KVAR needed to improve the power factor from 68 percent to 85 percent may be calculated as follows:

$$350 \text{ KW} \times 0.458 = 160 \text{ KVAR}$$

Table 6-4. KW Multipliers to Determine Capacitor Kilovars Required for Power Factor Correction (courtesy of Sprague Electric Company).

Orig-inal Power Factor	CORRECTED POWER FACTORS																				
	0.80	0.81	0.82	0.83	0.84	0.85	0.86	0.97	0.88	0.89	0.90	0.91	0.92	0.93	0.94	0.95	0.96	0.97	0.98	0.99	1.0
0.50	0.982	1.008	1.034	1.060	1.086	1.112	1.139	1.165	1.192	1.220	1.248	1.276	1.306	1.337	1.369	1.403	1.440	1.481	1.529	1.589	1.732
0.51	0.937	0.962	0.989	1.015	1.041	1.067	1.094	1.120	1.147	1.175	1.203	1.231	1.261	1.292	1.324	1.358	1.395	1.436	1.484	1.544	1.687
0.52	0.893	0.919	0.945	0.971	0.997	1.023	1.050	1.076	1.103	1.131	1.159	1.187	1.217	1.248	1.280	1.314	1.351	1.392	1.440	1.500	1.643
0.53	0.850	0.876	0.902	0.928	0.954	0.980	1.007	1.033	1.060	1.088	1.116	1.144	1.174	1.205	1.237	1.271	1.308	1.349	1.397	1.457	1.600
0.54	0.809	0.835	0.861	0.887	0.913	0.939	0.966	0.992	1.019	1.047	1.075	1.103	1.133	1.164	1.196	1.230	1.267	1.308	1.356	1.416	1.559
0.55	0.769	0.795	0.821	0.847	0.873	0.899	0.926	0.952	0.979	1.007	1.035	1.063	1.093	1.124	1.156	1.190	1.227	1.268	1.316	1.376	1.519
0.56	0.730	0.756	0.782	0.808	0.834	0.860	0.887	0.913	0.940	0.968	0.996	1.024	1.054	1.085	1.117	1.151	1.188	1.229	1.277	1.337	1.480
0.57	0.692	0.718	0.744	0.770	0.796	0.822	0.849	0.875	0.902	0.930	0.958	0.986	1.016	1.047	1.079	1.113	1.150	1.191	1.239	1.299	1.442
0.58	0.655	0.681	0.707	0.733	0.759	0.785	0.812	0.838	0.865	0.893	0.921	0.949	0.979	1.010	1.042	1.076	1.113	1.154	1.202	1/262	1.405
0.59	0.619	0.645	0.671	0.697	0.723	0.749	0.776	0.802	0.829	0.857	0.885	0.913	0.943	0.974	1.006	1.040	1.077	1.118	1.166	1.226	1.369
0.60	0.583	0.609	0.661	0.687	0.713	0.740	0.766	0.793	0.821	0.849	0.877	0.907	0.938	0.970	1.004	1.041	1.082	1.130	1.190	1.333	1.333
0.61	0.549	0.575	0.601	0.627	0.653	0.679	0.706	0.732	0.759	0.787	0.815	0.843	0.873	0.904	0.936	0.970	1.007	1.048	1.096	1.156	1.299
0.62	0.516	0.542	0.568	0.594	0.620	0.646	0.673	0.699	0.726	0.754	0.782	0.810	0.840	0.871	0.903	0.937	0.974	1.015	1.063	1.123	1.266
0.63	0.483	0.509	0.535	0.561	0.587	0.613	0.640	0.666	0.693	0.721	0.749	0.777	0.807	0.838	0.870	0.904	0.941	0.982	1.030	1.090	1.233
0.64	0.451	0.474	0.503	0.529	0.555	0.581	0.608	0.634	0.661	0.689	0.717	0.745	0.775	0.806	0.838	0.872	0.909	0.950	0.998	1.068	1.201
0.65	0.419	0.445	0.471	0.497	0.523	0.549	0.576	0.602	0.629	0.657	0.685	0.713	0.743	0.774	0.806	0.840	0.877	0.918	0.966	1.026	1.169
0.66	0.388	0.414	0.440	0.466	0.492	0.518	0.545	0.571	0.598	0.626	0.654	0.682	0.712	0.743	0.775	0.809	0.846	0.887	0.935	0.995	1.138
0.67	0.358	0.384	0.410	0.436	0.462	0.488	0.515	0.541	0.568	0.596	0.624	0.652	0.713	0.745	0.779	0.816	0.857	0.905	0.965	0.965	1.108
0.68	0.328	0.354	0.380	0.406	0.432	0.458	0.485	0.511	0.538	0.566	0.594	0.622	0.652	0.683	0.715	0.749	0.786	0.827	0.875	0.935	1.078
0.69	0.299	0.325	0.351	0.377	0.403	0.429	0.456	0.482	0.509	0.537	0.565	0.593	0.623	0623	0.654	0.686	0.720	0.757	0.798	0.846	0.906
0.70	0.270	0.296	0.322	0.348	0.374	0.400	0.427	0.453	0.480	0.508	0.536	0.564	0.594	0.625	0.657	0.691	0.728	0.769	0.817	0.877	1.020

	Col 1	Col 2	Col 3	Col 4	Col 5	Col 6	Col 7	Col 8	Col 9	Col 10	Col 11	Col 12	Col 13	Col 14	Col 15	Col 16	Col 17	Col 18	Col 19	Col 20	Col 21
0.71	0.242	0.268	0.294	0.320	0.346	0.372	0.399	0.425	0.452	0.480	0.508	0.536	0.566	0.597	0.629	0.663	0.700	0.741	0.789	0.849	0.992
0.72	0.214	0.240	0.266	0.292	0.318	0.344	0.371	0.397	0.424	0.452	0.480	0.508	0.538	0.569	0.601	0.635	0.672	0.713	0.761	0.821	0.964
0.73	0.186	0.212	0.238	0.264	0.290	0.316	0.343	0.369	0.396	0.424	0.452	0.480	0.510	0.541	0.573	0.607	0.644	0.685	0.733	0.793	0.936
0.74	0.159	0.185	0.211	0.237	0.263	0.289	0.316	0.342	0.369	0.397	0.425	0.453	0.483	0.514	0.546	0.580	0.617	0.658	0.706	0.766	0.909
0.75	0.132	0.158	0.184	0.210	0.236	0.262	02.89	0.315	0.342	0.370	0.398	0.426	0.456	0.487	0.519	0.553	0.590	0.631	0.679	0.739	0.882
0.76	0.105	0.131	0.157	0.183	0.209	0.235	0.262	0.288	0.315	0.343	0.371	0.399	0.429	0.460	0.492	0.526	0.563	0.604	0.652	0.712	0.855
0.77	0.079	0.105	0.131	0.157	0.183	0.209	0.236	0.262	0.289	0.317	0.345	0.373	0.403	0.434	0.466	0.500	0.537	0.578	0.626	0.685	0.829
0.78	0.052	0.078	0.104	0.130	0.156	0.182	0.209	0.235	0.262	0.290	0.235	0.262	0.290	0.318	0.346	0.376	0.407	0.439	0.473	0.510	0.551
0.79	0.026	0.052	0.078	0.104	0.130	0.156	0.183	0.209	0.236	0.264	0.292	0.320	0.350	0.381	0.413	0.447	0.484	0.525	0.573	0.633	0.776
0.80	0.000	0.026	0.052	0.078	0.104	0.130	0.157	0.183	0.210	0.238	0.266	0.294	0.324	0.355	0.387	0.421	0.458	0.499	0.547	0.609	0.750
8.81		0.000	0.026	0.052	0.078	0.104	0.131	0.157	0.184	0.212	0.240	0.268	0.298	03.29	0.361	0.395	0.432	0.473	0.521	0.581	0.724
0.82			0.000	0.026	0.052	0.078	0.105	0.131	0.158	0.186	0.214	0.242	0.272	0.303	0.335	0.369	0.406	0.447	0.495	0.555	0.698
0.83				0.000	0.026	0.052	0.079	0.105	0.132	0.160	0.188	0.216	0.188	0.309	0.343	0.380	0.421	0.469	0.529	0.529	0.672
0.84					0.000	0.026	0.053	0.079	0.106	0.134	0.162	0.190	0.220	0.251	0.283	0.317	0.354	0.395	0.443	0.503	0.646
0.85						0.000	0.027	0.053	0.080	0.108	0.136	0.164	0.194	0.225	0.257	0.291	0.328	0.369	0.417	0.477	0.620
0.86							0.000	0.026	0.053	0.081	0.109	0.137	0.167	0.198	0.230	0.264	0.301	0.342	0.390	0.450	0.593
0.87								0.000	0.027	0.055	0.083	0.111	0.141	0.172	0.204	0.238	0.275	0.316	0.364	0.424	0.567
0.88									0.000	0.028	0.056	0.084	0.114	0.145	0.177	0.211	0.248	0.289	0.337	0.397	0.540
0.89										0.000	0.028	0.056	0.086	0.117	0.149	0.183	0.220	0.261	0.309	0.369	0.512
0.90											0.000	0.028	0.058	0.089	0.121	0.155	0.192	0.233	0.281	0.341	0.484
0.91												0.000	0.030	0.061	0.093	0.127	0.174	0.205	0.253	0.313	0.456
0.92													0.000	0.031	0.063	0.097	0.134	0.175	0.223	0.283	0.426
0.93														0.000	0.032	0.066	0.103	0.144	0.192	0.252	0.395
0.94															0.000	0.034	0.071	0.112	0.160	0.220	0.363
0.95															0.000	0.034	0.071	0.112	0.160	0.220	0.363
0.95																0.000	0.037	0.079	0.126	0.186	0.329
0.96																	0.000	0.041	0.089	0.149	0.292
0.97																		0.000	0.048	0.108	0.251
0.98																			0.000	0.060	0.203
0.99																				0.000	0.143
																					0.000

Assuming that the installed cost of static capacitors is $10.00 per KVAR, the 160 KVAR capacitor above would cost $1,600.00. But the $156.28/month saving on the electric bill pays for this in less than 11 months. After the first 11 months, the monthly savings continue at no further costs for correction. If the power factor was improved above the 85 percent used in the example above, the billing demand charges would be further reduced.

The cost of $10.00 per kilovar is based on the use of 480-600 volt capacitors. For 240-volt units, the cost would be about $20.00 per kilovar. But even for the more expensive installation, a substantial over-all saving would be achieved.

It is suggested that you consult with your power company to establish what savings you will obtain in your area. Because of the wide variety of power factor clauses in existence it is impossible to present them all.

IMPROVED ELECTRICAL EFFICIENCY

Other advantages of corrected power factor are concerned with the better performance obtained from electrical equipment when no longer burdened with excess reactive power.

Released System Capacity

The reactive power used by inductive circuits consists of a reactive current (also called "wattless current," "magnetizing current," or "non-working current") multiplied by the supply-line voltage. Total reactive power (and current) increases, while power factor decreases, when the number of inductive devices that require reactive power is increased. Each inductive device added to the system contributes to the reactive power requirements of the system.

When power factor is improved, the amount of reactive current that formerly flowed through transformers, feeders, panels, and cables is reduced. Power factor correction capacitors, installed directly at the terminals of inductive loads such as motors, generate most or all of the reactive power needed to set up the magnetic fields of the motors, and thus reduce or eliminate the need to supply this power from the distribution system.

For example, if four motors operate at a power factor of 75 percent, power factor correction to 95 percent will release enough system capacity to accommodate an additional motor of the same size.

Where transformers and circuits are being overloaded, power capacitors installed at the various sources of inductive reactive

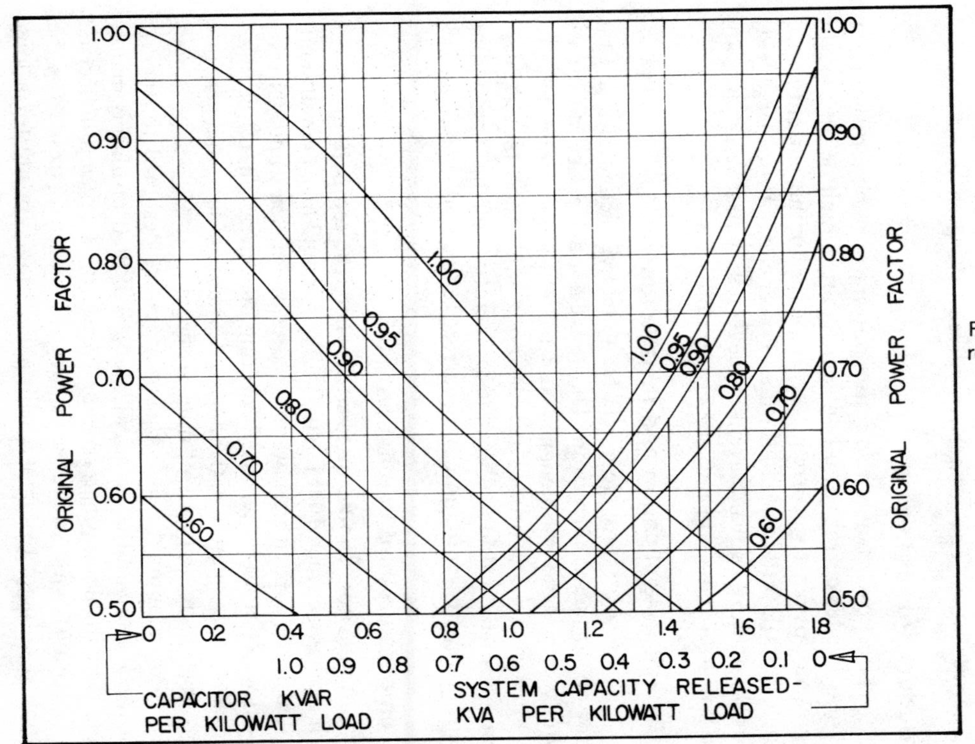

Fig. 6-18. How system capacity is released as power factor is improved.

load can release system capacity or relieve the overload to allow service to added equipments (See Fig. 6-18). The installation of power capacitors may, in some instances, eliminate the need for installing larger power transformers, rewiring a plant, or possibly both.

Improved Voltage Conditions

Low power factor can reduce plant voltage when kilovar current is drawn from the feeder line. As power factor decreases, total line current increases (from greater reactive current), causing greater voltage drops across line resistances. This is because the voltage drop in a line equals the current flowing multiplied by the line resistance. The greater the current, the greater the drop.

Reduce Power Losses

Low power factor also causes power losses in the distribution lines. Feeder current at low power factor is high because of the presence of reactive current. Any reduction in this current results in lower kilowatt losses in the line. Power capacitors, by reducing or eliminating kilovar current in the feeders, can save a significant amount of money in reduced electric bills.

Capacitor Application and Installation

Capacitors provide a reliable, effective, economical means for power factor correction. They have a long service life and require almost no maintenance. But to get the maximum benefits out of capacitors, proper installation is quite important. Among points to be considered are the temperature of the surrounding air, adequate ventilation, and correct fusing and switching.

Temperature and Ventilation

Capacitors should be located where the temperature of the air does not exceed 104° F (40° C). They should also be well ventilated since capacitors always operate at full load and generate heat which must be removed for long life operation. (Other electrical devices operate with varying amounts of load and generate proportionate amounts of internal heat).

For economic reasons, capacitors are designed to operate near the limits of the electrical strength of their insulating materials. The insulating material in a capacitor is called a dielectric, and it must withstand a great deal of electrical stress from the applied voltage. Combined with a high temperature for a long

Table 6-5. Recommended Wire, Switch and Fuse Sizes for Use With Individual Capacitors (courtesy of Sprague Electric Company).

KVAR	Rated Current*	Wire Size** Type R or Equal	Fuse Amperes	Switch Smperes	Rated Current*	Wire Size** Type R or Equal	Fuse Amperes	Switch Amperes	Rated Current*	Wire Size** Type R or Equal	Fuse Amperes	Switch Amperes
240 VOLT SERVICE												
0.5	2.08	14	6	30	—	—	—	—	1.20	14	3	30
1.0	4.17	14	10	30	2.08	14	6	30	2.41	14	6	30
1.5	6.25	14	15	30	3.12	14	6	30	3.61	14	6	30
2.0	8.33	14	15	30	4.17	14	10	30	4.81	14	10	30
2.5	10.4	14	20	30	5.21	14	10	30	6.01	14	10	30
3	12.5	12	25	30	6.25	14	15	30	7.22	14	15	30
4	16.7	10	30	30	8.33	14	15	30	9.62	14	20	30
5	20.8	10	35	60	10.4	12	20	30	12.0	12	20	30
6	25.0	8	45	60	12.5	10	25	30	14.4	12	25	30
7.5	31.2	6	60	60	15.6	10	30	30	18.0	10	30	30
8	33.3	6	60	60	16.7	10	30	30	19.2	10	35	60
10	41.7	4	70	100	20.8	8	35	60	24.1	8	40	60
12.5	52.1	3	90	100	26.0	6	45	60	30.1	6	50	60
15	62.5	2	110	200	31.2	6	60	60	36.1	6	60	60
480 VOLT SERVICE												
1.0	2.08	14	6	30	—	—	—	—	1.20	14	3	30
1.5	3.12	14	6	30	—	—	—	—	1.80	14	3	30
2.0	4.17	14	10	30	2.08	14	6	30	2.41	14	6	30
2.5	5.21	14	10	30	—	—	—	—	3.00	14	6	30
3.0	6.25	14	15	30	3.12	14	6	30	3.61	14	6	30
4	8.33	14	15	30	4.17	14	10	30	4.81	14	10	30
5	10.4	14	20	30	5.21	14	10	30	6.02	14	10	30
6	12.5	12	25	30	6.25	14	15	30	7.22	14	15	30
7.5	15.6	10	30	30	7.80	14	15	30	9.02	14	15	30
8	16.7	10	30	30	8.33	14	15	30	9.62	14	20	30
10	20.8	10	35	60	10.4	12	20	30	12.0	12	20	30
12	25.0	8	45	60	12.5	10	25	30	14.4	12	25	30
12.5	26.0	8	45	60	—	—	—	—	15.0	10	25	30
15	31.2	6	60	60	15.6	10	30	30	18.0	10	30	30
20	41.7	4	70	100	—	—	—	—	24.1	8	40	60
600 VOLT SERVICE												
1.0	1.67	14	3	30	—	—	—	—	.96	14	3	30
1.5	2.50	14	6	30	—	—	—	—	1.44	14	3	30
2	3.33	14	6	30	1.67	14	3	30	1.92	14	6	30
2.5	4.16	14	10	30	—	—	—	—	2.41	14	6	30
3	5.00	14	10	30	2.50	14	6	30	2.89	14	6	30
4	6.64	14	15	30	3.33	14	6	30	3.85	14	10	30
5	8.33	14	15	30	4.16	14	10	30	4.81	14	10	30
6	10.0	14	20	30	5.00	14	10	30	5.77	14	10	30
7.5	12.5	12	25	30	6.25	14	15	30	7.22	14	15	30
8	13.3	12	25	30	6.66	14	15	30	7.69	14	15	30
10	16.7	10	30	30	8.33	14	15	30	9.62	14	20	30
12	20.0	10	35	60	10.0	12	20	30	11.5	12	20	30
12.5	20.8	10	35	60	—	—	—	—	12.0	12	20	30
15	25.0	8	45	60	12.5	10	25	30	14.4	12	25	30
20	33.3	6	60	60	—	—	—	—	19.2	10	35	60

*Rated current based on operation at rated frequency, voltage, KVAR

**For other wire types, consult National Electrical Code.

period of time, high electrical stress can cause the dielectric to weaken or fail. Freely circulating air is most important to hold the temperature down, thereby extending the life of the capacitor unit.

If at all avoidable, capacitors should not be installed in small rooms near radiators or other types of heating units, where ventilation is restricted, or in outdoor areas where heat from the direct rays of the sun might raise the capacitor temperature excessively. The capacitor case temperature should not rise above 131° F (55° C) under normal operating conditions—that is, at the nameplate voltage and frequency value for which the capacitor is designed. If capacitors are operated at overloads for short periods, the case temperature should not exceed 158° F. (70° C).

Temperature rise depends on the capacitor tolerance, line frequency and operating voltage. Temperature as a function of frequency need not be considered, since the line frequency is constant in modern power systems. (The rated frequency of the capacitor, stamped on the nameplate, must of course be the same as the line frequency of the circuit). Overheating at a normal operating voltage is practically impossible, but when the voltage exceeds 110 percent of the capacitor rating, damage may occur. A 10 percent over-voltage means a 21 percent rise in KVAR; together with the KVAR tolerance of 15 percent, the maximum operating conditions may be exceeded. When the operating voltage is 10 percent or more above the capacitor's rated voltage, the line voltage should be reduced or the capacitors should be switched off the line during light-load periods.

It is also important to avoid mounting capacitors too closely together, since the heat from one will affect the next. Mounting methods provided by the capacitor manufacturer reduce the danger from this factor.

Fusing

Power capacitors are provided with fuses to protect distribution systems in case of an internal short circuit. These fuses are rated from 165 percent to 250 percent of the rated kilovar current to allow for maximum operating conditions plus momentary surges of current. Adjacent equipment requires protection from a shorted capacitor mainly because of the danger of a burst capacitor case.

Fuses disconnect a shorted capacitor from the circuit before gas pressures, built up within the capacitor as a result of the breakdown, are severe enough to cause the case seams to break.

Table 6-6. Recommended Wire, Switch and Fuse Sizes for Use With Rock Type Capacitors (Courtesy of Sprague Electric Company).

KVAR	SINGLE-PHASE				THREE-PHASE			
	RATED CURRENT*	WIRE SIZE** TYPE AVA OR EQUAL	FUSE AMPERES	SWITCH AMPERES	RATED CURRENT*	WIRE SIZE** TYPE AVA OR EQUAL	FUSE AMPERES	SWITCH AMPERES
240 VOLT SERVICE								
20	83.3	3	150	200	48.1	6	80	100
25	104	1	175	200	60.2	4	100	100
30	125	0	225	400	72.2	4	125	200
40	167	000	300	400	96.2	2	175	200
45	188	0000	350	400	108	1	200	200
60	250	300M	450	600	144	00	250	400
75	312	500M	600	600	180	000	300	400
80	333	500M	600	600	192	0000	350	400
90	375	600M	800	800	216	250M	400	400
100	417	750M	800	800	241	300M	400	400
105	438	800M	800	800	253	300M	450	600
120	500	1000M	1000	1200	289	350M	600	600
135	562	1500M	1000	1200	325	500M	600	600
150	—	—	—	—	361	600M	600	600
180	—	—	—	—	434	(2) 250M	800	800
240	—	—	—	—	578	(2) 350M	1000	1200
270	—	—	—	—	675	(3) 300M	1200	1200
360	—	—	—	—	868	(3) 400M	—	—
480 VOLT SERVICE								
25	52.1	6	90	100	30.1	10	50	60
30	62.5	4	110	200	36.1	8	60	60
40	83.3	3	150	200	48.1	6	80	100
45	93.8	2	175	200	54.1	6	90	100
60	125	0	225	400	72.2	4	125	200
75	156	00	300	400	90.2	2	150	200
80	167	000	300	400	96.2	2	175	200
90	188	0000	350	400	108	1	200	200
100	208	250M	350	400	120	0	200	200
105	219	250M	400	400	126	0	225	400
120	250	300M	450	600	144	00	250	400
135	281	350M	500	600	162	000	300	400
140	292	400M	500	600	168	000	300	400
150	312	500M	600	600	180	000	300	400
160	333	500M	600	600	192	0000	350	400
240	—	—	—	—	288	(2) 00	600	600
320	—	—	—	—	384	(2) 0000	800	800
360	—	—	—	—	432	(3) 00	800	800
480	—	—	—	—	576	(3) 0000	1000	1200
600 VOLT SERVICE								
25	41.6	8	70	100	24.1	10	40	60
30	50.0	6	90	100	28.9	10	50	60
40	66.6	4	110	200	38.5	8	70	100
45	75.0	4	125	200	43.3	8	80	100
60	100	2	175	200	57.7	6	100	100
75	125	0	225	400	72.2	4	125	200
80	133	0	225	400	77.0	4	150	200
90	151	00	250	400	86.6	3	150	200
100	167	000	300	400	96.2	2	175	200
105	175	000	300	400	[illegible]	1	175	200
120	200	0000	350	400	115	1	200	200
135	225	250M	400	400	130	0	225	400
140	233	250M	400	400	135	0	225	400
150	250	300M	450	600	144	00	250	400
160	266	350M	450	600	154	00	300	400
240	—	—	—	—	230	(2) 1	500	600
320	—	—	—	—	308	(2) 00	600	600
360	—	—	—	—	345	(3) 1	600	600
480	—	—	—	—	460	(3) 00	800	800

*Rated current based on operation at rated frequency, voltage, and KVAR
**For other wire types, consult National Electrical Code.

Sprague's Unipak® capacitors consist of *internally fused* unit cells enclosed in a single outer container. One advantage of this fusing-packaging method is that the internal fuses will remove any defective cells from the circuit, while permitting the rest of the capacitor unit cells to continue in operation.

Conductor Current Ratings

Conductors used with capacitors should have a current rating of 135 percent of the minimum rated current of the capacitor, to allow for the possiblity of maximum operating conditions (see Tables 6-5 and 6-6).

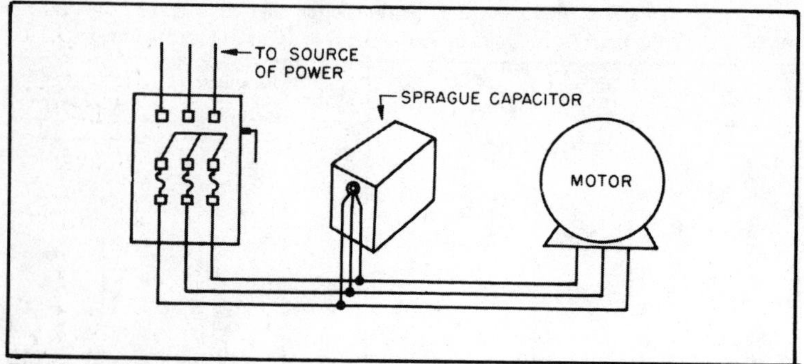

Fig. 6-19. Capacitors connected directly to motor and automatically removed from system when motor is shut down. (courtesy of Sprague Electric Company).

Switching

The National Electrical Code requires that power capacitor installations be equipped with a disconnecting means to permit their removal from the circuit during light loads, or during equipment maintenance periods. Switches used for this purpose should have a current rating of at least 165 percent of the rated current of the capacitors which they serve. See Figs. 6-19 and 6-20.

When capacitors operate with induction motors, installation of the capacitor unit on the load side of the motor starter eliminates the need for both switches and fuses. The saving thus gained is one of the reasons for the "correction-at-the-load" technique pioneered by Sprague.

Courtesy of Sprague Electric Company

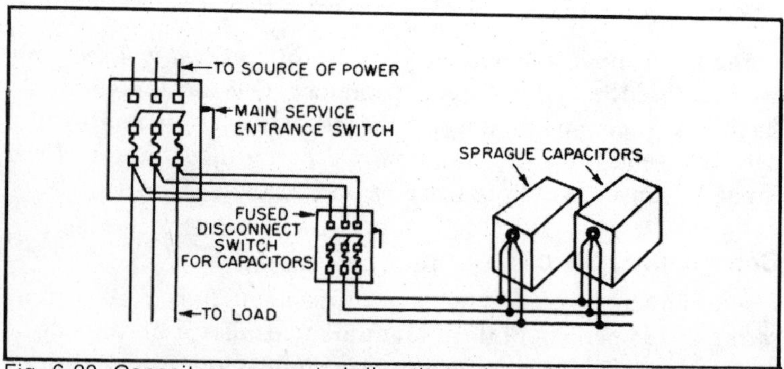

Fig. 6-20. Capacitors connected directly to the switchboard or at distribution or load centers (courtesy of Sprague Electric Company).

Fig. 6-21. Measuring voltage with a clamp-on volt/amp/ohmmeter (courtesy of AMPROBE Instrument).

CONDUCTING THE ELECTRICAL POWER SURVEY

The following individual test instructions should be conducted by appropriately trained personnel with adequate electrical background. The personnel conducting the tests as required for data collection should first acquaint themselves with the instruments they will accordingly be using, and ascertain that the instruments will be operating in voltages and current values well within their respective range of maximums. It is preferred practice to utilize instruments that will provide meter readings at "mid-scale" points of their intended maximums. Attention should be given that the instruments will not adversely load the circuit or affect additional instrument errors. It is desirable to begin the survey with the assurance of a recently calibrated instrument. Many types of meter movements may, after a lengthy time out of service, become inaccurate to sufficient degrees that would abort the effective collection of data. It is recommended that the plants' voltmeters, ammeters, ohmmeters and wattmeters be shop calibrated by an approved organization (Figs. 6-21 through 6-23).

Comments on each of the electrical instruments, with a schematic of each, and their schematically represented manner of hookup is also provided to eliminate any possiblity of erroneous readings by improper connections. Particular considerations must be given when metering polyphase circuits to assure that the phases so metered are in proper electrical sequence for power measurement.

The data and values collected by the tests specified will then be applied to the formulas as given in Table 6-1 and in the survey form itself. All of the formulas and data required for their computations may be obtained by the deployment of range suitable voltmeters, ammeters and a wattmeter. The engineer may avail himself with an oscillosope, which will indicate directly the phase synchronism and near accurate phase shift of polyphase systems.

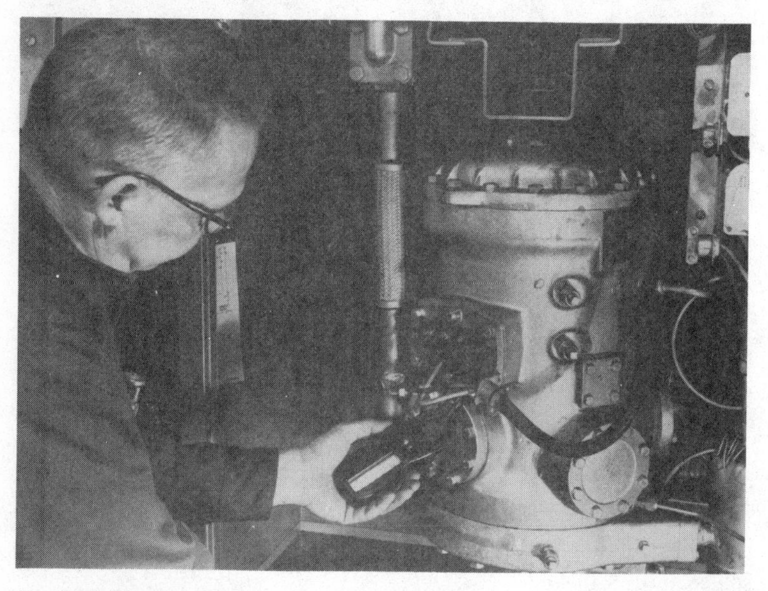

Fig. 6-22. Measuring current with a clamp-on volt/amp/ohmmeter (courtesy of AMPROBE Instrument).

VOLTMETER

The *voltmeter* is an instrument for measuring the electromotive force or potential difference of a circuit (or any part of a circuit), which will therefore indicate the actual voltage drop or difference in potential between two points (Fig. 6-24). The voltmeter will indicate the potential difference or driving EMF between these two points.

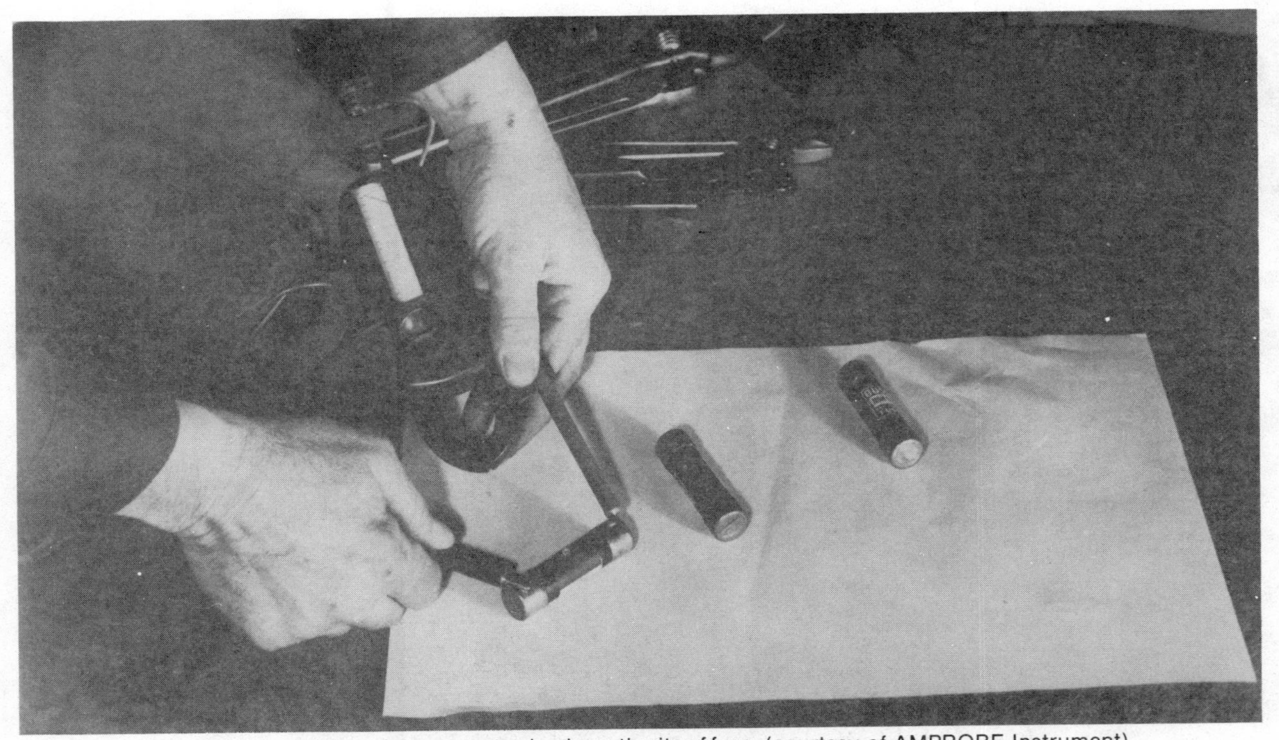

Fig. 6-23. Using a clamp-on volt/amp/ohmmeter to check continuity of fuse (courtesy of AMPROBE Instrument).

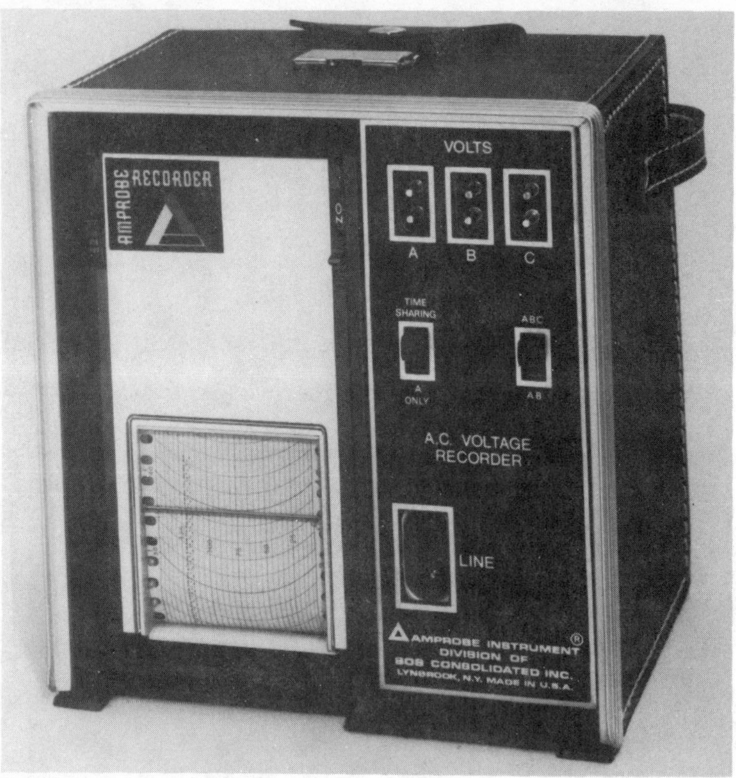

Fig. 6-24. Voltage recorder (courtesy of AMPROBE Instrument).

If we were to place the two probes of a voltmeter across a circuit that has an originating EMF of 220 volts, our reading will be 220 volts between the two hot legs of the lines. However, were we to probe only one of the two lines and the other to ground (assuming ground to be neutral or O volts), we could find a marked difference above or below one-half of the nominal voltage we would expect to find. Again, the voltmeter will measure differences in a circuit's voltage as created by the driving EMF. The addition of all the voltage drops, or potential differences, as collected about the circuit, proceeding from beginning to end, will equal the total EMF or potential difference across the circuit as a whole. This phenomenon is known as *Kirchhoff's Law* and is utilized in all electrical computations involving voltages.

CONNECTION FOR MEASUREMENT

The voltmeter itself may be of the electrodynamic, permanent magnet-moving coil, electrostatic or moving iron varieties.

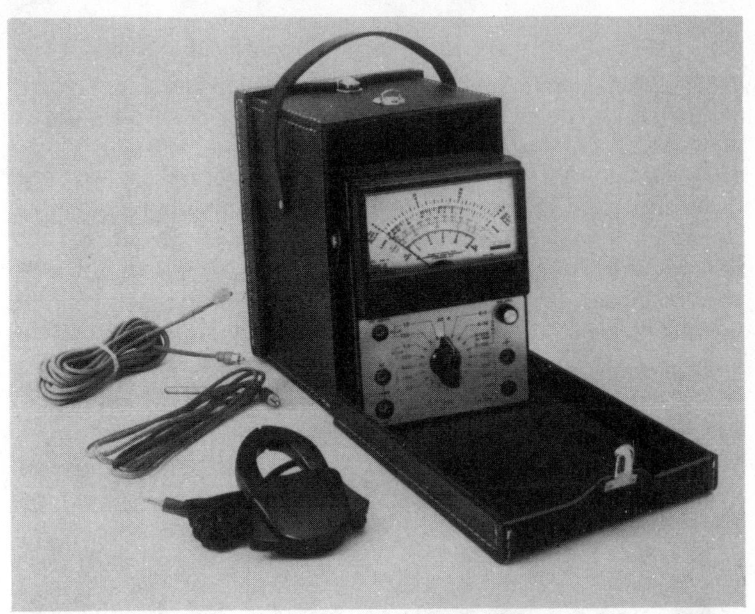

Fig. 6-25. Multi-meter (courtesy of AMPROBE Instrument).

The meter is placed across the line in that it must measure a difference in force, do not put or attempt to put a voltmeter in series with the circuit. This will immediately either ruin the meter or cause damage to the circuit.

Before placing the voltmeter on the line, examine the dial indicator setting of the meter. Make sure that your meter is set for the measurement of voltage, and for the proper ac or dc type of EMF. Most multi-meters must be set for both types of current and expected maximum values (Figs. 6-25 through 6-28). The meter should then be examined to make sure that the maximum value of potential difference you will be measuring is at least less than the range that the meter is set for. Many meters have a multiplier range to extend the usefulness of the meter in taking higher measurements than the meter movement itself is capable of handling. Care must be exercised to make certain that there is enough resistance in series with the meter movement to protect the movement from damage. Figure 6-29 is a schematic diagram of a voltmeter placed across a circuit and shows the resistors employed to protect the meter movement. This resistance multiplying arrangement on most meters can be adjusted outside of the meter by merely rotating the selector switch to a setting that will be higher than the voltage expected to be encountered in the Circuit being measured. If in doubt, set the meter to the highest voltage setting and progressively work on down to a setting that will put the meter approximately at mid-scale position.

Illustrated in Fig. 6-29 is a voltmeter of the permanent magnet variety. The line voltage is 400 volts. The proper resistance selection is had by placing enough resistance in series with the meter armature to permit it to withstand a maximum voltage of 500 volts. If we were to put the meter end of the probe into the next lower tap point of resistance of 350 volts, we would undoubtably burn out the movement. By selecting too high a set point, such as putting the probe into the 750 volt tap, it would result in

Fig. 6-26. Model AM-1A multi-meter (courtesy of AMPROBE Instrument).

Fig. 6-27. Model Am-2A multi-meter (courtesy of AMPROBE Instrument).

too high a resistance. Select a voltage setting that will place the needle of the meter at as close to meter scale mid-point position as possible. At this position the meter should be at its most sensitive position.

SHUNT-TYPE AMMETER

The shunt-type ammeter, as illustrated in Fig. 6-30, is representative of the present day ammeter. See Figs. 6-31 and 6-32 for other current recorders. It can be of any type movement with a rectifier in series with the meter if required for ac use. Formerly, because the ammeter was placed in series, the circuit had to be

Fig. 6-28. Model AM-3 multi-meter (courtesy of AMPROBE Instrument).

interrupted and the probes placed on each side of the break in order for the entire current to pass through the meter movement. With the advent of the shunt-type selector ammeter, the shunt as selected will permit, by adherence to basic parallel flow laws, only a portion of the entire current to flow into and through the entire meter movement. The shunt to be selected will of course be selected to accommodate the amount of current to be measured and

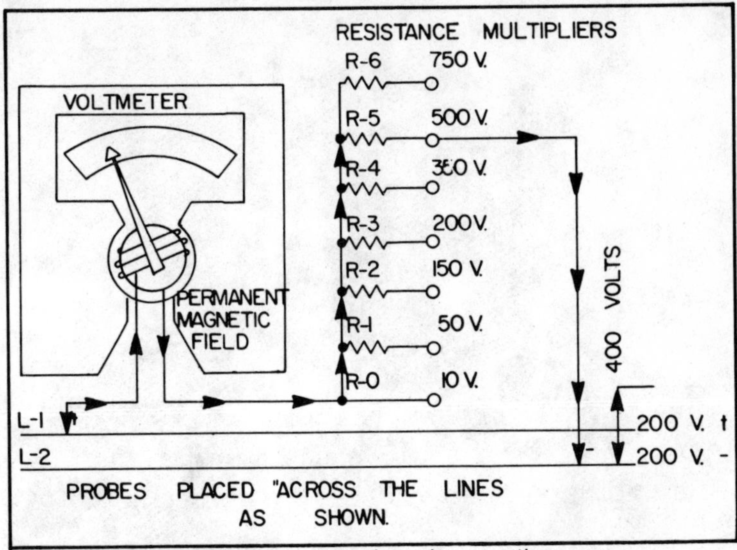

Fig. 6-29. Typical voltmeter schematic and connection.

passed through the meter movement. The heavier or more conductive the shunt is will naturally permit a larger amount of cur-

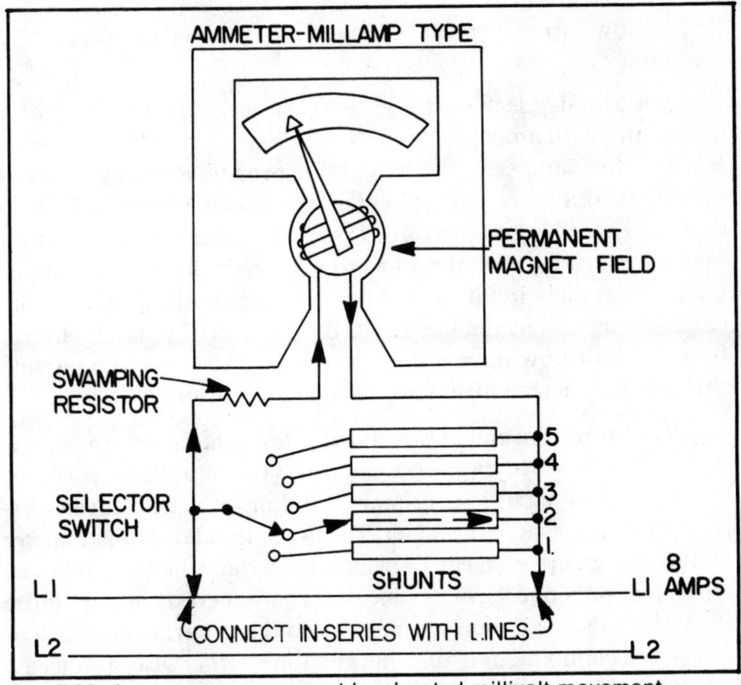

Fig. 6-30. Ammeter measurement by shunted millivolt movement.

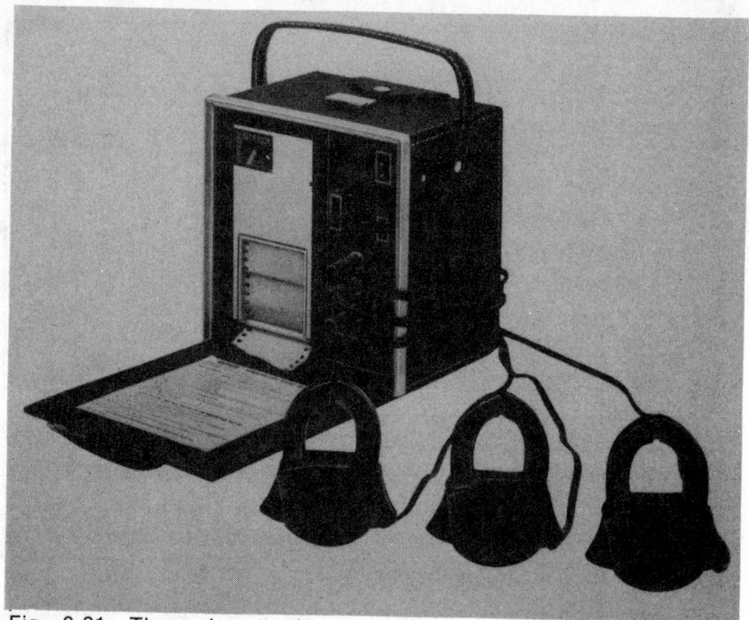

Fig. 6-31. Three-phase volt-amp recorder (courtesy of AMPROBE Instrument).

rent to flow through the parallel portion of the circuit created by the shunt.

The shunt selected in Fig. 6-30 can measure a current of a maximum of 10 amperes. Because the circuit is conducting a current of only 8 amperes, the selection is proper. Refering again to Kirchhoff's Law, the current will be divided between this deliberately "shunted" circuit provided for measurement. Since we have, for example, a current flow of 8 amperes in the line, we might put down a parallel circuit for the current to pass through of equal conductance as has the line itself. By the laws of parallel current flow, we will now have a current of half the line value passing through the meter movement, or a current of only 4 ohms.

The more convenient method of using a clamp-on ammeter is accomplished by the exploitation of the laws of induced currents. The clamp, as such, is a primary winding of a transformer arrangement internal to the meter. The use of the clamp-on ammeter will simplify current measurements. Be certain that you take the reading of only one leg of the line. If the power cord contains more than one power leg, then you must of course separate the legs and attach the clamp on only one line at a time. Otherwise, a canceled effect would result.

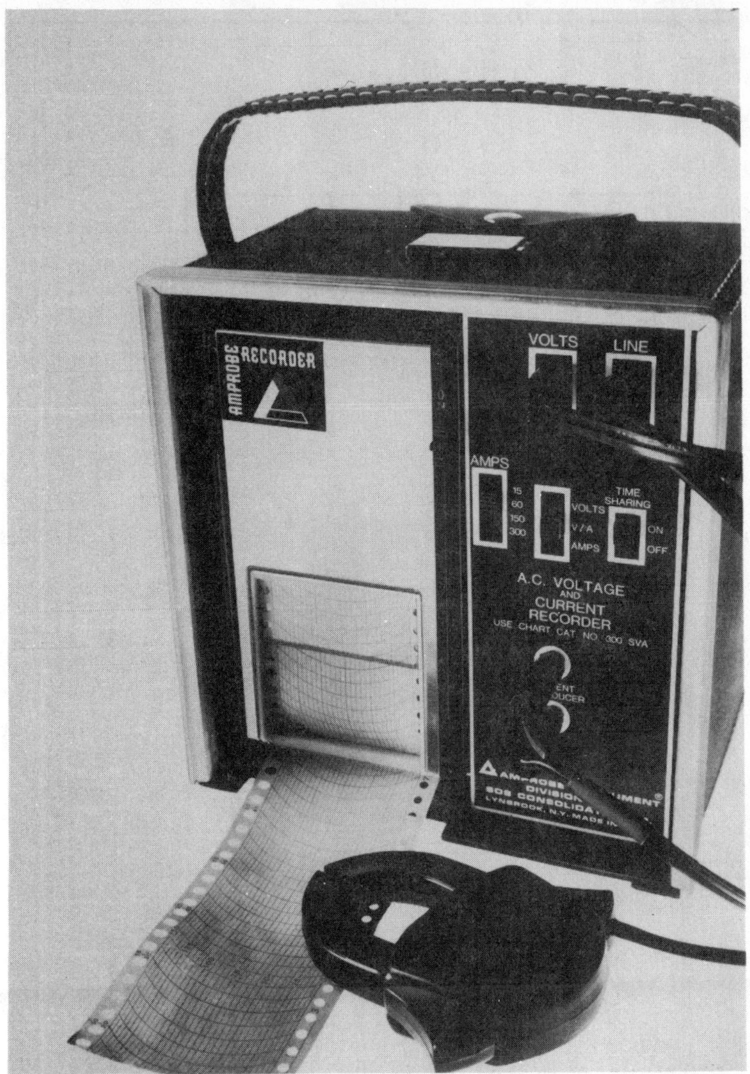

Fig. 6-32. Voltage and amperage recorder (courtesy of AMPROBE Instrument).

OHMMETER

The *ohmmeter*, is a self-contained instrument for the measurement of the resistance of a circuit (Fig. 6-33). The resistance thus indicated is the ohmic resistance if it is a dc circuit. It is not to be confused with the impedance, as would be the case with ac circuits. Essentially, the ohmmeter is an ammeter of very sensitive movement and response. Generally the meter will be a milli-

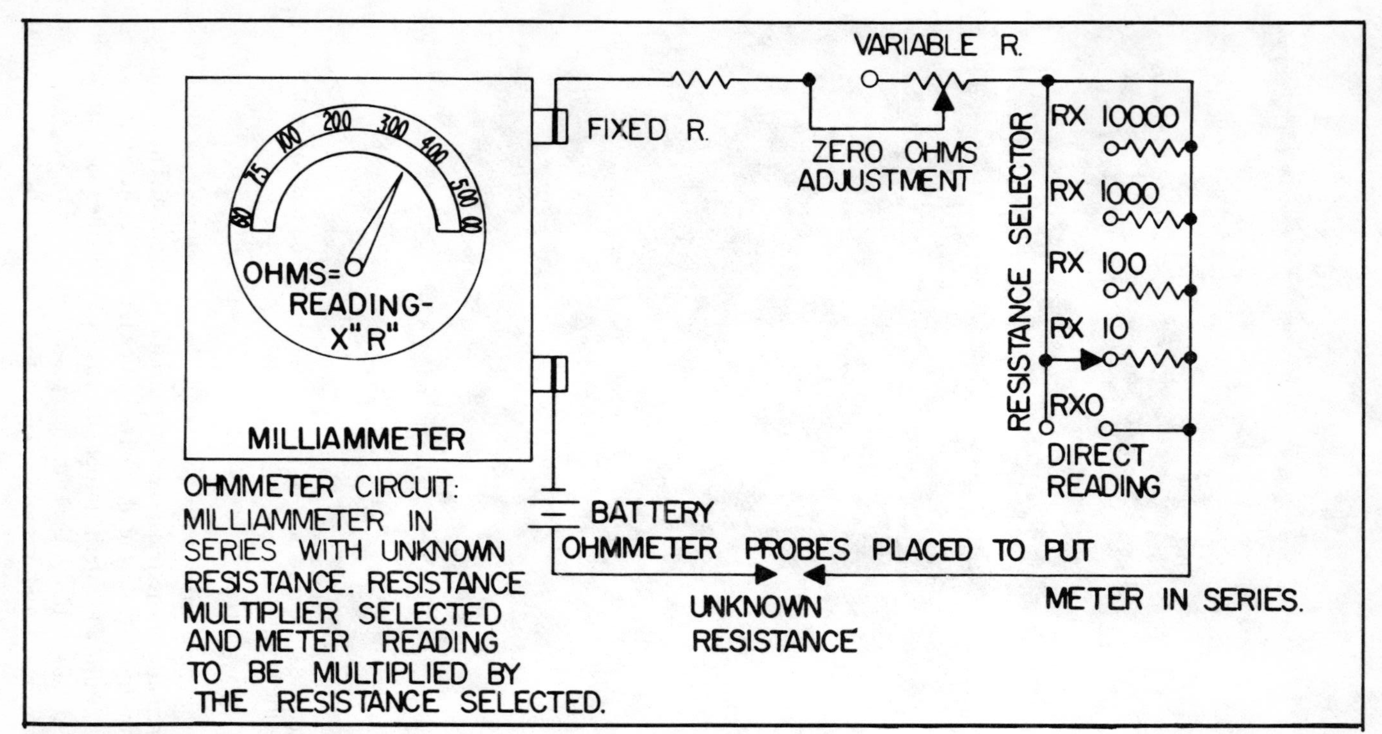

Fig. 6-33. An ohmmeter connected in circuit.

ammeter, thereby requiring only a thousandth part of an ampere of current to cause full swing of the meter.

The ohmmeter is connected in electrical series with the resistance to be determined. Make sure the power to the entire circuit is off, and all capacitors and inductors have either been discharged. Or discharge them to ground before connecting the ohmmeter to the circuit. The ohmmeter contains its own independent driving EMF, usually from a renewable dry cell battery. If the ohmmeter is of the plug-in type, it will be powered by the current from another line through a stepdown transformer. When a battery is used, a zero adjustment will have to be made before every reading to compensate for battery depletion. This is done by touching the two probes together and adjusting the controller until the meter needle swings all the way to the zero point of the scale. If a sharp swing to the zero point can not be had, then the battery should be replaced. These batteries exhaust themselves quite rapidly in ohmmeters.

The ohmmeter is placed in series with the resistance to be determined. The meter face is observed to read a value near midpoint. Either the resistance (in a multiplier circuit) is decreased to show a readable indication or the battery is boosted by placing in more EMF in series. The infinity sign— ∞ —indicates that the resistance is beyond the measuring value of the battery and/or internal resistance of the ohmmeter/resistance circuit.

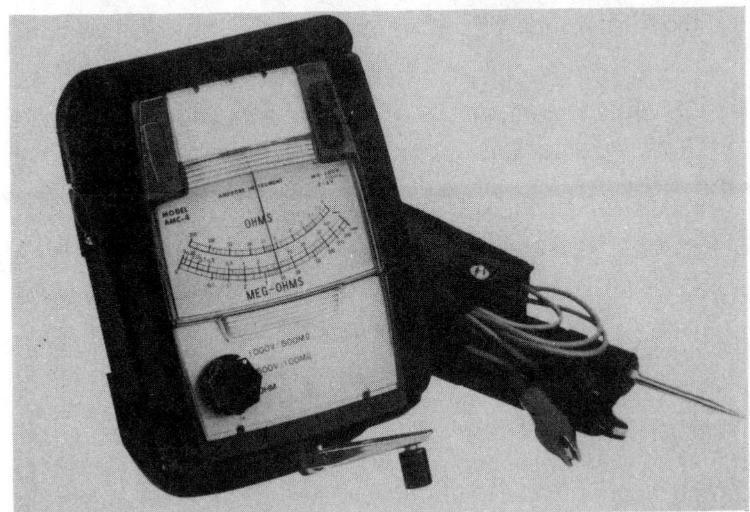

Fig. 6-34. A model AMC-4 megohmmeter (courtesy of AMPROBE Instrument).

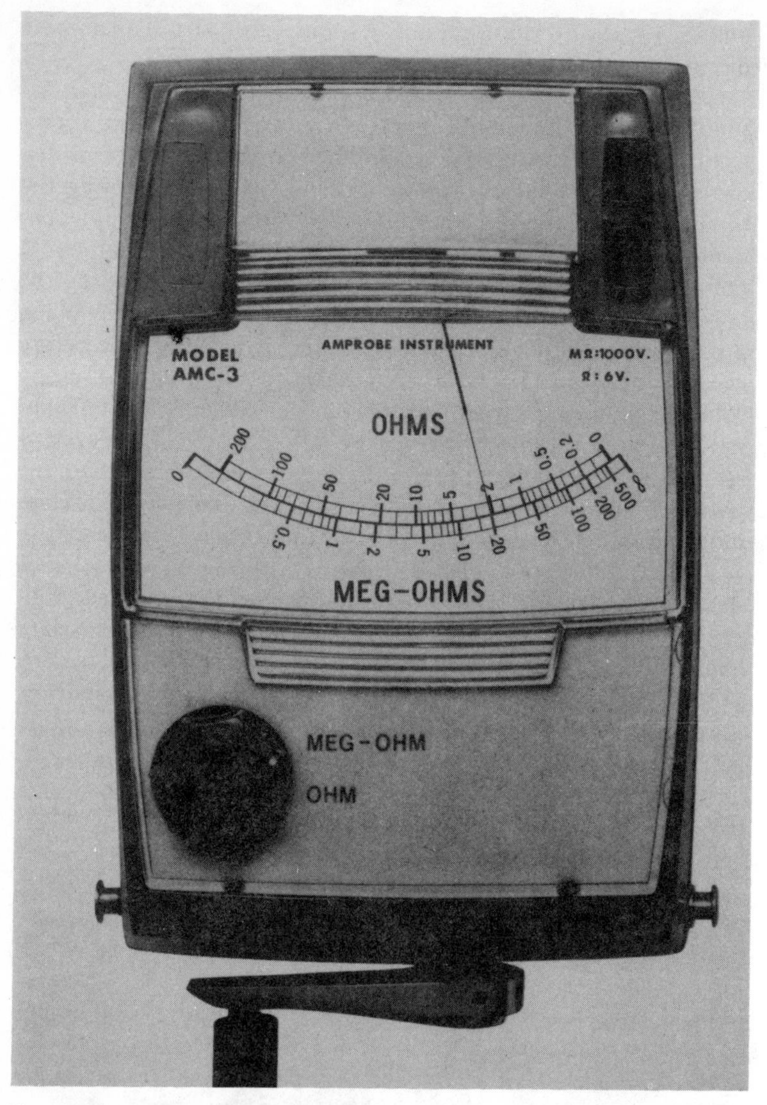

Fig. 6-35. A model AMC-3 megohmmeter (courtesy of AMPROBE Instrument).

MEGOHMMETER

The *megohmmeter* is an instrument capable, by its self-contained high voltage supply, of measuring resistance values far greater than the *ohmmeter* is able to achieve.(Figs. 6-34 through 6-37). The megohmmeter is principally used in taking insulation

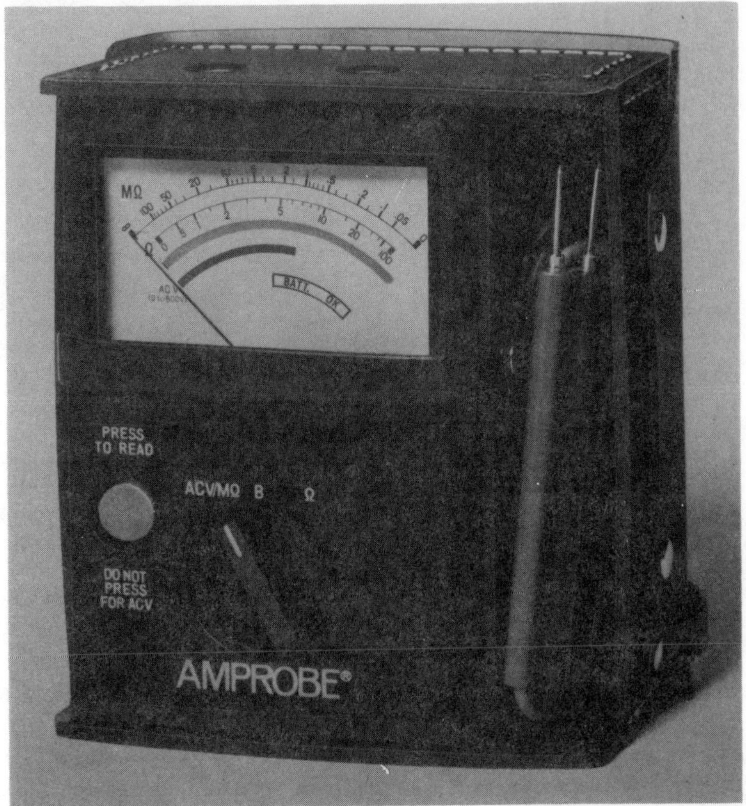

Fig. 6-36. A model AMB-1 megohmmeter (courtesy of AMPROBE Instrument).

resistance measurements, which requires the application of great internal voltage. Most insulation resistances are over 5,000 volts. The megohmmeter is very capable of indicating resistances of up to and in excess of several thousand million ohms. The megohmmeter is of use in locating shorts and high resistance breakdowns in motors, generators and transformers.

The megohmmeter is similar to the ohmmeter, because of the necessity of higher internal voltages. The battery is replaced by the higher driving EMF supply. Hand-cranked dc generators are often incorporated into the megohmmeter, and the unit is cranked until the ultimate voltage is generated to cause breakdown. The resistance is at that value noted. More modern variations have power packs to facilitate the cranking necessary with manual models; however, the exercise made available should be appreciated by most engineers. In addition, the manually cranked model

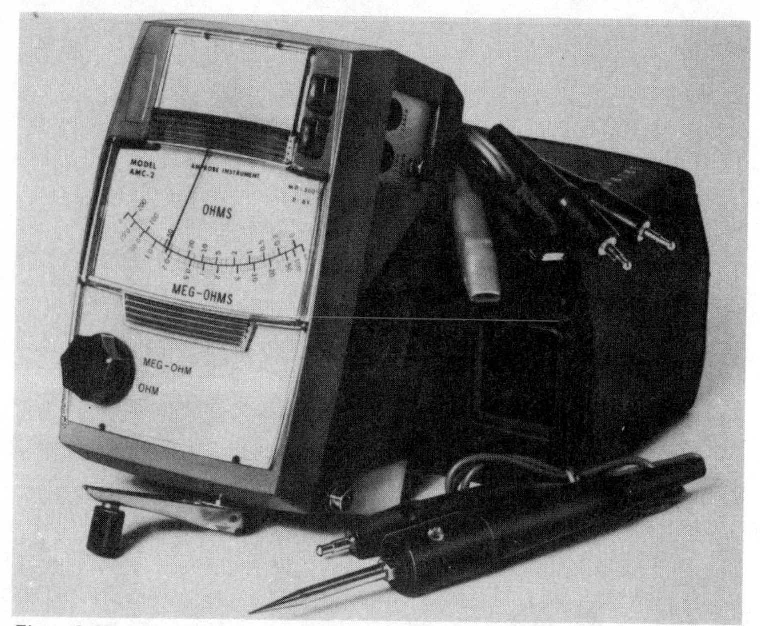

Fig. 6-37. A model AMC-2 ohmmeter (courtesy of AMPROBE Instrument).

is less subject to problems and requires no power packs or current. It may be used virtually any place for field measurements.

WATTMETER

The *wattmeter* is in reality a combination voltmeter and ammeter. The wattmeter as illustrated in Fig. 6-38 may be used either on single-phase ac circuits or on a dc circuit. The internal meter movement is of the electro-dynamic variety and thus will respond to the reversals of an ac circuit as though it were a dc source.

In selecting a wattmeter for power evaluation, particular attention should be given to the maximum current (in amperes) expected to be passing through the meter movement. The actual true power may be very small and well within the scale deflection due to low power factor conditions, but the actual current may be excessive.

By studying the wattmeter schematic (Fig. 6-38), it will be noted that the armature or moving coil is connected in shunt, or in parallel with the load. The field or stator winding is in series with the load current. Thus, the moving coil is actually voltage responsive. The field or stator windings are current responsive.

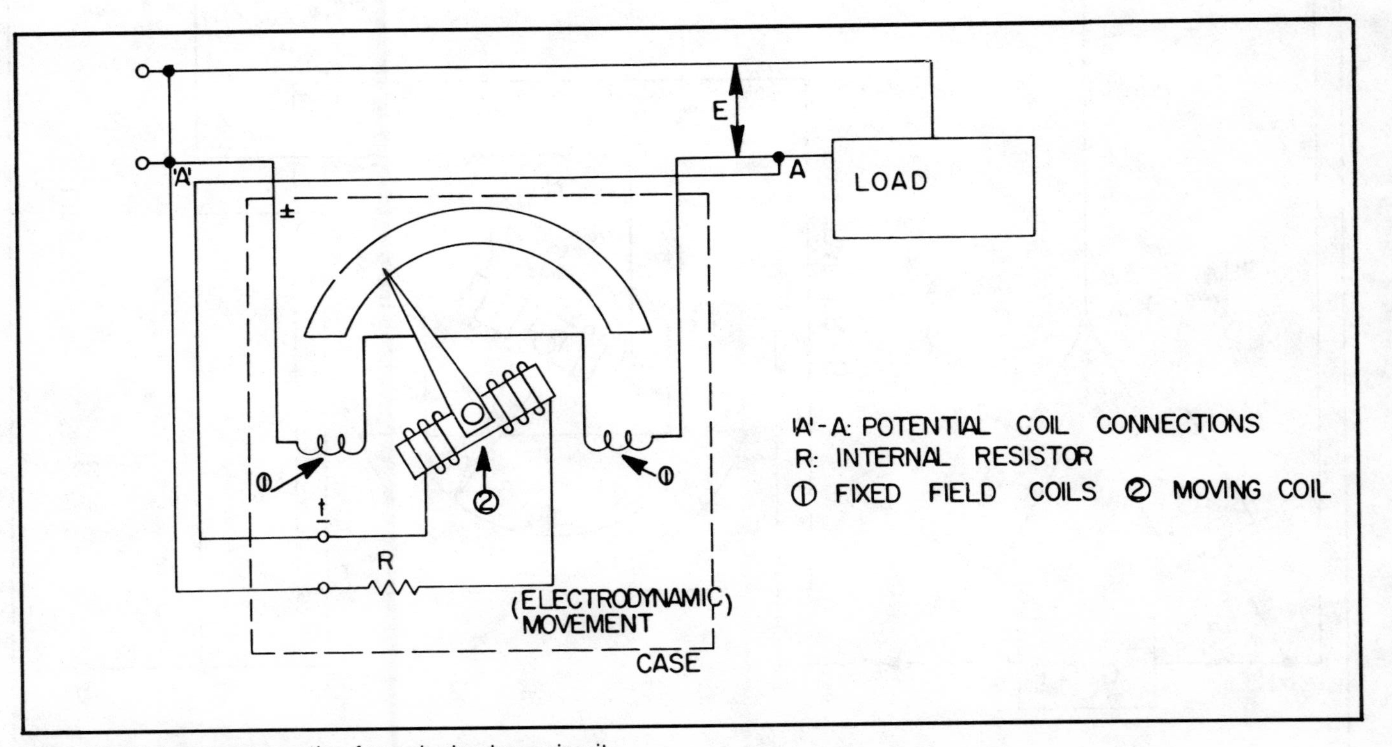

Fig. 6-38. Wattmeter connection for a single-phase circuit.

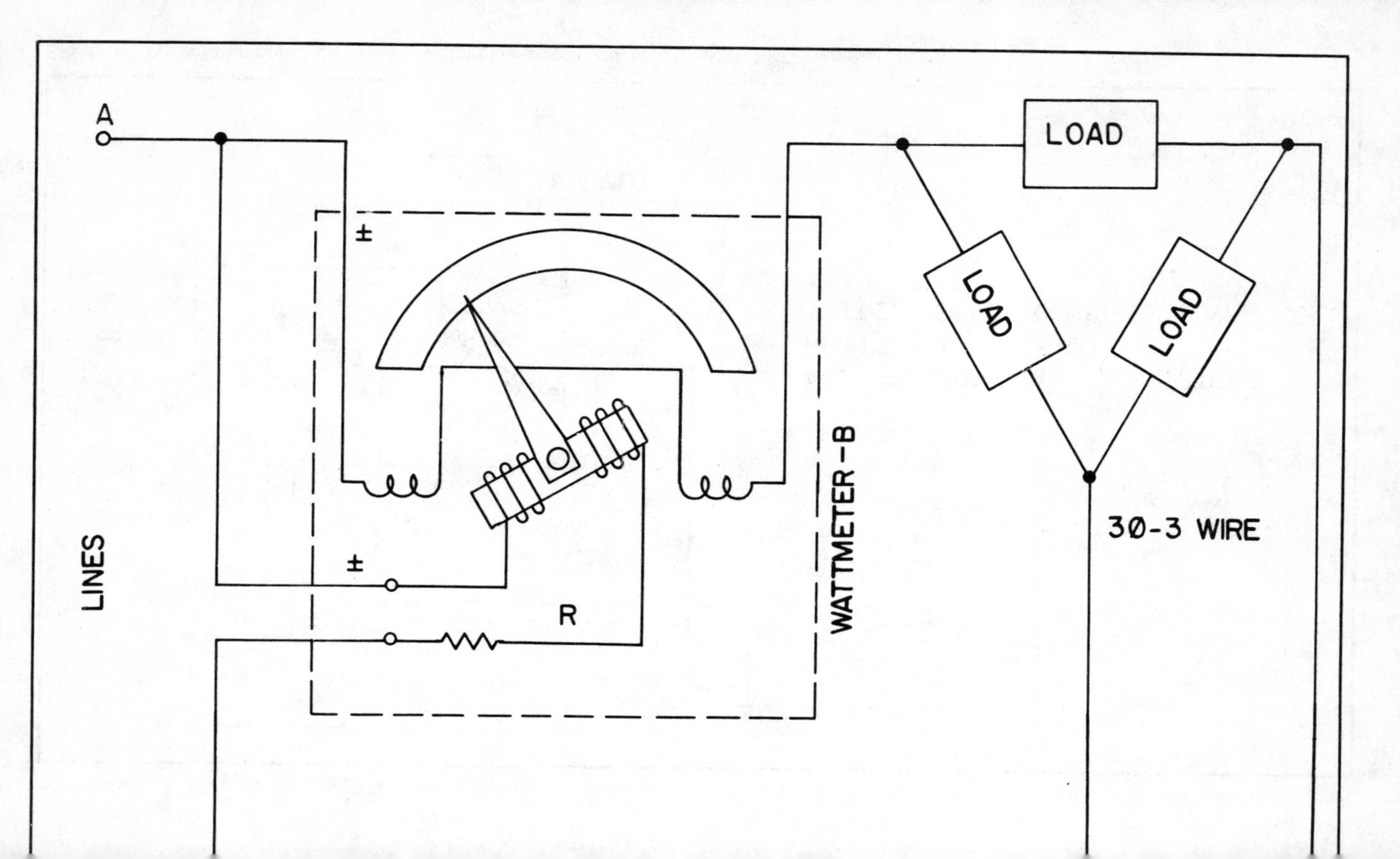
A
LOAD
LOAD
LOAD
±
±
R
WATTMETER-B
3Ø-3 WIRE
LINES

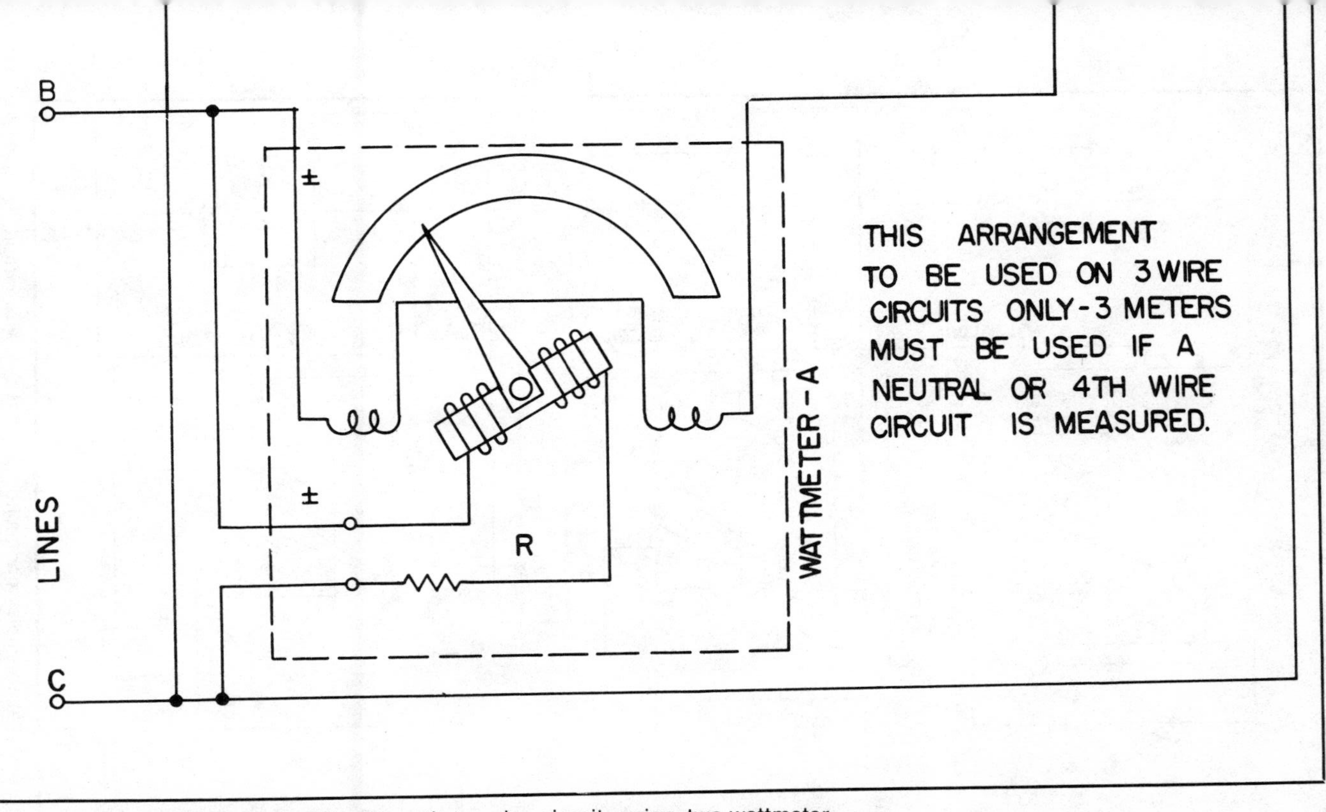

Fig. 6-39. A wattmeter connection for a three wire circuit, using two wattmeter.

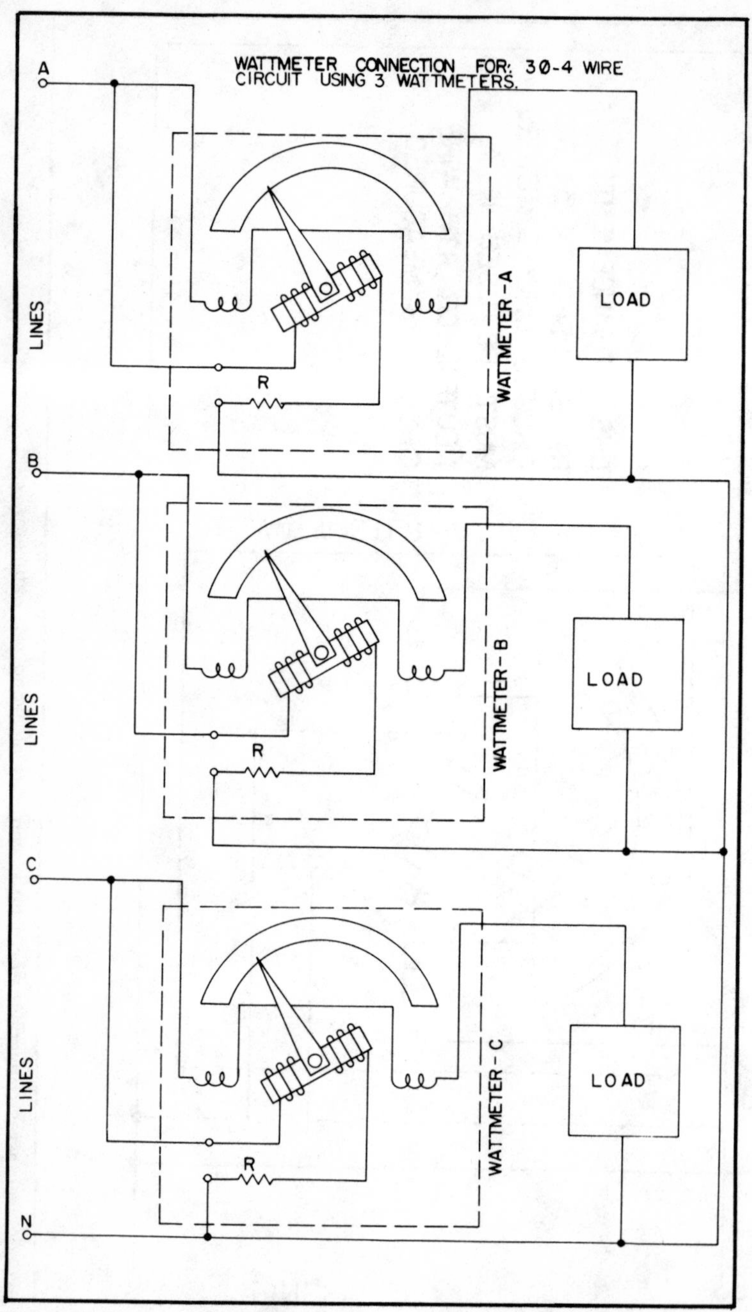

Fig. 6-40. A wattmeter connection for a 3 0-4 wire circuit using three wattmeters.

Because the actual simultaneous magnetic force so exerted is somewhat of an electrical compromise, the meter will indicate a concerted, algebraic response to the influences of both values.

The wattmeter as illustrated in Fig. 6-38 is for the application of either dc or single-phase hookups. What do we do when we wish to test a two-phase or three-phase circuit? Use two or more wattmeters. The actual manner of including two or more wattmeters to a polyphase system is clearly shown in Figs. 6-39 through 6-41. The methods of connecting the wattmeters to the line must be carried out as shown to assure correct algebraic or vectorial collection, as is required by virtue of the polyphase generation and power effects brought about by the various phase and phase wiring systems.

POWER FACTOR TEST

The power factor of any ac circuit, regardless of the number or wiring configuration, may be computed very easily by taking a wattmeter reading and a voltage and amperage reading of the circuit. The readings are then inserted into the required formulas, and the computations completed to indicate the plant's power factor. The entire plant power factor may be taken from the service entrance on the plant's side of the transformer. Or the individual motors and generators may be independently tested. The independently collected power factors may not be added as such to compute the plant's total power factor because of a number of additional influences. These two computations will then have to be considered separately.

The power factor may also be obtained directly from a power factor meter. The plant reactance is measured by a *varactance meter*. Since these meters are a part of the designed plant instrumentation, their individual functions in collecting survey data will require no explanation. Their respective values will then be read directly from the meter scale and applied to the survey.

A power factor meter is shown illustrating its internal coil arrangement and scale face in Fig. 6-42. It will be observed that unity or zero point is at center or mid point scale, with *lag* as indicative of inductive reactance. Lead is capacitive reactance falling to either side of the unity or center point. Quite frequently the power factor meter will be utilized in conjunction with a *varmeter*, which indicates the total effect of reactance directly. The varmeter will then express the percentage of inactive or non-power load that is nonetheless included as a burden to the system.

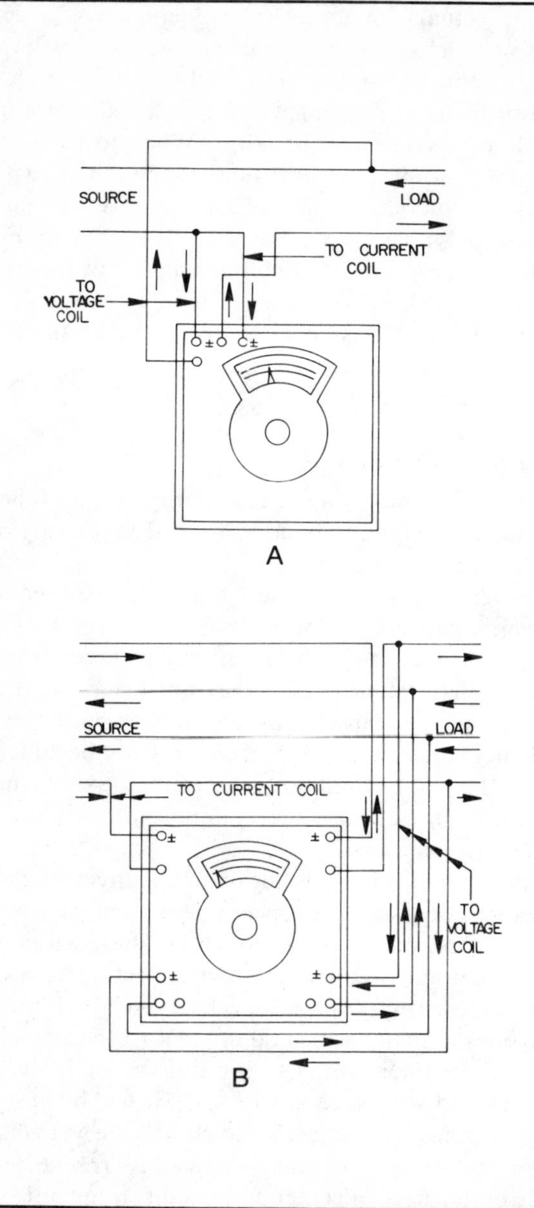

Fig. 6-41. (A) Connections for measuring watts in single-phase ac circuit. (B) Connections for measuring watts in a two-phase, four-circuit; (C) Connections for measuring watts in two-phase and three-phase, three wire ac circuits; (D) Connections for measuring watts in three-phase, four wire ac circuits.

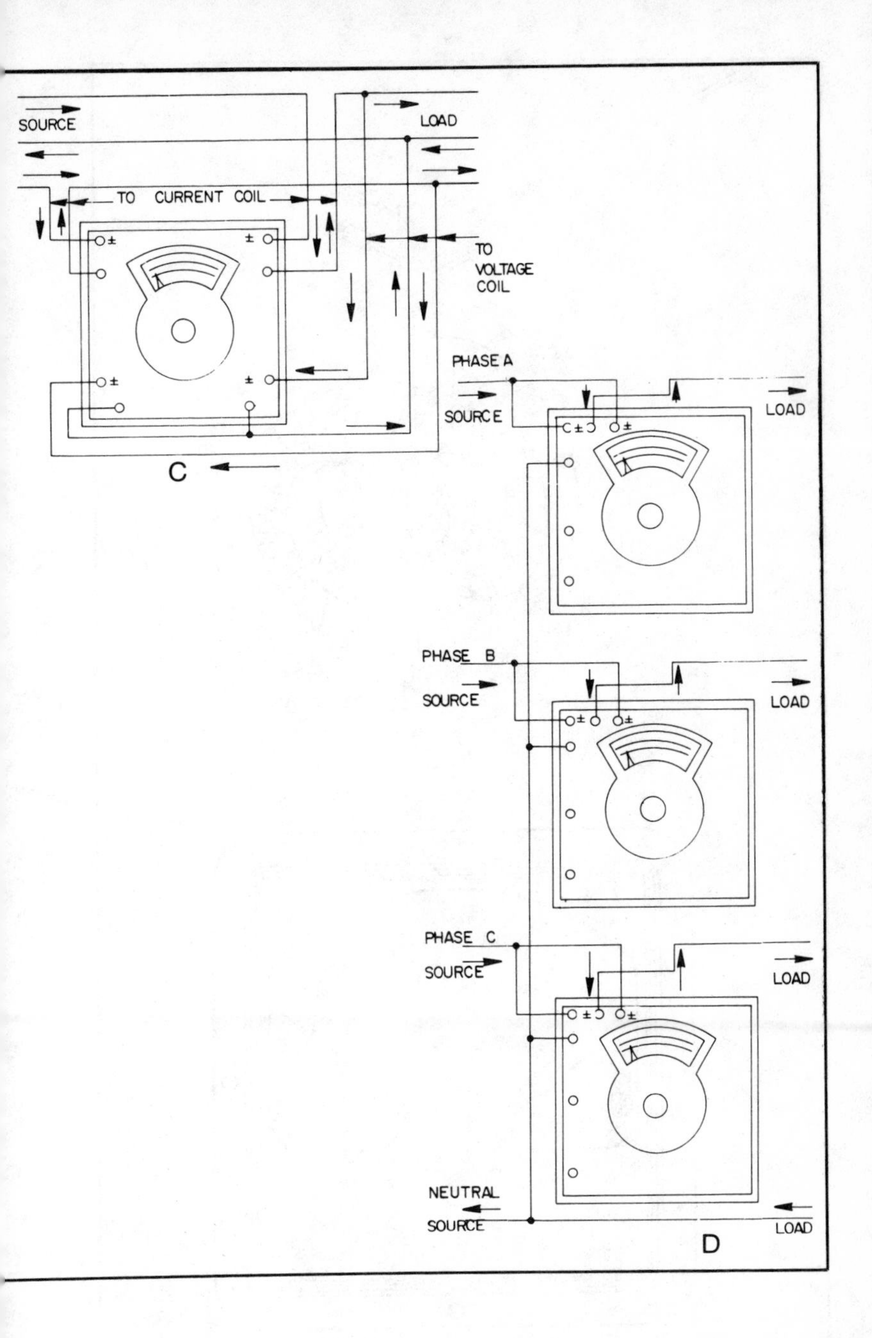
SOURCE
LOAD
TO CURRENT COIL
TO
VOLTAGE
COIL
C
PHASE A
SOURCE
LOAD
PHASE B
SOURCE
LOAD
PHASE C
SOURCE
LOAD
NEUTRAL
SOURCE
LOAD
D

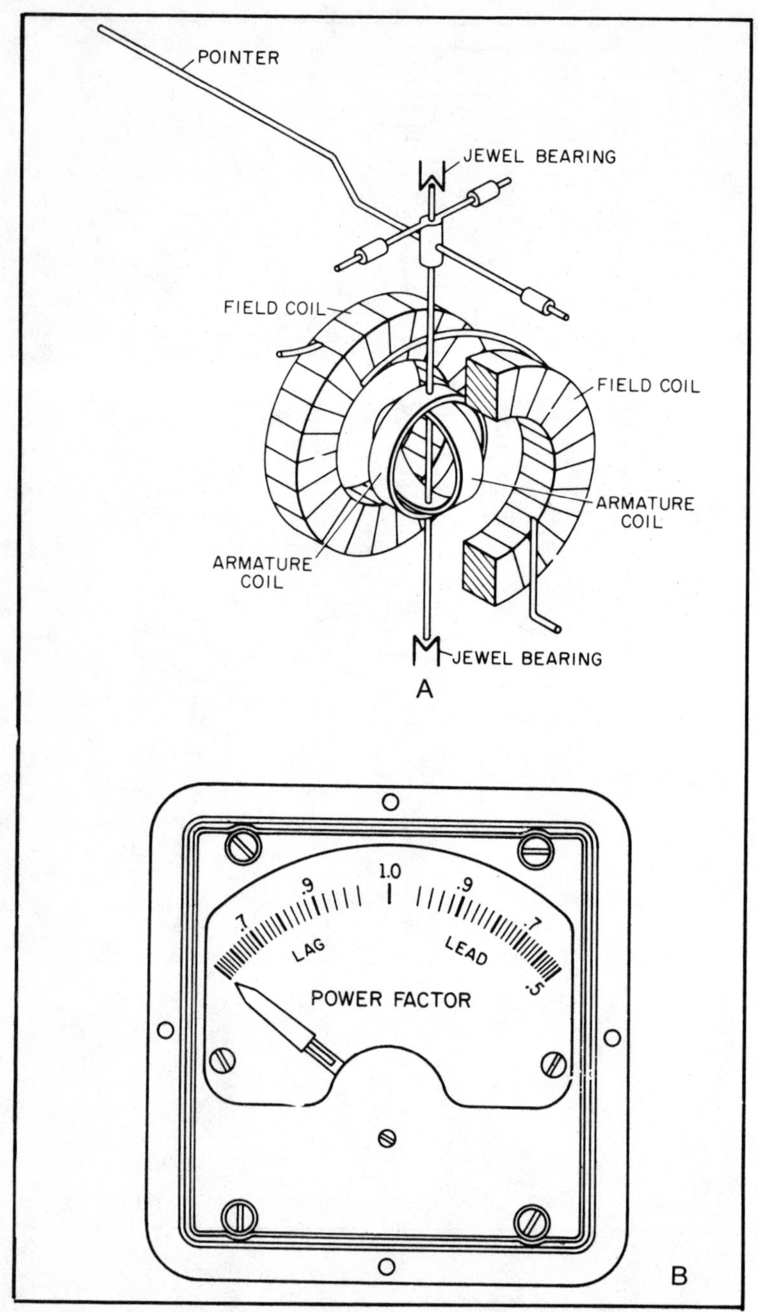

Fig. 6-42. (A) Crossed coil power factor meter. (B) Single-phase power factor meter.

POWER FACTOR DETERMINATION FOR POLYPHASE SYSTEMS

The determination of power factor (PF) for the single-phase system is accomplished by merely taking the wattmeter reading, which will express true power, and dividing this reading by the collected readings of the amperage (I), and the voltage (E). The formula is expressed as:

$$\text{power factor} = \frac{\text{true power (by wattmeter)}}{\text{apparent power (volts} \times \text{amps)}} = \text{percent}$$

The situation is identical in theory with polyphase systems. Because the polyphase system is being fed by more than one power circuit, the formulas as given must be assumed to require all of the power supplied to the system. In cases of two-phase, three wire systems, the true power as recorded by the wattmeter will naturally necessitate the power supplied by both phases. It will then be divided by the apparent power multiplied by the two phase, three wire factor which is 1.41. The formula then becomes:

$$\text{PF} = \frac{\text{watts (phase \#1)} + \text{(phase \#2)}}{\text{volts} \times \text{amperes} \times 1.41}$$

For two-phase, four wire arrangements, the formula is the same as that for the three wire system. But the phase factor two is accordingly substituted:

$$\text{two-phase, four wire PF} = \frac{\text{watts (phase \#1)} + \text{(phase \#2)}}{\text{volts} \times \text{Amperes} \times 2}$$

The formula for three-phase systems is the addition of its three power fed circuits, (Figs. 6-43 and 6-44). The three-phase factor of 1.73 is added:

$$\text{three-phase PF} = \frac{\text{watts (phase \#1)} + \text{(phase \#2)} + \text{(phase \#3)}}{\text{volts} \times \text{amperes} \times 1.73}$$

The use of three wattmeters will be required when a three-phase, four wire system is to be studied. Two wattmeters may be used in a three-phase, three wire system. The formula will only be altered to adjust the numerator portion to accommodate the addition (or alebraic collection) of the readings of the two wattmeters. The phase factor of 1.73 will remain unchanged.

POWER FACTOR DETERMINATION AT GENERATOR TERMINALS

Where the density or uniformity of the current flow as supplied to each circuit of a polyphase system may be in question, a more analytical approach may be done by considering the amperage (I) consumption of each circuit independently, as pre-

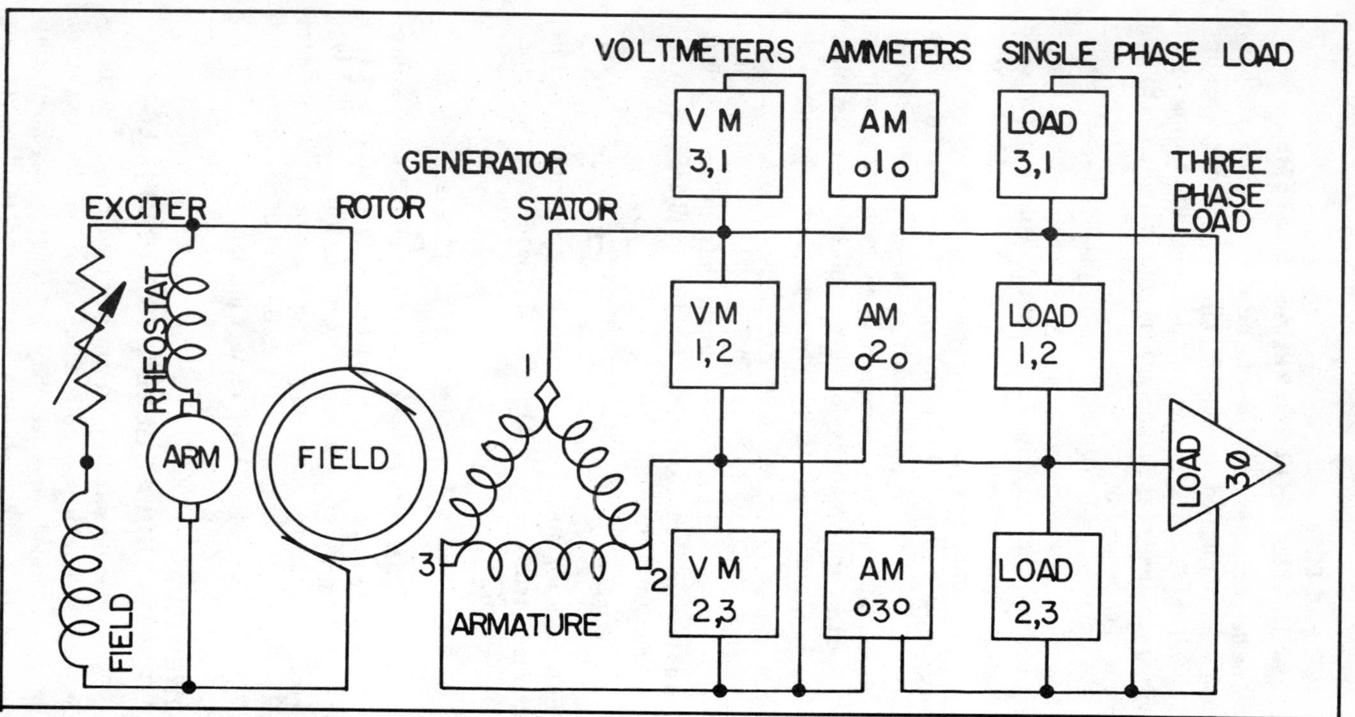

Fig. 6-43. Three wire, three-phase generator.

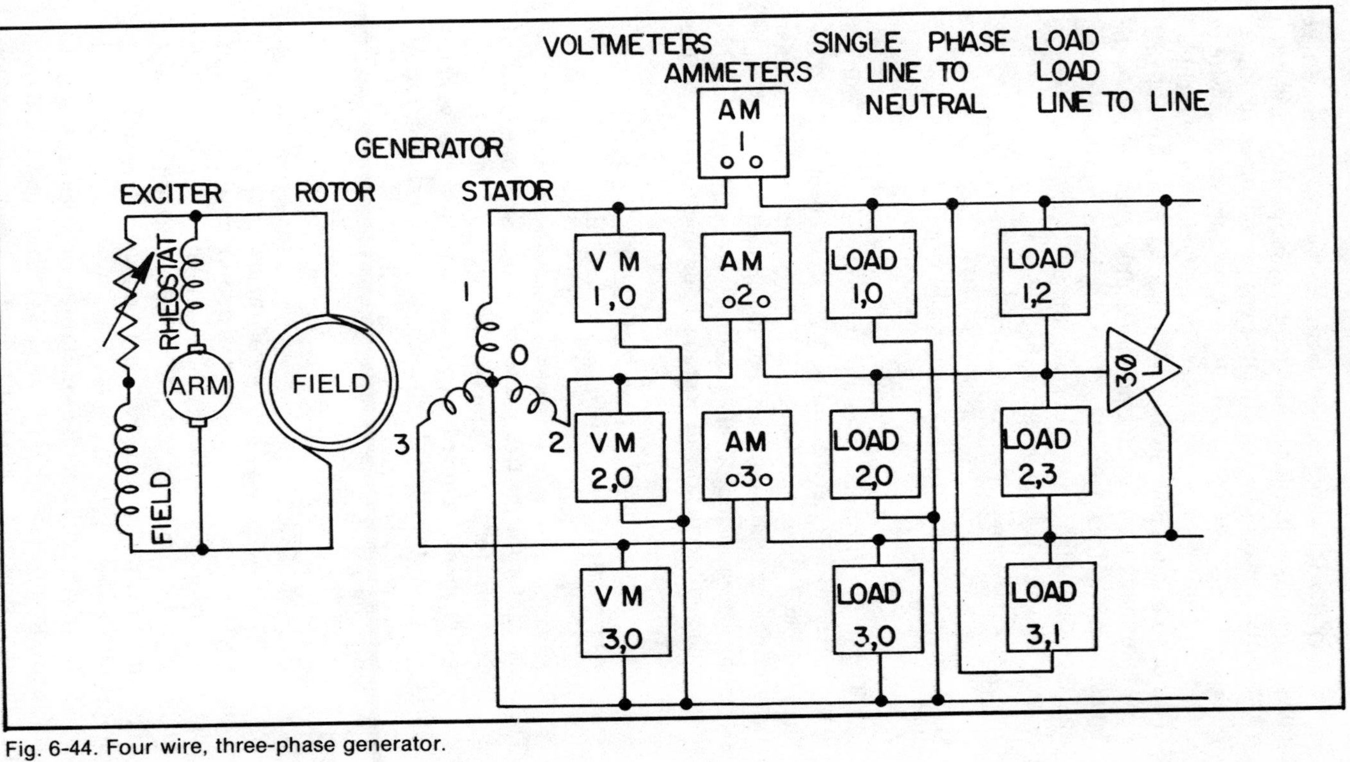

Fig. 6-44. Four wire, three-phase generator.

scribed in the following power factor formulas. The total true power consumption will be collected by appropriately connected wattmeters, and will be the effective power consumed for three-phase, three wire three-phase, four wire, two-phase, three wire or two-phase, four wire. The pertinent phase factor will be included in the denominator portion of the formula, such as 1.41, 1.73 or 2. As with all other formulas, the power factor so obtained will be the "phase angle" of the system (cos Ø). It may be used in determining the system's reactive power consumption in VARs, (volt amperes reactance) which is expressed by the formula:

$$\text{Reactive power in VARs} = \text{Sin}\,Ø \times \text{apparent power (VA)}$$

$$\text{power factor } (\cos Ø) \text{ (for } 3Ø \text{ systems)} = \frac{\text{total true power consumed by system (by wattmeter)}}{(I^1 + I^2 + I^3)\dfrac{\text{terminal volts } (*)}{\text{(phase factor)}}}$$

(*)-average sum of terminal voltage

The above formula is for the three-phase system, and the phase factor to be inserted will be 1.73. It will be noted that the three independent current values "I" are required for this formula. The above consideration may be equally applied to the two-phase system by merely including but two current values, "I^1+I^2," and the appropriate phase factor, depending upon the system's wiring characteristic.

$$\text{power factor } (\cos Ø) \text{ (for } 2Ø \text{ systems)} = \frac{\text{total true power consumed by system (by wattmeter)}}{(I^1+I^2)\dfrac{\text{terminal volts } (*)}{\text{(phase factor)}}}$$

(*)-average sum of terminal voltage

The phase factor for the two-phase, three wire system is 1.41 and the phase factor for the two-phase four wire system is 2.

For a properly balanced load demand, the terminal voltages will be all equal to within 5 percent of each other. If a serious disparity between terminals exists, then redistribution of the load will be mandatory.

POWER FORMULAS

The resolution of problems involving *true power* in respects to the generation of a sinusoidal wave output may be trigonometrically arrived at by considering the interrelationship by both the *apparent power* and the *reactive power*. This relationship is trigonometrically depicted by the triangulation technique as shown in Fig. 6-45.

True power is the actual amount of electrical power consumed to produce actual work, as to overcome resistance from conductors. It is thus converted to heat in the conductors and

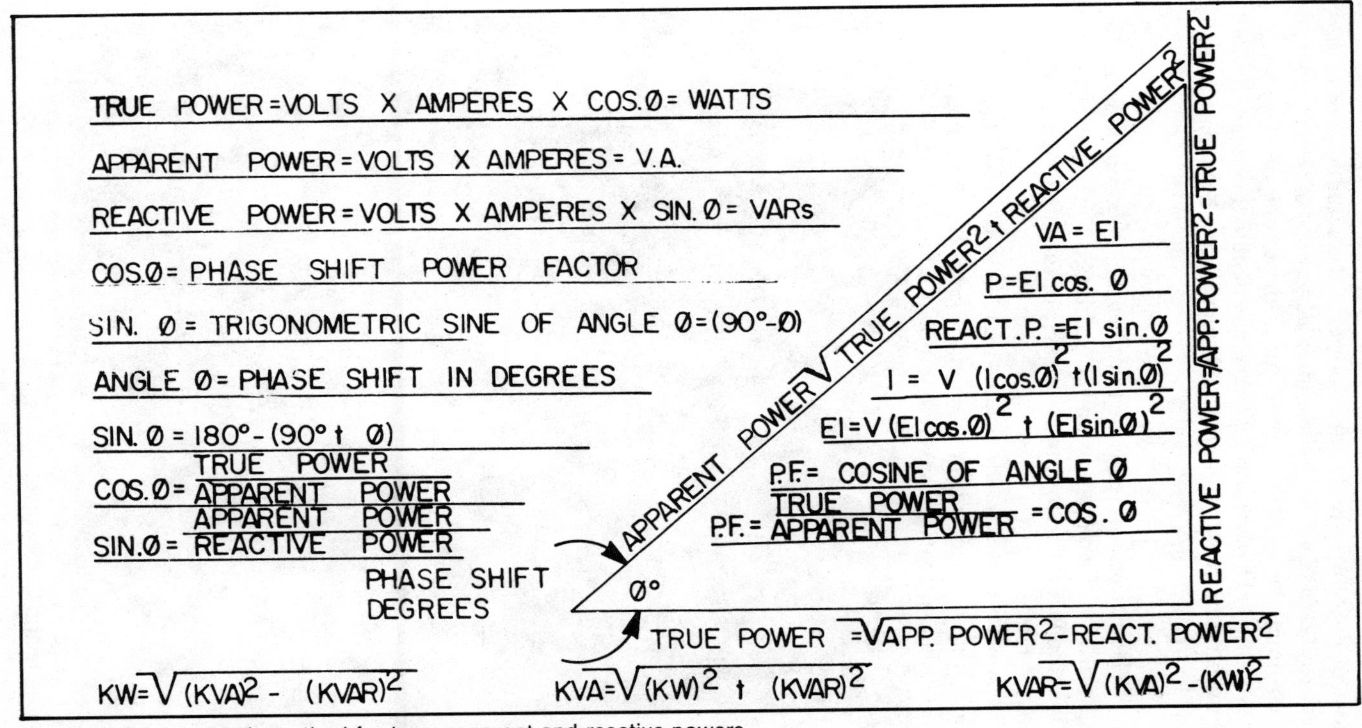

Fig. 6-45. Trigonometric method for true, apparent and reactive powers.

core of the windings. The true power is the actually present current that is in phase with the impressed voltage of the circuit, and accordingly equals the current times the cosine of the angle of phase shift. The true power is also equivalent to the square rooft of the apparent power squared, subtracted by the square of the reactive power in VARS.

$$\text{True Power} = \sqrt{(V.A.)^2(VAR)^2)}$$

The true power, referring to our earlier computations, is also equivalent to the apparent power multiplied by the power factor. This is in fact the trigonometric cosine of the angle created by the phase shift.

Apparent power is that power which is assumed, because of measured or indicated presence in the circuit, to be consumed by the circuit. Apparent power is obtained by the direct multiplica-

Fig. 6-46. A 345-kilovolt high voltage transformer (courtesy of Westinghouse Electric Corporation).

Fig. 6-47. A 525-kilovolt auto transformer (courtesy of Westinghouse Electric Corporation).

tion of the volts by the amperes flowing in the circuit. It is also equivalent to the square root of the sum of the true power square and the reactive power squared.

$$\text{apparent power} = \sqrt{\text{true power}^2 + \text{reactive power}^2}$$

Apparent power may also be obtained from the following formulas:

$$\text{apparent power} = \sqrt{(\text{I}\cos\phi)^2 + (\text{I}\sin.\phi)^2} \quad \text{apparent power} = \sqrt{(\text{EI}\cos\phi)^2 + (\text{EI}\sin.\phi)^2}$$

It will be seen that the reactive power is actually the sine of the angle of phase shift or degrees displacement. The reactive value in VARs may then be obtained by consulting a trigonometric table of sines, or by the sine scale on an appropriate slide rule (Table 6-7). The apparent power is the combined vectorial collected sums of the true power that may actually be realized as work performing power (active component) and the inactive component required for the magnetization or charging of circuit capacitors which, although is not actually consumed, is neverthe-

Table 6-7. Sine Table of Power Factor Losses Expressed as Percent Reactive Power.

Where the Sine of angle -θ= (90-cos. θ); Reactive Power= sin-θ (x) E (x) I

Sine of Phase-Shift=(90-cosθ) Degrees	Reactive Power-% Percent	Sine of Phase-Shift=(90-cosθ) Degrees	Reactive Power-% Percent	Sine of Phase-Shift=(90-cosθ) Degrees	Reactive Power-% Percent
0	0.0000	31	.5150	61	.8746
1	.0175	32	.5299	62	.8829
2	.0349	33	.5446	63	.8910
3	.0523	34	.5592	64	.8988
4	.0698	35	.5736	65	.9063
5	.0872	36	.5878	66	.9135
6	.1045	37	.6018	67	.9205
7	.1219	38	.6157	68	.9272
8	.1392	39	.6293	69	.9336
9	.1564	40	.6428	70	.9397
10	.1736	41	.6561	71	.9455
11	.1908	42	.6691	72	.9511
12	.2079	43	.6820	73	.9563
13	.2250	44	.6947	74	.9613
14	.2419	45	.7071	75	.9659
15	.2588	46	.7193	76	.9703
16	.2756	47	.7314	77	.9744
17	.2924	48	.7431	78	.9781
18	.3090	49	.7547	79	.9816
19	.3256	50	.7660	80	.9848
20	.3420	51	.7771	81	.9877
21	.3584	52	.7880	82	.9903
22	.3746	53	.7986	83	.9925
23	.3907	54	.8090	84	.9945
24	.4067	55	.8192	85	.9962
25	.4226	56	.8290	86	.9976
26	.4384	57	.8387	87	.9986
27	.4540	58	.8480	88	.9994
28	.4695	59	.8572	89	.9998
29	.4848	60	.8660	90	1.0000
30	.5000				

SINE of Phase Shift (deg. θ) = (90- cosine of angle)

Reactive Power = Sin-θ E.I. = VARs; $\text{Sin-}\theta = \frac{\text{Apparent Power}}{\textit{Reactive Power}}$; $\text{Cos-}\theta = \frac{\text{True Power}}{\textit{Apparent Power}}$

Note: When the Power Factor is known, the trigonometric Cosine value in degrees is
Reactive Power Apparent Power

NOTE: When the Power Factor is known, the trigonometric Cosine value in degrees is subtracted from 90, and the difference in degrees is the Sine-θ.

Fig. 6-48. This auto transformer weighs more than 200 tons (courtesy of Westinghouse Electric Corporation).

less supplied to the electrical devices. It must be carried in the utility companies' supply lines. transformers and included in the generated load of the supplying generators.

Reactive power is also known as the inactive component of the total load carried in the system's circuitry. It manifests itself as the out of phase power and may be expressed by the following trigonometric relationships:

$$\text{reactive power (in VARs)} = \sqrt{(\text{apparent power})^2 - (\text{true power})^2}$$

or

$$\text{reactive power (in VARs)} = \text{volt-amperes} \; (\times) \; \text{sine } O\text{; where } O = \text{phase shift}$$

The reactive power is that power carried by the circuit as required for such purposes as magnetizing the inductive devices or charg-

Fig. 6-49. A 235-ton exciter transformer (courtesy of Westinghouse Electric Corporation).

ing the capacitors. It is returned to the circuit when the ac cycle is reversed. Thus, it is only "held" by the reactance offered by the

Fig. 6-50. An 1100-kilovolt transformer (courtesy of Westinghouse Electric Corporation).

Fig. 6-51. A new addition to the Westinghouse fleet of Schnabel railroad cars. This car can be used to ship giant transformers (courtesy of Westinghouse Electric Corporation).

devices as mentioned. But the current required for this non-consumed power must nevetheless be kept in the circuit. It will be

Table 6-8. Electrical Survey Form: Part B.

PART-"B"

ELECTRICAL SURVEY
GENERATORS, TRANSFORMERS, & MOTORS

Plant Load ________ KVA or DCW

Plant "PF" ________ %

KVA Supplied ________ KVA Utilized ________ Trans. Eff.: ________ %

Transformer Efficiency (at Rated Full Load) $= \dfrac{\text{Output}}{\text{OUTPUT + LOSSES}}$ (X) 100 Losses: Core, Copper, Hysteresis, Heat

Transformer Voltage Regulation percent $\% = \dfrac{\text{(No Load Sec. Voltage)} - \text{(Full Load Sec. Voltage)}}{\text{Full Load Secondary Voltage}}$ (X) 100

Transformer Resistance expressed-% KW $= \dfrac{\text{Copper Loss In KW's}}{\text{Rated Output-KW's}}$ TRANS.-"I^2R" LOSS ________ %

Transformer Impedance $= \sqrt{(\text{percent resistance})^2 + (\text{percent reactance})^2}$ = "Z" ________ %

Electric Generator Efficiency/Mech. HP $= \dfrac{\text{KVA (or D.C.KW's) xPFx(Phase Factor)}}{\text{Prime Mover HP x .746 (as applied)}}$ (X) 100=Eff. ________ %

Prime Mover HP/KW Conversion Index $= \dfrac{\text{KVA (or KW-DC)xPFx(Phase Factor)}}{\text{Eff. of Generator x .746}}$ = H.P./KW ________ -Index

Individual Motor Efficiency $= \dfrac{\text{Horsepower at shaft (x) .746}}{\text{Volts(x) Amps(x) PF(x) (Phase Factor)}}$ = Motor Eff. ________ %
To be collected for each Motor used-Non-Additive

Note - Phase Factors: D.C. = 1.0; Single 0=1.0; 20–3 wire=1.41; 20-4wire=2.0; 30= 1.73

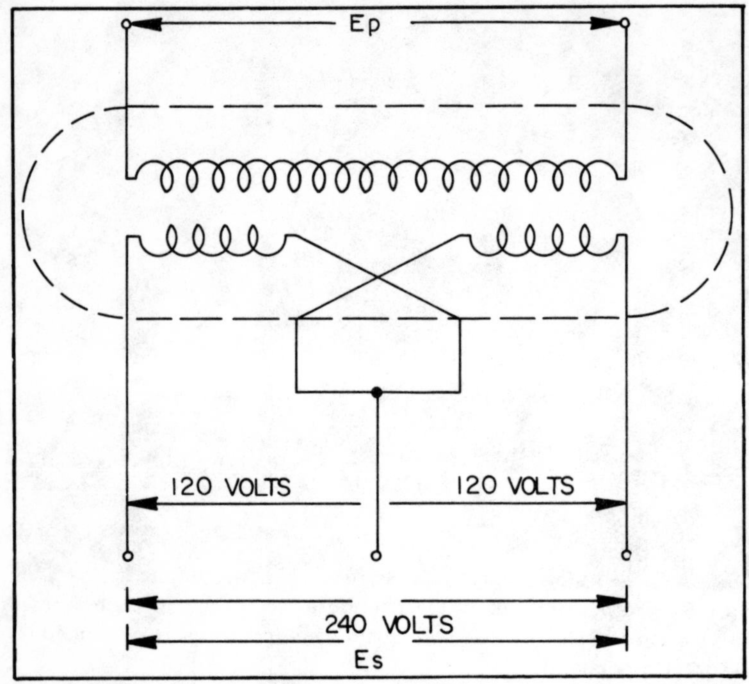

Fig. 6-52. A single-phase transformer with two secondaries describing basic transformer voltage, current and power ratios.

recalled that although the current as necesitated by this reactive current is not in itself utilized, its loss is felt by the waste it offers to the current carrying conductors. The waste is felt as the square of the extra current.

Reactive power may be measured directly by a varmeter, which electrically is similar to a wattmeter but has its potential coil arranged 90 degrees out of phase with the actual voltage. The phase angle measured will then be 90-ϕ, so that the meter will then measure EI (×) cos (90-ϕ).

PLANT TRANSFORMERS

The transformers of interest to the plant engineer will be the distribution transformer and the individual power transformers at the particular demand points as located throughout the plant. See Figs. 6-46 through 6-51 There are a number of variations that the transformer may assume, in respect to its manner of winding or technique in geometrically utilizing the windings. The transformer may be of the single polyphase type, where all of the

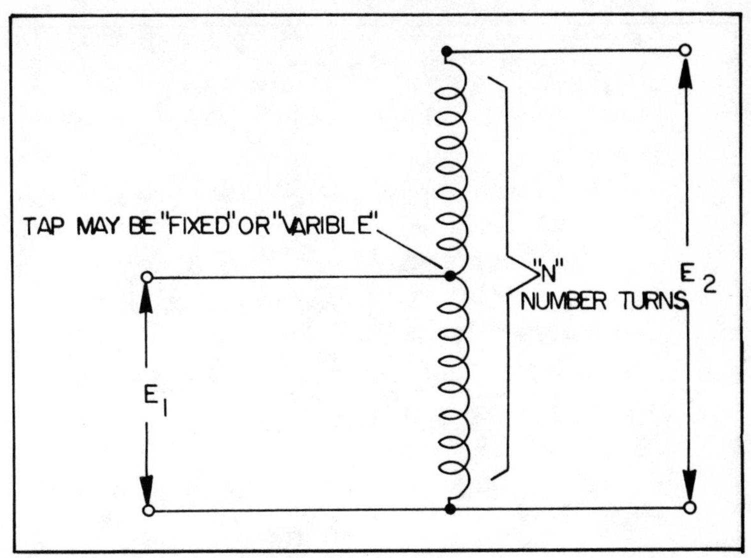

Fig. 6-53. Schematic of an auto transformer.

individual phase windings are mechanically wound about one common core. Or the transformer may exist as a battery of individually, mechanically separated transformers, connected electrically to render the same effect as one large unit. The use of separate or bank transformers is more expensive initially. But a bank transformer offers the advantage that if replacement becomes necessary to just one coil, or one phase, then the entire transformer does not require overhaul. The one damaged core and phase may be removed from the line, another added, and the power restored.

The varieties of phase windings are chiefly the delta, wye (or star), the open delta and "scott" (or T) configurations. These variations may further be modified or rearranged in the transformer circuitry to provide the following arrangements and their respective electrical advantages: delta-delta, wye-wye, delta-wye; and wye-delta; the open delta arrangement is really an emergency remedy. If one phase of a three-phase transformer were to burn out, the one phase of the transformer is disconnected, the original primary is fed as before, and the secondary load is drawn off at only 58 percent of the original capacity. The scott or "T" technique is merely a "tap" configuration whereby the primary of a three-phase transformer is configuratively changed to permit the use of two-phase secondary supply. There are, of course, additional transformer variations as designed for special use purposes. Be-

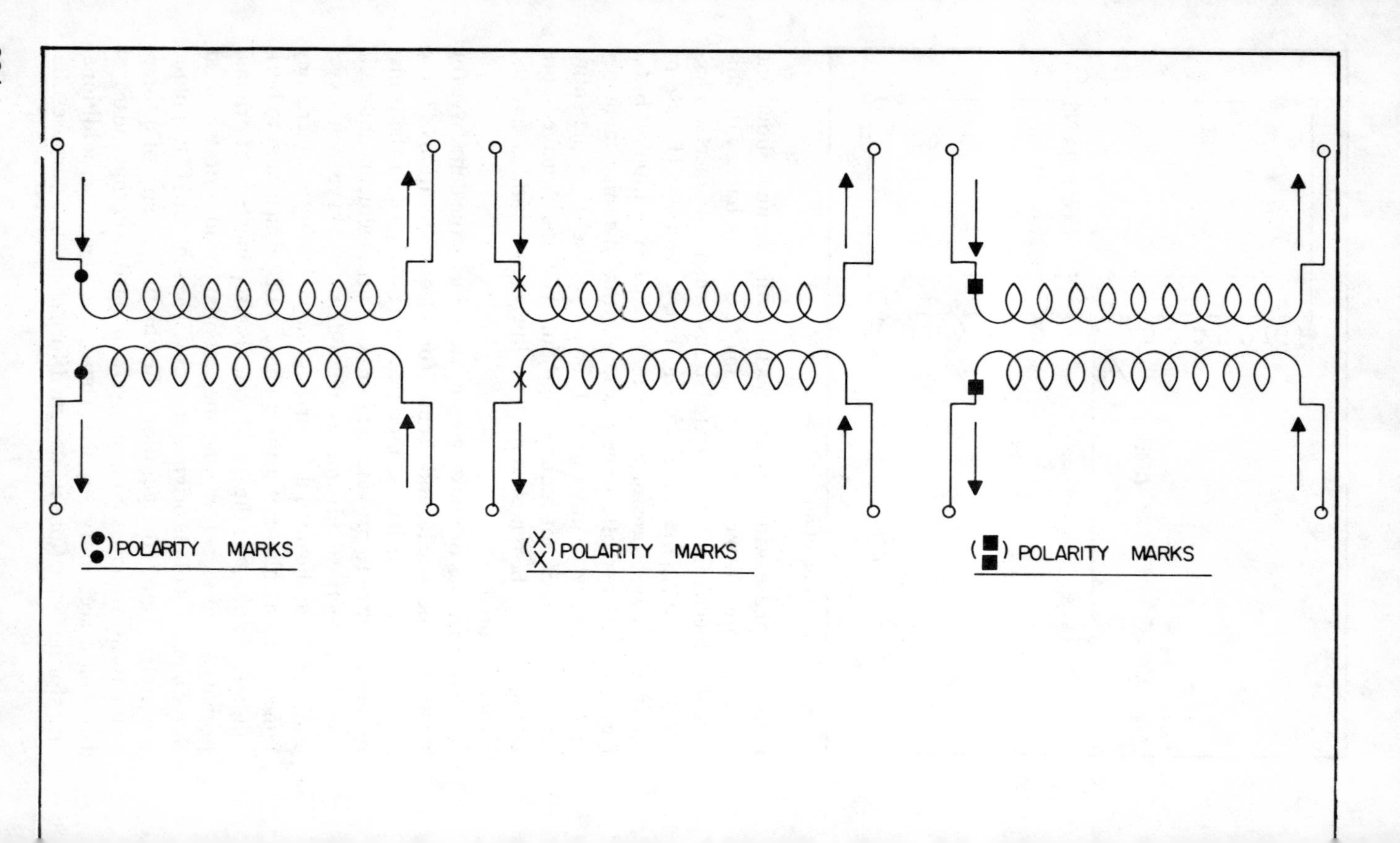
POLARITY MARKS
(X X) POLARITY MARKS
POLARITY MARKS

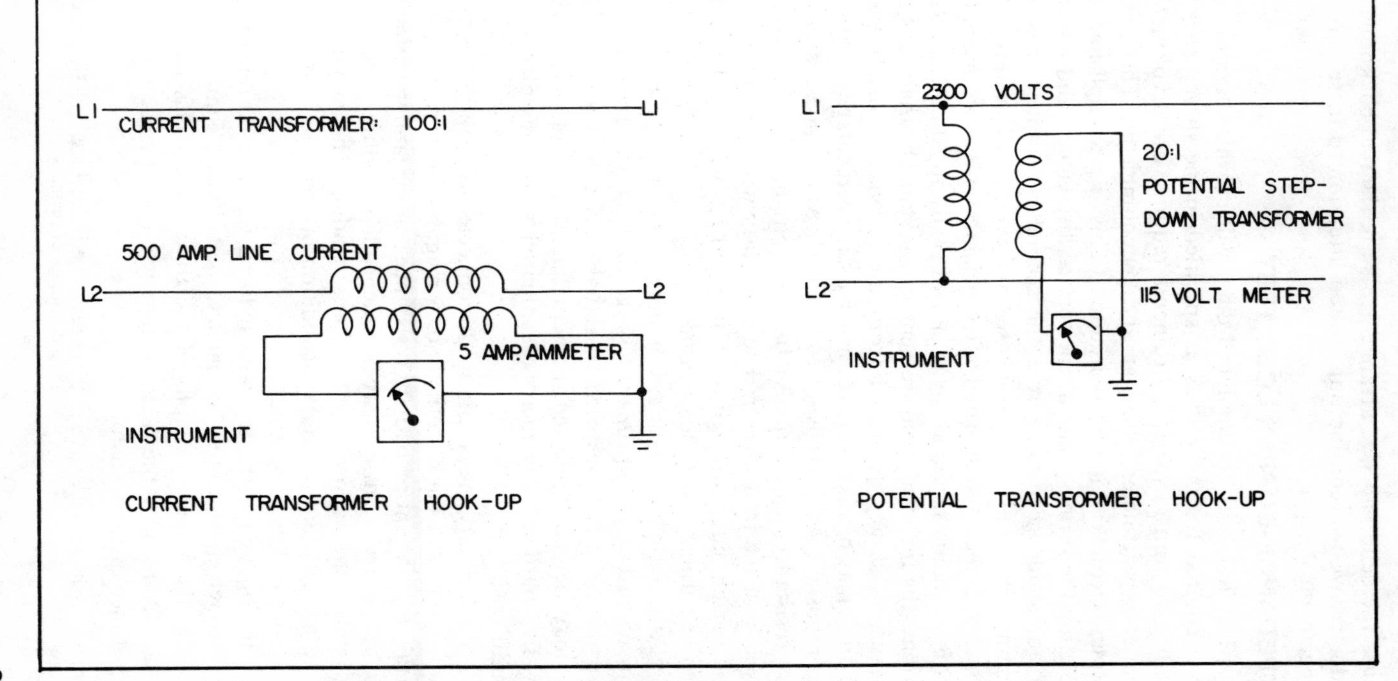

Fig. 6-54. Instrument transformer hookups with polarity identification.

cause the delta and wye arrangements are the real backbone of industry, we will confine our limited attention to these two arrangements.

IMPEDANCE-COPPER LOSS TEST

This test, with its combined data, will provide the information required for those tests as specified in the survey part B, relating to transformers. Instruments required are a voltmeter, ammeter, wattmeter, rheostat and megohmmeter. The transformer secondary is short-circuited. This is accomplished by isolating the primary side at the service disconnects and placing an adjustable rheostat across the secondary output, or three separate rheostats in three-phase unit type transformers. The primary is then reconnected after placing no-load equivalence upon secondary terminals by means of rheostat. At no-load conditions, the included measurements are taken. The rheostat's resistance is slowly decreased from the secondary. Incremental readings may be taken and recorded at each percentage of equivalent rheostat load point as selected. The rheostat is now placed on zero resistance, which equals a full short circuit upon the secondary side of the transformer. The secondary will now have a full load current circulating through it. Readings will now be taken for the purpose of full load values.

At any time of testing, it is only necessary to admit to the primary just enough primary voltage to equal rated values. The transformer is then restored to actual service connections. The test may then be duplicated and actual plant loads of reactive and resistive burdens may then be evaluated at various plant load conditions.

The primary may again be disconnected, and the entire secondary replaced back on the line. A megohmmeter reading may be taken of the entire load, for resistance reading, and the primary and secondary windings may be independently "ohmed" to make an initial winding-insulation resistance reading. This may periodically be repeated to ascertain condition of breakdown insulation resistance by "meg" readings.

In conducting the test with induced full load currents passing through the "shorted" secondary, the voltmeter connected will indicate the impedance volts of the transformer. This indicated voltage divided by the rated voltage will give the percentage of impedance of the transformer:

$$\text{percent impedance} = \frac{\text{voltage actually required to affect full load current-sec.}}{\text{rated voltage of primary}}$$

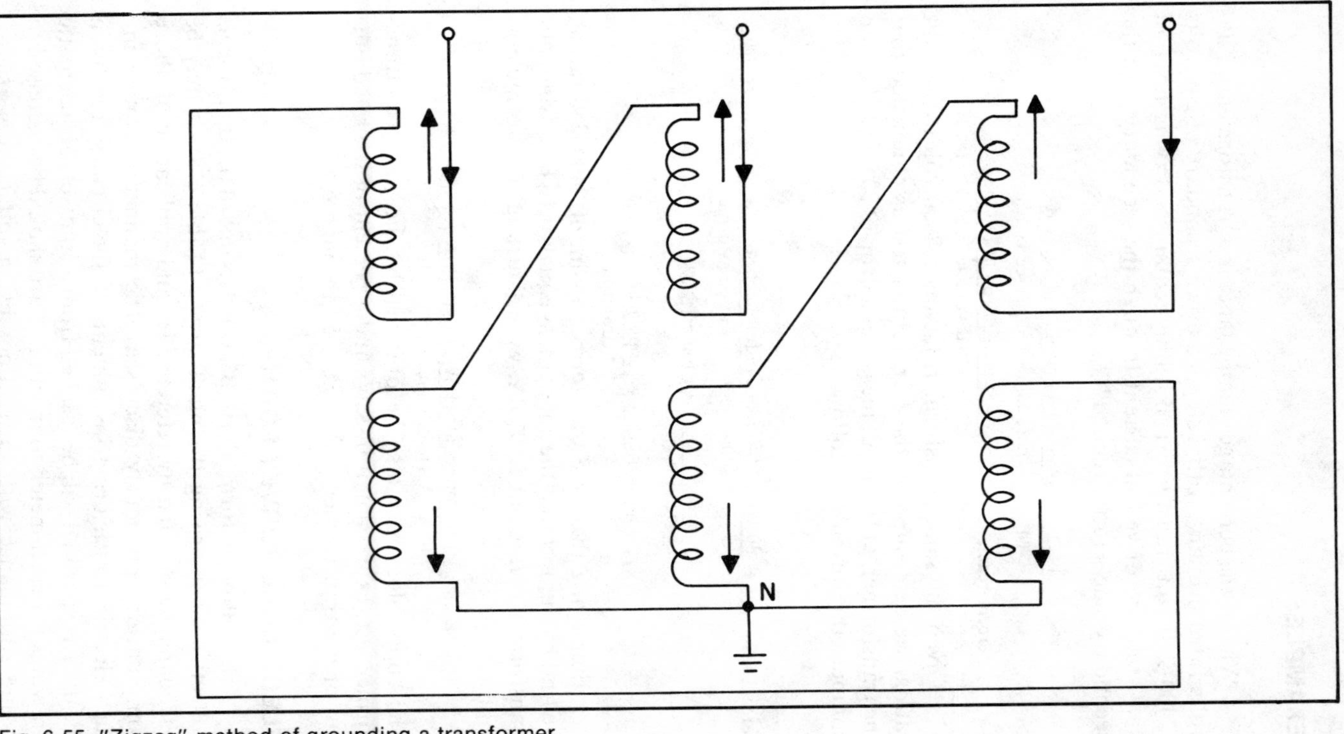

Fig. 6-55. "Zigzag" method of grounding a transformer.

EXAMPLES

What is the percentage of impedance in a transformer of 2400 volts rated primary, with a stepped down secondary of 240 volts (2400/240 volt transformer) if it only requires a driving EMF of 48.72 volts to drive full current through the secondary with the secondary side short circuited?

$$\frac{\text{48.72 volts (required for "short" full sec. current)}}{\text{2400 volts as rated for operational sec. current}}$$

$$\text{percent reactance} = \frac{\text{reactive voltage drop at full load}}{\text{rated voltage}} (\times)\ 100$$

Given the same transformer, having a 2400 volt to 240 volt rating, we determine by the transformer test as conducted that the transformer, at full load, measures a drop of 30 volts. We then compute the formula as follows:

$$\text{percent reactance} = \frac{\text{30 volts (difference in full load-no load voltages)}}{\text{2400 rated volts of transformer}} (\times)\ 100 =$$

$$\frac{\text{30 Volts}}{\text{2400 Volts}} (\times)\ 100 = 1.25\%$$

$$\text{percent resistance} = \frac{\text{copper loss}}{\text{rated output}} (\times)\ 100$$

Employing the same transformer again, of 2400/240 rating, we find by our test that the copper loss equals 0.16 KW. Since our transformer is rated at 10 KVA, we will arrange our formula as follows:

$$\%\ \text{resistance} = \frac{\text{0.16 copper loss-KW}}{\text{10 KVA Trans. rating}} (\times)\ 100 = \frac{0.16}{10} (\times)\ 100 = 1.6\%$$

The copper loss may also be obtained by the "shorted" transformer test, expressing it in KWs and a percent of the rated output of the transformer. See Figs. 6-52 through 6-55.

ELECTRICAL SURVEY FORM

The following form, Part-"B", which is one of the three forms collectively comprising the plant survey (Table 6-8). It may be adequately completed by reference to, and application of, those computations previously discussed. The manner of connecting such electrical measuring instruments to the circuitry to obtain the data as specified will be recalled from a study of the schematic drawings for each meter connection. Care should be exercised to assure compliance with instructions by manufacturers' recommendations to avoid danger to the technician, the circuitry under tests and the instruments employed. It is again urged that the tecnician reacquaint himself with all vital instructions as provided with all electrical instrumentation before conducting tests upon live circuitry.

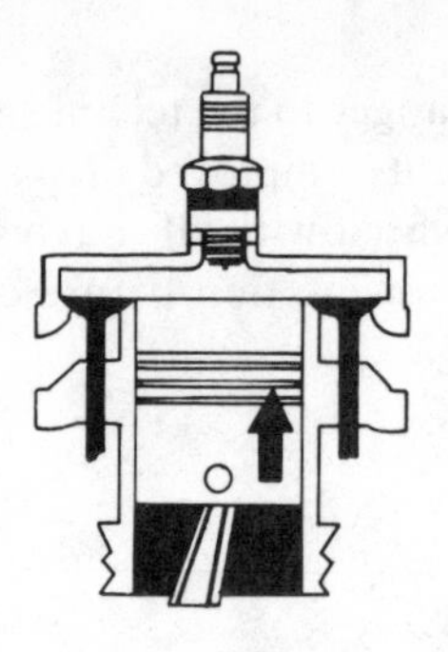

Chapter 7
Prime Movers

Throughout history, the genius of man has been illustrated by an endless and diverse agglomeration of means and utilities for harnessing the kinetic and latent forces of natural power sources. Many such contrivances were short lived, such as the Newcommen and watt condensing steam engine and the tesla turbine. Others, through modification and scientific adaptations have survived the colorful chronology of man's aspirations for mechanical supremacy and mastery over his environmental elements.

PERPETUAL MOTION MACHINE

In the early 1800s the French Academy of Science had, with resignation, closed its academic doors to any further investigations and pursuits to the development of perpetual motion, and the devices which were facetiously attributed to function by means of perpetual motion power. Man's futile attempts to utilize perpetual motion was for centuries only surpassed by his blinded and vain efforts in alchemy, the science primarily founded for the purpose of changing the base metals of lead and iron into gold. Mechanical perfection in the form of a perfect 100 percent efficient machine was unknown to nature and therefore unattainable to mankind.

Although a perpetual motion machine of 100 percent efficiency is certainly impossible there are many machines and devices that can function with near total efficiency, such as the electric transformer which operates at above 95 percent efficiency. The electric motor which is about 80 to 90 percent efficient on the average, and the newly perfected magnetohydrodynamic generator (MHD) has reached the practical apex of efficiency in its exclusive province as an electric converter. All of those devices mentioned derive their significantly high/efficiencies because of the direct and uncomplicated "one step" system involving the input and output of supplied energy.

COMPUTING EFFICIENCY

In contrast to the simplicity and virtual advantages gained by such uncomplicated devices of electrical nature, we have the comparatively complex and interdependent parts of the fossil fuel generating system. To say the least, that system is little more efficient then the windmill technique of grinding grist. The actual thermal efficiency or gross ability for even the best power plant's output-input ratio is staggeringly low, the average being about 18 to 25 percent. This figure is the "gross" product of all the combined efficiencies starting from the combustion chamber of the boiler and ending at the transmission transformer. The figure is arrived at by actually applying the Btus in the fuel delivered by the KVA power available at the secondary bushings.

Because of the combined effect of corporate efficiencies, we will exact a more equitable and realistic figure of efficiency by evaluating each integral part of the power plant. We will not collect a total or "raw to finished" index of ability in respect to the power plant as an energy converting entity. If we were to stand back and view the plant as a homogeneous piece of machinery, I feel it would not only be an erroneous perspective but one which would perhaps be penalizing specific high efficiency devices through the incorporated association with devices of less efficient nature.

Thus, we will not then be penalizing an efficient package boiler of 85 percent efficiency when connected to an antiquated slide valve engine, where the engine is perhaps but 16 percent efficient. On the other side of the coin, we could not serve justice by considering the combined efficiency of a boiler-engine unit, when we pipe a 68-year-old boiler to a new turbine. This collective method of computations will leave us in no intelligent position to effectively administer corrective measures, which is exactly why this manual and its related three-part systematized procedure is arranged as it is—to conduct specific thermal, electrical and mechanical data and apply that data to those specific and respective parts of the plant where it will be most effective.

We will be evaluating the *prime mover* which, depending on the particular plant, may be a diesel bank, an entire turbine deck, a gas or hydraulic turbine or even a water wheel. More important than the specific type of prime mover is its efficiency. When we consider the prime mover, we will limit our scope to the machine itself, from throttle to discharge port. We will then consider only the energy of the fluid or medium entering the point of admission to the prime mover. We terminate our investigation at the point where it leaves the machine, collecting only the work performed or energy expended between the points of admission.

This suggestive formula for limited energy computation is very simple for a single turbine. We simply subtract the heat energy (Q) leaving the turbine from that heat energy in the steam it had upon entering. The difference is the heat absorbed or expended in spinning the rotor through the period of time that the *dynamometer* or *prony brake* was attached to the shaft. The same basic formula could also in theory be applied to any prime mover, with very elementary revision. In the final analysis the efficiency of any prime mover or energy converting device may be literally summed up to expressed the following. Efficiency equals energy in mechanical power produced by the machine divided by the energy applied to the machine (in equal units of measurements). Expressing this algebraically we have:

$$\text{efficiency} = \frac{\text{energy output}}{\text{power input}} = \%$$

We must of course expand on this formula in applying it to the particular demand characteristics of the individual prime mover. As presented, the basic formula remains—input divided by output equals efficiency.

POWER

Power is defined by many authorities as the expenditure of energy, in both quantitative and qualitative dimensions. Translating this to more practical terms, we may redefine power as the production of a work converting force or effort, maintained for a controllable period of time. The most used application of power is in reference to *horsepower* or mechanical power. In the late 1700s much confusion was prevalent regarding the actual power producing ability of the newly developed steam engines or condensing engines. Many mine owners and textile manufacturers wishing to consider the adoption of such reliable mechanical power needed information in respect to the replacement value of a steam engine, as compared to the power requirements of their particular operation. There was really no accepted or established conversion constant to equate the rules of force distance moved and time in respect to units of deliverable power.

It was left to James Watt, a noted engineer and practical scientist, to provide the required answer to the question of power and its measurement. Since James Watt was in the business of building steam engines and experimenting on everything else that moved or could be made to turn on a bearing, he then gave the world the most well known and universally adopted rendition of the term power—horsepower. Since horses were the main source of mechanical effort in Watt's day, it was only natural for Watt to express the equivalent power in terms of a horse's worth.

After such deliberation, Watt made an arrangement of pullies and rigging to accommodate the typical horse to be harnessed. Since the average speed of a typical draft horse was 2.5 miles an hour, or 220 feet per minute, Watt computed this data and arrived at the figure of 22000 foot pounds per minute, adding another 50 percent for lost motion and rigging losses. He then presented to the technological world the horsepower equivalent, the work performed by the average draft horse, at 33,000 ft. lbs per minute.

AVOIDABLE VERSUS UNAVOIDABLE LOSSES IN PRIME MOVERS

All machines are susceptible to three primary areas of efficiency areas of efficiency loss. These include design limitations *internal* or *frictional losses* and *contributory losses.*

Design Limitations

If we had a machine that could function with no loss or inherent resistance, we would have a perpetual motion machine. Once started, it would continue to run until enough counter force equaling the initial force was applied to brake the movement. The foremost limiting factor is the machine's inability to convert or utilize all of the energy input as imparted to the machine and return this as mechanical output. This inherent limitation we shall ascribe to design limitations. Considering a machine which was to literally "float" in a cradle of lubricant of no resistance, then all of the loss or difference between power applied and power delivered would express this design loss.

To exemplify this situation, let us consider a few representative prime movers. In a hydraulic turbine, we find by formula that the theoretical horsepower of a hydraulic turbine equals the actual water delivered in dimensions of force (weight or head of water), the distance moved (volume per minute) and the time (in minutes). This combined power is then divided by 33,000. Regardless of how much water will be flowing we cannot expect to convert it all to mechanical power, even with as simple a machine as a turbine or *pelton* rotor. The rotor is subject to what is called *slip*. If a theoretical amount of water was to flow through and create 100 mechanical horsepower, and by a dynamometer or prony brake we find only 95 horsepower at the shaft, then this particularly designed rotor has a design limitation of 5 percent and a mechanical efficiency of 95 percent. Obviously, there is nothing the engineer can do to increase the output above 95 percent of the input. If we were to find later that our output for identical input data has decreased to 88 percent, then we have 95 percent - 88 percent equals 7 percent additional loss to investigate. Of this 7 percnt additional loss, a certain unavoidable portion will represent internal friction. The balance of internal friction is then appropriately deemed avoidable and thus the engineer's responsibility to correct or show why it cannot be eliminated.

In evaluating the *reciprocating* type prime mover, which chiefly is comprised by the diesel, The otto cycle engine, the stirling, and the steam engine, our index of design limitations will be proportionately larger. This is due primarily to the greater weight and mass of the moving parts, admission and discharge port resistances, time allotted for expansion of gases before they must be discharged from the combustion cylinder, the mechanical resistance caused by the numerous interrelated functional parts, and the complexities caused by the multi-directional distribution of internal forces created within the prime mover during normal power generation. The design limitation factor peculiar to reciprocating engines is best expressed by the percentage given when the indicated horsepower given by the prime mover is divided by the actual available thermal energy provided for its utility. By formula:

$$\text{design limitation} = \frac{\text{indicated horse power (IHP)}}{\text{available energy in equiv. power}}$$

$$= \text{percent loss}$$

$$\text{design limitation} = \frac{\text{500 indicated horse power}}{\text{Btus delivered } (\times)\ 0.0236} = \frac{\text{500 IHP}}{\text{1,500 equiv. units}}$$

$$.333 \text{ percent design loss}$$

Internal or Frictional Losses

These losses are necessary evils that escape no machine or mechanical device, regardless of how superior the lubrication system or how few the moving parts. Internal friction will be an inherent factor in prime mover losses. The percent efficiency or amount of actual frictional horsepower of the prime mover can be near accurately determined in two principal ways. First, the indicated horse power will be subtracted by the actual brake or dynamometer horsepower. This will provide the horsepower required by the prime mover itself in overcoming its own resistance. The percent loss may be found by dividing the brake horsepower by the indicated horsepower.

$$\text{frictional hp} = \frac{\text{brake hp developed}}{\text{indicated horsepower}} = \text{percent loss}$$

The frictional or internal horsepower may be directly determined by connecting a driving motor to the output shaft of the prime mover. The prime mover's lubrication system is engaged to normal operating specifications. The driving motor is brought up to the prime mover's rated shaft speed to provide a very close simulated required horsepower. The horse power of the driving machine is then recorded, or its equivalent power is converted to mechanical power. The resulting power consumed will then be a near approximation to the prime mover's frictionally required horsepower. It may be expressed as an index or as a percentage of efficiency loss. Divide the indicated horsepower by the frictional horsepower required by the driving motor to attain its rated shaft or flywheel speed. There are, of course, other factors to be entertained, such as the increase in internal resistance upon thrust and main bearings with the increase in load and other factors which will occur only under actual load conditions. This test and the data resulting from it will quite adequately serve our purposes.

This method of obtaining frictional horsepower may be applied to essentially any prime mover. Every prime mover has to have some form of output shaft. With a little creative ingenuity, an appropriate drive motor testing arrangement may be had. The favored driving motor is the dc series wound motor with a fully adjustable rheostat and recently calibrated ammeter to indicate precise electrical power requirements:

$$\underset{\text{(frictional hp)}}{\text{driving horsepower}} = \frac{\text{watts power consumed at prime mover Rpm}}{\text{driving motor efficiency } (\times)\ 746}$$

Contributory Losses

These losses contribute to output deficiencies relating to improper operating procedures, maladjustment of critical parts, uncorrected wear and misalignment of bearing and journal surfaces, and about a million other contributing factors. When added up, these factors account for a large chunk of needlessly expended lost motion, heat producing friction and clearance seizing effects—all of which are avoidable. These losses may be substantially reduced by maintaining all adjustments, such as valve clearances and linkage distances to factory specifications. Provide the proper grade and weight of lubricating oil. Change the oil and flush all debris from the engine's oil delivery network. Exercise intelligent adherence to all manufacturers' recommendations and have a periodic preventive maintenance procedure established by a qualified mechanic or technician.

SUMMARY OF EFFICIENCY LOSSES

The three areas of representable losses are expressible as a proportion of efficiency lost. Each loss contributes its individual and distinguishable percentage of that deficiency peculiar to each and every type of prime mover. Two of the three losses, namely design limitation loss and frictional internal resistance loss, are basically fixed constants. All factors being under ideal requirements, they will remain an inherent part of the prime mover and must be accepted as unavoidable loss. All losses above and in excess of these two losses must then be avoidable losses and should be investigated and corrected by whatever remedies needed.

The first avenue of correction for avoidable or contributory losses should be a through understanding of the prime mover per the manufacturer's manual. Have a factory-trained service and operating technician in for an in-plant evaluation and instruction session, with all engineers and mechanics present. The nominal fee charged by the manufacturer for such service will more than be offset by the savings in otherwise lost fuel economy.

Many leading companies such as Fairbanks Morse, Westinghouse, General Motors, Caterpillar, General Electric and the Bosch Company, just to mention a few, offer regularly scheduled service and operation classes—at little or no charge to the companies who wish to send their personnel there for valuable instruction and updates on equipment innovations. It would certainly be worthwhile to contact the manufacturer of your main equipment and inquire about their training programs available.

THERMAL HORSEPOWER

The preceding discussions of power has treated the term horsepower as a purely mechanical aspect. This is generally, true, as horsepower is in fact universally measured in manifestations of so much work performed by a quantity of force moving through a given distance in a defined period of time. This is the most direct technique of metering or testing the machine or device for its equivalent work producing ability.

The mechanical efficiency of the prime mover engine, or machine, is the actual measured horsepower as delivered at the shaft or flywheel divided by the engine's indicated horsepower.

$$\text{mechanical efficiency (percent)} = \frac{\text{actual shaft or flywheel horsepower (*)}}{\text{indicated horse power}}$$

(*)as tested by the prony brake or dynamometer method

Usually this mechanical horsepower will be nearly or about 90 percent. The only deficiencies are the engine's internal or frictional resistance and the normally less than optimum maladjustments and imperfections in operational techniques.

All prime movers of the thermal classifications, either of the internal or external combustion varieties, may be computed to determine the thermal efficiency. Merely ascertain by easily obtained records and instrumentation the actual amount of fuel or combustible delivered to the combustion chamber. Thus, if we were to deliver a given unit of combustible to our engine to convert the combustible to an equivalent amount of mechanical energy, we would then need to consider the following related essentials. One British thermal unit (Btu) equals the mechanical expenditure of 778 foot pounds of work performed, and vice versa. One mechanical horsepower, as expounded by Watt, equals the expenditure of 33,000 foot pounds of physical work performed per minute, or 550 ft./lbs. second.

That 778 ft./lbs. is exactly equal to and replaceable (theoretically) by 1 Btu of released heat energy. To translate this to horsepower, we must include the additional time element, be it either the second, minute or hour. Our second consideration will be represented by how much we are doing in how long a period of time--such as producing 33,000 ft./lbs. of work in a minute, or 550 ft./lbs. of work each second, both equaling 1 horsepower. By applying both preceding points to our observations, we may rightfully deduce that we may evaluate an actual mechanical production of power, in this case the equivalent horsepower, by applying the fuel or combustible so consumed to effect the developed horsepower.

We agreed that 778 ft./lbs. of work equals 1 Btu. We further learned that if we were to continuously effect the output of 33,000 such foot pounds of work for the duration of one minute, or 550 ft./lbs. of work for one second, then we would be producing one horsepower of mechanical power. By dividing the horsepower equivalence, 33,000 by the Btu/ft. lb. equivalence of 778, we will have a quotient of 42.42. This tells us that if we were to supply 42.42 Btus of heat in one minute to a prime mover of 100 percent efficiency, then we would have in return for this minutes worth of applied thermal energy one mechanical horsepower. Let's expand upon this cross-application. We can revert the division process:

$$\frac{778\text{-(Btu/ft. lb. conversion constant)}}{33{,}000\text{-(the mechanical hp/min)}} = 0.0236$$

The resulting quotient tells us accordingly that 1 Btu per minute is the equivalent of one mechanical horsepower. Advancing this illustration further, we find that by simply multiplying the "factor," 0.0236, by the amount of fuel or combustible fired per minute, by the calorific value of the combustible so fired per each minute, we may obtain the theoretical thermal efficiency of the engine or prime mover under study.

Testing a Diesel

Testing a very hypothetical diesel of 100 brake horsepower (BHP), we find that the diesel consumes 0.6 lbs. of #2 diesel fuel per minute to develop the 100 BHP. Number 2 fuel oil has a calorific value of approximately 19,500 Btus/lb. The total amount of Btus delivered will then be 0.6 × 19,500 = 11,700 Btus. For the 100 BHP developed, a heat release or supply of 11,700 Btus is needed which gives us a ratio of 100/11,700. This may further be reduced to express the resultant of 117 Btus/BHP/minute.

The theoretically attainable horsepower may be ascertained by multiplying the consumed Btus by the factor 0.0236 (for consumption per minute). We find that for the same 117 Btus fired per minute as needed by our "test subject," a diesel of 100 percent efficiency could produce 117 × 0.0236 = 2.76 horsepower for the very same 117 Btus of heat applied. We could handle the matter by simply dividing the established 42.42 Btus/theoretical horsepower by the wasteful 117 Btus as delivered to our hypothetical diesel. We arrive at the efficiency quotient of 0.36. This indicates that our diesel is only 36.2 percent as efficient as it could theoretically be, if it were to convert every Btu of heat supplied to its combustion chamber into mechanical horsepower at its output shaft.

Establish an Index

Although our pure theoretical formulations and conversion factors are truly academically abstract in their practical applications, by having the knowledge and the working comprehension of true efficiency we, as an engineering/scientific community, may employ these interchangeable factors to studies of thermal and mechanical efficiency as to circumscribe our parameters of desirable and attainable values. We do so if for no other purpose than to establish an index of what actually is indeed possible. We cannot, in any intelligent stretch of the imagination, expect to meet the 100 percent point of efficiency, not even by the use of nuclear reaction by fissionable materials. Contrary to popular belief, the nuclear reactor is not even as efficient as some of our worse misrepresentations of prime movers because of the greater temperature differences and losses involved. We can and must, however, strive to meet the greatest possible degree of proximity and never cease to question ourselves with each successive advance and attainment. Is this our ultimate optimum point? Can we exact from our combustible another one-half of a percent? Could we expect to do so at a future date with more perfected correlated devices and materials?

All heat engines, be they of the internal or external combustion types or of the displacement or acceleration classes, may be evaluated by the thermal/mechanical formula. This heat input to horsepower output situation may be expressed as such:

$$\text{thermal/mechanical efficiency} = \frac{\text{actual horse power at shaft per minute (BHP)}}{\text{Btus supplied/min.} \ (\times)\ 0.0236 \quad (*)}$$

(*) The actual thermal horsepower per minute, as discussed, equals the amount of Btus released to the prime mover times the factor--0.0236.

This particular formula, which may be used universally upon any of the related engines and prime movers, is given in survey/form, part C, for the gas turbine. I recommend the exclusive adaptation of this formula for all prime movers and heat engines.

PRIME MOVER CLASSIFICATIONS

A prime mover may be defined as a device for the conversion or translation of power, from one genetic form of energy to that of another. Thus, a diesel engine, a steam or gas turbine, a fuel cell,

Fig. 7-1. Here is an automatic plant powered by a windwill to produce nitrogen fertilizer from air and water (courtesy of United States Department of Energy).

a windmill or water turbine are such devices or implements that are capable of converting the primary energy applied to a secondary or alternate form of energy (Figs. 7-1 through 7-3). This is usually manifested as mechanical output by rotational power as transmitted to a shaft. The primary energy may be in any number of forms: the heat released by the combustion of fossilized fuels or nuclear fuels, the kinetic energy of the wind, or the rushing torrents of high pressure water flowing through the delivery tubes supplying a hydraulic turbine (Fig. 7-4). We may also consider the more direct conversion of primary fossilized fuel (in liquid form), as it is injected into the combustion chamber of a diesel or otto cycle engine or into the "combustor can" of a gas turbine.

The manner of conversion, or technique of utilizing the pri-

Fig. 7-2. The Sandia Laboratories in Albuquerque, New Mexico, have filled patent applications for improvements in the vertical axis wind turbine seen here (courtesy of United States Department of Energy).

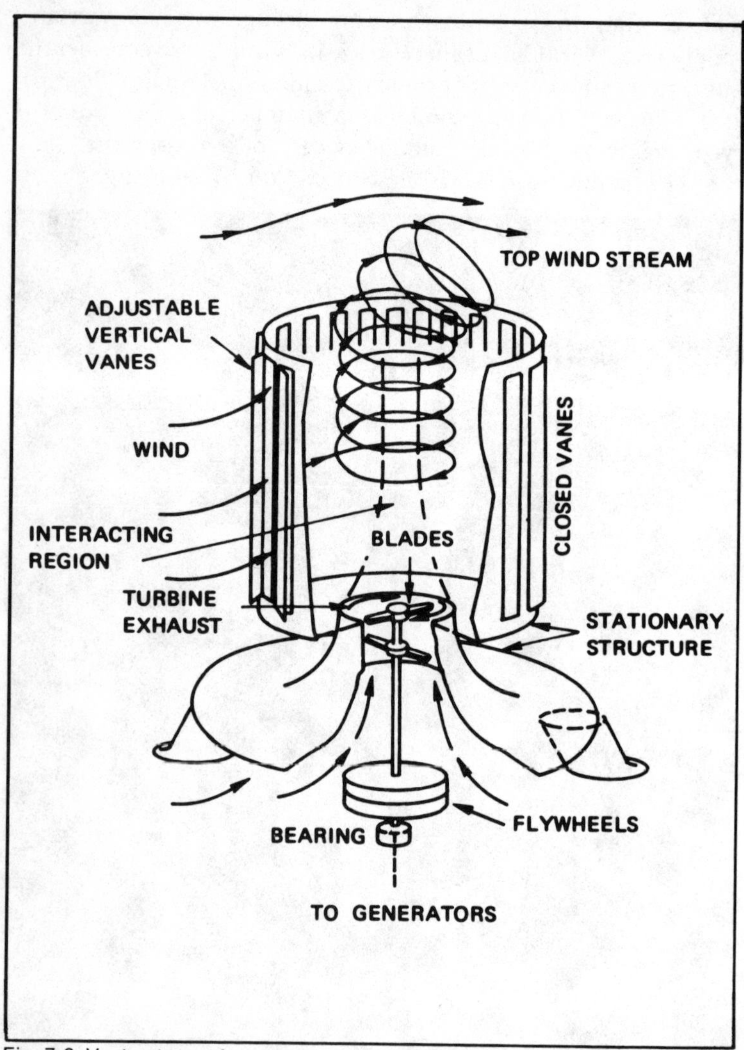

Fig. 7-3. Vortex tower for omnidirectional winds (courtesy of United States Department of Energy).

mary application of energy, is accomplished in one of two distinct function—either by the affects of displacements of the energy bearing medium or by the acceleration of this energy bearing medium. Our prime movers are thus either of the displacement or acceleration variety.

Displacement Prime Movers

These devices are chiefly reciprocating engines such as the

Fig. 7-4. A giant nuclear turbine generator (courtesy of Westinghouse Electric Corporation).

diesel, otto, stirling and steam reciprocating engines. If the horsepower computation as required by its respective formula necessitates the use of mean effective pressure (MEP) as a factor in the expression, then the prime mover may be classified as the displacement type of prime mover.

The MEP is an inherent aspect peculiar to only the displacement varieties, by virtue of the physical arrangement of the device's technique of mechanical reaction or response to absorbing the "fluid's" energy and rendering this absorbed energy to the shaft. In all cases this will be accomplished by absorbing the energy force of the displacement fluid, or power medium, and converting this power from the effective area of the piston face, in a lineal direction. By use of connecting rods and an offset crank, or multiple of cranks, the rectilineal motion is translated to that of rotary motion.

Formerly, the steam engine was designed by virtue of its intended purpose to act without the rotational effect as provided by a crank. In most imaginable situations. the motion required by all of today's technologically perfected machinery will necessitate that the prime mover supply a rotational shaft-delivered effort.

The formula accordingly required by all displacement type prime movers is as follows:

$$\text{mechanical horsepower} = \frac{\text{MEP }(\times)L(\times)A(\times)N(\times)K}{33{,}000}$$

MEP= Mean effective pressure in pounds per sq. inch
"L"=length of stroke-feet
"A"=area of piston-sq. in.
"N"=number of rpms
"K"=number of power strokes per single revolution.

Acceleration Type Prime Movers

These prime movers are grouped as per the physical type of their working medium or "fluid." Acceleration devices include the hydraulic turbine, wind turbine and Pelton wheel. The applicable formula for the hydraulic turbine is as follows:

$$\text{hydraulic turbine HP maximum attainable} = \frac{Q(\times)\ H\ (\times)\ \text{efficiency}}{8.8}$$

where: Q=rate of fluid displacement in cu. ft./sec.

The formula for thermodynamic acceleration type turbines, such as the steam or gas turbine, considers the thermal or heat differences and relates the difference, or heat absorbed, as the primary energy supplied for prime mover utilization. Expressed algebraicaly as follows:

$$\text{thermal turbine maximum attainble HP} = \frac{\text{heat absorbed by turbine}/T(\times)\ 778}{33{,}000}$$

The factors for steam turbines are identified as heat absorbed or "Q" = the difference from (T-1)-(T-2), where T-1 is enthalpy of steam at the throttle, or admission port. T-2 is the enthalpy of heat content of the steam per pound as it leaves the prime mover. The Q value for points T-1 and T-2 may be found by the use of a steam table or mollier chart. Since one Btu is equivalent to 778 ft. lbs. of mechanical work, the total heat absorbed multiplied by the conversion constant (778) will equal the effective gross work performed or available. When divided by the horsepower constant of 33,000, this will produce the mechanical horsepower equivalence for the prime mover per minute.

The thermal horsepower formula may be applied to both steam and gas turbines. In computing the pertinent values of

Table 7-1. Performance and Dimensional Data Related to Displacement Prime Movers.

Prime Mover Type:	HP:	RPM's:	Stroke:	Ratio:	MEP-psi:	Efficiency:
Steam Reciprocator	25-500	100-300	6-24"	0.8-1.2	50-100	0.5-0.8(*)
Otto Cycle Engine	10-300	2000-4000	3-8"	0.9-1.1	50-150	0.4-0.6
Diesel, Low Speed	100-5000	100-300	10-24"	0.8-1.1	40-80	0.4-0.8
Diesel, High Speed	25-1000	1500-2000	3-6	0.8-1.0	50-100	0.4-0.6

NOTE: MEP is Brake MEP, psi; Ratio is Bore to Stroke; Efficiency - Indicator Card Factor; (or Air Card Standard) (*) = Logarithmic Standard.

Table 7-2. Performane and Dimensional Data Related to Acceleration Type Prime Movers.

Type P.M.	Rating-KW's	Stages	Head-pressure	Temp. "F."	rpms	Eff.% (★★)
Pelton Wheel	1000-200,000	1	500-5000 ft.	#	100-1200	.75-.85%
Francis Turbine	1000-200,000	1	50-1000 ft.	#	72-360	.80-.90%
Kaplan Propeller	5000-200,000	1	20-100 ft.	#	72-180	.80-.90%
Steam Turbine (Small size)(*)	100-5000	1-12	100-400 psi	400-700	1800-10,000	.50-.80%
Steam Turbine (Large size)(*)	100,000-1,000,000	20-50	1400-4000 psi	900-1100	1800-3600 3600	.80-.90%
Gas Turbine	500-20,000	10-20	70-100 psi	1200-1500	3600-10,000	.80-.90%

(#): Ambient temperature. (*): Condensing exhaust plants. (★★ Mechanical Efficiency.

fuel combustion as is required for the gas turbine, it will be more expedient to employ the following formulas:

the maximum horsepower attainable thermodynamically = calorific value of fuel fired (×) 0.0236

The formula for determining the efficiency by such thermodynamic values is:

$$\text{gas turbine efficiency per fuel consumption} = \frac{\text{actual shaft horsepower by prony brake}}{\text{calorific value of lbs. fuel fired } (\times)\ 0.0236} = \text{percent eff.}$$

An electrical dynamometer may be used in place of the prony brake, in this as well as all computations where the developed shaft horsepower is to be determined. The contant .0236 is obtained by dividing 778 by 33,000. See Tables 7-1 and 7-2 for date on displacement and acceleration prime movers.

ANALYSIS OF PRIME MOVERS

In Table 7-3 is a comparative analysis of today's prime movers, categorizing them according to their abilities in respect to actual power output. They are listed in order of their thermal superiority.

Table 7-3 cites the related prime movers according to their total degree of operational efficiency and in net terms of thermal utility. To be more descriptive, let us consider the diesel in contrast to the steam turbine, although it can be argued that the incidental losses of the diesel are greater than are those for the turbine, the diesel has a superior advantage in actual delivered horsepower per quantity of fuel delivered. This contrast becomes less distinct when we evaluate other prime movers. Although the prime movers may be at first glance quite efficient in power extracted by initial conversion, the same prime mover may be less effective in providing for reclaimable heat, which could just as economically be utilized in the plant for such purposes as boiler feedwater heating, process heating, comfort heating, etc. Also, the degree of additional heat reclaiming and exchanging equipment must be kept in mind.

Table 7-3. Comparative Analysis of Thermal Distribution.

Prime Mover: Cycle Classification	Shaft Output	Reclaimable Losses-Max.	Unreclaimable Losses-Min.	Incidental Losses-Avg.
The Stirling Engine	40-45%	40-50%	10-15%	5-7.5%
The Diesel Cycle	30-40%	30-40%	20-30%	8-12%
The Gas Turbine	15-22%	40-45%	30-40%	15-10%
The Otto Cycle	18-22%	15-25%	35-50%	15-20%
The Steam Turbine	15-20%	20-35%	20-45%	3-8%
The Steam Reciprocator	10-18%	20-35%	30-60%	10-15%

We could also compare the suggested superiority of the otto cycle's shaft output of 18-22 percent against the 15-22 percent as averaged by the gas turbine. Because of the inherent simplicity offered by the relative absence of complex accessories of the turbine, the turbine's incidental losses are correspondingly less. Another advantage with the gas turbine is the greater adaptation of heat reclaiming devices, which by virtue of the gas turbines' conversional mode may be advantageously affixed to the turbine's exhaust system to reclaim much otherwise spent energy. This extensive heat absorption idea may not be as effectively applied to the otto cycle because of the greater back pressure such devices would create upon the exhaust system, thus reducing the otto cycle's mean effective pressure.

Those prime movers of the reciprocating varieties all indicate a relatively higher incidental loss ratio than their acceleration counterparts. This is explained by considering the additional mechanical forces and resistance that the engine must overcome pursuant to its mode of operation. The steam turbine enjoys the advantage of requiring no such element to sustain its own operation. However, the steam turbine incurs more efficiency losses.

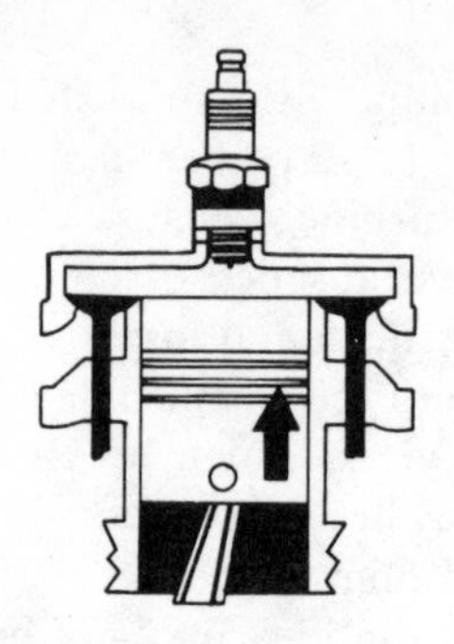

Chapter 8
Internal Combustion Engines

The diesel cycle is one which may be completed in either two or four strokes per power impulse. The diesel cycle differs from the more common otto cycle in that it may commence its power stroke without the aid of an ignition system. In addition, the diesel requires no carburetor carburet the combustible fuel/air mixture. Because the diesel utilizes very high compression ratios, which fall in the neighborhood of from 12:1 to 20:1 (as compared to the 4:1 to 10:1 ratios used by the otto cycle), the compression will raise the temperature of the inducted or "charged" air to approximately 1,000 degrees F. Thus, the temperature of the air is more than adequate to ignite and foster combustion when the fuel oil is injected to the combustion chamber, without the aid of a spark plug or separate ignition. This ignitionless process described is for the pure diesel only, as some modifications and variations of the diesel do in fact require an igniter or "punk stick."

DIESEL CYCLE

Fundamentally, the power process is as follows. The Piston, at bottom dead center BDC, has drawn in from the atmosphere a sufficient volume of air to fill its cylindrical volume. The piston will now proceed upward to top dead center (TDC), compressing the captured air charge, by virtue of the admission valves closing until the piston has arrived at TDC. The compressed gas is now approximately 1,000 degrees in temperature.

With the piston at TDC, fuel oil is then forced into the enclosed and highly heated combustion chamber by the fuel injection system. The fuel being injected into the chamber will immedi-

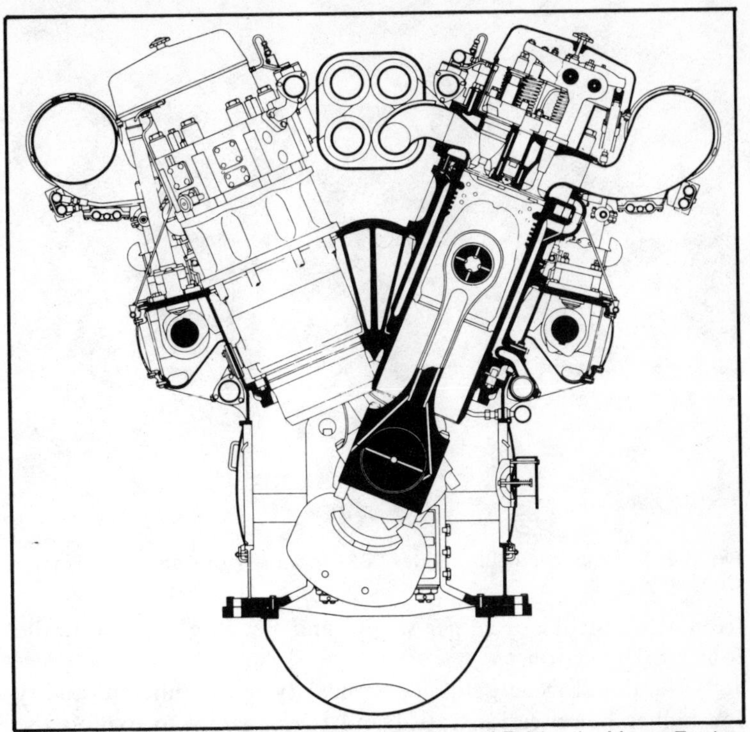
Fig. 8-1. A four stroke diesel engine (courtesy of Fairbanks Morse Engine Division, Colt Industries).

ately ignite. Combustion will progress as long as fuel is injected to the chamber. This usually is accomplished within 10 percent of the power stroke. This stage of the injection and combustion of fuel with the resultant pressure buildup upon the piston face, causing its forced movement, is called the *power stroke*.

The last stroke of the four cycle system is the *exhaust stroke*. The piston, after completing its power stroke at BDC, then returns to TDC. With the exhaust valves open, it will force the spent products of combustion out of the cylinder and into the exhaust piping, arriving back at TDC ready to induct another fresh air charge into its induction stroke by closing the exhaust valves and opening the admission or intake valves.

This power process may be completed in abbreviated form by the two stroke cycle, which is not as efficient as the more mechanically perfected four stroke cycle. The two stroke engine is quite often used in automotive and compacted installations where size is premium over actual efficiency. More horsepower may be had

Fig. 8-2. Marine propulsion model D339 (courtesy of Caterpillar Tractor Company).

from a two stroke cycle per weight and size as compared to the four stroke version.

The diesel cycle gains its popularity by its inherent ability for higher compression ratios and the capacity to exploit the higher heat content (calorific values) of heavier carbon-based fuels, which are substantially less expensive than the higher octane

Fig. 8-3 Marine propulsion model D339 with reduction transmission drive (courtesy of Caterpillar Tractor Company).

Fig. 8-4. Stationary 2000 KVA model D399 generator set (courtesy of Caterpillar Tractor Company).

and more sophisticated, volatile gasoline mixtures as is required by the otto cycle. In Table 8-1 are thermal losses incurred by both the diesel and otto cycle engines pursuant to allocation of operational percentages.

Fig. 8-5. Another view of the model D399 (courtesy of Caterpillar Tractor Company).

Table 8-1. Thermal Losses Incurred by the Diesel and Otto Cycle Engines.

Loss Classification	Diesel Engine (Injection)	Otto Cycle (Mixture)
Developed Shaft Output	33%	20%
Exhaust Losses	33%	40%
Cooling System Losses	33%	40%
Incidental Losses	01%	0-1%
Total Input	100%	100%
Avg. Thermal Efficiency	30-40%	20-26%

Fig. 8-6. This 14 cylinder Colt-Pielstick diesel engine is being prepared for final testing (courtesy of Fairbanks Morse Engine Division, Colt Industries).

In addition to the ability to utilize the heavier and more economical fuels, the diesel further offers the advantages of a higher maximum figure of attainable horsepower per cylinder and lower crankshaft and consequent piston speeds, resulting in longer operating life between overhauls. Higher mean effective pressures developed in the combustion chamber and cylinder of the diesel will yield a higher indicated horsepower. The diesel is available in either single or double acting power strokes and double opposed piston type arrangements to offer higher hp per size and weight. The diesel has single or dual fuel use applications. Both four and two cycle systems are available, with and without supercharging. See Figs. 8-1 through 8-21. Table 8-2 is a representative performance guide for typical applications.

Fig. 8-7. A stationary 2000 KVA generator set (courtesy of Caterpillar Tractor Company).

Because the diesel cycle is of the displacement variety, the horsepower formulation will adhere to the basic indicated horsepower computations as given. The mean effective pressure is taken by an engine indicator. The internal or frictional resistance will be found to be proportionately higher for this type of prime mover, so surprise should not be given when final net deductions are arrived at. See Figs. 8-22 and 8-23.

$$\text{diesel horsepower} = \frac{M.E.P.(\times)L.A.N.k}{33{,}000}$$

where: MEP= Mean effective pressure
L=length of Stroke in feet
A=area of piston face
N=number of rpms
k=number of power strokes, or "acting strokes"/revolution

note: "k"=1.0 for two cycle engines.
"k"=0.5 for four cycle engines.
"k"=2.0 for opposed piston engines.

OTTO CYCLE

N.A. Otto built a highly successful working engine based upon the sequence of operations as proposed by Beau deRochas in 1862. In essence the engine had four independent functions. First was the *induction stroke*, whereupon a gaseous mixture of combustible nature was drawn into the combustion chamber/cylinder by the downward motion of the piston in a direction towards bottom dead center. Second was the *compression stroke*, which was affected by having the piston return from BDC to TDC with all valves closed. Third, the *power stroke* was accomplished by igniting the compressed and combustible mixture upon arrival

Fig. 8-8. Colt-Pielstick generating set for nuclear standby installation (courtesy of Fairbanks Morse Engine Division, Colt Industries).

Fig. 8-9. This engine is ready to roll (courtesy of Fairbanks Morse Engine Division, Colt Industries).

Fig. 8-10. Colt-Pielstick engine generating set on factory test prior to shipment and installation in a large industrial plant (courtesy of Fairbanks Morse Engine Division, Colt Industries).

at TDC by an electric spark induction discharge. This caused the piston to return by such imposed pressure back down to BDC while transmitting its developed force (pressure resulting from combustion times area of piston face) to its connected crank and attached flywheel. Fourth was the exhaust stroke, being affected by opening of the exhaust valves and returning the piston back up to TDC (Fig. 8-24). The spent products of combustion were expelled from the cylinder through the opened exhaust valves and into the atmosphere, preparing the process for the subsequent recycling as required by the next induction stroke.

Originally the otto cycle was comprised of four strokes to the power impulse, but this has been abbreviated to two strokes in the two cycle engines by the appropriate incorporation of functions and a rearrangement of sequences (Fig. 8-2). In most cases the

Table 8-2. Representative Performance Guide for Typical Engine Applications.

TYPE OF APPLICATION:	H.P.:	M.E.P.:	Weight- lb/HP:	Comp. ratio:	Eff."T"
Air Injection Engine,	300-500	50-85	25-200	12:1-15:1	30-35%
Solid Injection Engine,	20-300	75-125	7-25	12:1-15:1	25-30%
Railroad Diesel-Elect.,	200-2500	60-90	10-40	12:1-15:1	30-35%
Stationary, Unsuperchgd.,	50-2500	70-80	10-100	12:1-15:1	30-35%
Stationary, Supercharged,	60-4000	110-125	7.5-80	10:1-13:1	30-40%
Dual Fuel, Average,	55-3000	80-135	7.5-100	10:1-15:1	30-40%

Fig. 8-11. Skid mounted turbocharged opposed diesel engine generating set (courtesy of Fairbanks Morse Engine Division, Colt Industries).

Fig. 8-12. Diesel generating set for marine propulsion system (courtesy of Fairbanks Morse Engine Division, Colt Industries)

Fig. 8-13. Cutaway of turbocharged opposed piston diesel (courtesy of Fairbanks Morse Engine Division, Colt Industries).

abbreviation to the two cycle sequence causes a reduction of actual single impulse power, but this is overcome by the increased power strokes per minute to provide for a comparable horsepower rating with somewhat less efficiency.

The Otto Cycle is almost universally utilized for automotive applications, and was once the power supply of all mobile transportation. The appearance of the otto cycle is rapidly decreasing in marine and small stationary power plants, being replaced by the diesel with its less expensive fuel requirement. Because of the otto cycle's exhaust pollution and expensive high volatile fuel necessities, it is now even being replaced in the automobile by the high speed diesel. The otto cycle was the first real internal combustion engine to be introduced to the engineering world. Unless its operation is modified to accommodate less expensive fuels and its combustion improved to assure complete and contaminate free combustion, then the age of the otto cycle may be nearing its completion.

Figure 8-26 illustrates the effect of compression ratio upon the otto cycle's thermal efficiency. It must of course be annotated that the compression of the inducted fuel/air mixture is halted by the mixture's tendency to preignite at certain pressures. This prevents the otto cycle from realizing its full and true theoretical ability to become a more effective and efficient prime mover. The preignition propensity has been kept in check by adding certain anti-preignition compounds containing lead which, when separ-

Fig. 8-14. Skid mounted turbocharged opposed piston diesel engine generating set—power generation application—on a rail car for shipment and installation in Ecuador (courtesy of Fairbanks Morse Engine Division, Colt Industries).

Fig. 8-15. Dual-fuel spark gas engine(courtesy of Fairbanks Morse Engine Division, Colt Industries).

ated in the products of combustion in this system's exhaust, will present a very dangerous ecological contaminate. Thus, it is the

Fig. 8-16. Testing diesel power plants by hydraulic dynamometer (courtesy of Fairbanks Morse Engine Division, Colt Industries).

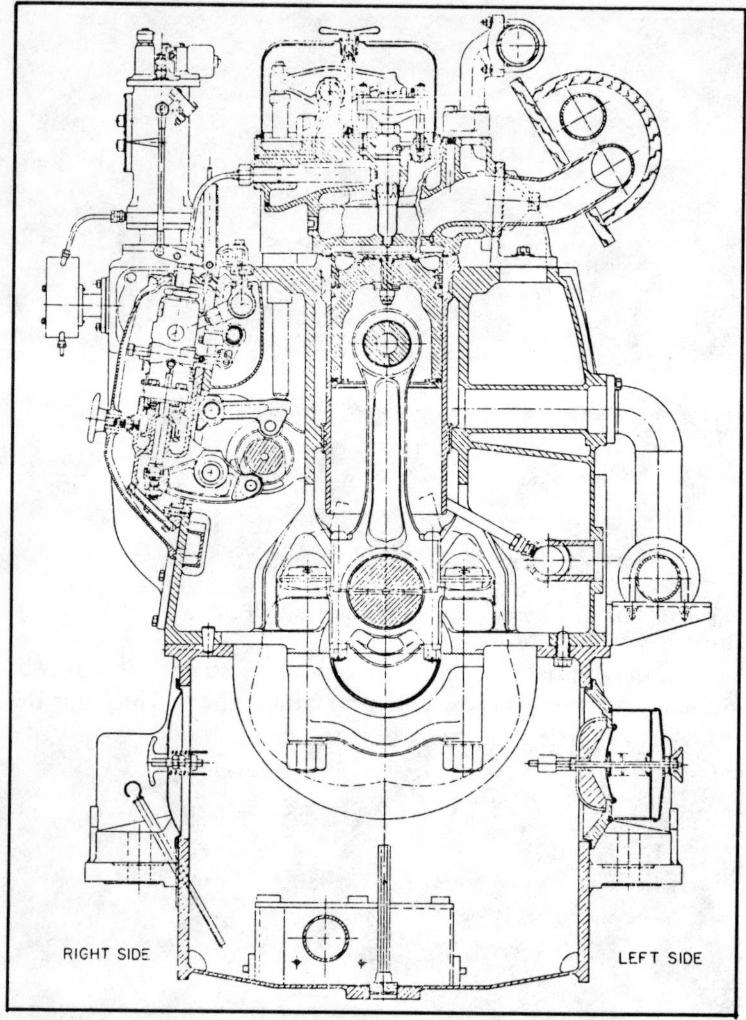

Fig. 8-17. Cross section through a 6-251 engine, looking from the free end (courtesy of Alco Power Inc.).

very technique employed to enhance the otto cycle's efficiency (increased compressability of fuel charge) that causes the engine to be a damaging device to our ecological life support system.

Since the otto cycle is a true displacement type prime mover, the system is computed by the displacement formula:

$$\text{otto cycle horsepower (indicated)} = \frac{M.E.P.(\times)L.A.N.k.}{33{,}000}$$

where: MEP=Mean effective pressure in psi.
L=length of stroke in ft.
A=area of piston face
N=number of rpms
k=number of power strokes per revolution/piston

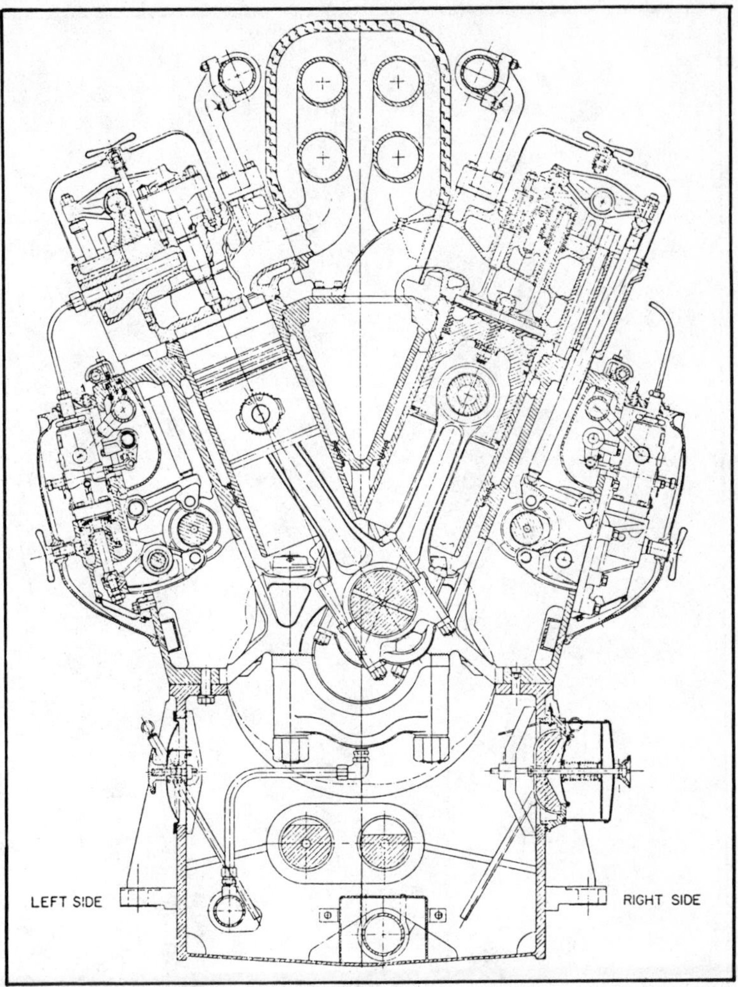

Fig. 8-18. Cross section through 8-Vee-251 engine, looking from the drive end (courtesy of Alco Power Inc.).

GAS TURBINE

The *gas turbine* is relatively new in the sense of its application as a self-contained prime mover for power generation purposes (Figs. 8-27 through 8-37). Its operating, or functional characteristics, are very similar to the steam turbine, which is an external combution engine. The similarities parallel in functional respects to the manner by which the kinetic energy, resulting from the high temperature and high pressure gas, passes through the blading or rotor.

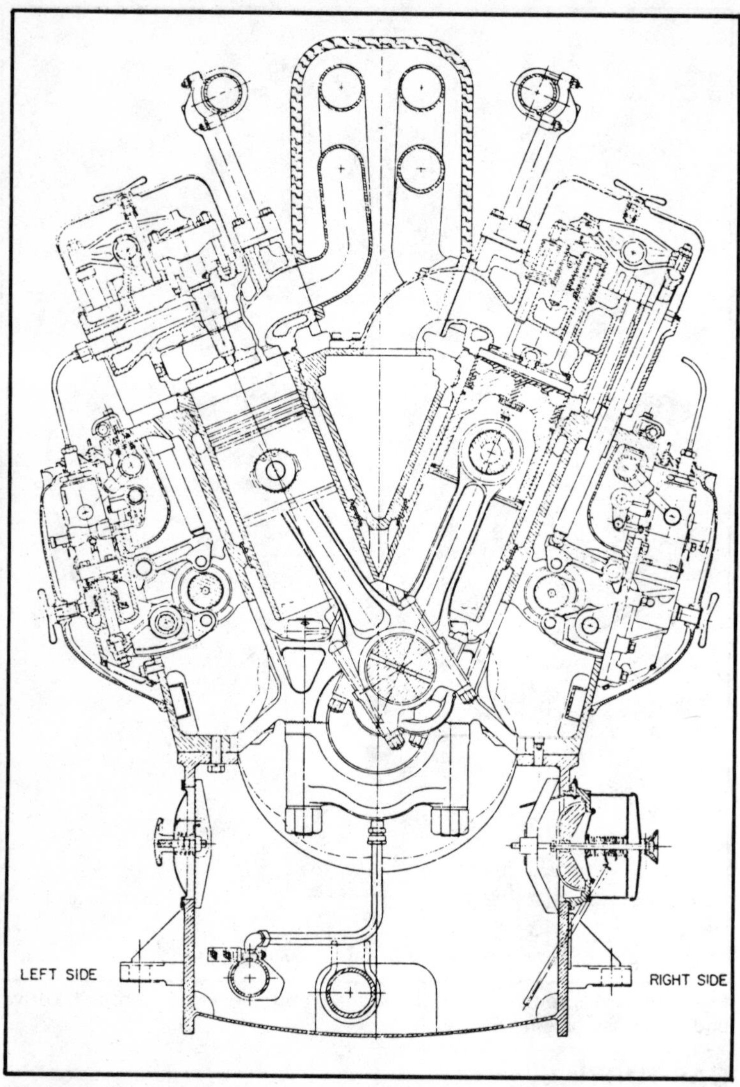

Fig. 8-19. Cross section through 12-Vee-251 engine, looking from the drive end (courtesy of Alco Power Inc.).

Basically, the gas turbine differs from the steam turbine by its self-contained energy fluid generation system. The gas turbine creates its own working medium of gas by the actual combustion process of fuel and air supply. The point of combustion is the combustion chamber from where the expanded and heated gas originates. The air supply is admitted to the combustion chamber

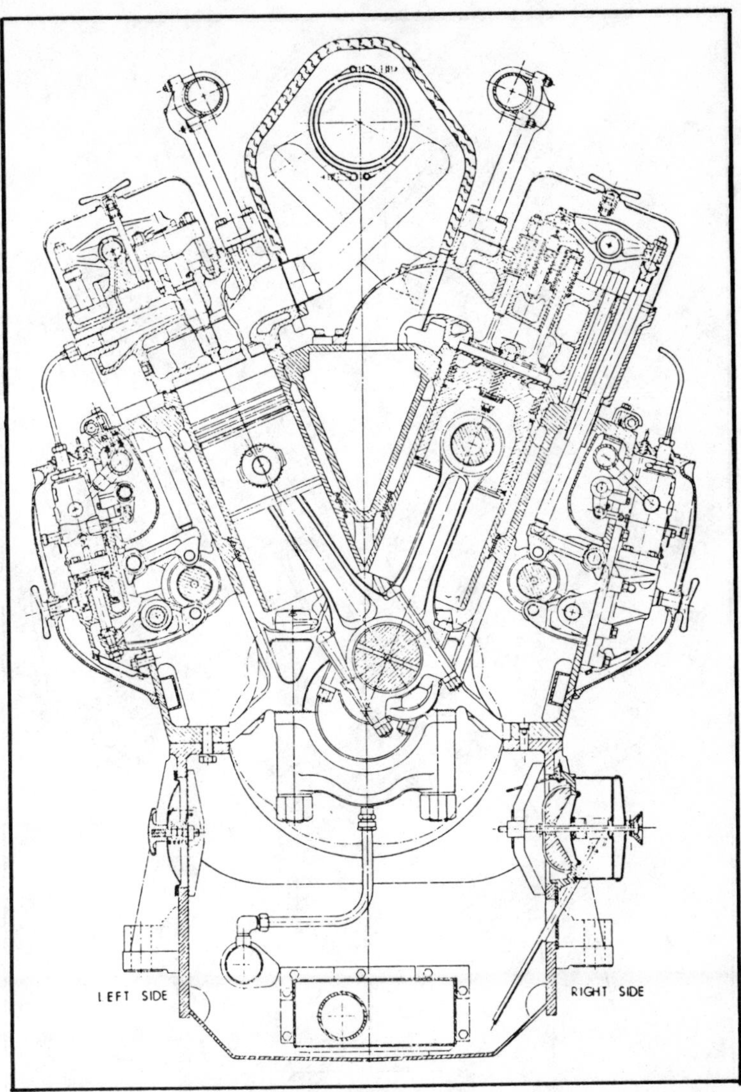

Fig. 8-20. Cross section through 16-Vee-251 engine, looking from the drive end (courtesy of Alco Power Inc.).

under high pressure delivery, as provided in a compressed high pressure volume by the air compressor. This is directly attached to the turbine's drive shaft and for all purposes may be considered almost integral to the turbine itself. The air compressor turns with the bladed rotor, usually at the same speed, unless geared for higher delivery rates. Because the turbine must have compressed

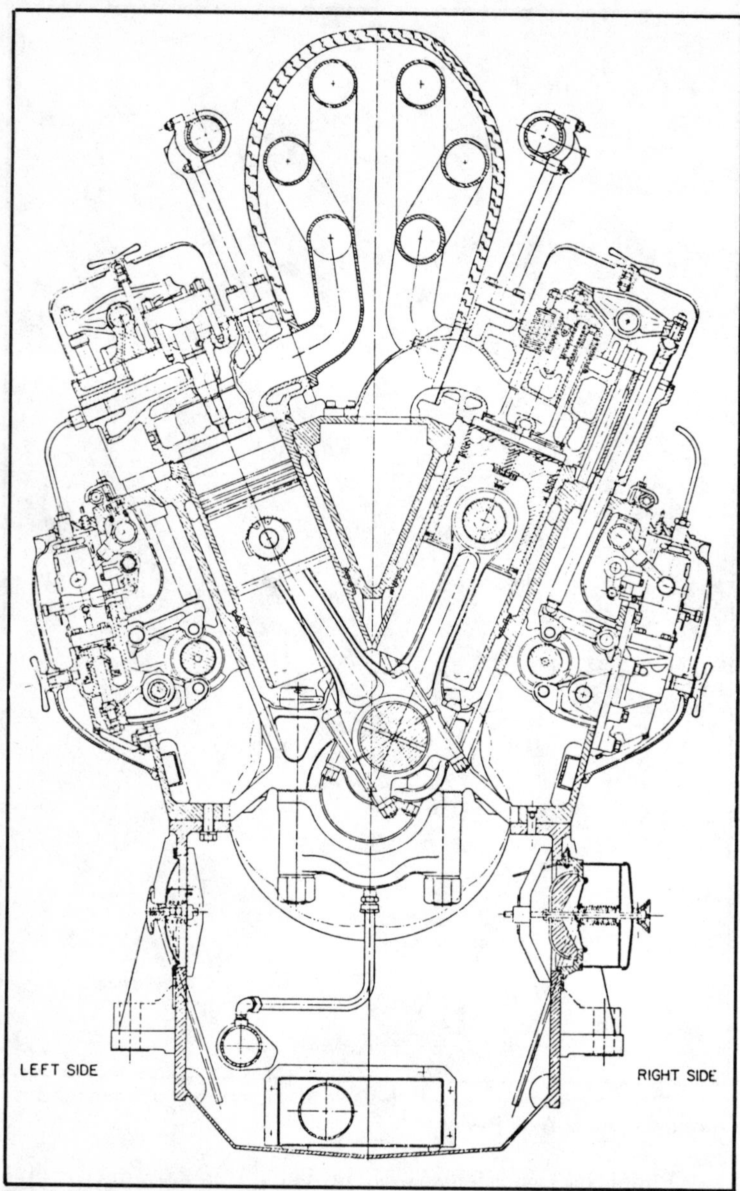

Fig. 8-21. Cross section through 18-Vee-251 engine, looking from the drive end (courtesy of Alco Power Inc.).

air for its combustion charge, the compressor then becomes a dependent and inseparable related function of the turbine proper The compressor adds to the overall compactness of the turbine

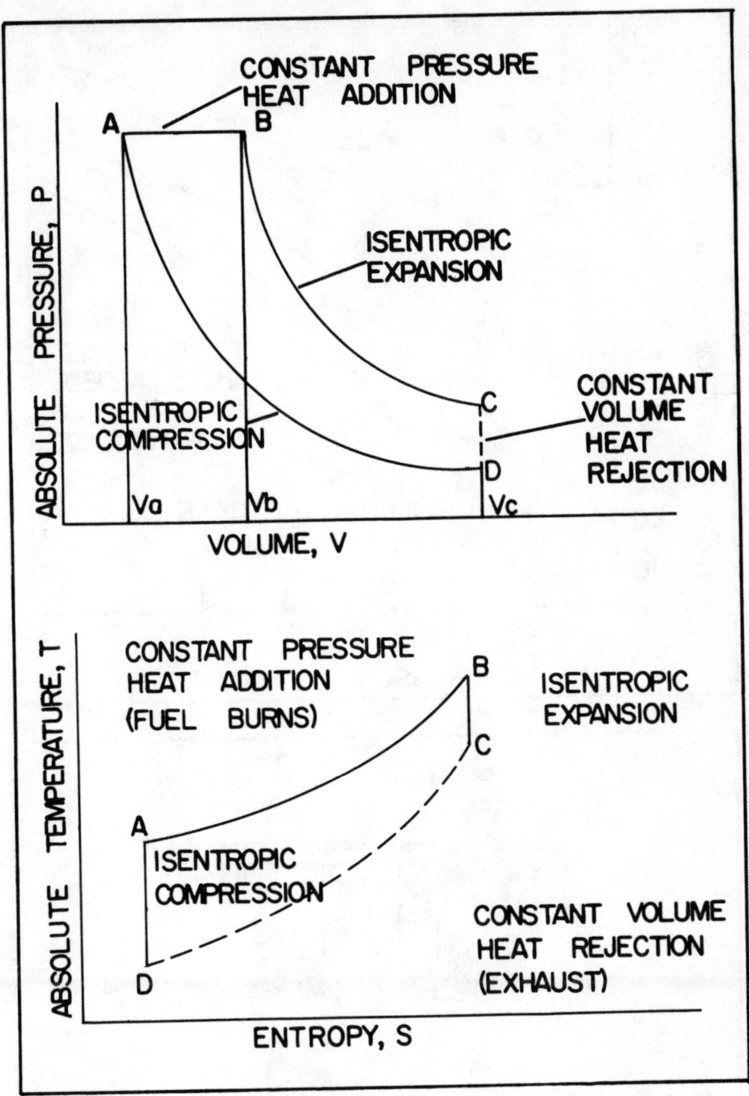

Fig. 8-22. Ideal diesel cycle with pressure-volume and temperature-entropy bases.

unit, and is a permanent and indispensable factor of efficiency in loss to the turbine in terms of deducted shaft horsepower.

While the compressed air charge is being delivered to the combustor, a metered supply of fuel is also being admitted to maintain combustion therein. The combustor serves both as a combustion chamber in itself and as the preliminary point of the

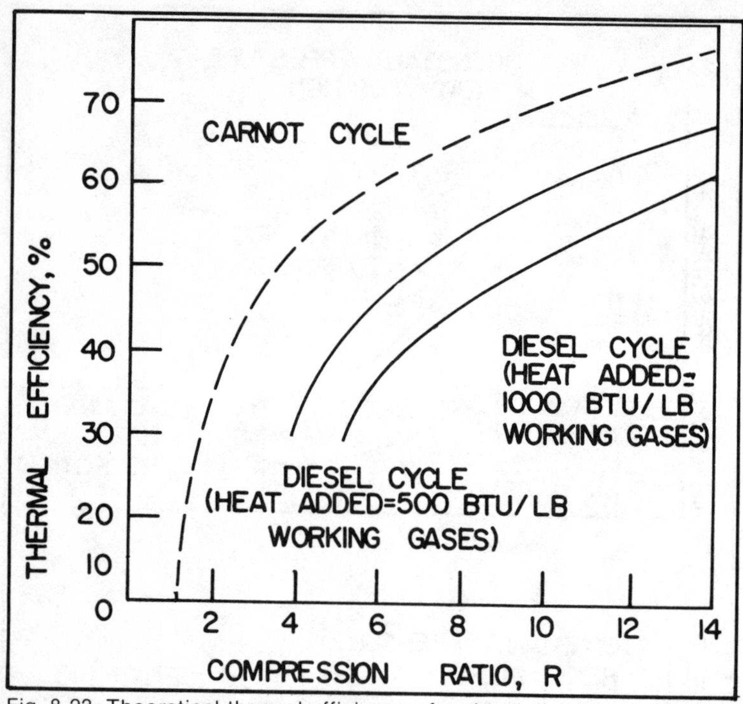

Fig. 8-23. Theoretical thermal efficiency of an ideal diesel cycle.

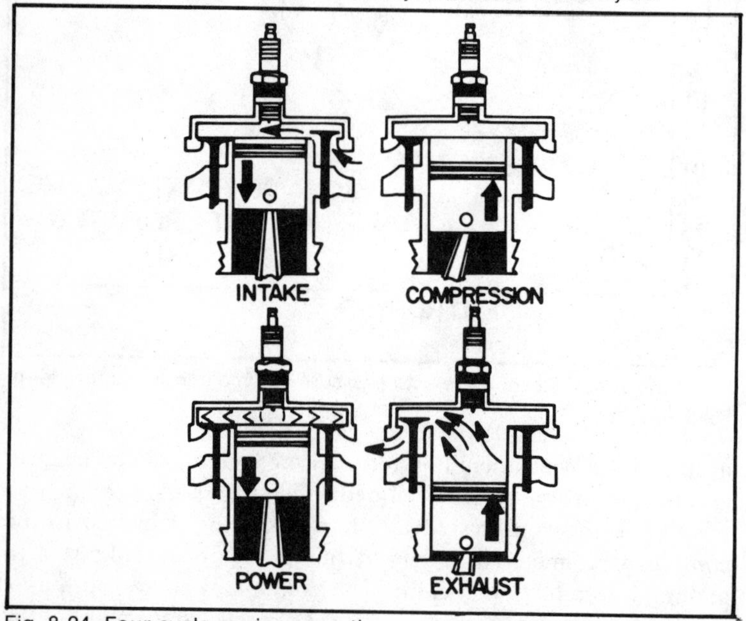

Fig. 8-24. Four cycle engine operation.

Fig.8-25. Two cycle engine operation.

expansion or energy conversion staged progression. Combustion is accomplished under better than average efficiency and completeness before being released by the combustor into the admission nozzles of the turbine rotor's blading and guide vane system. It is the combustor's duty to execute complete, efficient and clean combustion of the gases before releasing the products of such combustion to the turbine. If complete and particle free combustion is not achieved, the rotor and guide vane blading will very shortly become clogged and corroded. A very time consuming and expensive overhaul will be necessary.

Thermal Efficiency

Following pressure and temperature let down and the consequent exchange of energy, the spend gases are then discharged from the last stage of rotor blading into either the atmosphere or an appropriately designed heat absorber or exchanger for further heat absorption. The exiting gases, having been reduced from intial inlet temperatures of from 2,500 to 3,200 degrees F., may be

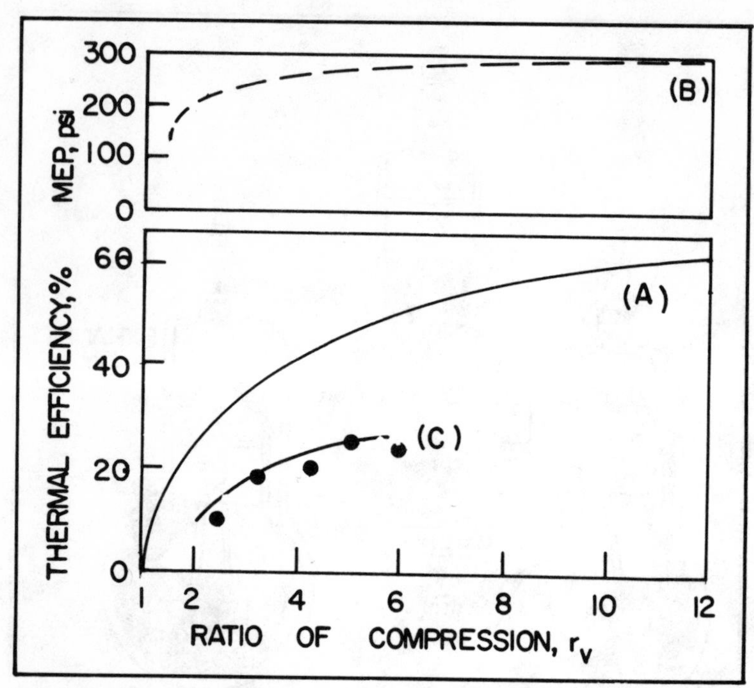

Fig. 8-26. Effect of compression ratio on the thermal efficiency and mean effective pressure of the otto cycle. Curve (a) shows thermal efficiency of air standard cycle. Curve (b) shows means effective pressure of air standard cycle. Curve (c) shows thermal efficiency of an actual engine.

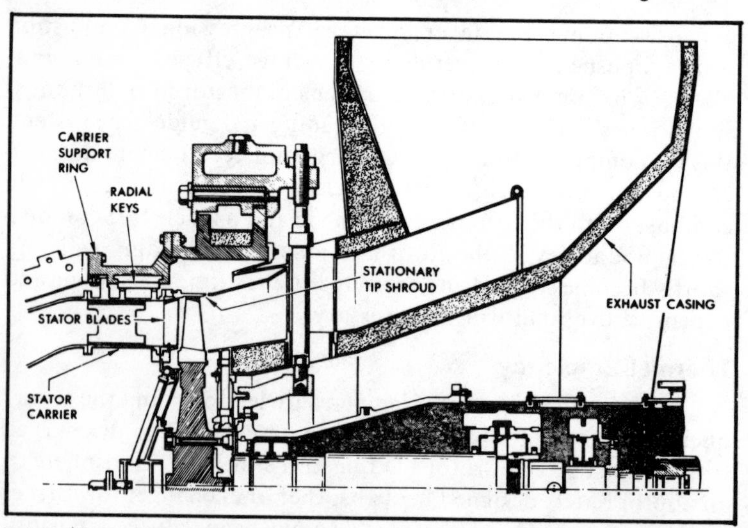

Fig. 8-27. Cross section of PT 125 power turbine (courtesy of Dresser Clark Division of Dresser Industries).

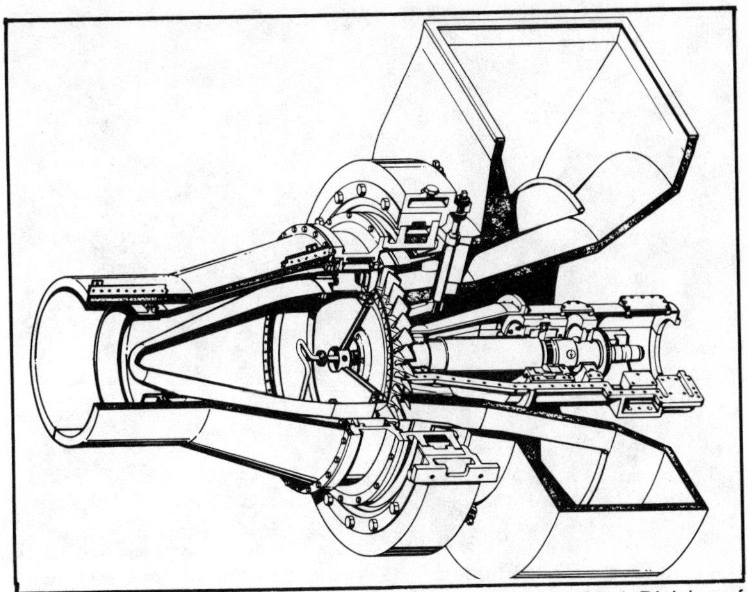

Fig. 8-28. Cutaway of a gas turbine (courtesy of Dresser Clark Division of Dresser Industries).

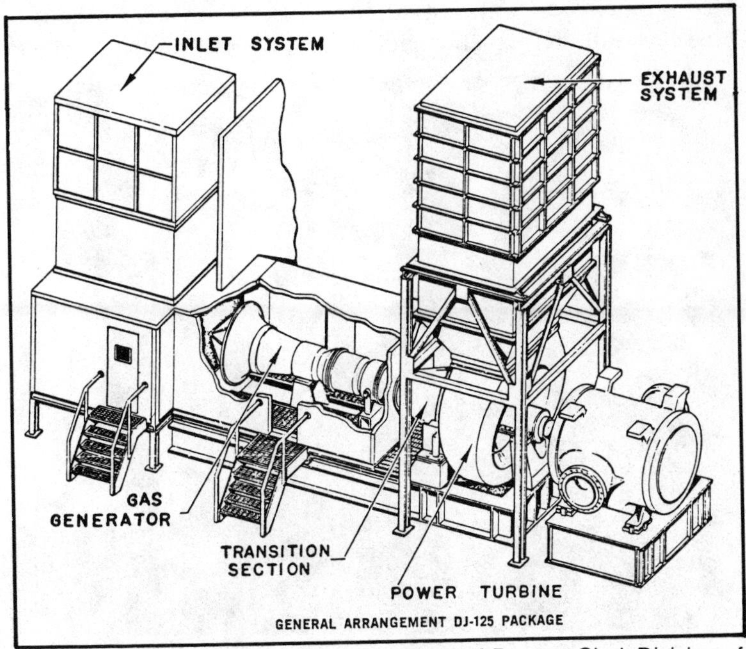

Fig. 8-29. Power plant gas turbine (courtesy of Dresser Clark Division of Dresser Industries).

Fig. 8-30. Model DJ-125 turbine installataion (courtesy of Dresser Clark of Dresser Industries).

further reduced by the application and exchange function of such incorporated heat utilizing additions as will enchance the turbine's overall thermal efficiency. Unless the discharged products

Fig. 8-31. A rotor on a balancing machine (courtesy of Dresser Clark Division of Dresser Industries).

Fig. 8-32. Model DJ 270 turbine frame and rotor, showing first stage disc and blades (courtesy of Dresser Clark Division of Dresser Industries).

of combustion are effectively and intelligently processed for further heat extraction, the turbine will not equal the diesel in its net efficiency as a prime mover. Typical thermal efficiencies are in the range of from 18 to 22 percent.

In general, there are certain distinct advantages and assets enjoyed by the employment of gas turbines for prime mover purposes. The turbine costs considerably more to build because of the precision machining and the expense of the special heat resistant alloy steels and materials, but this is offset by the minimal number of parts, chiefly the reduction of main bearings involved. The typical "V-12" reciprocating engine has approximately 90 moving

Fig. 8-33. Model DJ-270 turbine rotor pedestal and bearing housings (courtesy of Dresser Clark Division of Dresser Industries).

or sliding members which must be lubricated, and they collectively add to the gross frictional deficiency of the engine. The turbine for the same shaft horsepower output will have only four matching bearings of contacting surfaces.

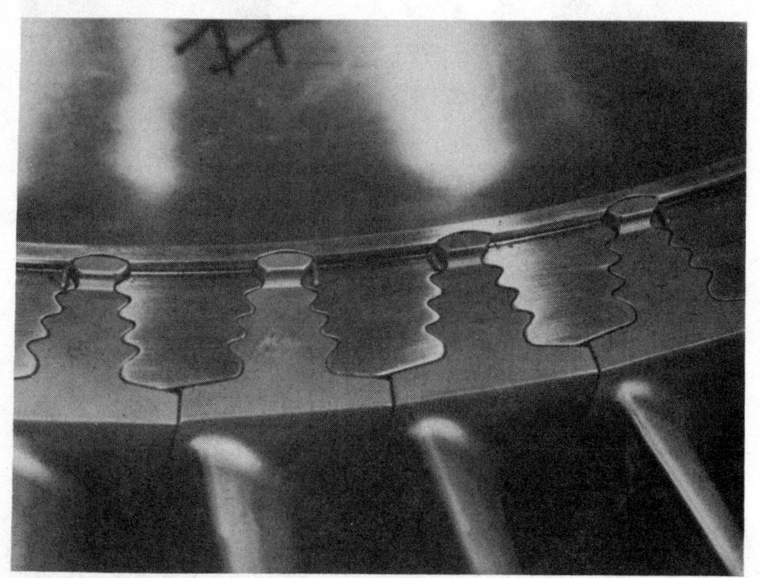

Fig. 8-34. Fir tree configuration on blade roots (courtesy of Dresser Clark Division of Dresser Industries).

Fig. 8-35. A model DJ-50 turbine (courtesy of Dresser Clark Division of Dresser Industries).

Mechanical Efficiency

The turbine excels over reciprocating engines in actual extraction of energy as applied to the output shaft. It makes better use of the force or released power as given by combustion because

Fig. 8-36. Installation of Garrett IE-990 gas turbine engine into model DJ-50 turbine (courtesy of Dresser Clark Division of Dresser Industries).

Fig. 8-37. (A) Model DJ-270 turbine on test stand. (B) Customers are looking at an older model 307 gas turbine rotor. (C) DJ—270 enclosure. (D) Model DJ-270 first stage rotor disc and blades (courtesy of Dresser Clark Division of Dresser Industries).

of its blading design. The blading design, and thus the character of the turbine, may be of either the impulse or of the reactance variety. The weight per shaft horsepower is considerably less than for reciprocating types. Because of the inherent simplicity and lack of multidirectional forces incidental to its operation, the turbine has superior balance properties. The turbine requires for less warmup time in comparison to its closest competitor, the diesel. The lubrication system for the turbine is absolutely unmatched, offering complete control and direct command of both the lubricant's pressure, temperature and cleanliness. Typically, the turbine's lubrication system may be entirely changed filtered, chemcially tested and renewed while the turbine is in full power operation. There is no sump built internal to the turbine, as is required by the reciprocator through the deployment of its crankcase or oil pan. The turbine is far more reliable and has a longer life than its reciprocating relatives. It may operate on less expensive fuels.

The gas turbine has no real upper limit or ceiling for its horsepower rating. Even the standard models may be boosted from their nominal ratings by simply converting from the standard one combustor can to multiple can type cumbustors. The volume of the products of combustion will increase. The efficiency will quite naturally decrease unless additional exit gas heat exchanging devices are contingently added or expanded. Theoretically, the gas turbine may be actually forced to 200 percent to 800 percent of its factory rating for temporary power outputs to accommodate load demands. Presently over 95 percent of the Nation's electrical power is generated by either the steam or gas turbine. The turbine's efficiency formula, as appearing on survey form part C, is as follows:

$$\text{gas turbine efficiency} = \frac{\text{actual shaft horsepower developed}}{\text{heat value of fuel fixed } (\times)\ 0.0236} = \text{percent} \_\_\_\_ \text{percent}$$

The gas turbine's thermal efficiency may be computed by the basic formula:

$$\text{percent thermal efficiency} = \frac{(\text{"T-1"}) - (\text{"T-2"})}{(\text{"T-1"})}$$

where "T-1"=combustion temperature
"T-2"=exiting gas temperature

The following is a discussion of the gas turbine from the United States Department of Energy.

Gas Turbines For Efficient Power Generation

The United States uses today one-fourth of its energy to produce electricity for homes, offices and factories. By 1990, nearly half of our energy may be used to generate electric power. Keeping up with growing electrical demand could require an enormous investment, averaging over $20 billion a year, in new generating plants.

Engineers at the United States Department of Energy (DOE) have been studying many alternatives for meeting electrical demands efficiently and ecomonically. The gas turbine engine, which can be used alone in small dispersed generating plants or for supplemental generating capacity at large central power stations, is a promising choice. But the gas turbine depends on fuels that are becoming more scarce,and it can waste up to 80 percent of the energy in the fuel it uses. Research at DOE aims to make today's turbine engines run more efficiently on those scarce fuels and to develop improved turbines that will operate efficiently on the synthetic fuels that will one day replace oil and natural gas.

Business and industry need two kinds of energy, electricity and direct heat for space heating, drying operations, and process steam. They generally satisfy these needs by buying electricity from central utilities (which throw away their excess heat energy) and by burning fuel in their own furnaces on site to produce heat.

Though power plants have been designed to produce the maximum electrical energy, no generating process is 100 percent efficient in using the energy from burning fuel. When electricity is produced at central power plants, leftover heat energy, often equal to more than 60 percent of the fuel energy used, is exhausted to the environment.

Studies have shown that small power generators which simultaneously produce, or cogenerate electricity and heat right on a commercial or industrial site can cost less to build, make better use of fuel, and do less environmental damage than large central nuclear or fossil plants which generally cannot make as effective use of their waste heat energy (Fig. 8-38). Congeneration systems would be ideal for industrial parks or other small "communities" like colleges or large shoping centers with a mix of facilities requiring both electrical power and process heat energy. requiring both electrical power and process heat energy.

Cogeneration theoretically could save over a half million barrels of fuel per day in just three of the country's most energy-intensive industries; petroleum refining, chemicals, and paper and pulp processing. Industrial gas turbines are the engines most widely used for cogeneration now, particularly in the chemical industry where high electrical and process heat demands have long made on-site power generation practical. Gas turbines will be an important option for cogeneration installations for more efficient power production in other industries as well.

Providing Peak Power at Utilities

In addition to generating power at small dispersed power

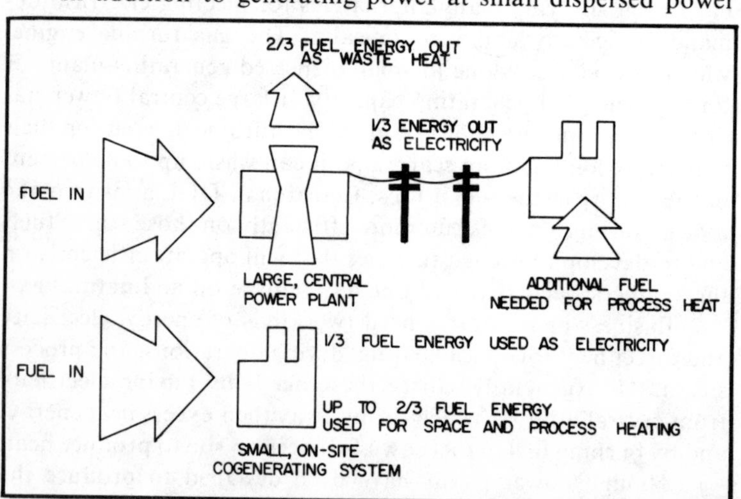

Fig. 8-38. Cogenerating electricity and heat makes better use of scarce fuels.

stations, gas turbines can help larger centralized power plants meet high electrical demands. Electric utilities provide a continuing flow of electricity to meet customer needs. But at certain times of the day and week, like the middle of a working day when every machine, light, and air conditioner in business and industry is running at full power. electrical demand reaches a peak. Gas turbines can start up quickly to help utilities respond to this peak demand.

Gas turbines can also be produced cheaply and quickly. They have a low initial cost of about $250 per kilowatt and can be installed and operating in less than 3 years, a very short lead time for generating equipment. In comparison, coal-fired steam plants cost about $850 per kilowatt and take 5-10 years to plan and build. Nuclear power plants cost an estimated $1100 per kilowatt and require 10-13 years lead time.

Power companies are understandably reluctant to invest large sums of money in extra generating capacity far in advance when recent projections of electrical demand have proven to be too high. Underused generating capacity means higher generating costs and increased costs to the consumer. But postponing the decision to build a coal or nuclear plant with a long lead time could mean not having enough generating capacity to meet future peaks. Power plant managers know that if they do not have reliable backup generators to pick up peak load, power outages could bring large areas of the U.S. to a virtual halt.

The gas turbine has become a popular backup engine for peak demand because it is quick starting, because it is cheap, and because its short lead time permits a utility some flexibility in planning for future power loads.

Improving Gas Turbine Efficiency

Though the initial cost of a gas turbine is low, the turbine's low efficiency makes its operating costs high. The average gas turbine being built today converts little more than one-fourth of the heat energy of its fuel into electricity, making it too expensive to run more than the 1500-2000 hours per year it is needed to meet peak demands (Fig. 8-39).

Gas turbine exhaust is generally still well over 1000° F. when it leaves the engine, hot enough to do a lot more work if that waste heat could be harnessed. A boiler installed in the exhaust steam can "recover" the waste heat energy. Water in the boiler, turned to steam by the 1000° F exhaust, can drive a steam turbine to generate electricity. Such a *combined cycle* arrangement using both a

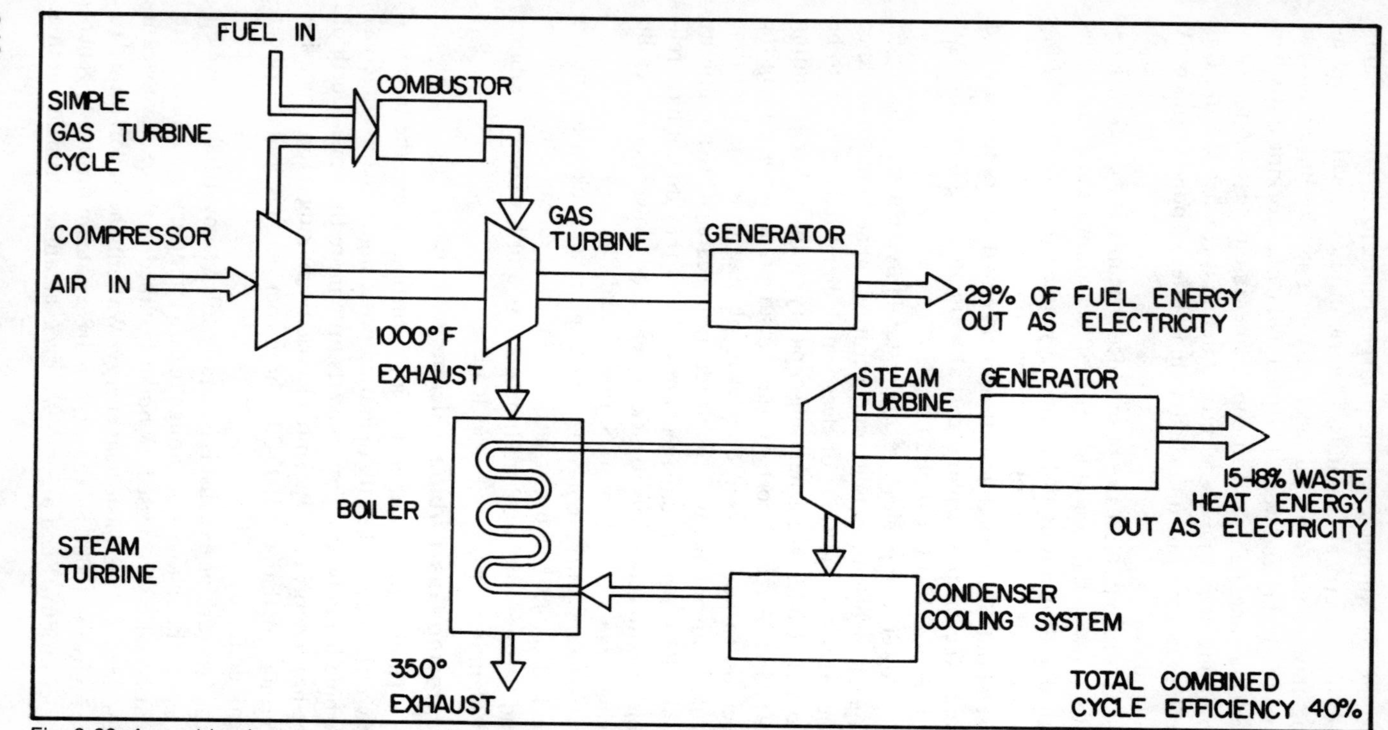

Fig. 8-39. A combined gas and steam turbine cycle increases electricity-producing efficiency.

gas turbine and a steam turbine system, will produce one-third more power than the simple gas turbine cycle, with no increase in fuel.

Increasing gas turbine efficiency from about 29 percent to 40 percent by adding a steam turbine cycle means gas turbines can be used economically for more than just 2000 hours of peak load duty a year. Combined cycle systems could be used to meet intermediate (2000-5000 hours per year) and base loads (over 5000 hours) as well.

Uses for Combined Cycle Systems

Where power plants cannot be converted to burn cheap, abundant coal because of environmental constraints or where fuels like natural gas are plentiful locally, as in the Southwest, combined cycle power plants could provide more economical and efficient baseload power generation than conventional oil and gas-burning systems.

Where converting an older, inefficient oil-burning system to coal would be possible but not economically or environmentally feasible, the life of the old system might be extended by "repowering" with a combined cycle system. This would mean installing a gas turbine to replace the plant's oil-burning system and replacing the old steam boiler with a steam boiler in the turbine exhaust steam. The remainder of the old steam system and generating equipment could still be used, making the conversion relatively cheap compared to constructing a new plant from the ground up.

Besides being economical, a repowered plant would be more efficient, using as much as 20 percent less fuel than the conventional oil-fired plant. It is estimated that twenty thousand or more megawatts of the nation's old, inefficient oil-burning capacity cannot be converted to coal for economic or environmental reasons. If those plants were repowered with combined cycle systems, the greater efficiency made possible by the gas turbine's waste heat recovery system could save over 150,000 barrels of oil per day.

How A Gas Turbine Works

A gas turbine is a relatively simple engine with four main parts:

☐ A compressor to suck in air from the atmosphere and pump it up to a high pressure

☐ A combustor to mix and burn fuel with the compressed air, creating a high temperature gas

□ Two turbines which are rotated by the force of expanding hot combustion gas rushing over their blade surfaces (Fig. 8-40).

One of the spinning turbines transfers energy along a compressor drive shaft. About half of the output of a gas turbine engine thus goes to compress air for its own operation. The power turbine transfers the rest of the available energy along a shaft to drive a generator or to provide the mechanical power to pump oil and gas through a pipeline.

Fueling Gas Turbine Systems

Because they can be used to cogenerate electricity and heat energy or, in combination with steam turbines, simply to generate electricity efficiently, gas turbine would appear to be the ideal power generators to stretch dwindling fossil fuel supplies until synthetic fuels from coal and oil shale become available. But gas turbines today require clean light distillate oil or natural gas for fuel, the same fuels needed to heat homes and run factories. Though gas turbines might be more efficient than other electrical generating systems, utilities must avoid adding generating capacity based on declining petroleum and gas fuels, except where coal is not environmentally practical or readily available.

How Research Can Help

Engineers at DOE are trying to resolve the gas turbine fuel problem. Working with industry, they hope to develop gas turbines which can operate efficiently on lower grade, more readily available residual oils and fuels made from agricultural and urban waste products.

The overhaul and maintenance costs of a gas turbine depend on the fuel used. The poorer the fuel, the lower the turbine reliability, the more frequent the overhaul, and the greater the operating costs. Run on distillate fuel oil, a turbine can last 30,000-50,000 hours; on natural gas, 100,000 hours. But when today's gas turbines burn the more abundant residual fuels, turbine life can be as short as 2000-5000 hours before overhaul is necessary.

Low-grade fuels burn hotter and contain more contaminants than do light distillate oil and natural gas. DOE is developing improved turbine combustors and blades with coatings that can withstand the hot, corrosive gases resulting from burning low-grade fuels. Improved cooling systems being developed should also increase gas turbine durability when hot-burning fuels are used. A new method of fabricating turbine blades will allow chan-

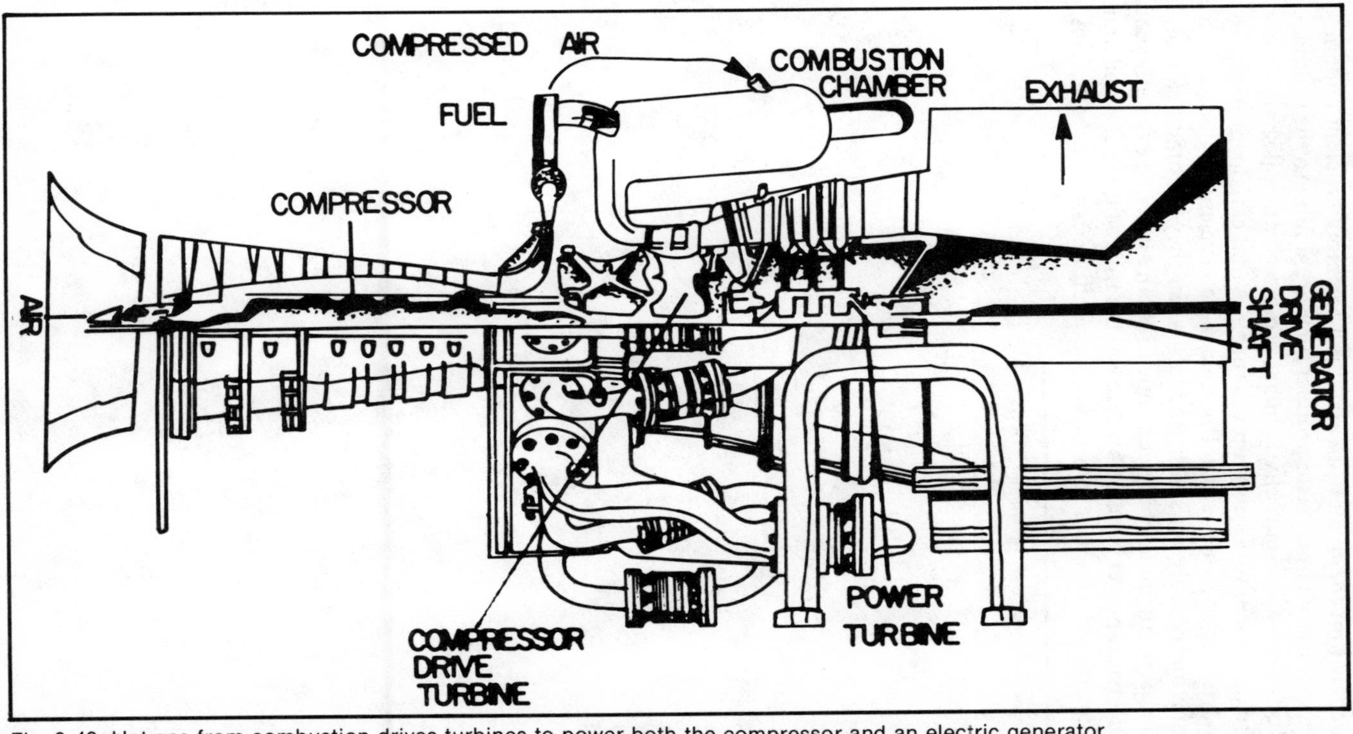

Fig. 8-40. Hot gas from combustion drives turbines to power both the compressor and an electric generator.

nels within the blades to carry fluid for more efficient cooling.

Residual and waste product fuels are similar to the synthetic fuels that will eventually be made from coal and oil shale. Work being done now to develop gas turbines that can operate reliably on low-grade fuels should permit these efficient engines also to operate reliably on synthetic fuels when they become available. Thus gas turbines could continue to be an alternative for efficient electrical power generation through the time of dwindling oil and natural gas supplies to a time of more abundant synthetic fuels.

Courtesy of United States Department of Energy

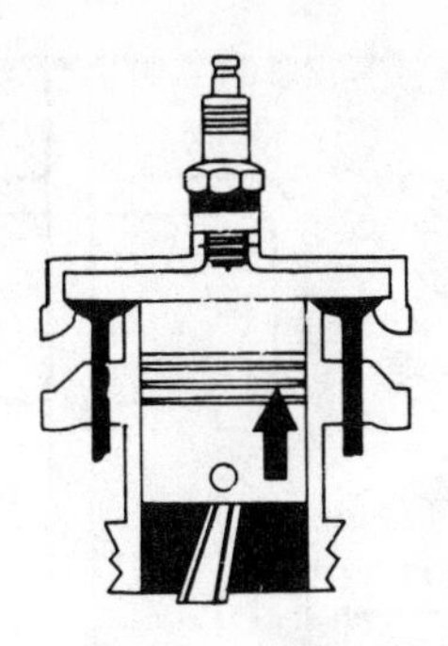

Chapter 9
External Combustion Engines

We owe much of our nation's engineering progress to the *reciprocating steam engine.* The steam engine, was first introduced to clear the English coal mines from the influx of water. Basically, the reciprocator is a pressure-actuated piston, free to respond to forces at either end of its travel. It transmits its motion to the flywheel by means of a rotational crank or offset. The admission and discharge of steam is accomplished by mechanically arranged valves and linkage. Although attached directly to the crankshaft. The valves are capable of being manipulated while under load to compensate for load changes. Figures 9-1 through 9-3 illustrate the function of the valves.

RECIPROCATING STEAM ENGINE

It was the problem of controlling the coordinated flow of steam to and from the piston-housed cylinder that presented the greatest obstacle in perfecting the steam engine to behave as a self-contained and dependable source of power. At first, the control of steam to the power cylinder was manually admitted and released by opening and closing of hand valves. It is customarily believed that around 1713 a boy named Humphery Potter, whose boring and repetitious job it was to operate these valves, tied cords and fixed catches on the valves and beams so that the engine operated automatically. Thus, the first automatic and self-contained prime mover appeared on the scene.

The effectiveness of the steam engine and its inherent efficiency as an economical prime mover have been influenced mostly

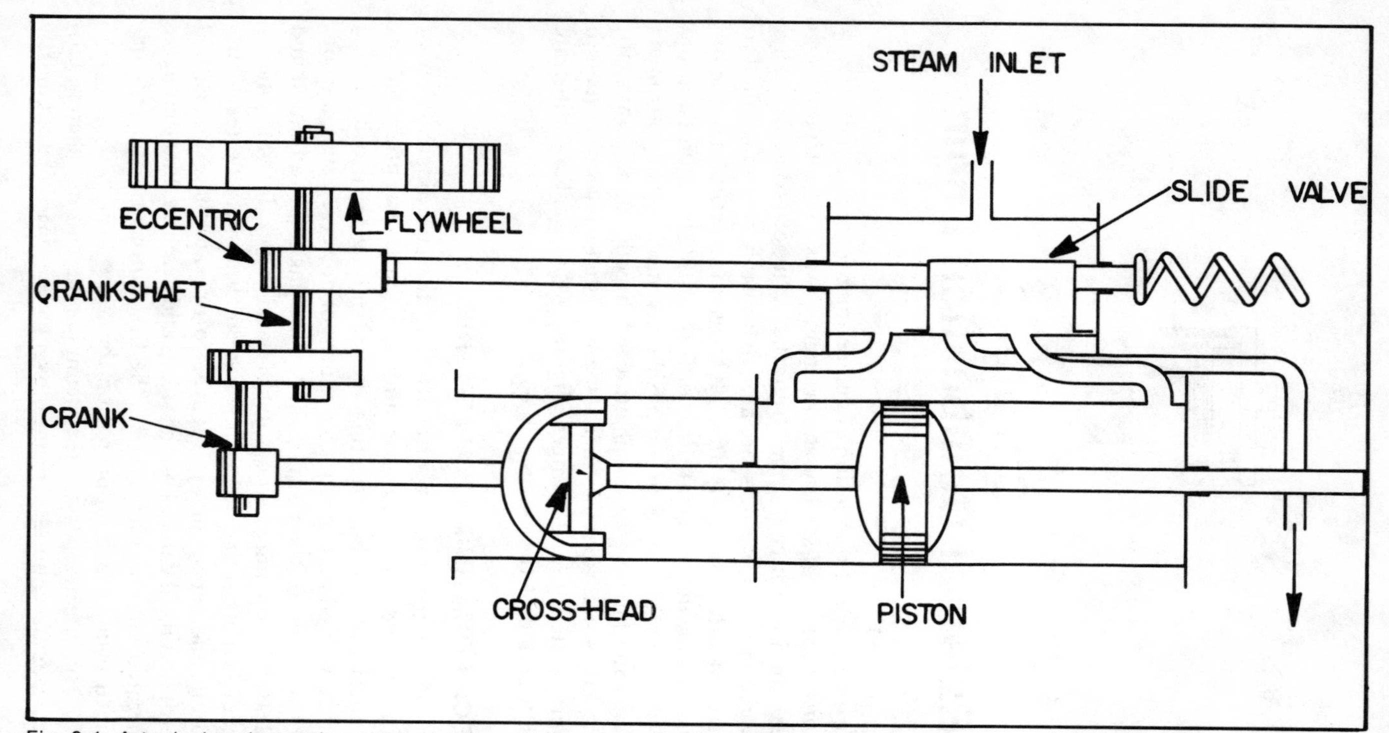

Fig. 9-1. A typical reciprocating steam engine valve controls the flow of steam. Shown are related parts of the engine.

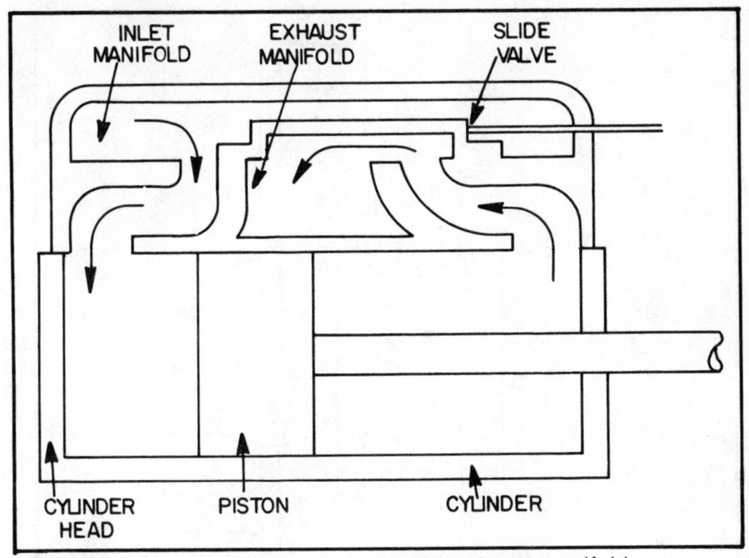

Fig. 9-2. The slide valve is located near the exhaust manifold.

by the development of the valve. The first and still practical valves were very primitive "flat valves" of either the "D" or "B" varieties. These flat valves limited the efficiency and speed of the engine by their lack of response and adjustment sensitivity. This deficiency has in large part been overcome by the use of balanced flat valves, piston valves and the very superior Corliss valve. The advent of the modern valve permitted use of higher pressure steam and high degrees of superheated steam, with accompanying response and "cutoff" control unequaled by the basic "D" and "B" valves.

The power output of the engine was first controlled by throttling the pressure to the engine. This resulted in a very wasteful account of the steam over any considerable load variation. By the adaption of the "cutoff control," the pressure at admission was kept constant. The steam was shut off from the engine at the valve by causing the valve to cover the admission port earlier in its travel. Figures 9-4 through 9-9 show the steam valves and the indicator cards obtainable from their expansive use. The first real cutoff control was the *Stephenson link*. This arrangement consisted of two eccentrics fitted on the crankshaft. An eccentric rod was connected to each eccentric and terminated at a curved link or quadrant. The quadrant was then attached by suitable linkage to the valve spindle by a block, and to the engine frame with an intervening control handle. The link could be used to regulate the

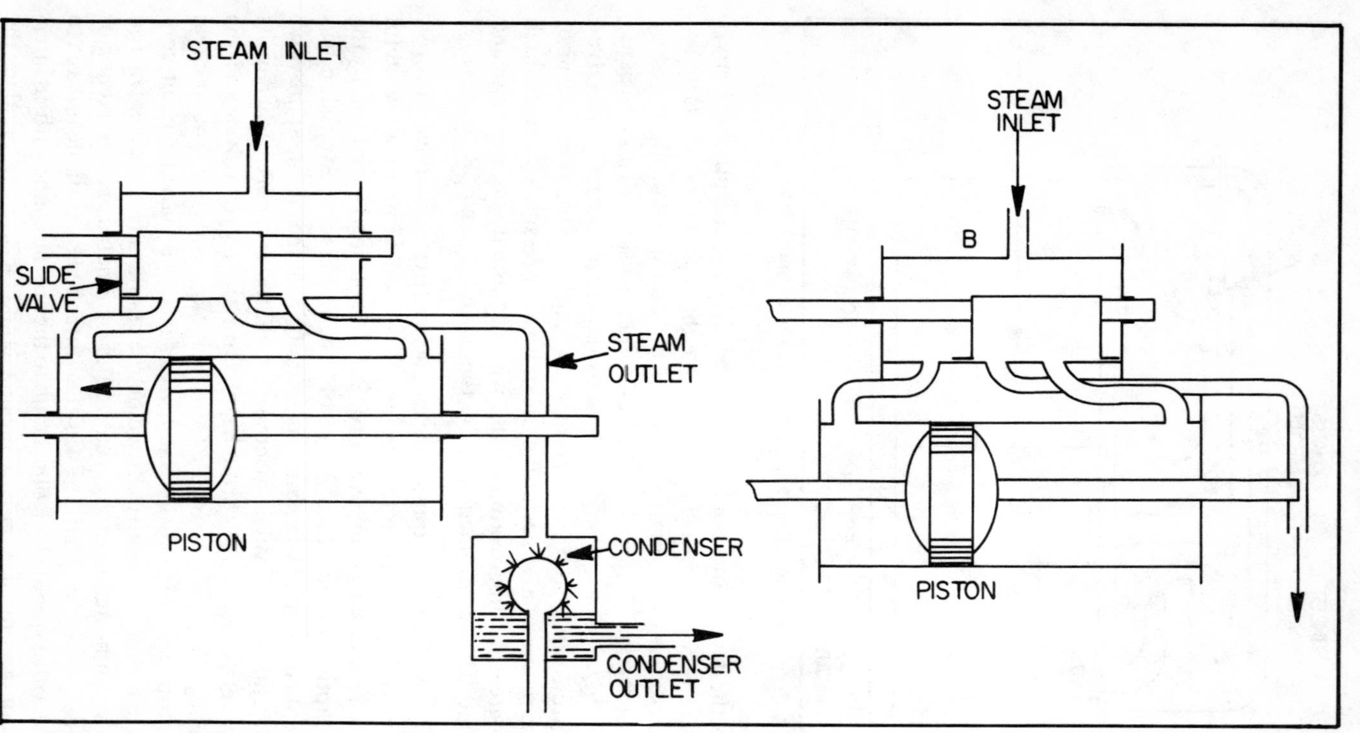

Fig. 9-3. Valve control mechanism.

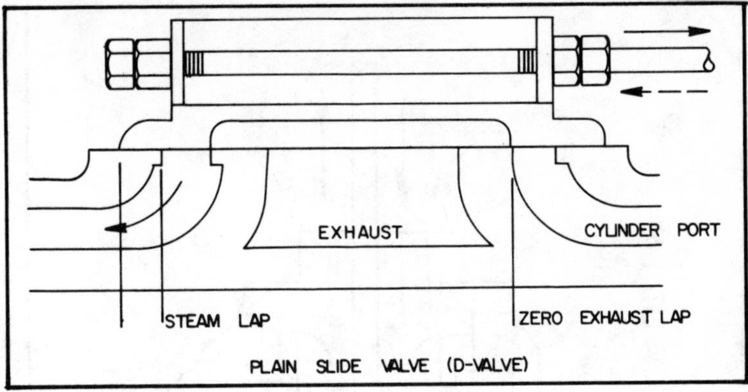

Fig. 9-4. "D" type slide valve.

position or percent of cutoff as well as reverse the rotational direction of the engine.

The efficiency of the steam reciprocator may be evaluated from both the thermal and mechanical aspects. Because the reciprocator is a displacement type prime mover, it will subscribe to the computational factors of the standard horsepower formula. Being a heat engine of the external combustion variety, its efficiency may be computed by the formula relating to heat engines:

$$\text{the mechanical horsepower (indicated horsepower)} = \frac{\text{mean effective pressure } (\times) \text{ L.A.N.k.}}{33{,}000}$$

where: MEP = average pressure as taken by an "indicator card", in psi
L=length of piston travel in feet
A=area of piston face, in square inches
* N=number of power impulses completed per minute (rpms)
k=number of power strokes (double acting, single acting, etc.)
(*) "N" also must express the number of expansion stages in a multiple expansion type engine.

The thermal efficiency of the reciprocator may also be computed as a basic heat engine prime mover where Q equals the heat in Btus per pound of steam consumed, being the difference between the heat content of the steam/lb. at admission to that of the steam at exhaust pressures.

$$\text{efficiency} = \frac{(\text{"T-1"})-(\text{"T-2"})}{(\text{"T-1"})} \text{ or } \frac{(\text{Q-1})-(\text{Q-2})}{(\text{Q-1})} = \% \text{ "Q" utilized.}$$

This may be also equated to express the thermal difference, or the heat consumed by the heat actually applied as:

$$\text{thermal/mechanical efficiency} = \frac{\text{actual shaft horsepower developed } (*)}{\text{Btus absorbed (Q-1)-(Q-2) } (\times)\ 0.0236\ **}$$

(*) The shaft horse power may be obtained by the prony brake or dynamometer.
(**) The figure 0.0236 is obtained by dividing 778 by 33,000.

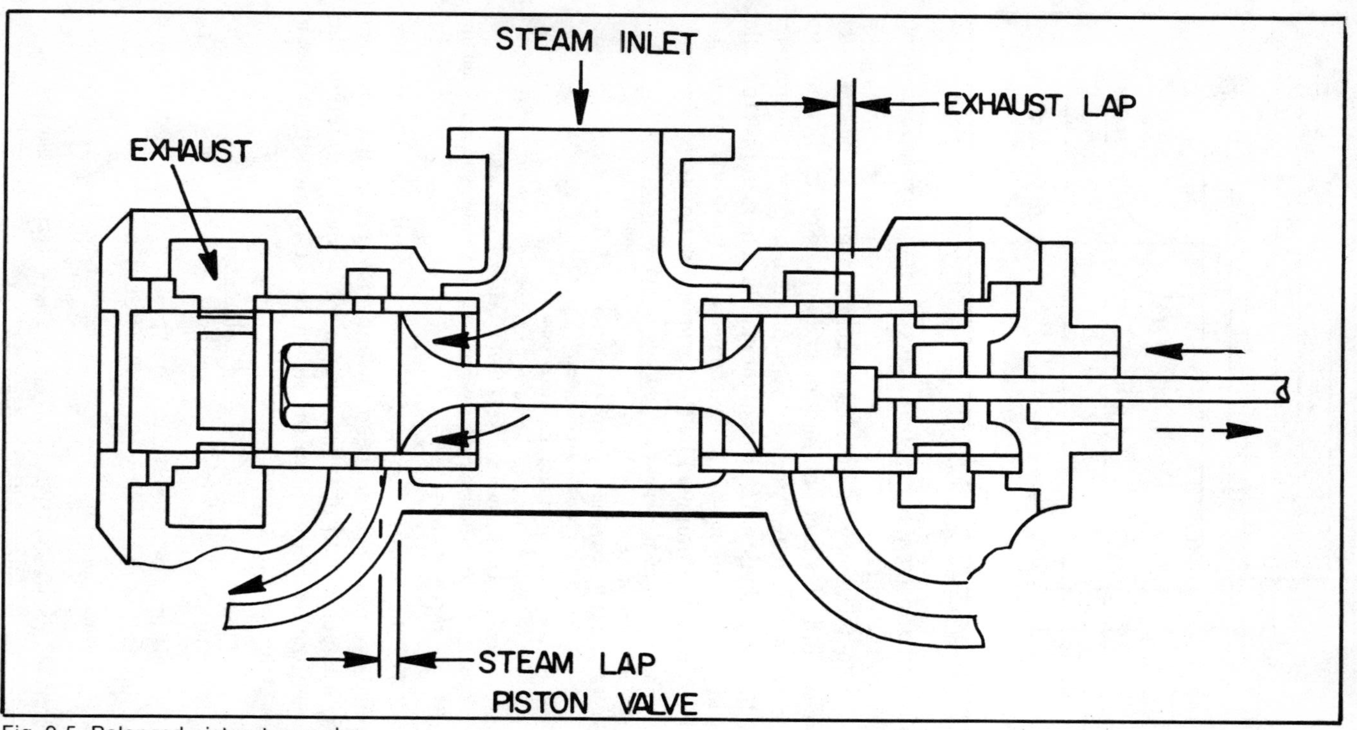

Fig. 9-5. Balanced piston type valve.

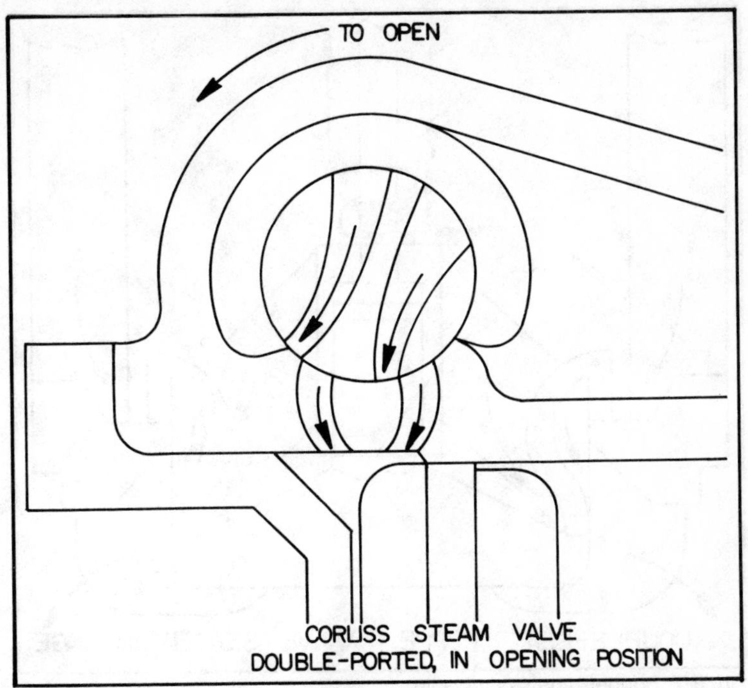

Fig. 9-6. Corliss type steam valve.

Generally, the net thermal horsepower of the reciprocator will be found to be disappointingly low. In the best of plants, a figure in excess of 18 percent is considered exceptional, with efficiencies of 10 to 12 percent as average. Customarily the mechancial IHP formula is used for computations regarding the reciprocator. Being a heat engine, the thermal formula is also applicable.

STEAM TURBINE

This type of acceleration prime mover has all but totally replaced the steam reciprocator as a driving device for our nation's electrical power generation. Its applications have greatly been multiplied in the marine engineering fields as an excellent constant speed propulsion system when geared to the propeller shafting of even the largest of ships. It requires very little adjustment or attention while in operation, due chiefly to its inherent simplicity and virtual absence of any interdependent parts.

The steam turbine may be of the *impulse* or *reactance type*, or an optimum combination of both. Quite frequently the turbine will be of the multi-stage expansion and/or pressure types, and

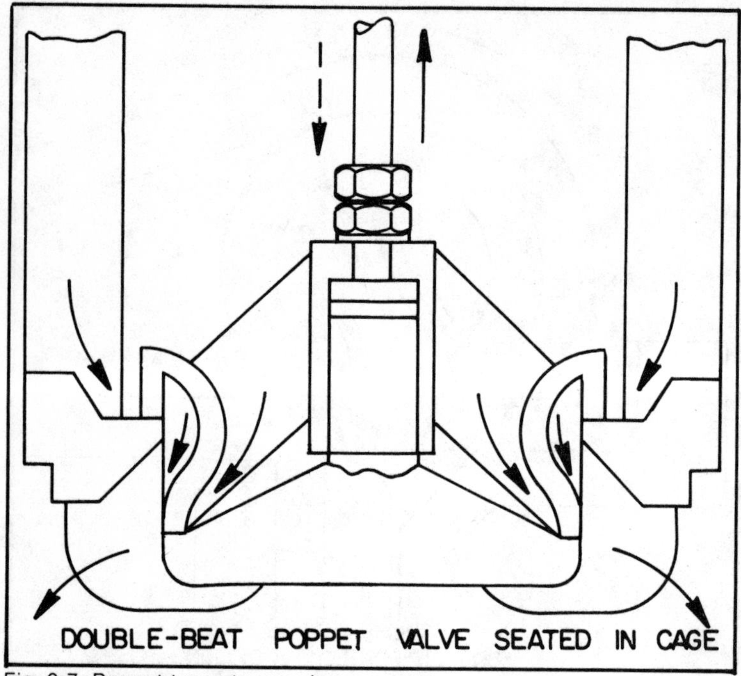

Fig. 9-7. Poppet type steam valve.

may further be compounded to incorporate the most desirablc properties of each function. In the *impulse* type of turbine, the steam is delivered to the initial expansion nozzles expansion takes place. The expanded, and thus high velocity, steam is then directed through appropriately designed guide vanes and applies its "impulse" upon the moving blades of the rotor. This velocity-created expansion process may be carried out repetitively throughout any number of stages, all the way through the turbine until exhausted to either a condenser under vacuum or out to the atmosphere or into a final heat exchanger.

The reactance type of turbine makes use of the reactive effort of the steam's "kickoff" reaction effort when expanded in the moving rotor blades. The blading of the reaction type rotor is designed so that the exiting cross section is larger than the entrance area. The steam is allowed to expand in the rotor blading and exert its force upon the blading as it "kicksoff" into the lesser pressure stage in front of this stage of the rotor. The successive stages of the turbine casing are made progressively larger to purposely accommodate the much larger volume as occupied by the expanded steam, and to reduce the impeding back pressure that

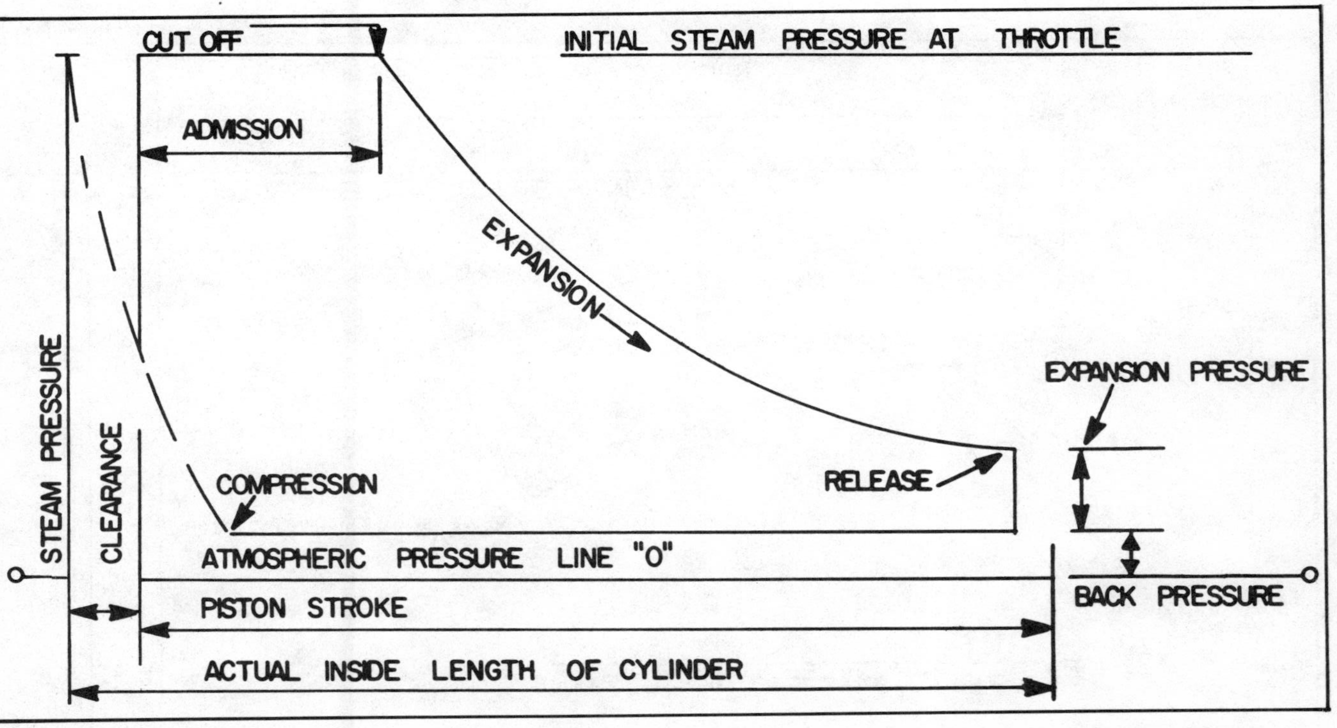

Fig. 9-8. Theoretical indicator card.

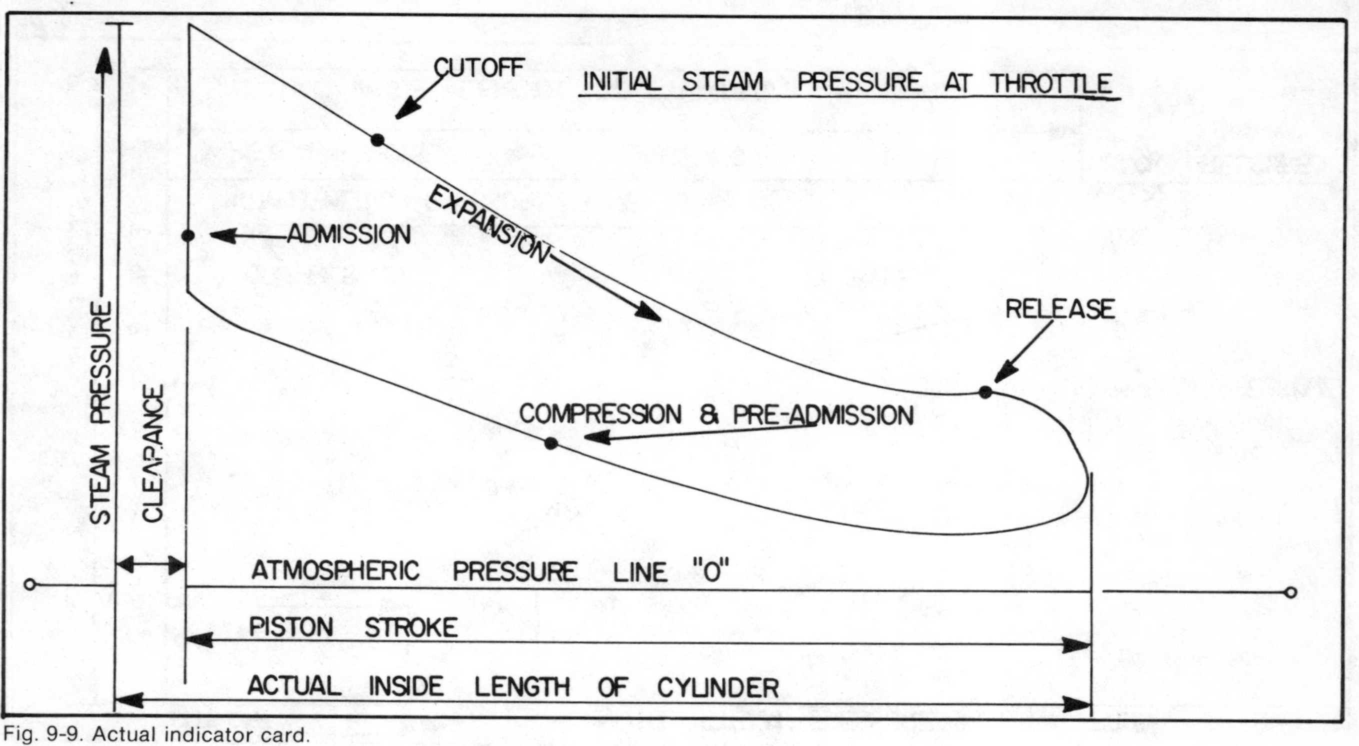

Fig. 9-9. Actual indicator card.

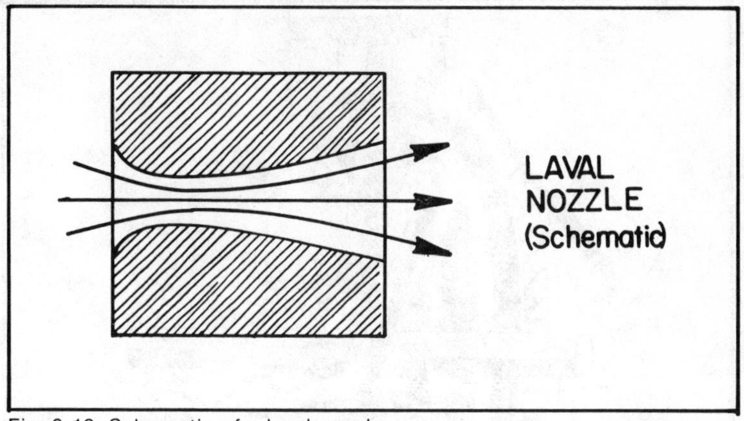

Fig. 9-10. Schematic of a laval nozzle.

would otherwise prevent the necessary pressure difference to affect the continued expansion from pressure stage to pressure stage.

Figures 9-10 through 9-14 illustrate diagrammatically the nozzle, rotor and stator blading and a cross sectional view of a typical condensing type turbine. The turbine is "staged" to extract the maximum latently stored potential energy from the steam as possible. Were the entire pressure to velocity transaction to be completed in one stage or between just one rotor and stator diaphragm the circumference, and consequently the diameter of

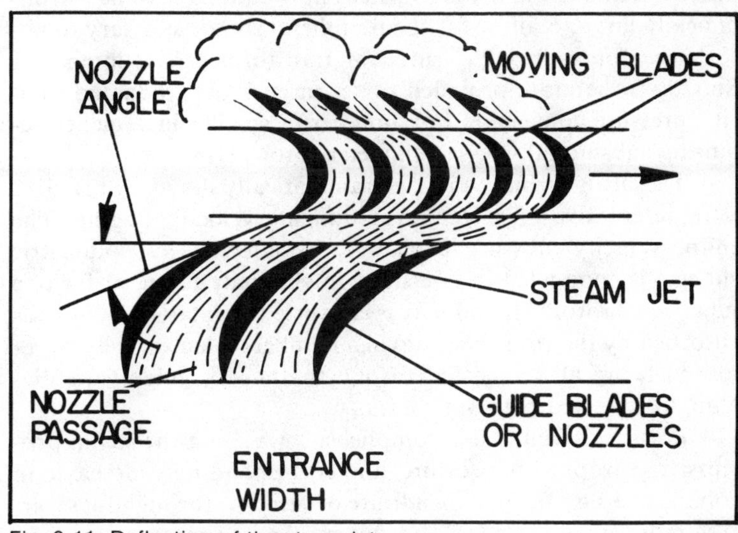

Fig. 9-11. Deflection of the steam jet.

Fig. 9-12. Drive of a wheel with rotor blades.

the rotor, and the entire turbine casing would have to be astronomically large to allow for the peripheral speed necessary to absorb the "once-through" surge of transformed kinetic energy. Staging is therefore provided to accomplish the graduated velocity/pressure conversion in a number of smaller increments, requiring substantially lesser diameter rotor peripheries.

The curtis turbine, as is diagrammatically shown by Fig. 9-13 is representative of a technique utilizing velocity staging. The entire velocity/pressure exchange, as effected by volumetric change, is conducted in one stage to eliminate the need for one super-sized rotor. The velocity is converted to rotational work (as absorbed by the rotor blading) in a number of successively larger rotors, being all connected to one shaft, with guide vane diaphragms inserted between rotors.

Pressure staging is accomplished by having the steam progressively dropped in pressure, and correspondingly increased in volume at a significant expenditure of velocity, throughout several or more pressure stages. The successive pressure stages are compensated for by the increased volume to be handled by their increased diameter, as the same amount (weight) of steam must pass through the entire length of the turbine. Because of the decreased pressure, a proportionately larger volume and rotor size must be anticipated. The pressure reduction between stages may also be achieved by the use of the *extraction* type of turbine, so-called because of the pressure drop created by having the steam from certain selected points extracted or tapped off from the turbine and then used to satisfy either process or industrial plant requirements. The extraction turbine may be considered as a feeder supply for the various process demands, while collaterally exploiting the pressure reduction caused by this "extraction" to facilitate

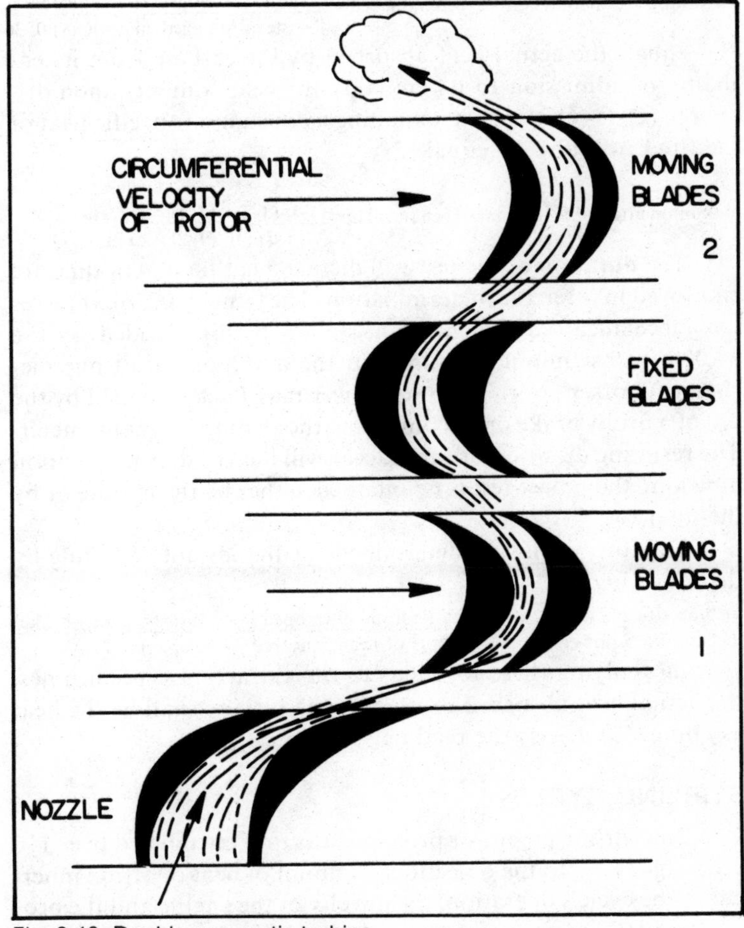

Fig. 9-13. Double row curtis turbine.

a reduced pressure area from one side of the stage to the next—with a power absorbing rotor in between these stages.

The two principal classes of turbines, the reaction and the impulse types, were introduced by DeLaval (impulse) and Parson (reaction). Generally, the impulse type is employed for high shaft speeds and the reaction types are used for high torque and lower shaft speeds. The Westinghouse turbines are an optimum combination of both types, compounded to provide the compromised benefits as offered by each (Figs. 9-15 through 9-17).

The acceleration formula and its modified derivatives are used in computing the turbine's horsepower. The maximum attainable horsepower is given as:

$$\text{maximum attainable thermal/mechanical hp} = \frac{\text{shaft horsepower } (*) \text{ delivered}}{\text{lbs. steam supplied/min}(\times)Q(\times).0236}$$

"Q" equals the actual heat absorbed by the turbine from its enthalpy of admission to its enthalpy (or heat content) upon discharge (Q-1)-(Q-2) which, assuming a constant of specific heat of the fluid, will also be equal to:

$$\text{maximum attainable thermal/mechanical hp} = \frac{(T\text{-}1)\text{-}(T\text{-}2)}{(T\text{-}1)}$$

where:
(T-1) = entrance "Q"
(T-2) = exiting "Q"

The duration of the test will dictate what factors of time are employed in the actual determination. The figure 0.0236 expresses the mechanical equivalent of heat, 778 ft. lbs. divided by the 33,000 ft. lbs./minute required for the development of one mechanical horsepower. The horsepower may be determined by the use of a prony brake or by connecting the shaft to a dynamometer. The resulting duration or time factor will then be dependent upon how long the power test is maintained, either by the minute or by the hour.

The torque may be determined for the instant of testing by the following:

$$\frac{\text{turbine efficiency}}{\text{thermal horsepower}} = \frac{\text{total shaft torque developed in ft. lbs. (by prony brake)}}{\text{Btus of heat delivered to turbine } (\times)\ 778}$$

In applying these formulas to the extraction type turbines, the actual heat absorbed or used by the turbine shall be the heat as computed across the blading.

STIRLING CYCLE

The stirling engine or principle was first introduced over 116 years ago. Due to the objectional amount of heat rejection inherent to the cycle's operation, the novelty of the engine and the profuse availability of petroleum fuels for the wanton internal com-

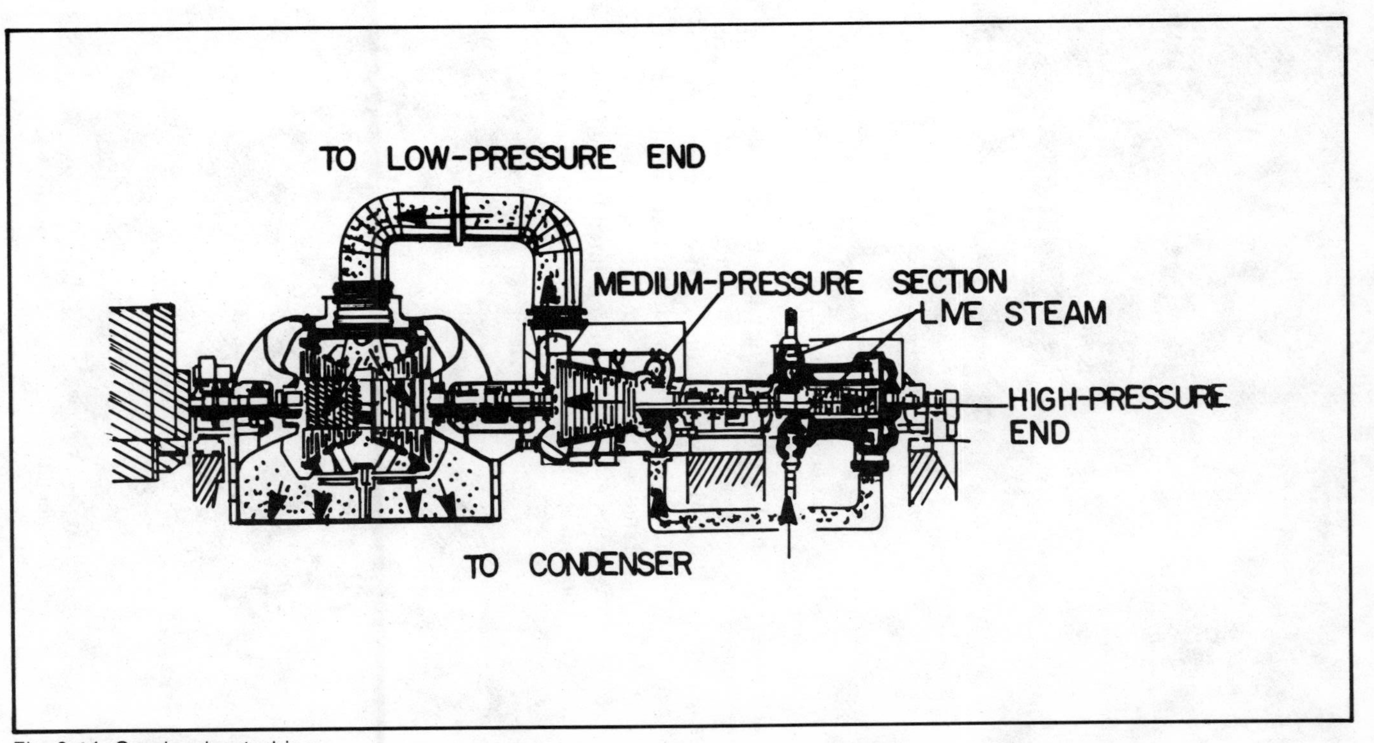

Fig. 9-14. Condensing turbine.

Fig. 9-15. A worker is observing a dynamometer (courtesy of Westinghouse Electric Corporation).

bustion engine, it has only been in the past ten years that attention has focused on this external combustion engine of the displacement variety. The stirling's cycle allows efficiencies in excess of 45 percent, which accordingly places it in a class by itself. Further, the stirling engine may operate on a multitude of fuels, both of the fossilized varieties as well as heat generated through suitable exchanges from nuclear reactors 2nd solar heaters. Because the

Fig. 9-16. View of a steam turbine test facility (courtesy of Westinghouse Electric Corporation).

engineering community will no doubt be witnessing the rebirth of this novel entity, it should then receive advance consideration as a duly deserving prime mover. The following abbreviated description of the cycle in theory, although certainly not of instructional conclusiveness, should serve to whet the appetitie of the reader for a more extensive familiarity with this "different" but potentially superior inovation.

Fig. 9-17. A megawatt turbine generator (courtesy of Westinghouse Electric Company).

Theory of Operation

First and foremost, the stirling engine is a heat engine and falls under the classification of an external combustion displacement type, of reciprocating to rotary drive. It is essentially, but not necessarily, of the self-contained package type unit. The engine may be utilized for mobile or automotive applications, in addition to stationary, marine and railroad utility. Because of the engine's somewhat objectionable weight to horsepower ratio and the superiority offered by the jet and turbojet, it is doubtful that we may witness an Airborne stirling.

The stirling engine as manufactured by the Philips stirling engine works of Eindhoven, Netherlands, and the engine/generator set as produced by the General Motors Corporation of America will be studied herein. The "cycle" as such can best be described as occurring in a succession of "phases" Figures 9-18 through 9-20 show the respective components and "phase" positions. It will be noted that the "phases" correspon to some degree with the related

Fig. 9-18. Philips stirling engine with generator to demonstrate its multi-fuel capacity (courtesy of Philips Research Laboratories).

functions of the steam reciprocator. The reader should acquaint himself with an identification of the engine's component parts (Fig. 9-19) before continuing.

The sequence of events or "phases" is as follows (Fig. 9-20). With the piston in its lowest position (phase 1) and the displacer in its extreme highest point, all of the gas is then contained in the cold space. With the displacer still remaining in its highest position, the piston now moves up to compress the gas at its low temperature (phase 2). The piston in its highest or top dead center position remains stationary. The displacer has moved down to its "near bottom" position while "displacing" the cold compressed gas from the cold space (or chamber) to the hot space (or chamber), absorbing heat from both the regenerator and heater (phase 3). The hot gas, after having been passed through the heater, has now quite subtantially expanded, pushing both the piston and the displacer to the bottom dead center positions. The displacer is now in proper position to rise again. The sequence of events will repeat themselves, as long as heat is applied and removed (phase 4).

Figure 9-21 diagrams the events of the "double acting" version of the stirling cycle. As illustrated, this is a four cylinder model. Each cylinder has a hot expansion chamber at the top of

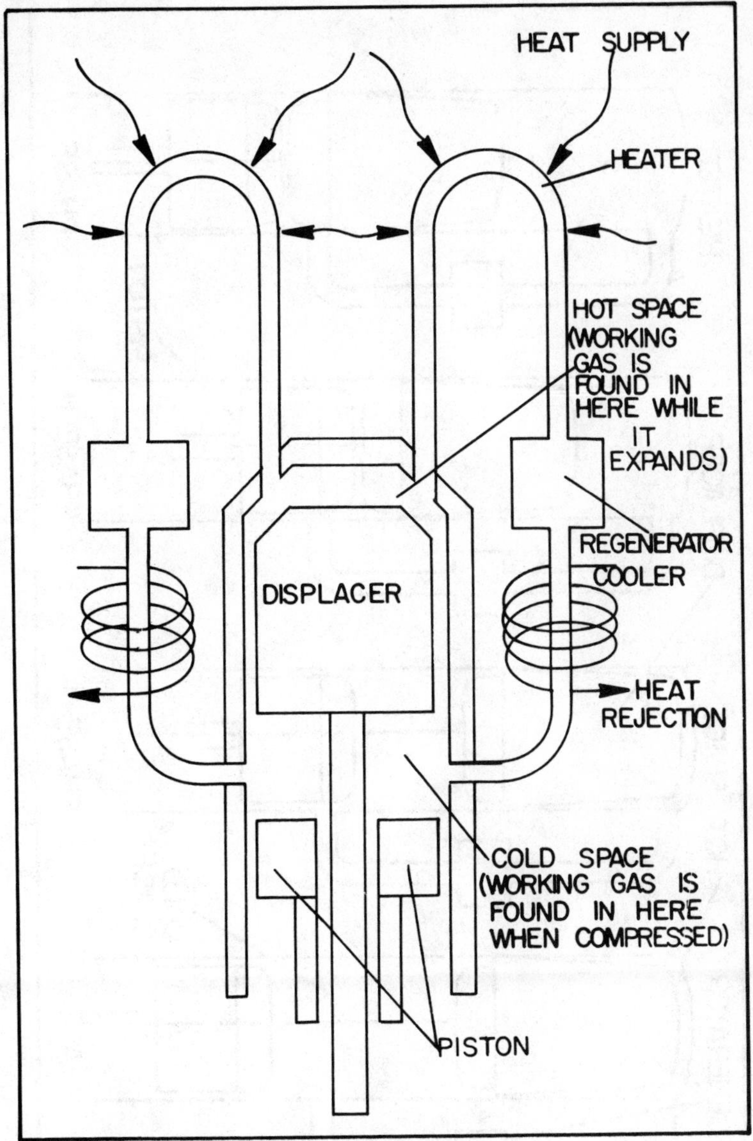

Fig. 9-19. Principle of the stirling displacer type engine.

each cylinder and a cold compression chamber at the bottom of each respective enclosure. The cranks are set 90 degrees apart in this system to facilitate the necessary "phase displacement" required for the cycle's inherent principle of heat transfer and consequent power producing expansion and contraction effects.

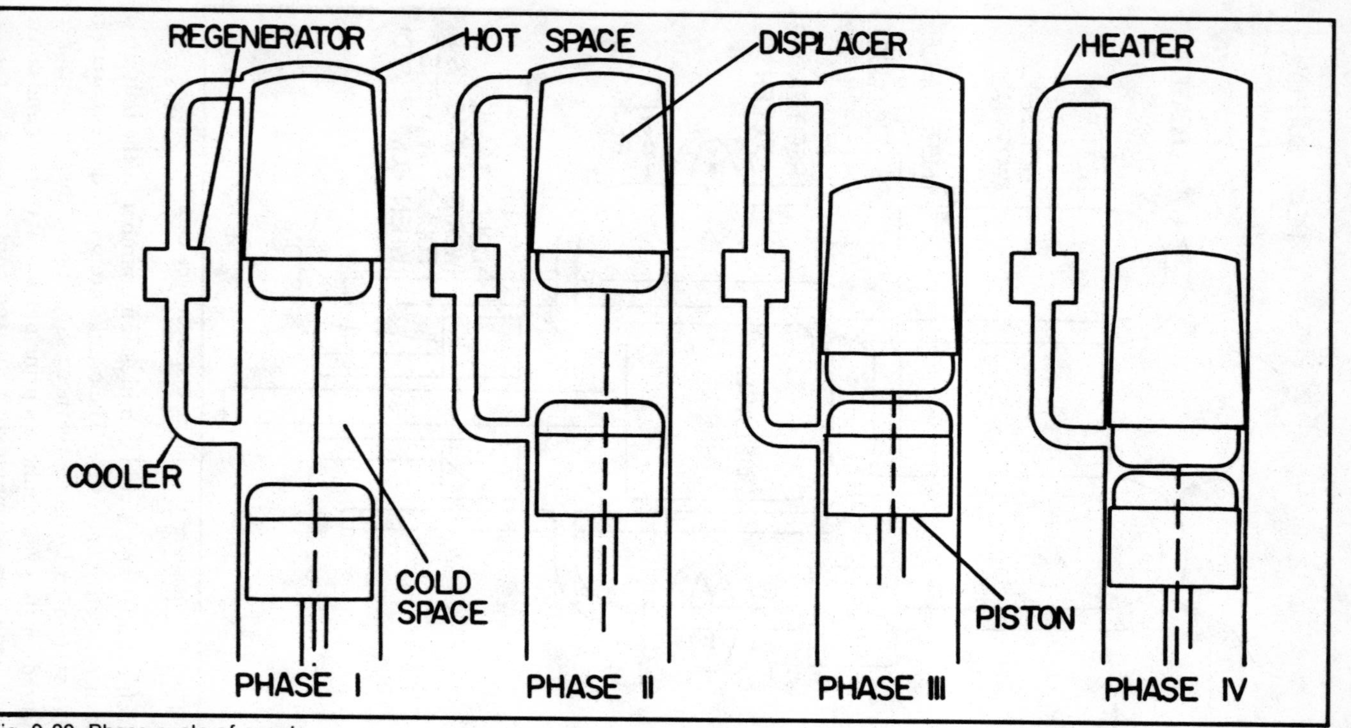

Fig. 9-20. Phase cycle of events.

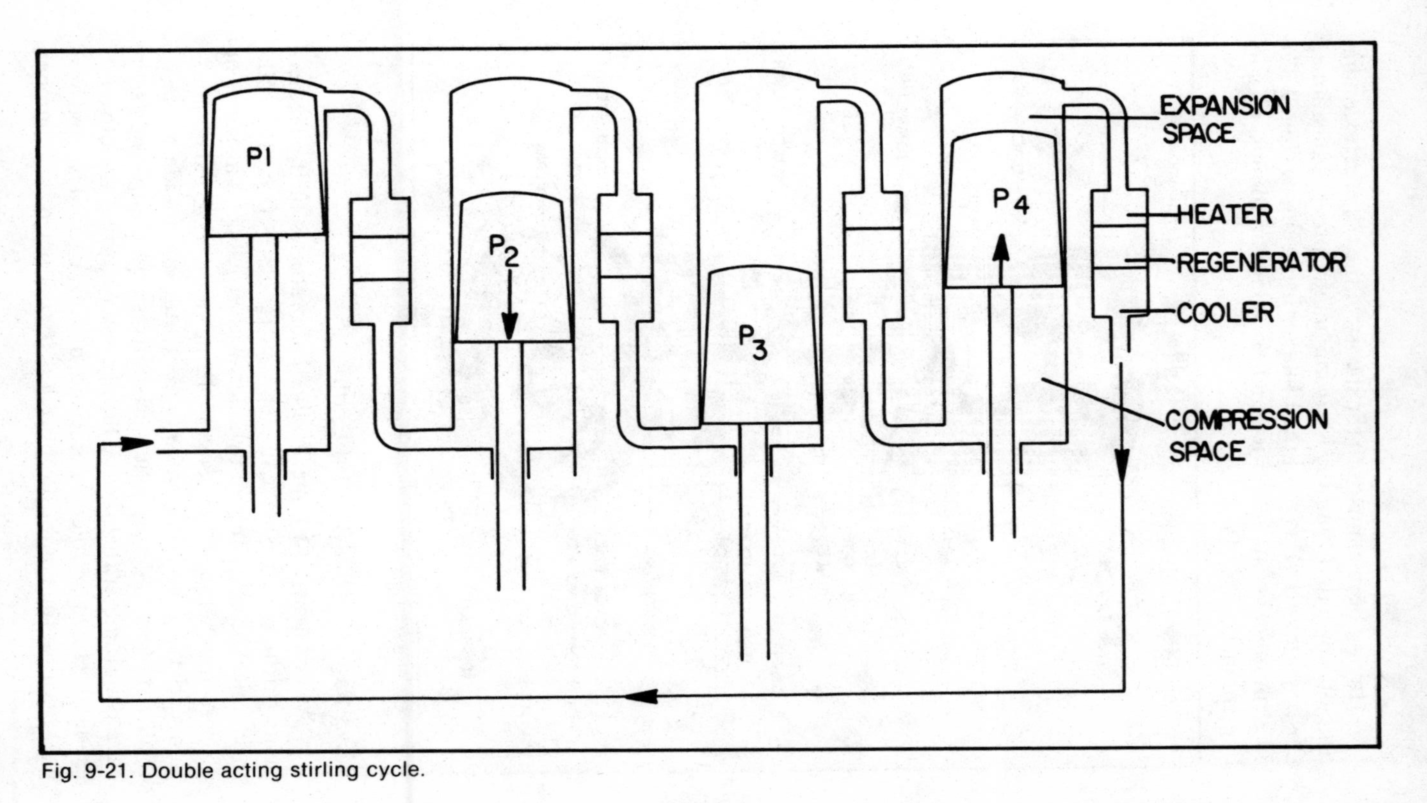

Fig. 9-21. Double acting stirling cycle.

Having been exposed to the theory of the cycle, an examination of an actual cutaway drawing might prove educational. Observe an internal view of the Philips stirling engine (Fig. 9-22). The components are identified and are common to most all stirling engines.

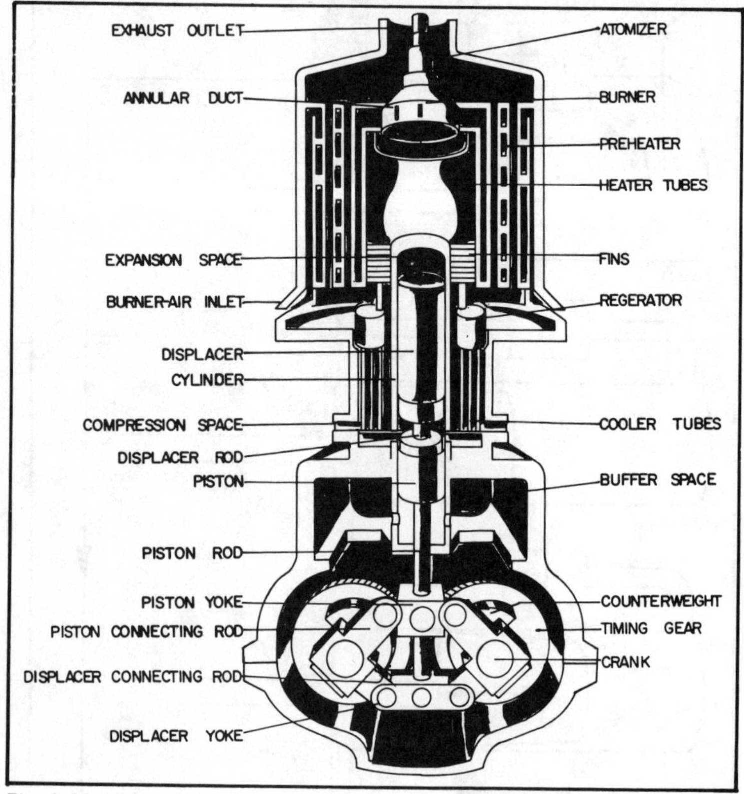

Fig. 9-22. View of the stirling engine.

Advantages of the Stirling Engine

As constructed by the Philips Laboratories in the Netherlands, stirling engines have horsepower ranges of from 10 to 500 mechanical horsepower per cylinder, with efficiencies in excess of 45 percent (thermal). The engine, being of the external combustion variety, enjoys the advantage of utilizing a wide range of fosillized fuels and synthetic derivitatives and compounds such as: alcohol, diesel fuel, heavy lubricating oil, salad oil, olive oil, crude unrefined petroleum, propane, butane, natural gas and biogas. Because the combustion process is conducted in a chamber

Fig. 9-23. 40 KW stirling automotive engine under test in laboratory (courtesy of United Stirling Company, Sweden).

other than the actual piston expansion cylinder, many otherwise unusable low grade hydrocarbon based fuels may be fired, with no adverse effects to the cleanliness of the cylinder walls or piston.

Approximately 40 percent of the heat produced by combustion is effectively employed for the purpose of actual power development. Exhaust and radiation do not exceed a mere 10

Fig. 9-24. 40 KW engine in an automobile (courtesy of United Stirling Company, Sweden).

Fig. 9-25. Closeup view of a 40 KW stirling engine (courtesy of United Stirling Company, Sweden).

Fig. 9-26. 65 KW Stirling engine in a 6-ton truck (courtesy of United Stirling Company, Sweden).

percent. However, a suitably designed radiator for the functional heat rejection must be used to handle approximately 50 percent of the heat so required for dissipation. A large radiator must be considered if this engine is to be successfully applied for automotive purposes. The efficiency depends very little upon actual load situations. The temperature of the heater component is maintained almost constant. Power regulation is affected by pressure of the working fluid (heated air) or alternate gas, such as helium, hydrogen or nitrogen.

Because of the extensive and independent control of the external combustion process, the products of combustion may

Fig. 9-27. 40 KW stirling engine with transmission for automotive use (courtesy of United Stirling Company, Sweden).

Fig. 9-28. 75 KW stirling engine on test stand (courtesy of United Stirling Company, Sweden).

easily be monitored and rendered compatible to the environment. The contaminates may be separately removed and disposed of; this reduction of contaminates cannot be so easily accomplished with the conventional internal combustion engine. Any scrubbing of precipitator devices would tend to increase the back pressure of the engine and diminish its mean effective pressure. There will be a reduction in the engine's shaft horsepower.

Disadvantages of the Stirling Engine

The principal drawback of the stirling engine is the size of the radiator or heat exchanger necessary to expel the discharge phase heat from the system. This will be no problem for stationary or marine power plants, but could conceivably present a problem for automotive applications. In both the stationary and marine plants, the heat dissipation may quite readily be achieved by the use of a water-cooled heat exchanger or an intermediate binary vapor system. This would effectively remove the discharged heat and render it available for feed water heating and process heating The absorbed heat may just be discharged to the ambient by means of a cooling tower.

The situational remedy is not as easy to implement for vehicles or mobile land propulsion machines. A rather large radiator will be required, which must handle the heat load promptly and constantly under all conditions, even in a traffic standstill in the middle of summer. This relative disadvantage should in no sense be considered as an insurmountable problem, as there are a number of ingenious techniques which could be employed for its resolution. The stirling engine can literally open doors of efficiency once thought closed to both the internal and external combustion engines. This engine has indeed been introduced at a most opportune time in our campagin for efficiency and conversation.

The termal and thermal/mechanical efficiencies may be accordingly calculated as follows:

$$\text{Thermal Efficiency} \quad \frac{(T\text{-}1)-(T\text{-}2)}{(T\text{-}1)} \quad or \quad \frac{(Q\text{-}1)-(Q\text{-}2)}{(Q\text{-}1)}$$

Where: (T-1) =temperature of combustion gases as generated in in heat chamber

(T-2)=temperature of products of combustion at exiting point in system.

(Q-1)=enthalpy (heat content) of amount of gasses generated in heater.

(Q-2)=enthalpy of the gases leaving the system after regenerator.

$$\text{thermal/mechanical horse power} = \frac{\text{actual shaft horsepower (*)}}{\text{Btu's generated (x) 0.0236-per min.}}$$

(*) Note that either the prony brake or the dynamometer may be employed for testing.

The mechanical horsepower of the stirling cycle may be computed as:

$$\text{mechanical horsepower} = \frac{\text{MEP (x) L.A.N.K.}}{33{,}000}$$

Note: All factors are equal to the conventional displacement type prime mover consideration

HISTORY AND DEVELOPMENT OF THE STIRLING ENGINE

The idea of the stirling cycle was originally a variation of the basic vapor cycle principle. It was first introduced by the Scottish Minister Reverend Robert Stirling in the year 1816. At first, the constructed engine in its working entirety consisted of a brick furnace fired with coal, a metallic shell supplied with air from the outside, a movable displacer fitted tightly within the shell and a hollow tube connecting the displacer to the working cylinder. A piston was mechanically connected to the displacer shaft through a hole in the center of the piston. The piston was timely moved through its required cooperation with the displacer through means of "yoke" type of crankshaft.

In 1827, after years of dedicated work on his original version, Stirling introduced the *regenerator*. Using thin strips of metal, he

Fig. 9-29. 75 KW stirling engine for automotive or marine use (courtesy of United Stirling Company, Sweden).

successfully transferred heat from the exhaust section of the engine back to the heat chamber. The exhaust gases were cooled by assisting the process of heat rejection and increasing the total available heat added to the combustion process. It should also be mentioned that this system of regeneration or transfer of heat between stages or "phases" has been utilized in many unrelated branches of science and engineering, since its conceptual introduction by Stirling for the use in his engine.

The work and ideas of Stirling have excited many experimenters of the scientific community. One such engineer was John Ericsson, who built a monsterous stirling engine, which he called a "caloric," for the purpose of propelling his 2,200 ton ship. Due to imperfections of design, his caloric engine, although successful to the point of proving its workability, could not compete as economically as the steam engines. The project was laid aside when he was called to devote his energies to design the famous iron-clad war ship, monitor," for the United States Navy to use in the Civil War.

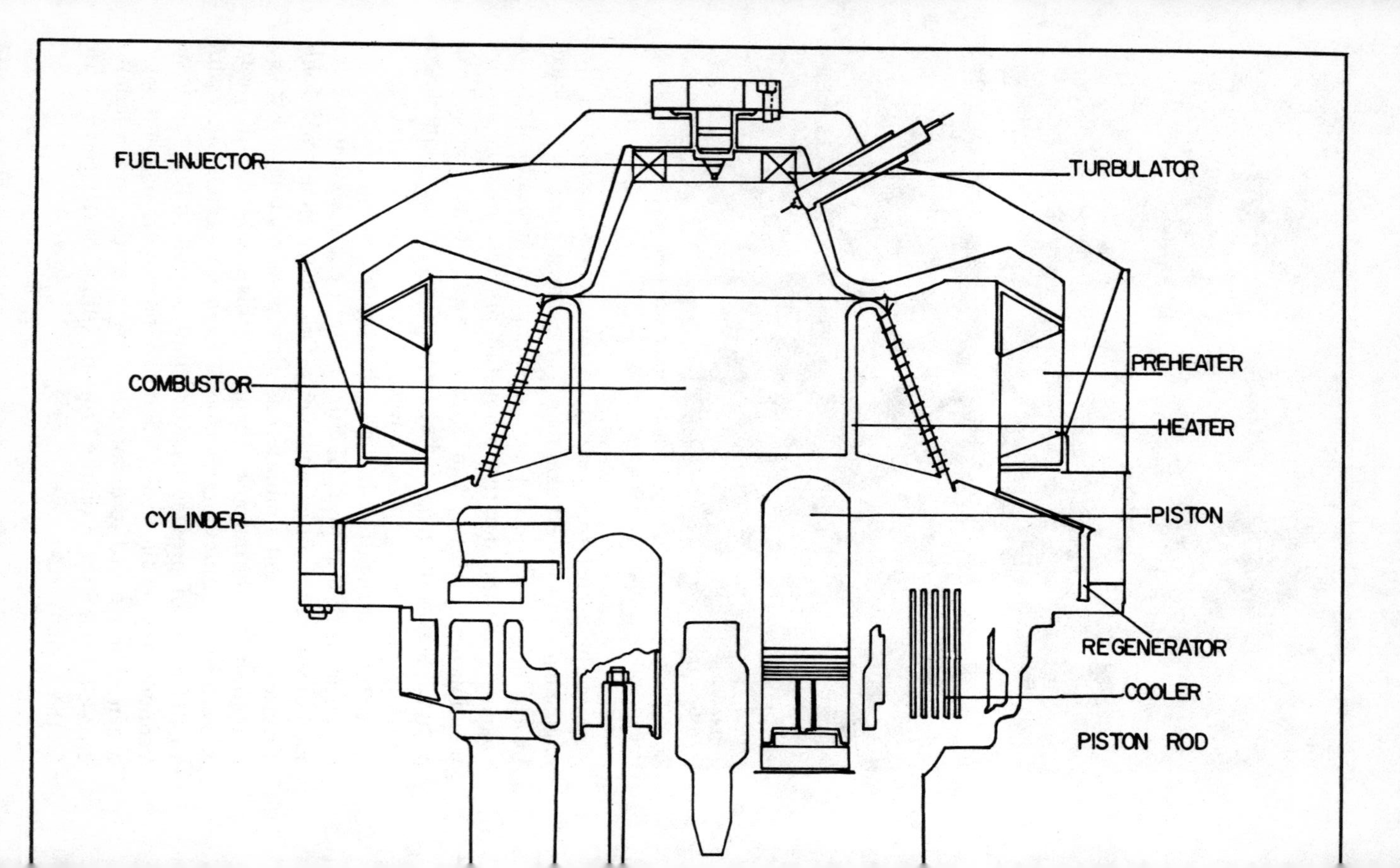
FUEL-INJECTOR
TURBULATOR
COMBUSTOR
PREHEATER
HEATER
CYLINDER
PISTON
REGENERATOR
COOLER
PISTON ROD

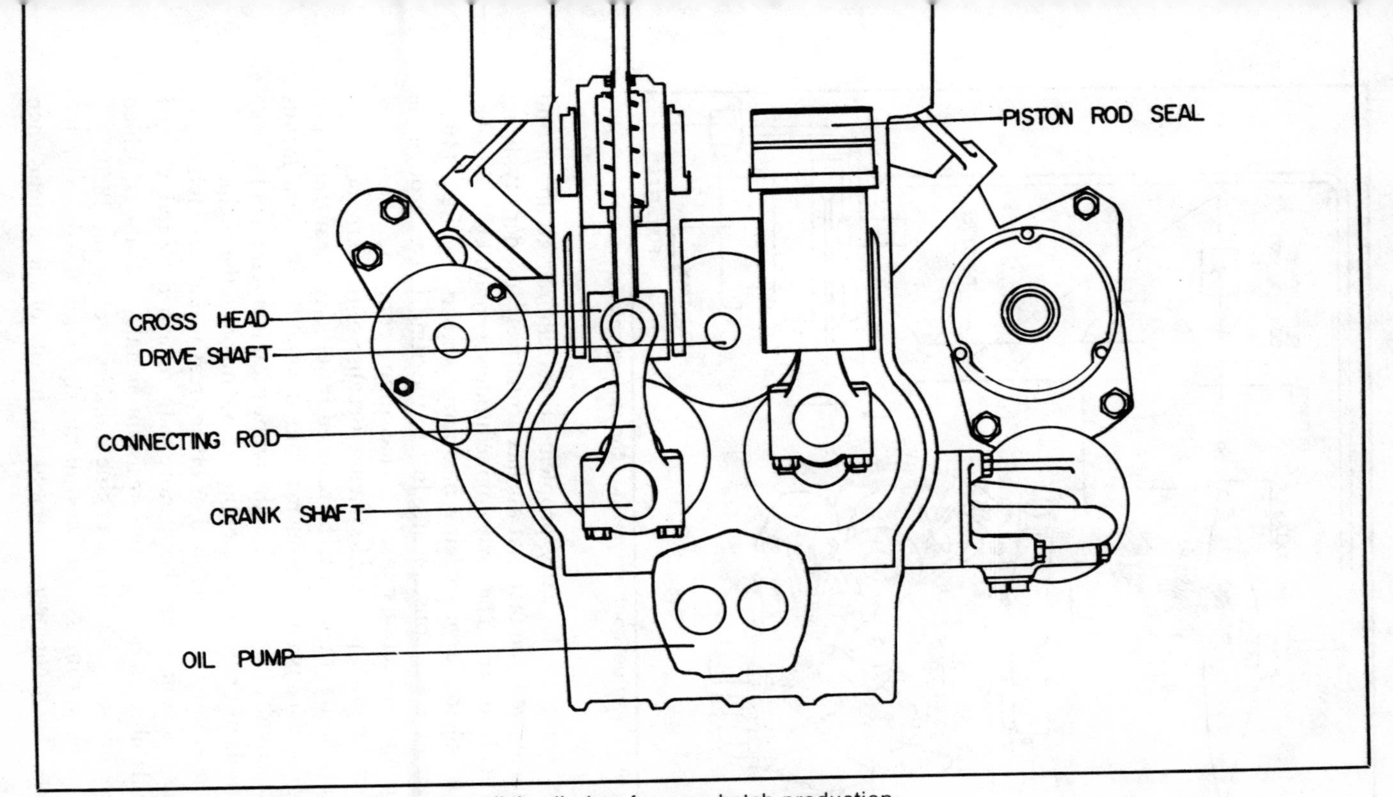

Fig. 9-30. Lastest twin-crank stirling has parallel cylinders for easy batch production.

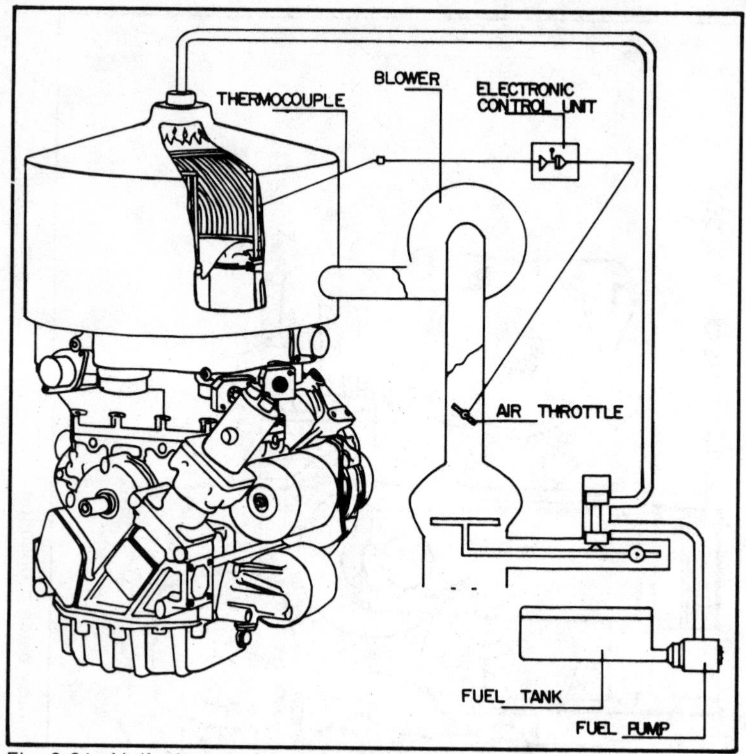

Fig. 9-31. Air/fuel control system of the stirling engine.

Some of the early problems that required attention and remedy were the fact that air has a relatively low density, being difficult to heat. The early engine models were bulky devices because of the large working medium chambers required. Much heated air was consequently lost to the atmosphere through the enormous rejection phase. This has almost completely been overcome by more perfected systems. Because of the extremely high temperatures that had to be reached for efficient heat transfer to the low density air working medium, it was difficult to find metals that could withstand the high combustion temperatures for long periods of uninterrupted exposure. Quite often the engine would keep running until its shell or other areas would just melt away. This, of course, has been more than remedied by today's vast availability of temperature-resistant metals and alloys, which were not available to the engine builders back in the 1850s.

The stirling system was all but forgotten for the past 100 years. Then engineers reviewed the merits of the system when

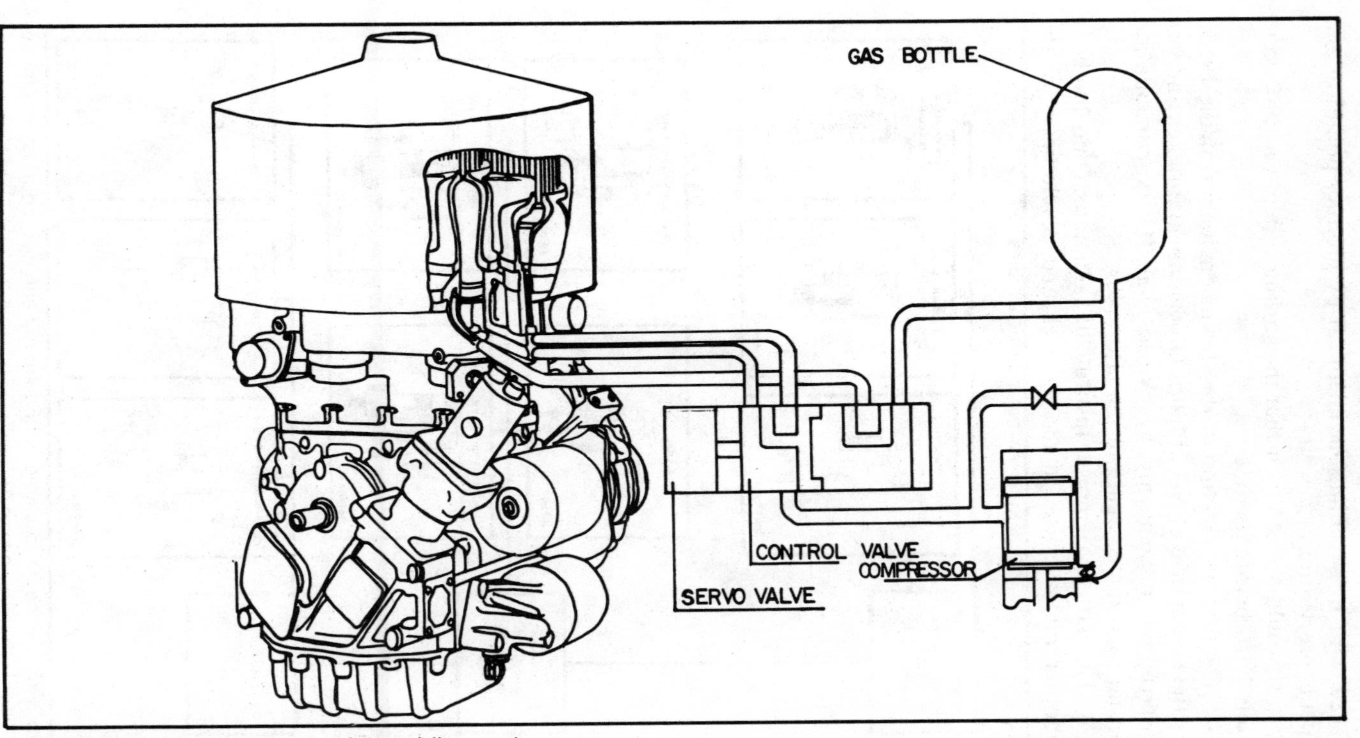

Fig. 9-32. Power control system of the stirling engine.

faced with noise pollution, chemical by-product pollution and fuel conservation.

It would appear then that the stirling system was unfortunately born before its time.

The first resurrection of the stirling engine was in the 1930s, by the Dutch firm of N.V. Philips Gloeilampenfabrieken and was interrupted momentarily by World War II. Experimentation promptly resumed again in the late 1940s.

The Battele Memorial Institute of Columbus, Ohio, has been another major center in the research and development of the

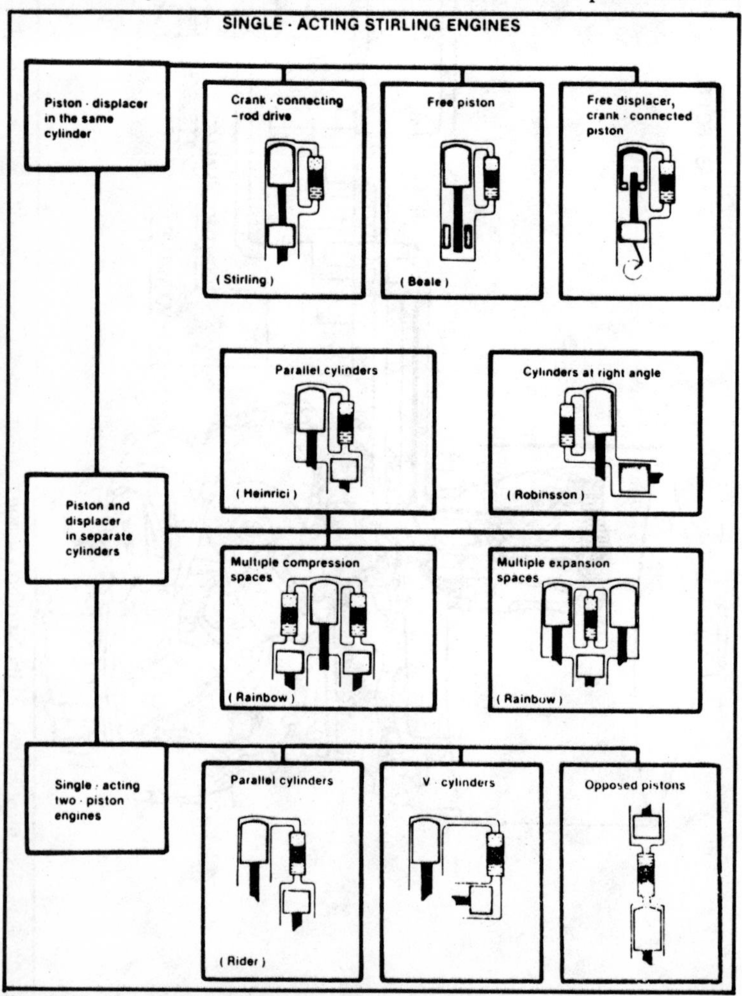

Fig. 9-33. Concepts in single acting stirling engines (courtesy of United Stirling Company, Sweden).

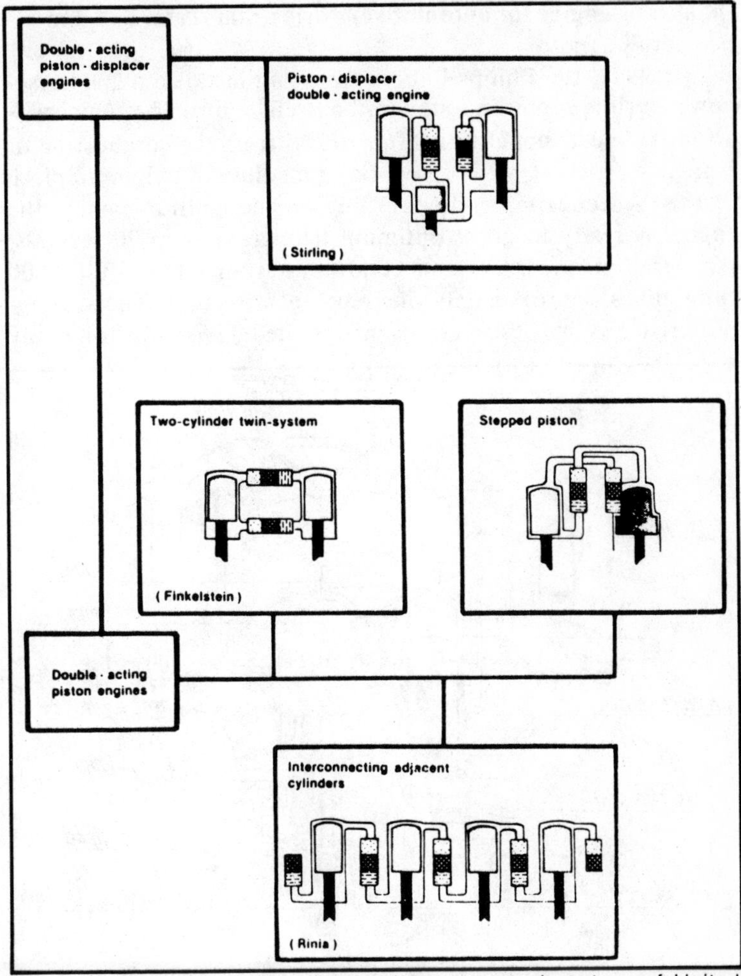

Fig. 9-34. Double acting stirling engine concepts (courtesy of United Stirling Company, Sweden).

stirling engine. Thus far, with efficiencies bettering 45 percent (thermal), such adaptations and modifications like substituting secondary gases such as nitrogen, helium, and hydrogen in place of air and designing the engine to operate from solar heat and nuclear energy have opened many possibilities for the stirling engine. There could not be so advantageously incorporated in other external or internal combustion engines. Many automobile manufacturers and governmental agencies are now allocating research departments for the exclusive purpose of perfecting the stirling engine for automotive, marine, stationary and domestic energy uses.

In 1971, the Philips Laboratories had installed a 200 horsepower stirling engine in a standard passenger bus. The engine was a four cylinder model which functioned from the combustion of regular #2 fuel oil, with the working medium of helium in place of the theoretical air. After turning on the ignition switch, the engine is ready to go at optimum temperatures in 20 seconds. Acceleration under no-load conditions from idle to full 3,000 rpms takes approximately one tenth of a second. The stirling-powered bus has thus far been operated favorably for many

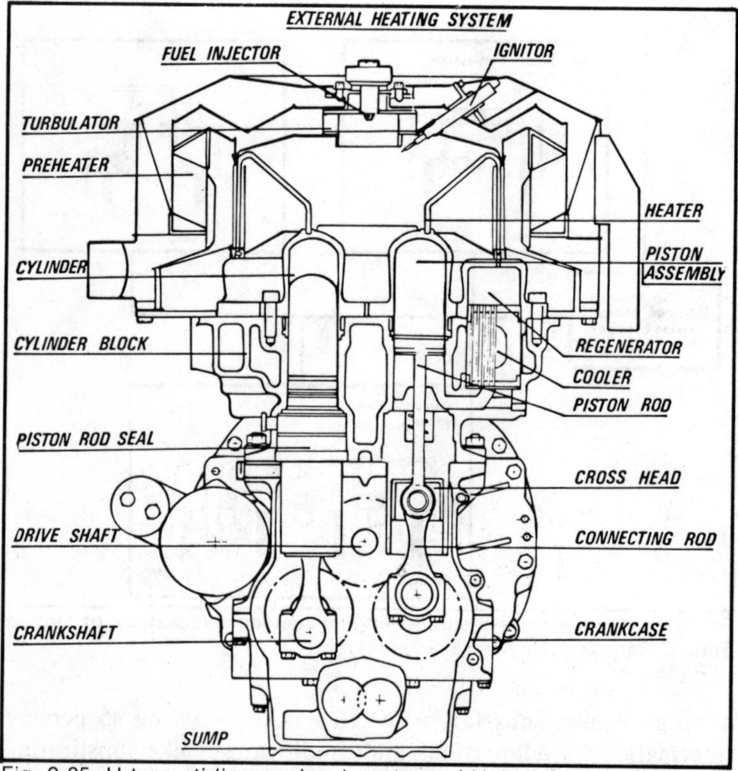

Fig. 9-35. U-type stirling engine (courtesy of United Stirling Company, Sweden).

thousands of miles through regular bus routes. Combustion is absolutely complete and the stirling makes almost no noise whatsoever. A running indicator must be connected to the system to signal that the engine is actually running.

Additional experiments have led to the multi-cylinder 400

horsepower stirling which, because of its interconnected piston arrangements, does not essentially require use of a separate displacer. The interrelated functions of the additional pistons force the required working fluid between its passages (Fig. 9-21). See Figs. 9-23 through 9-35 for photos and diagrams of stirling engines. The following information on the stirling engine comes from Philips Research Laboratories in the Netherlands.

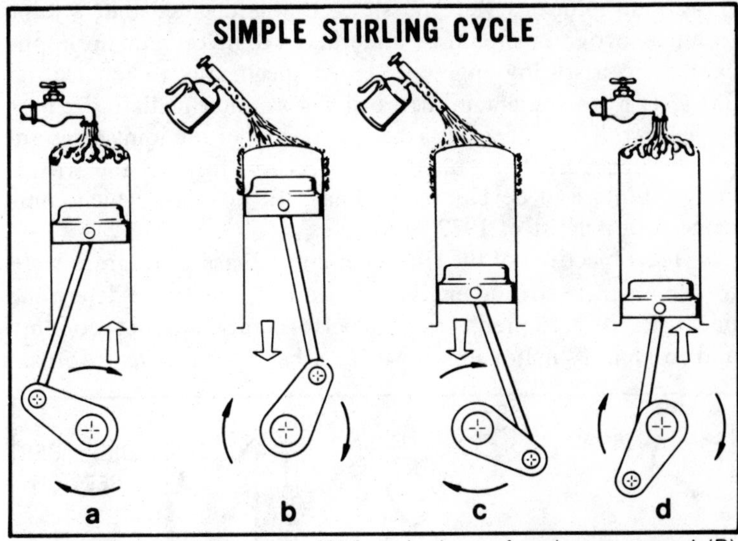

Fig. 9-36. Simple stirling cycle. (A) A cool volume of gas is compressed. (B) It is heated by an external heat source. (C) Expansion takes place. (D) The gas is cooled by an external cooling source (courtesy of Philips Research Laboratories).

INTRODUCTION TO THE STIRLING PHASE I PROGRAM

For the past several years Ford Motor Company has been active in a variety of programs to determine prospects of engines other than the conventional reciprocating piston engine for use in the automotive field.

In late 1970 a meeting was held with N.V. Philips of Holland to discuss their progress in development of stirling cycle engines. Philips is one of the largest corporations in the world and is a major producer of electrical and electronic equipment. Philips originally started work on the stirling engine as a power source for electrical generator sets. Since World War II, under the direction of Dr. R. J. Meijer, development of the stirling engine has continued with a variety of applications in mind ranging from torpedo propulsion and space power to boat and submarine power sources.

Until recently it had appeared that this type of engine was too heavy and complex for passenger car application and that NOx emissions were excessive. Philips Laboratories had been working actively to solve these problems and, at the time of their meeting with Ford in 1970, sufficient evidence was provided to show that solutions were possible. It was decided that a joint technical program should be undertaken to investigate the applicability of a stirling engine designed specifically to replace the 351 CID piston engine in the Ford Torino intermediate size passenger car. This paper describes the results of the joint program which subsequently resulted in a decision to continue with a design, build and development, Phase II, program which commenced in August of 1972.

The objectives of the stirling engine, Phase I, program were to demonstrate stirling engine emission capability, to determine packageability, to predict vehicle performance and fuel economy and to identify major unknowns as a basis for further efforts.

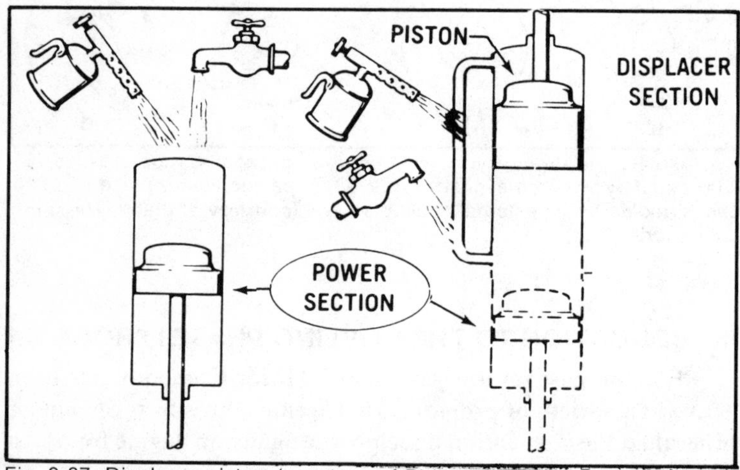

Fig. 9-37. Displacer piston (courtesy of Philips Research Laboratories).

HOW A STIRLING ENGINE WORKS

Before describing the results of the study, it may be helpful to explain how a stirling engine works. Conventional piston engines produce power by expanding a compressed and heated volume of air. A stirling engine also uses this principle, the major difference being the method by which the heat is added. In a conventional engine the heat is supplied by burning a quantity of fuel

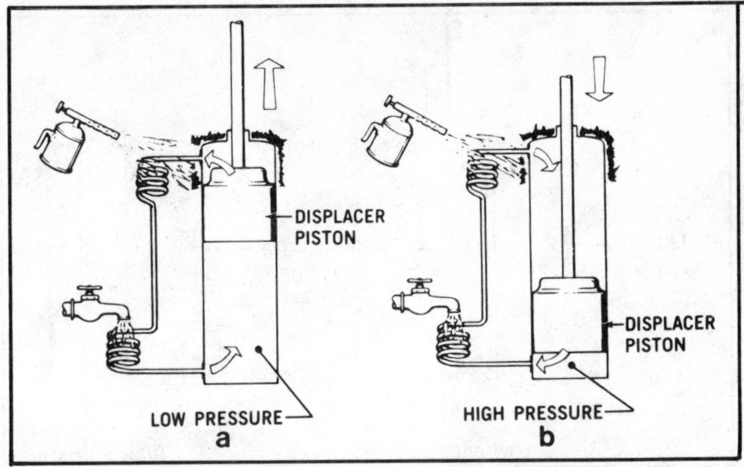

Fig. 9-38. How the displacer piston mechanism works. (A) Low pressure. (B) High pressure (courtesy of Philips Research Laboratories).

inside the chamber. In the stirling engine, the heat is added by an external flame through a heat exchanger (heater head) to the working gas inside the engine.

First a cool volume of gas, entrapped by a piston, is compressed (Figure 9-36A) and then heated by an external heat source (Figure 9-36B). As the gas heats, its pressure increases and the piston is driven downward to turn the crankshaft. After expansion (Figure 9-36C), the gas is cooled by an external cooling source (Figure 36-D). Its pressure decreases, and the gas is once again compressed. Since the pressure during the hot expansion is much higher than during the cool compression, there is a net work output from the engine. The complete cycle takes place in one revolution of the crankshaft as opposed to two revolutions required by conventional engines.

Since exchanging the heating and cooling sources is a cumbersome process, Robert Stirling, for whom the cycle is named, conceived a refinement to overcome this problem. His invention replaced the alternating use of hot and cold sources by addition of a mechanism called a displacer piston which serves to move the gas between a stationary hot chamber and a stationary cold chamber (Figure 9-37). This displacer type engine is described here as a convenience in explaining the principle of a stirling engine. However, a different configuration, the double acting engine, is superior for passenger car application.

The displacer piston mechanism allows the heating source to be stationary at one end of the cylinder and the cooling source

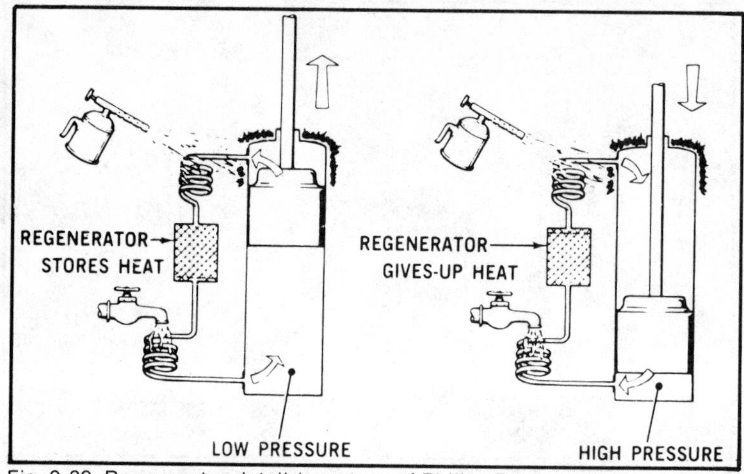

Fig. 9-39. Regenerator detail (courtesy of Philips Research Laboratories).

to be stationary at the other end (Figure 9-38). When the displacer piston moves upward (Figure 9-38A), the hot working gas from the upper portion of the cylinder is first moved through the heating coil or heater tubes. The gas then flows through the cooling coil where it is cooled until most of the working gas is in the cold section below the displacer piston. Because the gas is cool its pressure is low. Moving the piston downward (Figure 9-38B) forces the working gas back through the cooling coils and into the heater tubes where it is heated and forced into the hot section

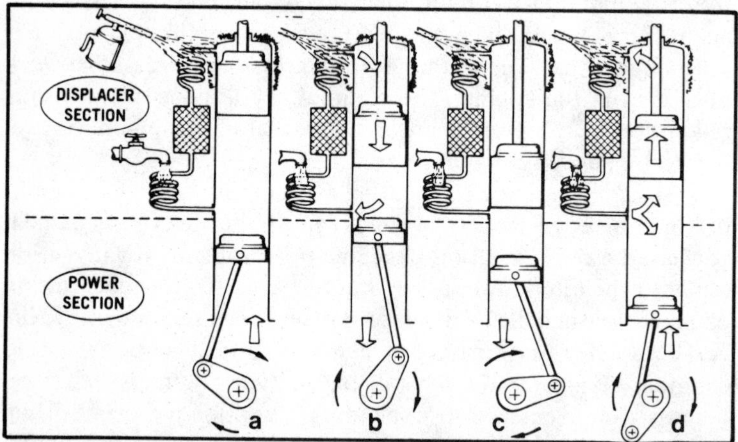

Fig. 9-40. Ideal stirling cycle. (A) Cooled gas is compressed by the power piston. (B) Gas is heated and its pressure is increased. (C) The hot gas has completed its heating cycle. (D) Displacer piston moves upward to force the working gas into the cool portion of the chamber (courtesy of Philips Research Laboratories).

above the displacer piston. Since the gas is hot, its pressure is high. There are no valves in the flow path, so that when the upper chamber is at high pressure, the lower chamber is also at high pressure.

One more addition is required to complete a practical stirling engine. The regenerator, as shown on the schematic (Figure 9-39) is located between the fixed heating and the cooling sources and stores otherwise wasted heat during the cooling process and permits recovery of the heat during the heating phase. This stored heat is equal to several times the heat added from the outside heat source.

Figure 9-40 shows the displacer section combined with the power section to form the basic stirling cycle power unit. Figure 9-40A shows the cooled gas being compressed by the power piston as in a conventional I.C. engine. In Figure 9-40B, the compressed

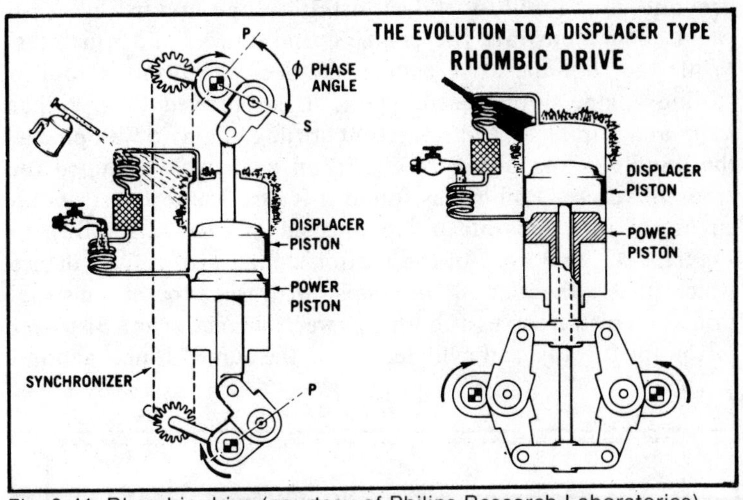

Fig. 9-41. Rhombic drive (courtesy of Philips Research Laboratories).

gas is being heated and its pressure increased because the displacer piston is moving a portion of the gas into the upper or hot part of the displacer section. The pressure increase is felt on the power piston. In Figure 9-40C the hot, high pressure gas has completed its heating cycle due to the descending displacer piston and the power piston has completed its power stroke driven by the high pressure gas. Figure 9-40D shows the displacer piston moving upward to force the working gas into the cool portion of the chamber, thus decreasing its pressure. The power piston is now ready to repeat the compression stroke and the cycle is completed.

The remaining mechanisms required to form a simple stirling cycle are those needed for driving the displacer piston at a fixed relationship to the power piston (90° out-of-phase) (Figure 9-41). This can be done by a crank and chain or preferably by an improved mechanism developed by Philips called the *rhombic drive*.

Since its invention in 1953, many rhombic drive engines have been built by Philips ranging from under 1 to over 300 horsepower. The rhombic engine, is a single cylinder engine. However, its characteristics are quite different from a single cylinder I.C. engine. Not only is the engine perfectly balanced but it has two power strokes per revolution and thus a one cylinder rhombic drive stirling engine has torque impulses comparable to a four-cylinder conventional I.C. engine.

Engines have also been installed in 5 kilowatt electrical power generators, in a multi-fuel demonstration unit and in a pleasure yacht to demonstrate the engine's smoothness and quietness. While the rhombic drive engine has been the primary tool in stirling engine development work, it has proved to be rather large and complex. A more recent stirling engine development. the double acting piston results in an engine better suited for automotive use. Philips has found it is possible to construct an engine of four separate interconnected cylinders as shown in Figure 9-42, and control the motion of the piston by a device which phases them at 90° intervals. With this type of construction, each piston serves as both a power, piston and as a displacer piston for the adjacent cylinder. Thus, the name "double acting."

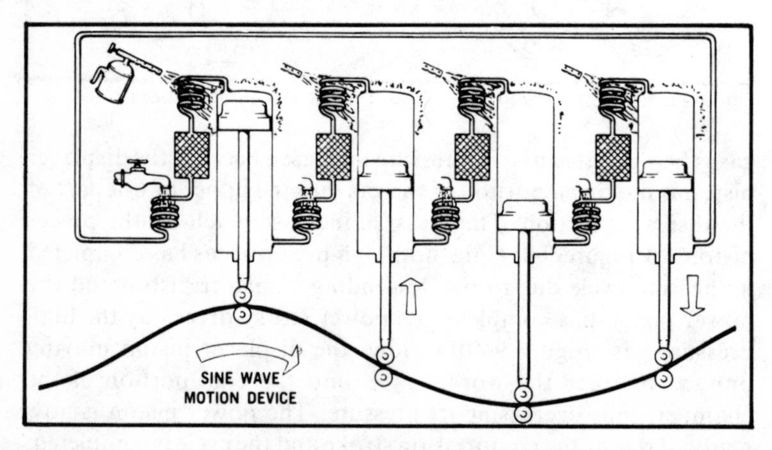

Fig. 9-42. Double acting piston (courtesy of Philips Research Laboratories).

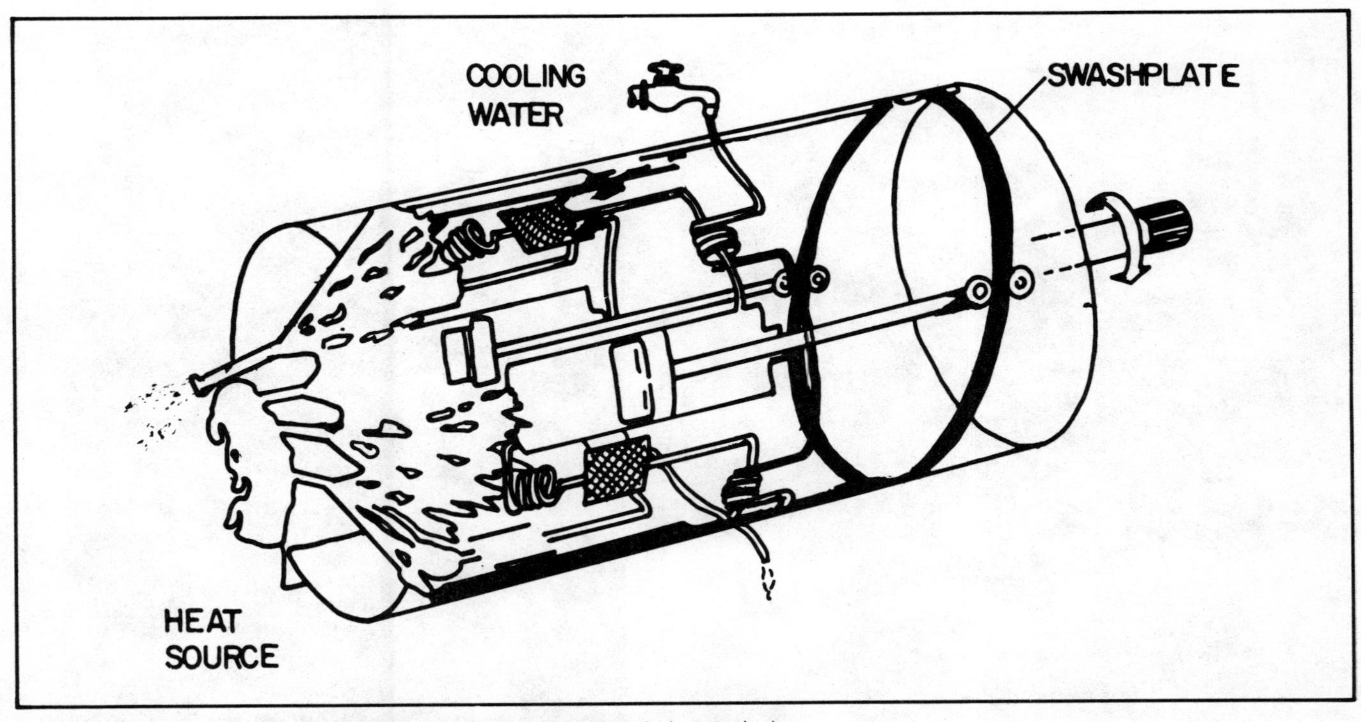

Fig. 9-43. Double acting engine (courtesy of Philips Research Laboratories).

Fig. 9-44. Generator set with stirling engine (courtesy of Philips Research Laboratories).

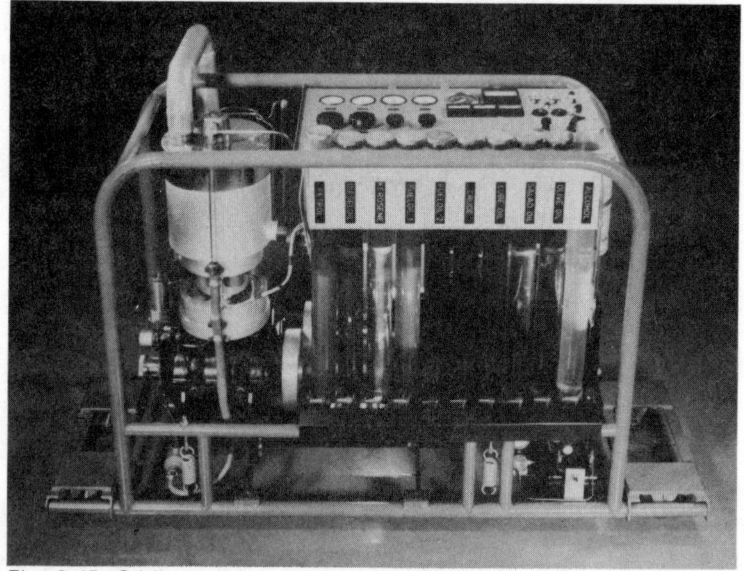

Fig. 9-45. Stirling engine with generator to demonstrate its multi-fuel capacity (courtesy of Philips Research Laboratories).

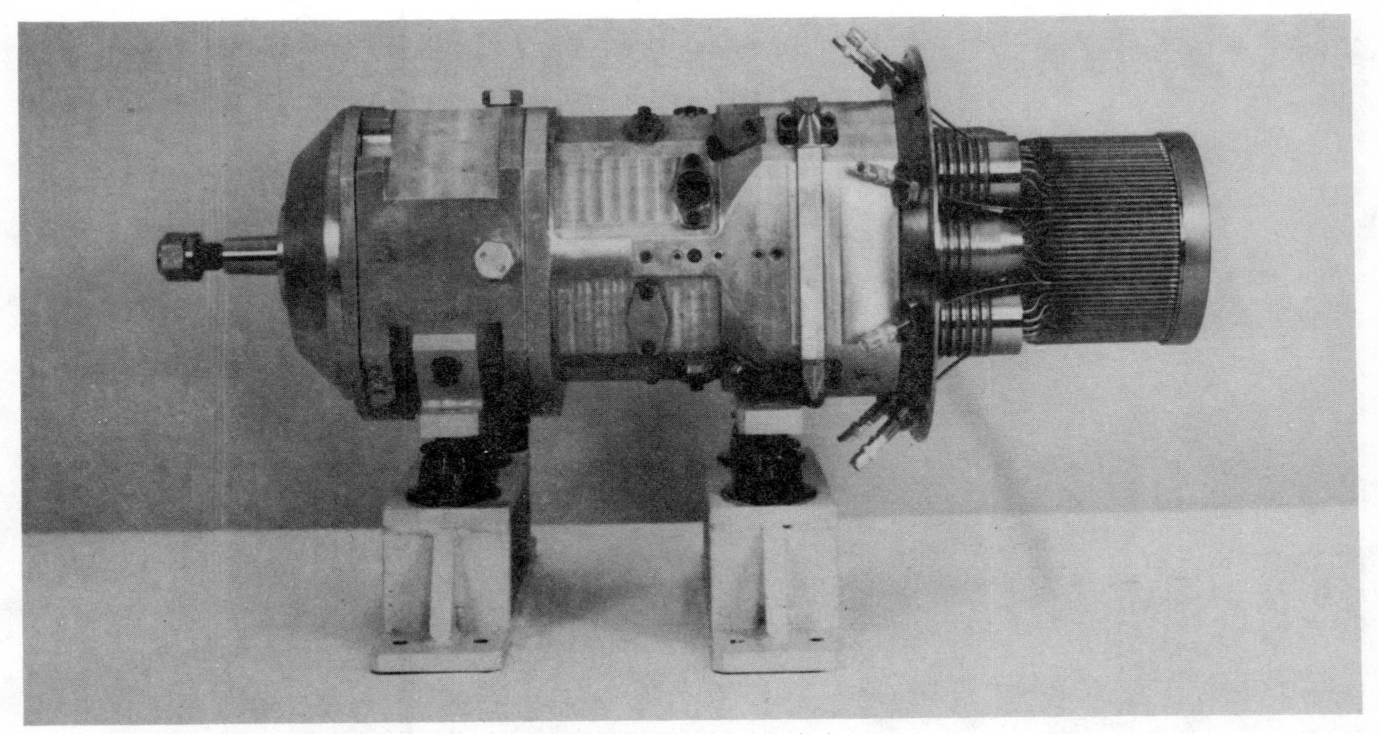

Fig. 9-46. A second generation stirling engine (courtesy of Philips Laboratories).

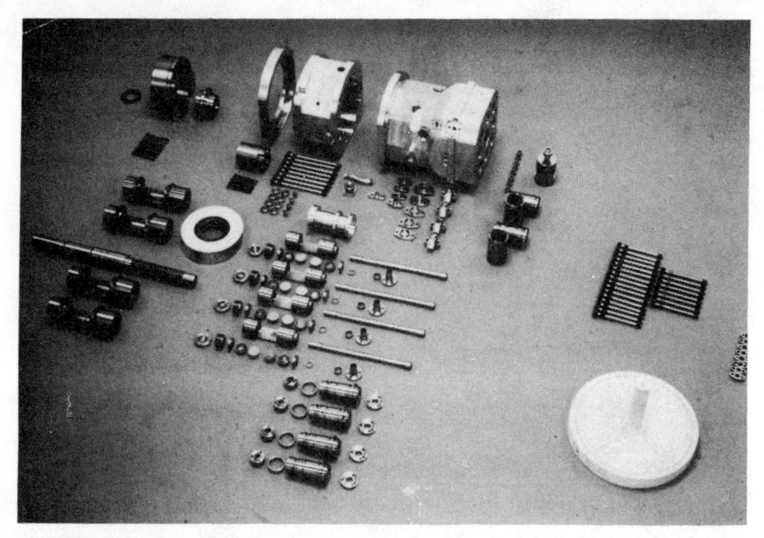

Fig. 9-47. A wide variety of engine parts (courtesy of Philips Research Laboratories).

By wrapping the double acting piston schematic into a cylinder, a compact swashplate engine is formed in which the four separate heating sources are grouped into one common source (Figure 9-43). The four cylinder swashplate engine is perfectly

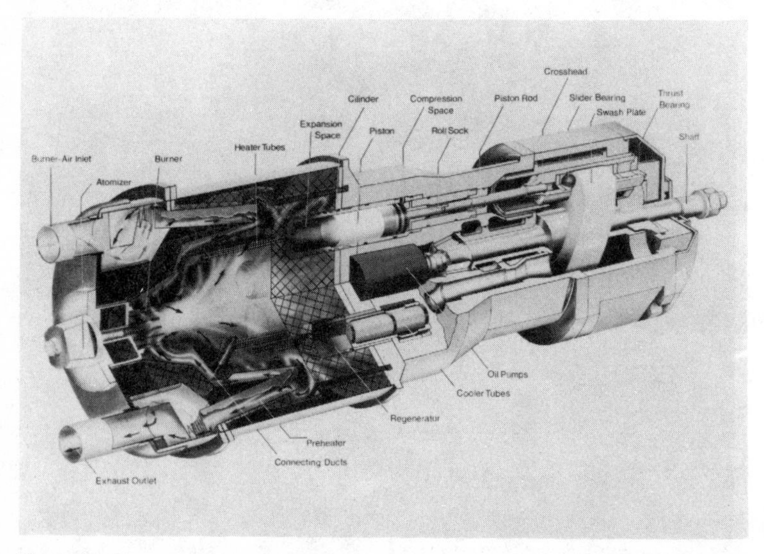

Fig. 9-48. Parts detail of the stirling #4-65 D. A. engine (courtesy of Philips Research Laboratories).

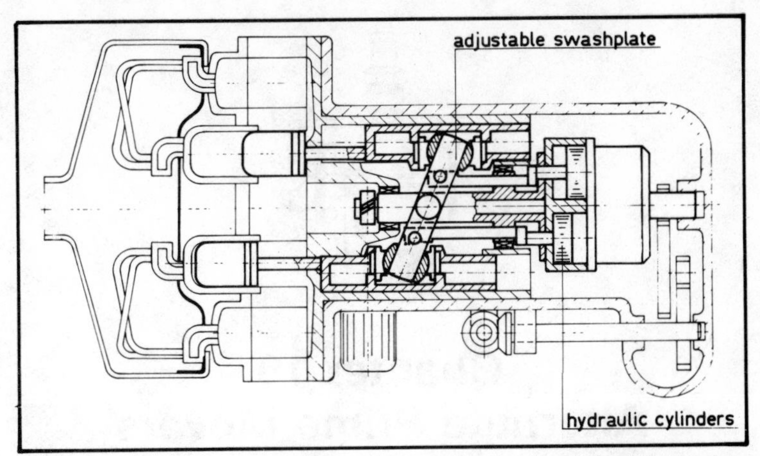

Fig. 9-49. Schematic diagram of a four cylinder double acting stirling engine with variable swash-plate drive. With the hydraulic pistions the position of the swash-plate can be adjusted (courtesy of Philips Research Laboratories).

balanced and has four torque impulses per revolution similar to an eight cylinder internal combustion engine. However, the magnitude of the impulses is much less than those of an eight cylinder I.C. engine. See Figs. 9-44 through 9-49.

Courtesy of Philips Research Laboratories

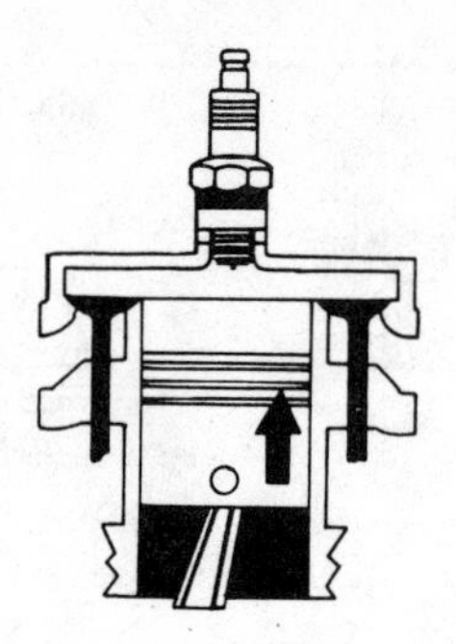

Chapter 10
Alternate Prime Movers

The most widely employed type of *hydraulic turbine* in use today is of the *Francis* or *Kaplan* types. These turbines are adaptable to essentially any type of head and volume conditions and are used

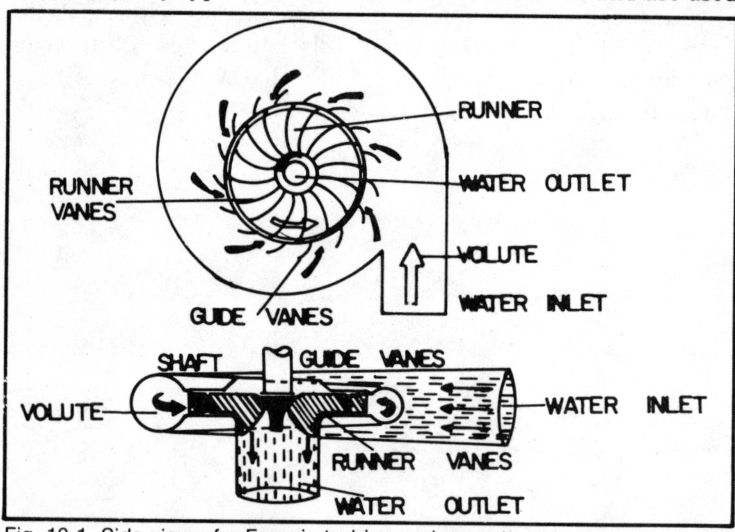

Fig. 10-1. Side view of a Francis turbine and a section of a Francis turbine.

extensively in the large hydroelectric plants in our nation's hydroelectric dams. In the hydraulic type of turbine, the divergence takes place inside the turbine proper. The divergence then is at right angles to the direction of entry, causing the "runner" or rotor to revolve about its shaft. Refer to Fig. 10-1 which describes the Francis type turbine.

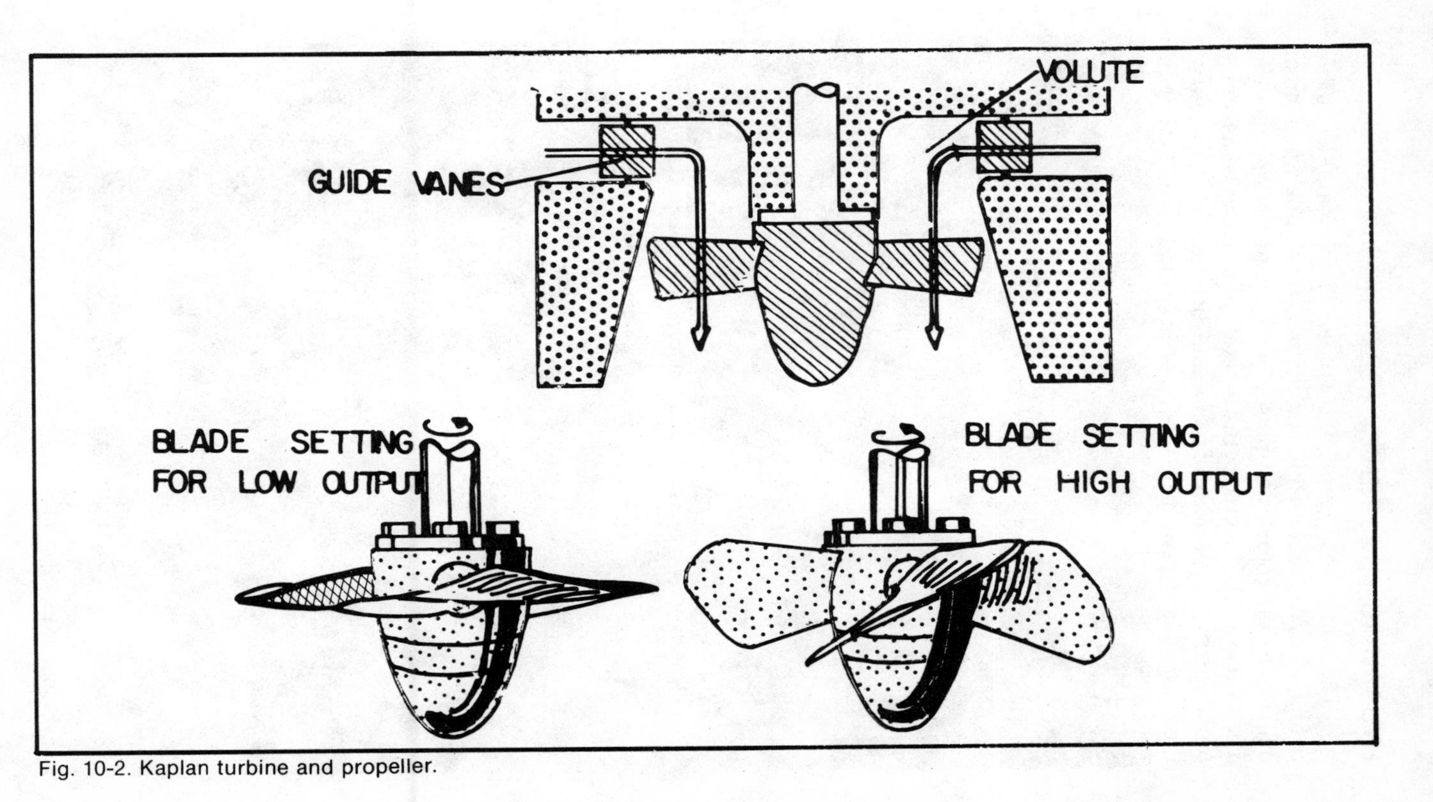

Fig. 10-2. Kaplan turbine and propeller.

HYDRAULIC TURBINE

The water first enters the volute chamber of the turbine casing. This volute is an annular channel, providing water stream delivery to the rotor at all points of its blading simultaneously. The water then is directed between fixed guide vanes which impart to the stream an optimum of computed direction, for final entry to the rotor. The stream, after having been steered through the guide

Fig. 10-3. Aerial view of Shasta Dam in California (courtesy of United States Department of Energy).

Fig. 10-4. Theodore Roosevelt Dam in Arizona (courtesy of United States Department of Energy).

Fig. 10-5. Glen Canyon Dam (courtesy of United States Department of Energy).

vanes, moves through the rotor in a radial direction and into its center. After imparting its kinetic energy to the rotor, the stream leaves through the center of the rotor and out into the tail piping.

The guide vanes of the Francis turbine are made adjustable to accommodate fluctuations in head pressure of the water supply and to handle load changes as required by the driven machine connected to its shaft, which in most cases is an electric generator. The guide vanes serve for the Francis turbine the same throttling control for shaft speed and torque regulation that the needle valve and deflector provides for the pelton wheel.

For very low heads and high flow rates, the Kaplan turbine will best be employed; this variation is also known as the propeller turbine (Fig. 10-21). The water enters the turbine casing laterally or parallel to the propeller axis and *transmits* its kinetic energy to the blades of the propeller. The blades are adjustable in their "pitch" or "bite angle" to accommodate load and water supply variations. Depending upon the design, the turbine may be composed of a volute section to redirect the water flow prior to admitting it through the propeller chamber. Or the water may be led directly through the propeller entirely in a parallel run from sluce gate to tailpipe without changing direction once.

Because of the characteristic of straight-through direction of water flow, these turbines are technically classified as axial-flow types. Even with the straight-through design, guide vanes are featured before the propeller to facilitate speed control. By decreasing the space between the vanes, the water flow may be throttled. This is done in addition to varying the angle or pitch of the blading.

Many low head dams, where the available flow or volume is sufficient to provide hydroelectric power, utilizing this form of turbine. The formulas for computing the hydraulic turbine's efficiency are contained in the part C survey form. The combined product of the weight of the water, the head pressure by which it is delivered at, and the displacement of the flow is divided by the horsepower constant of 33,000 to compute the turbine's theoretical available horsepower. Anything below this value is attributed

Fig. 10-6. Aerial view of Hoover Dam and Lake Mead (courtesy of United States Department of Energy).

Fig. 10-7. Artist's conception of a low head dam (courtesy of United States Department of Energy).

to percent "slip" and operational losses. The efficiency of these turbines is of average return. In considering the medium employed for conversion—water—they are an ideal source of power and electrical generation. See the hydroelectric dams in Figs. 10-3 through 10-7.

PELTON WHEEL

The pelton wheel, sometimes called the pelton turbine is a direct modification of the older water wheel. By virtue of its impulse buckets placed around its periphery, the pelton wheel converts the kinetic energy of the discharged water aimed at its circumference into direct rotary motion. It may be employed to drive machinery, generate electricity or drive any number of shaft-connected loads. The output of the pelton wheel is surprisingly efficient for its simplicity and greatly exceeds its parent water type wheels, having only radial paddles about its rim.

The pelton wheel functions by having a high velocity stream of water aimed at its properly positioned blades, which is at a line tangent to the wheel's radius. The buckets, as illustrated in Fig. 10-8 receive the stream of water and are thus acted upon by the impulse, or reaction to impingement of the high velocity jet of water. The release of the impinged water stream is affected by specially designed blades. The blades curvatures are so hydrodynamically designed as to provide the greatest conversion of the imparted kinetic energy, with the least resistance or other incidental impeding back pressures.

As one can imagine, if the load were suddenly removed from the shaft of this prime mover, the wheel speed would almost instantaneously reach the critical rim speed. The "G force" (centrifugal force) would cause the destruction of the entire wheel and a major catastrophe. To eliminate any problems resulting from load

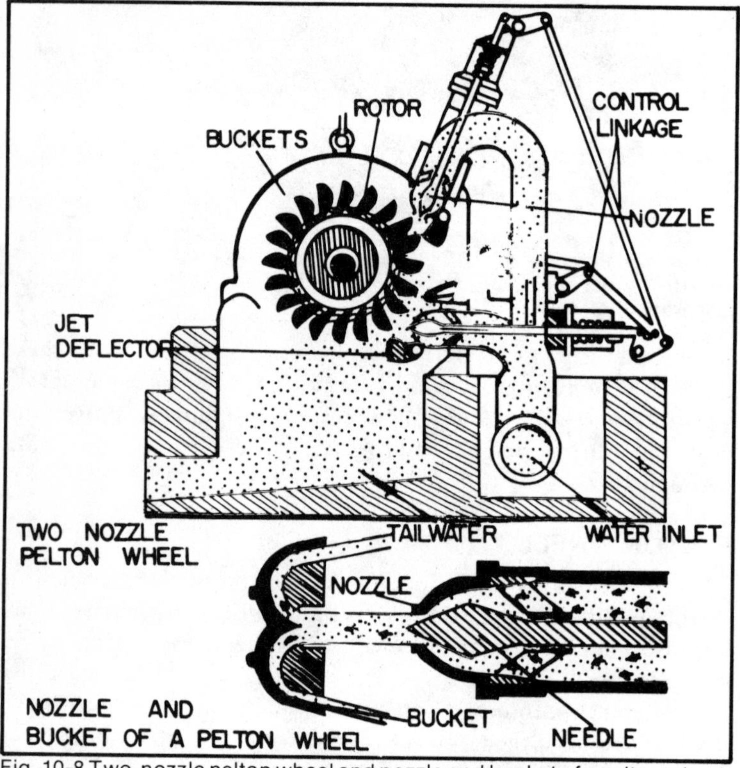

Fig. 10-8 Two-nozzle pelton wheel and nozzle and bucket of a pelton wheel.

fluctuations, needle flow valve arrangement is provided to throttle the flow of water to the buckets. The less the load demand, the more the needle valve will throttle the delivered stream and vice versa.

To further eliminate the possiblity of "water hammer," which would surely follow an abrupt and complete stoppage altogether of the water flow by the needle valve closing entirely, a deflecting or diverting control arrangement is added. This diverting action is accomplished by connecting the needle valves to the "deflectors" by such linkage as will permit the "first step" deflection of the

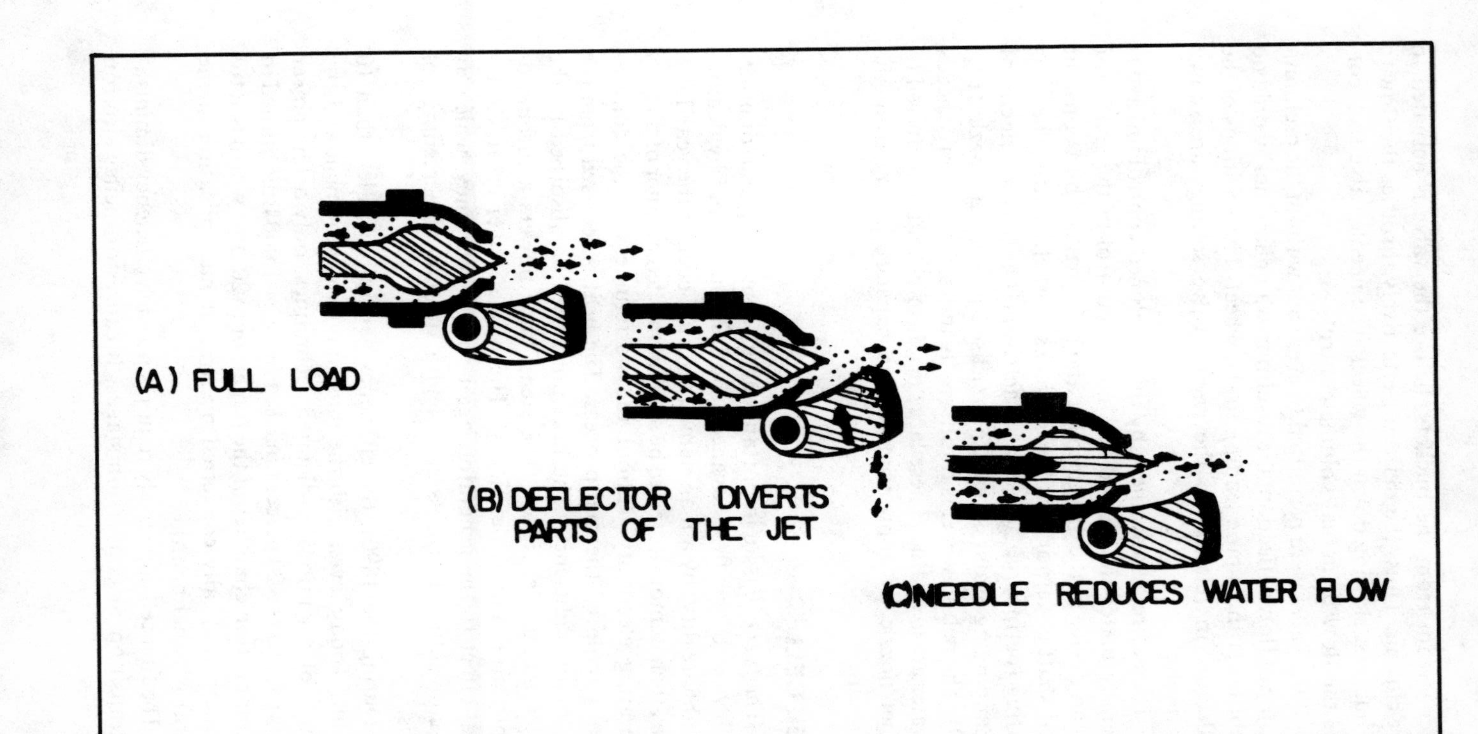

Fig. 10-9. Control of the jet of a pelton wheel.

water stream from the buckets, before the valve is actuated to start closing. Thus, the deflector effectively throttles the flow to the buckets, slowing down the wheel. The needle valve then corrects the flow by actual volume control.

If the valve were permitted to ever shut without the preliminary step of flow deflection, the resulting "water hammer" would no doubt burst the piping leading to the wheel. The positions of the "deflector" in relation to the needle valve settings are shown separately.

The entire arrangement by this combined control linkage is generally under the electro-mechanical control of the generator or load device at the driven shaft, and is "signalled" by a constant speed shaft governor. The wheel works most efficiently when it is rotating (peripherial speed) at exactly one-half of the velocity of the impinging water stream as released by the jet nozzle (Fig. 10-9). The pelton wheel is traditionally utilized where large heads of water are available. These large heads of water are delivered to the jet nozzle by appropriately high capacity piping from the reservoir.

FUEL CELL

The *fuel cell* is an important energy converting and/or utilizing device. It was first advanced, both in theory and by demonstratable reality, by a man named Ostwald back in the year 1894. It was then demonstrated to be an unparalleled manner of directly generating electricity without the wastefulness of the conventionally required thermal process. The heat is released from the fuel, minus deficiency. The heat must then be absorbed by the broiler, minus deficiency. The steam then powers a turbine or reciprocator, minus deficiency. The prime mover creates rotational effect to the generator, again minus deficiency, which by this conventional process can yield a substantial percentage of losses.

During the 1890s, the efficiency generally was less than 10 percent. Today, even with the most modern equipment and the greatest of scientifically dictated techniques, only a few power plants exceed the 20 percent mark. For every 1,000 Btus burned in the combustion chamber of the boiler, only 20 Btus worth is actually mathematically converted by straight multiplication at the generator panel board.

The fuel cell conversely requires no such efficiency diminishing transition or conversion stages. It causes or incites electron movement and thus current, directly, by chemical activity much like a primary cell, but with much greater power capacity. The actual efficiency of the fuel cell has reached the 90 percent mark, and this is 90 percent of the maximum attainable value of the cell ceiling of 1.23 volts. This efficiency is far superior to the average 20 percent of the thermal/mechanical/electrical generating efficiencies as realized to date. Only the cells themselves are required, obviating the steam generator, the prime mover and the generator.

The fuel cell will only provide direct current (dc), but may easily enough be converted to alternating current by either using an ac-dc converted, motor generator set or by employing an oscillator for low power electronics applications. This latest advance has been achieved with electrolyte temperatures only slightly above room temperature.

CELL THEORY AND CONSTRUCTION

The cell was constructed by applying the basic concepts of a carbon anode (+) and a liquefied silver cathode (-), with molten soda as the electrolyte. In essence, the carbon anode is also the combustible, or in this case the "fuel." Being contained in a positively atmospheric free environment, the electrolyte and silver cathode are then fired to a temperature of between 1000 to 1100 degrees Centigrade. The anode then enters the solution and releases carbon ions, each having a (4 positive charge). The liquid silver provides negatively charged ions of 0--(2 negative charge) receiving these "O--" ions from the continuously injected oxygen that is supplied to the electrolyte solution during electrochemical combustion. The conventional products of $C0^2$ result. The reaction is carried out as follows:

1-C++++ plus 2-0-- = 1-$C0^2$ plus heat.

For every carbon atom released to solution or converted, four electrons are given off to the carbon rod (anode) and four electrons are withdrawn from the oxygen cathode.

The foremost disadvantage of the related fuel cell is the high temperature of electrochemical reaction, which would result in a very short cell life. This situational handicap is remedied by the replacement of gases, such as hydrogen as a fuel or combustible in lieu of carbon. The hydrogen combustible allows for the generation of a current of densities up to 6.5 amperes/sq. in. at a reduced temperature of approximately 240 degrees Centigrade, but still with pressures in excess of 1000 psi, which necessitates special high strength containers.

The ionization of the gas fed to the cell is done by porous diffusion type electrodes of nickel composition. These electrodes are sintered components. On one side they are connected to the gas supply, and on the other side they are in direct contact with the electrolyte. The active region or chemically participating area is the *interface*, or boundary, of the "three phase effect" of gas, electrode and electrolyte.

In constructing this cell to meet the design intention of "extended boundary," all of the pores have a specially computed "optimum diameter" which provides for the absolute maximum utility of all areas per actual unit of length. In order to completely eliminate the possiblity of passage of unaffected gas, which could otherwise escape between the pores, each electrode is covered with a very fine pored outer layer in addition to the conventional single face composition. As a result of the high catalytic activity of the electrodes as comprised, the cell can be made to function with reasonable efficiency even at room temperature.

The dissolved fuel cell functions at room temperature and illustrates another method of constructional physics in basic cell character. In this cell, the oxygen electrodes (-) contain raney silver while the hydrogen electrodes (+) contain raney nickel as the added catalyst, enabling this unique variation to operate not only at room temperature but at atmospheric pressure too.

The electrodes described in the preceding example are known as "double-skeleton on catalyst electrodes." Because of their greater catalytic activity, they are able to dehydrate liquid organic fuels such as methanol. When using an alcohol fuel in solution in a potassium hydroxide base electrolyte, the fuel cell provides for a very efficient and desirable prime mover. The following information on the fuel cells is from the United States Department of Energy.

A NEW KIND OF POWER PLANT

The prototype of a new electric power option, a high efficiency, low maintenance, variable size environmentally safe, fuel cell generating system, is being designed and will be built and tested by United Technologies, with support from ERDA and the Electric Power Research Institute (EPRI).

The fuel cell concept itself is not new: such cells have already provided power for moon landings and, between 1971 and 1973, provided electric power to 50 apartment houses, commercial establishments, and small industrial buildings. What is new is an effort to capitalize on the fuel cells inherent flexibility safety, and efficiency by putting together a generating system that can use a variety of fuels to meet today's utility-scale power needs economically.

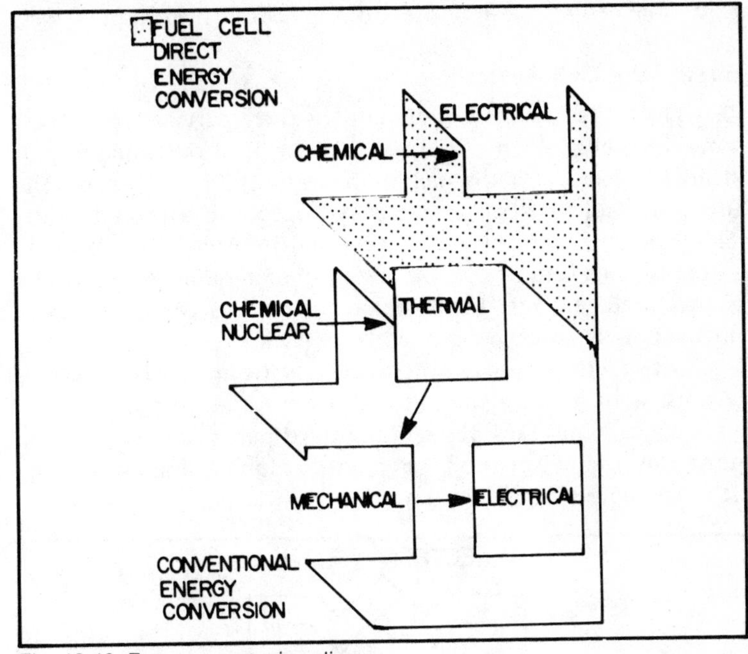

Fig. 10-10. Energy conversion diagram.

Fuel Cell Operation

In the combustion process used in conventional power plants, power generation requires three steps Fuel and air combine releasing their stored chemical energy as heat energy which is used to make steam. The mechanical energy of steam turning a turbine is converted to electric energy.

In the electrochemical process which takes place in a fuel cell, the chemical energy that bonds atoms of hydrogen (from a hydrocarbon fuel such as coal, oil or gas) and oxygen (from air) is converted directly to electrical energy (Fig. 10-10).

A fuel cell is a sandwich consisting of an anode, electrolyte, and cathode, much like a battery. Hydrogen-rich fuel is fed down the anode side of the cell, where the hydrogen loses its electrons, leaving the anode with a negative charge. Air is fed down the cathode side, where its oxygen picks up electrons, leaving the cathode with a positive charge. The excess electrons at the anode flow towards the cathode, creating electric power (Fig. 10-11). Meanwhile, hydrogen ions produced at the anode (when electrons are lost) and oxygen ions from the cathode migrate together in the electrolyte. When these ions combine, they form water, which leaves the cell as steam because of the heat of the cell processes.

Basic Fuel Cell System

The fuel cell generator system has three parts: a fuel processor, a fuel cell stack, and a power inverter. Hydrocarbon fuel and steam (recycled from the fuel cell operation) are first fed into the fuel processor and converted to hydrogen and carbon dioxide. This hydrogen-rich mixture is then fed into the fuel cell stack, where the fuel cells are piled one of top of another. As electricity is produced, exhaust from the stack includes carbon dioxide, nitrogen, and water, condensed from steam.

Direct current dc power electricity from fuel cell electrochemistry must be converted to ac power for utility applications. This conversion takes place in the third part of the system, the power inverter, which can convert large amounts of dc power to ac at nearly 96 percent efficiency (Fig. 10-12).

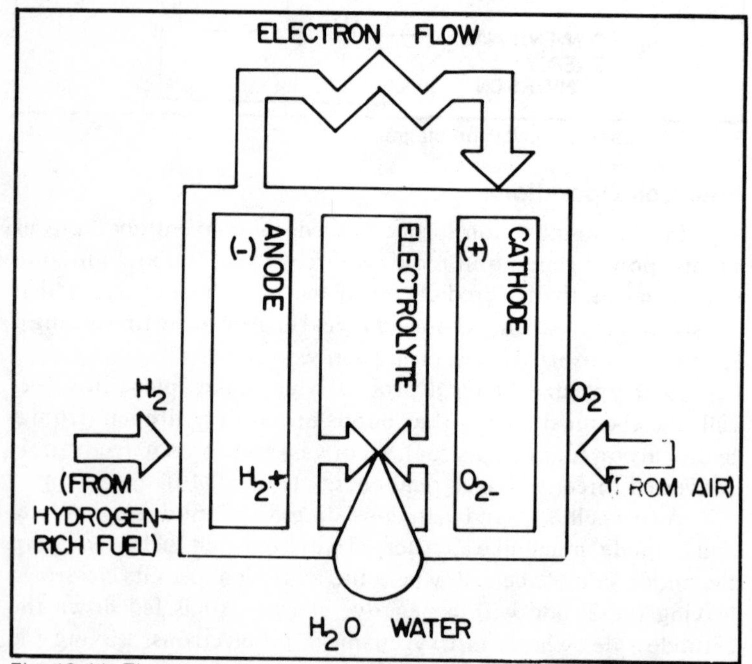

Fig. 10-11. Electron flow from anode to cathode.

The three segment fuel cell system can already produce electricity, from fuel to utility grid, at an efficiency of 38-40 percent, comparable to the best conventional combustion plant. Within 10 years, this efficiency is expected to be 50-55 percent. In addition, emissions from the generating process are well within present environmental limits, 10 times cleaner than the EPA requires.

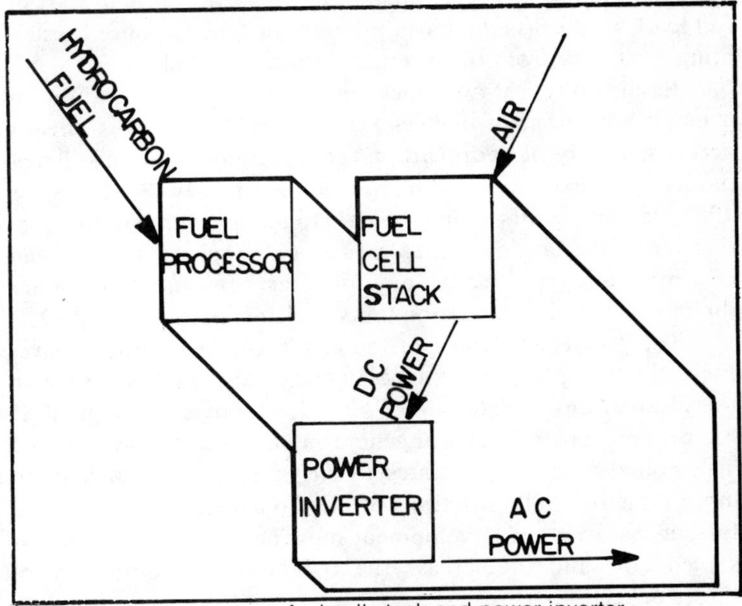

Fig. 10-12. Fuel processor, fuel cell stack and power inverter.

Advantages

In a time when power plant costs are high, construction lead times long, and future power demand uncertain, the flexibility of fuel cell systems is a major attraction. Like the fuel cells themselves, all three segments of a fuel cell system are modular and can be connected in parallel to meet additional power demand. Standardized fuel processors, fuel cell stacks, and power inverters could be factory built and trucked to the plant site. This modularity could mean lower cost, shorter plant construction lead time, and greater flexibility in the size of the plant.

Fuel cell systems, in various sizes, could serve at any point in a utility system, from a central power station to a user site. Siting close to point of use can reduce transmission losses (now about 8-10 percent of power, transmitted) and cut the costs for transmission lines, which add about $50 to $100 kW to the cost of present plant construction. Onsite use allows additional fuel savings through constructive use of system heat usually discarded in central generating systems.

Fuel cell system modularity also affects system efficiency. A large scale combustion plant is inefficient at part-load operation because some of its expensively-built capacity is not being used. Fuel cell plants, however, operate equally efficiently at part or full load. Fuel cells could also provide "instant response," generating electricity from the moment they are turned on.

Besides physical and operating flexibility fuel flexibility is a major advantage of fuel cell systems. Fuel processors already accept a variety of hydrocarbon fuels, including light distillates, natural gas, methanol, and high, medium-and low-Btu gases. By 1985, fuel processors should be able to accept synthetic fuel products from the nation's more plentiful coal supplies. Development of a more advanced technology for integrating fuel cell systems directly with coal gasification units is also expected during 1985.

The greater efficiency of the electrochemical process gives fuel cell power plants advantages in operating costs. By 1985, with the installation of 20,000 MW of fuel cell power, savings of $1 billion per year in electrical generating costs could be expected This dollar saving incorporates a yearly fuel saving equivalent to more than 100 million barrels of oil. To achieve these benefits, further research and development must be done to lower the installed cost and to increase the reliability and durability of these fuel cell systems.

The greater efficiency of the electrochemical process should give fuel cell power plants advantages in operating costs. By 1985, with the installation of 20,000 MW of fuel cell power, savings of $1 billion per year in electrical generating costs could be expected. This dollar saving incorporates a yearly fuel saving equivalent to more than 100 million barrels of oil. To achieve these benefits, further research and development must be done to lower the installed cost and to increase the reliability and durability of these fuel cell systems.

Environmental considerations like low water requirements, limited emissions, and quiet operation help make fuel cell plants an attractive power option. Where fossil fuel and nuclear plants require large quantities of water for cooling, fuel cells, which generate less heat, will be air-cooled by low speed fans. Because fuel cells can use a variety of hydrocarbon fuels, they share with conventional generating processes the environmental problems

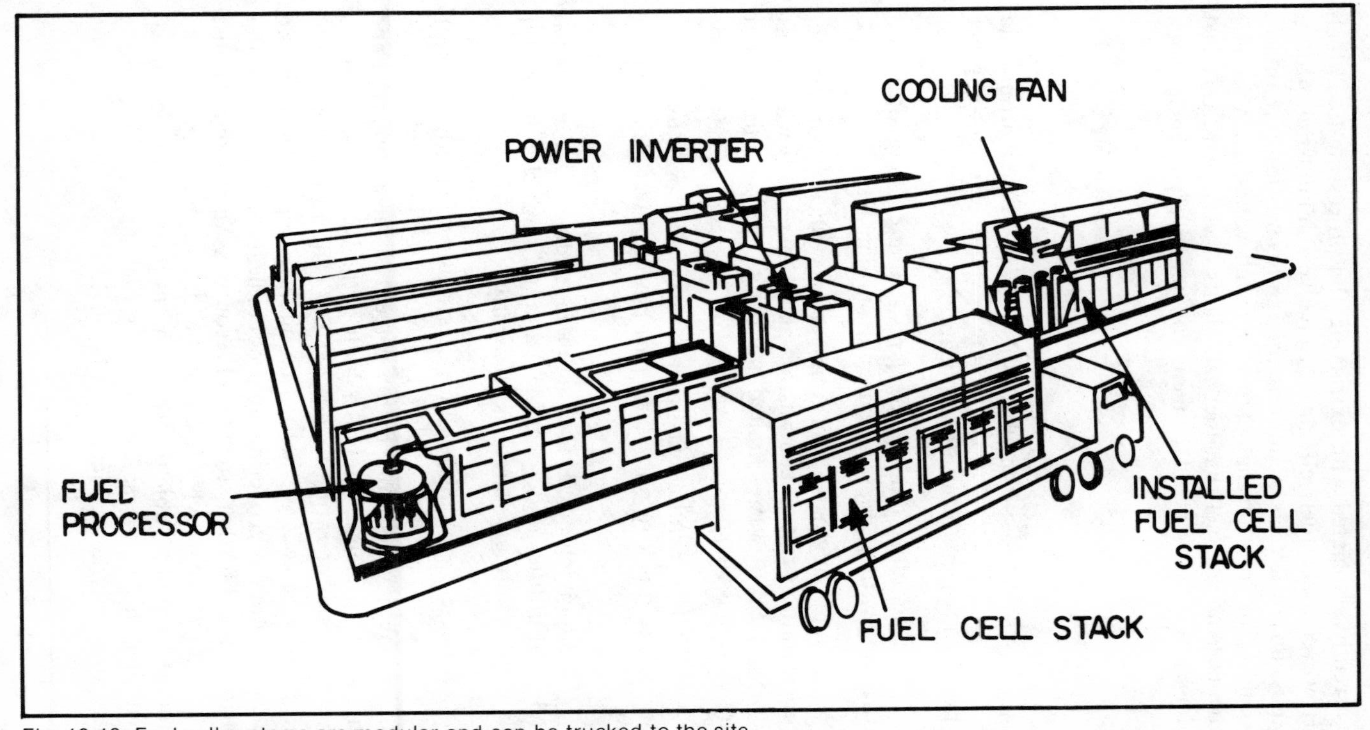

Fig. 10-13. Fuel cell systems are modular and can be trucked to the site.

currently associated with extracting and processing fossil fuels. However, since fuel cells do not involve a combustion process, emissions from their operations are significantly lower than emissions from conventional plants and well within EPA requirements. And the fact that fuel cell plants operate with very little noise also helps to make them environmentally "good neighbors."

Applications

The range of sizes, the modularity, and the environmental advantages make the fuel cell system a candidate for power generation in a variety of utility applications:

☐ Upgrading old urban plants, using existing sites more efficiently with decreased environmental impact.

☐ Supplying new generating capacity where environmental considerations restrict combustion plants (expecially when transmission right-of-way is limited and plants must be sited close to population areas).

☐ Complementing existing power systems' peak load-capacity, where quick response and part-power power efficiency are required.

☐ Supplying power for small and medium sized municipal and rural utilities under 100 MW, a range in which other power plant types cannot operate as efficiently.

Because they can produce electricity efficiently on both small and large scales, fuel cell systems are also candidates for onsite power generation. Eighty percent of the commercial and multi-unit residential buildings built in the U.S. annually have a maxmum power rating under 200 kW. Onsite fuel cells could save 25-30 percent of the fuel required to supply electricity to such buildings. Recovering by product heat for space and water heating could further stretch fuel resources. An apartment house study showed that fuel cells, coupled with heat pumps and thermal storage, should be able to provide all of the building's electrical and thermal requirements, with less fuel than was previously used to supply heat alone. See Fig. 10-13.

Courtesy of United States Department of Energy

COMPARISON OF THE INTERMITTENT AND CONTINUOUS COMBUSTION PROCESSES

The theories and discussions of combustion efficiency and the objectional by-products emitted from the spent products of fossilized fuel engine combustion have been a major topic of every forum of our engineering/scientific community. The theoretical expansiveness of the topic of pollution free combustion efficiency may be academically reduced to the primary aspect of just how, or by which, of two principal techniques the combustion processes are conducted by. First, we have the intermittent process, as utilized by the diesel and otto cycles as applied to the internal combustion engine. The second process or technique is represented by the *continuous process*, as utilized by the steam reciprocator, the steam turbine, (the boiler furnace serving as a continous combustion process), the gas turbine and the stirling engine.

Ideally, combustion should be affected by a continuous process, whereby the combustible and oxygen (from the air) are brought into intimate contact with each other, while in an environment of proper ignition temperature. There must be ample time to assure that all the elements will adequately combine into the ideal final products of combustion with the greatest exothermic release of heat. This returns us directly to the science and concepts of combustion. Let's deal again with the "three Ts of combustion; temperature, turbulence and time. We will study each "T" in its individual related role regarding the prime mover.

Temperature

The effect of temperature, or the mean effective molecular kinetic energy in any given mass or conglomeration of combustible elements, is of chief concern because of the directly proportional rate that chemical combinations or reactions proceed at in respect to their temperature. The higher the temperature, the faster will be the reaction rate. The lower the temperature, the slower or more retarded will the reaction rate be. Thus, the temperature is a decisive factor in the degree of chemical activity.

A certain minimum temperature must be achieved or provided for before any degree of combustion may be expected to be initiated. For all definitive purposes, this minimum temperature may be called the ignition point. If the combustible (as typified by the fuel fired in the prime mover) is introduced into the combustion chamber at a temperature less than the minimum ignition temperature of that required by the element, combustion will either be totally aborted or retarded and in complete. The expelled products of combustion will be little more changed than when the combustible was first introduced to the combustion chamber, at best comprising unburnt hydrocarbons, carbon monoxide, free oxygen and nitrogen, and a good deal of carbon particle soot.

If the temperature is too excessive, such as above a temperature of 3,000 degrees F., combustion will initially be accomplished with probable completeness, assuming a proper amount of oxygen and turbulence. However, too high of a temperature can be equally as adverse to the product pollution free combustion and the efficiency of the process. Let me exemplify this situation by citing the four most notorious contaminates as emitted from the combustion of refined gasoline.

Contaminates

First are *unburnt hydrocarbons* caused chiefly by incomplete combustion. There is insufficient time in the combustion chamber before discharge to the exhaust. Improper turbulence is caused by poorly designed combustion chamber. There is not an intense enough ignition temperature to promote instantaneous combustion. Also there is improper ignition timing of the electrical system.

Second is *carbon monoxide*, which is the unfulfilled $C0^2$ molecule. The partial combustion of carbon with its necessary two parts per mole of oxygen can be caused by low ignition temperature, a poorly designed combustion chamber or an insufficient oxygen supply from air charge starvation. The complete $C0^2$ molecule can contact a relatively cool area of the combustion chamber, releasing a portion of its heat. Or carbon monoxide (co) may be formed by the newly founded theory that the carbon dioxide might have reverted back to carbon monoxide by being exposed to *too high* of a combustion process temperature. The $C0^2$ molecule is subjected to temperatures in excess of 3,000 degrees and will undergo reverse combustion, as pronounced by the dissociation of the $C0^2$ back into carbon monoxide and an absorption of heat (endothermic reaction). The $C0^2$ will absorb 4,345 Btus/lb. of carbon dioxide reverting to carbon monoxide. This combustion reversal will rob heat from the combustion area and promote further adverse conditions. Also at this temperature (3,000 to 3,500 degrees), a good percentage of water vapor will dissociate back to hydrogen and oxygen. This reversed combustion is also of an endothermic nature and will exact an additional 61,100 Btus/lb. of water from the heat process.

The third contaminate is oxides *of nitrogen* (NO^2) which are created by high temperatures in the combustion space. The formation of this very objectional contaminate is markedly increased by the instantaneous peak temperatures and not so much by the average temperatures. Thus, the intermittent combustion process is especially prone to the creation of nitrous oxides. The peak temperatures, however brief, are sufficient enough in their short lived intensity to contribute the required temperature necessary to aid and abet the formation of this particularly detrimental pollutant. Notably, the continuous combustion process if not as disposed to the high temperature conditions as is peculiar to the intermittent process.

In the continuous process as exemplified by the boiler furnace in providing continous heat to the steam reciprocator and turbine, the independent gas turbine and the stirling cycle, the temperatures of combustion may be more constantly controlled to eliminate the temperatures that lead to nitrous oxide formation. The continous combustion process, by its controlled and heat regulated character, may be considered free from this particular contaminate, providing excessive temperatures are avoided. The intensity of continuous combustion must be kept at less than 3,000 degrees F. The problem of nitrous oxides is also encountered in steam generation. Through the use of proper lower peak temperature techniques, though, at least this contaminate may successfully be precluded from the stack gas analysis.

The fourth and last contaminate is lead, being directly derived from the admixture of tetraethyl lead as added to gasoline at a refinery to eliminate engine "knock." The presence of lead will only be effectively eliminated from the atmosphere when the use of leaden gasoline is terminated. At the date of this writing, there is no positive or effective method of removing this contaminate from the products of combustion of the gasoline engine. It is felt by governmental agencies that since the anti-pollution acts of the early 1970s, the use of leaden gasoline will in perhaps another 10 to 15 years be eliminated by the expired life span of the automobiles that require its use.

All of the four contaminates are peculiar to the gasoline otto cycle engine. Three contaminates, unburnt hydrocarbons, carbon monoxide and nitrous oxide, are peculiar to the internal combustion engine because of the engine's nature of intermittent combustion.

Turbulence

The intimate contact that is so imperative for the complete and prompt combustion of all participating elements of the combustion process is greatly and primarily influenced by the degree of turbulence. Turbulence is chiefly controlled by the physical design of the combustion chamber and by the manner and geopositional function of the injector in the diesel and fuel injected versions of the otto cycle. The adaptation of pre-combustion chambers and special piston head designs have aided the efforts of effective turbulence and consequent perfection of combustion. Through admixture of the combustible in its ideal vaporous form and intimate diffusion with the optimum amount of oxygen is of sacrificed at high speeds of the reciprocating engine, where there is a substantial and marked reduction in volumetric efficiency and functional ability.

This problem of lack of turbulence can be more realistically and immediately dealt with in cases of continuous combustion. The combustible and air charge may be made to combine in the vaporous and most optimum admixture even before the final designation for combustion is reached. Because the flow of combustible of fuel and air charge will be occurring at a steady rate, a very positive regulation of diffusion by turbulence or other more sophisiticated means may be assured. The continuous combustion process has by documented tests, in all instances. provided a more desirable and near optimum condition of homogeneity of air/combustible charge than could be performed by the intermittent cycle. Thus, the continuous combustion technique will have superior advantages in reducing or totally eliminating unburnt hydrocarbons and carbon monoxide by its inherent ability to provide for better turbulence or the balanced distribution of the combustible within intimate vaporous contact in the confines of the combustion chamber, prior to ignition.

Time

The duration or period for the completed combustion starting from the instant of entry, either as a fuel/air mixture in the tube of the carburetor or upon first meeting in the combustion chamber of the diesel, through the stage where diffusion is expected to be performed, will in most all cases be found to be functionally inadequate. The explosive mixture must now meet a relatively cool combustion chamber upon its compression stage, which will certainly work an adverse effect upon the next stage of action—combustion. Combustion must be initiated by properly timed ignition. It must be completed during and before the piston (which serves as the counter chamber wall of the combustion arena) reaches bottom dead center, or whatever selected point where the exhaust valves will be open and the "expected" completed combustible charge is subsequently discharged to the exhaust system. Quite frequently the charge will still be undergoing final combustion while halfway out of the exhaust manifold. This is frequently evidenced by the trail of flame issuing from the exhaust ports of such an intermittent cycle engine when the exhaust manifold is disconnected. Thus, it is quite obvious that the intermittent cycle fails the requirements for providing the essentially vital element of time for the proper and thorough combustion of its delivered combustible charge.

SUMMARY OF IMPORTANT POINTS

The favorable potentials of proper and thorough combustion are all but precluded by the use of the intermittent system of combustion. The short comings of the intermittent system are evidenced by the emission of contaminates (unburnt hydrocarbons, carbon monoxide and oxides of nitrogen) and, in the case of the otto cycle, the emission of lead from the use of leaden fuels. This evidenced character of pollutants in the intermittent system's products of combustion is largely due to the inherent inability to perform the essentials of combustion without the instantaneous peaked increase of temperature, which is responsible for the reversed combustion and dissociation of the elements and compounds as mentioned.

The promptness by which the intermittent system must perform the functions of fuel/air charge preparation prior to actual combustion usually is accomplished in an abrupt manner, as opposed to the more deliberate and controlled technique of the continuous combustion system. The temperature variation that the combustion chamber is exposed to by the swing from lower temperature induction stage to higher temperature combustion stage causes a drastic effect upon the combustible charge, in initial reaction upon first entry to the combustion chamber. The unevenly propagated temperature wave or wedge as created by the act of preliminary detonation of the fuel/air charge makes the even distribution of temperature-pressure energies a haphazard accomplishment at best. This is primarily why the efficiency of the engine falls off so markedly at higher speeds, in regard to its fuel consumption per developed shaft horsepower at higher rpms.

The continuous combustion mode offers all of the advantages of the intermittent system with but only one disadvantage. The continuous combustion system naturally requires a separate and distinct combustion chamber or heat source. It is not as self-contained as the intermittent system, which serves as its own combustion chamber or heat source. However, in the case of the stirling engine, this factor is not so pronounced. The combustion chamber is literally wrapped around the power cylinder and made for all practical purposes a part of the power unit.

The functions of the separate combustion chamber cannot be made quite so compact and self-contained in the considerations given to the system reciprocator or the steam turbine. The gas turbine is almost the perfect example of a self-contained power unit, in spite of the fact that it requires anywhere from one to several combustors. These are tailor-fitted about the turbine casing as to seemingly be an integral part of the power train itself.

In stationary and marine installations, good use may be made of the separability of the combustion chamber from that of the main power unit. A larger and more elaborate combustion process system may then be enjoyed. We may design our prime mover about the dictations of the power plant instead of the more frequently encountered situation where the power plant must be built around the dimensions and demands of the engine.

Although it can be argued by intermittent system advocates that the exhaust gas pollutants can be cleaned and rendered free of contaminates by the use of reactors and after burners, this is a ridiculous policy. At best the problem is only moved from one point of the system to another. This remedial technique necessitates an additional expenditure of either energy or fuel, or both, to produce a function of energy conversion that should rightfully have been performed by the combustion process before the combustion was considered completed. The continuous combustion process offers the opportunity to provide the time, turbulence and temperature under the optimum conditions that can only be effectively accomplished by the separate and independent combustion system, as associated with the continuous mode of heat generation and delivery.

Continuous combustion contributes to the stirling engines thermal efficiency of slightly more than 45 percent. The diesel cycle engine, because of its high compressibility factor and its higher relative mean effective pressure, has a thermal efficiency of from 30 to 40 percent. The otto cycle engine has an earned average of approximately 25 percent thermal efficiency. Both the gas and steam turbines have efficiencies of around 20 percent. The unsinkable steam reciprocator has "combined efficiency" of 12 to 16 percent.

EMISSION REDUCTION BY CONTINUOUS COMBUSTION PROCESSES

Considering the most prevalent contaminates in respect to environmental pollution, the most temperature-sensitive contaminates are oxides of nitrogen. Lead compounds have been deleted from this discussion, because we will be dispensing with the tetraethyl lead additives of gasoline in the very near future.

One recognized technique employed in the reduction of nitrous oxides is the "two stage" (off-stoichiometric) method, where completed combustion takes place in two distinct stages or transitions. The initial combustion stage is a "fuel-rich" environment, effecting combustion with a deliberate deficit of oxygen supply which will allow for the greatest amount of actual combustible oxidation. This is at the expense of depriving the "waiting" nitrogen molecules from any chance of combining with the scarce and already combined C0, $C0^2$ and $S0^2$ molecules. The second stage of the process is carried out in a more oxygen-available environment, but under even lesser temperatures than the first stage. The first stage has a temperature of less than 3000 degrees F. The second stage requires even less of an ignition temperature, 2700 degrees F.

NATIONAL LEGISLATION AFFECTING COMBUSTION EMISSIONS

The Clean Air Act of 1963 was the first of many federally initiated attempts for the cooperation of government and industry to investigate, research and promulgate environmentally oriented programs for the ultimate purification and preservation of our environment and atmosphere. The Air Quality Act of 1967 provided much needed federal control for the administration and enforcement of such programs. It empowered the Secretary of Health, Education and Welfare to aggressively impose federal regulations of upon states that either did not have much regulations or did not seem anxious to legislate or enact such safeguards on their own. The Act of 1967 more importantly authorized the

Major Classification:	Subclassification:	Members of Subclassifications:
Organic Gases	Hydrocarbons	Hexane, Benzene, Ethylene, Methane, Butane, Butadiene.
	Aldehydes/Ketones	Formaldehyde, Acetone.
	Other Organics	Chlorinated Hydrocarbons, Alcohols.
Inorganic Gases	Oxides of Nitrogen	Nitrogen Dioxide, Nitric Oxide
	Oxides of Sulfur	Sulfur Dioxide, Sulfur Trioxide.
	Carbon Monoxide	Carbon Monoxide.
	Misc. Inorganics	Hydrogen Sulfide, Ammonia, Chlorine.
Aerosols	Solid Particulate	Dust, Smoke, Soot, Fumes.
	Liquid Particulate	Oil Mist, Entrained Liquid Droplets.

Table 10-1. Major Contaminates.

Department of Health, Education and Welfare to seek court injunctions immediately where public health was felt to be jeopardized.

Most recently, the Clean Air Act of 1970 incorporated under one authority all or most of the provisions of the two preceding acts. It further mandated commissions to establish scientific minimums regarding contaminating pollutants and expanded on enforcement policies to include procedures for monitoring by federal inspectors to assure compliance by all power plants and industrial installations. Table 10-1 is a list lf major contaminates by categorized classification.

MEAN EFFECTIVE PRESSURE AND THE INDICATOR CARD

The term mean effective pressure denotes the actual pressure which imposes the driving force as applied to the piston face on any displacement type prime mover. Briefly, the MED is the pressure existing in the power cylinder or combustion chamber over the entire length of piston displacement, from top dead center, to the point near bottom dead center when the exhaust valves or discharge phase occurs. It is not sufficient enough for computational purposes to consider only the forward or positive driving pressures. The back pressure which acts in a subtractive manner against and in opposition to the movement of the power producing piston, and its direction, must also receive attention to ascertain the actual or usuable net effort.

The actual mean effective pressure is then "mean" pressure as evidenced by actually taking a pressure and stroke coordinated indicator card reading of the graphically obtained pressure/ power stroke or phase. Subtract from this the "mean" pressure of the opposing back pressure. Thus, the effective value of the power available will then appear as the mean effective pressure for that particular engine or device. Generally, in cases of internal combustion engines, the mean effective back pressure will be equal to the atmospheric pressure, as applied throughout the piston's entire stroke. In cases of external combustion engines of the displacement types (such as the reciprocating steam engine) which may be exhausting into a condenser, the mean back pressure will be an assistance to the power output. It will be of an additive value when applied to the gross mean effective pressure of the engine. In respect to the stirling engine, this back pressure may be of the subtractive nature. The back pressure may be above atmospheric, or above the final exhaust pressure of the power phase, for the working medium.

When an indicator card has been "taken" of the engine, its interpretation will provide the engineer with two evaluations.

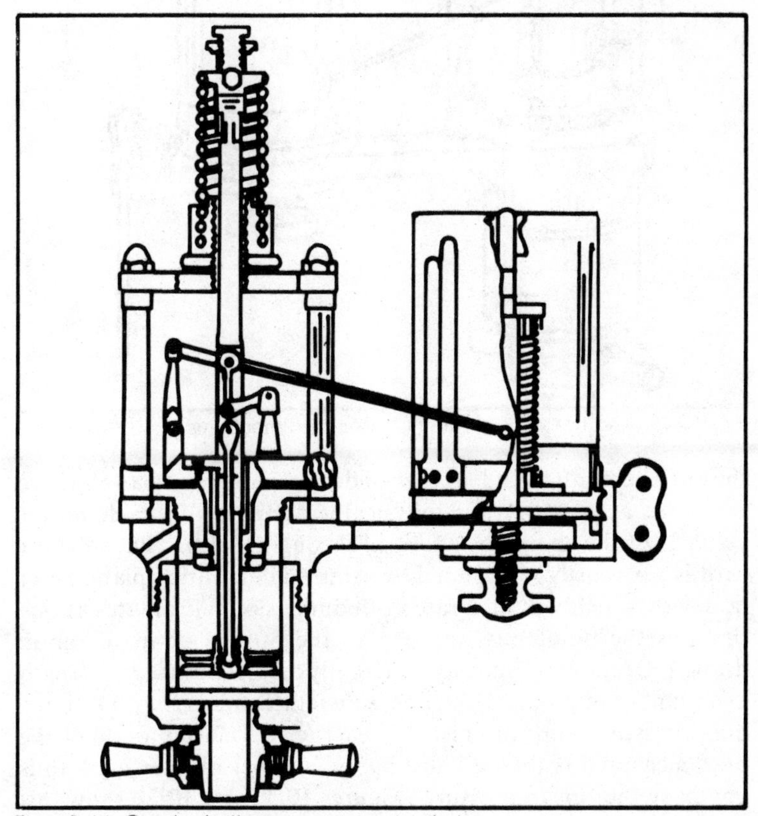

Fig. 10-14. Crosby indicator card mechanism.

First, the card will show the operation of the valves, and the prefered sequence of events occurring in the power cylinder, along with the evenness of the distribution of power between opposing piston efforts and/or the coordination of multi-expansion components of the engine. Second, the indicator card will enlighten

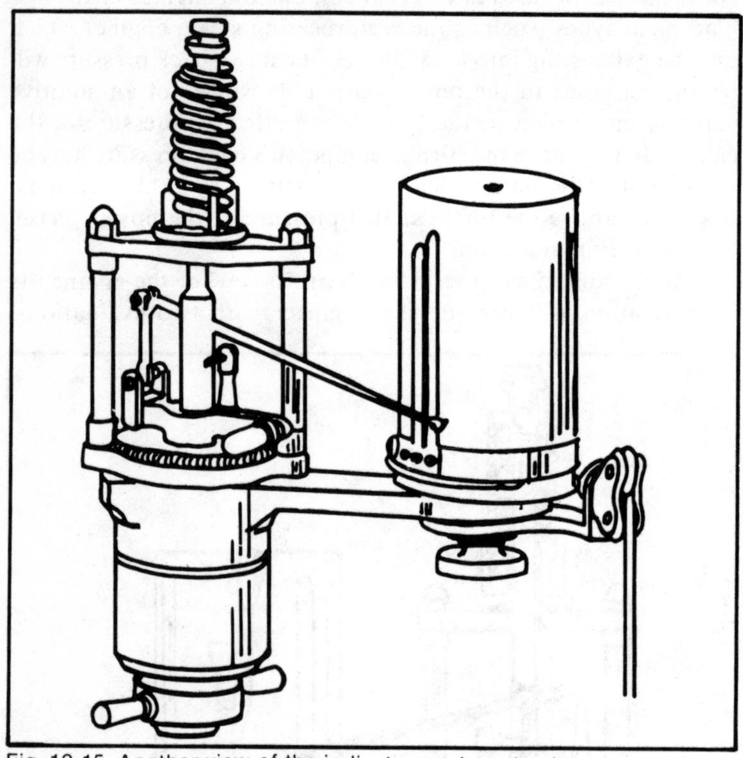

Fig. 10-15. Another view of the indicator card mechanism.

the engineer with the advantage of determining the mean effective pressure, as actually and physically occurring in each power cylinder of the engine (Figs. 10-14 through 10-18). The indicator card is very easily interpreted by using an integrating planometer (a device which will integrate both dimensional ordinates at one time, as the pointer is traced over the outline of the diagram drawn). Or the diagram may be actually divided through its length at a number of points. Each line is then measured from "O" to the high pressure point or crest, (at least ten). The sum of all of the lines measured is then divided by the number of lines used: thus you have the "mean pressure" Figures 10-19 and 10-20 show this method of obtaining the mean pressure of the card. If the actual

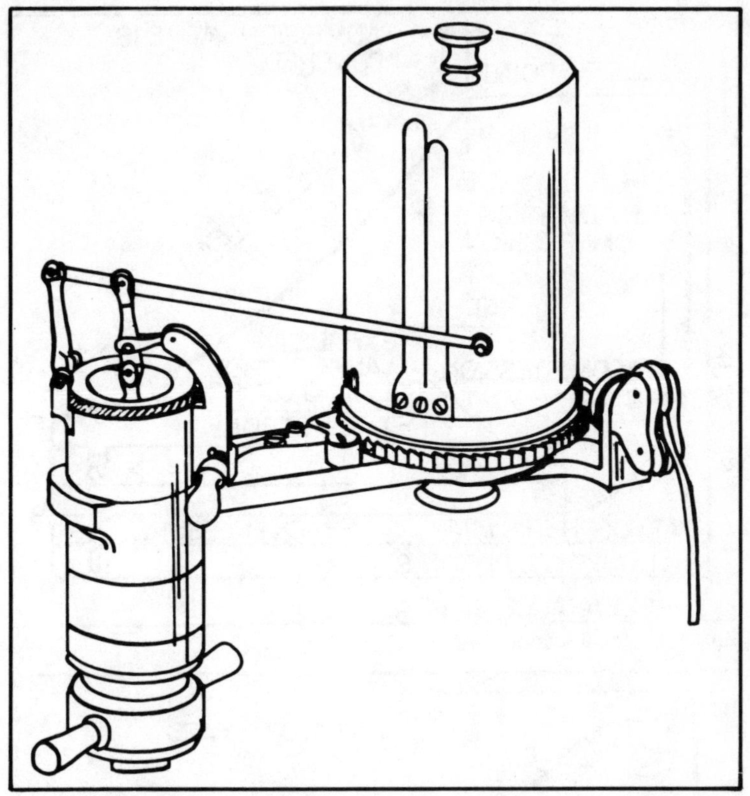

Fig. 10-16. Crosby steam engine indicator with a detent attachment and a drum 2 inches in diameter.

computed "area" of the card is determined, the MEP may be had by dividing the area by the stroke.

It goes without saying, that the engine, regardless of the number of cylinders, should be balanced, so that the load will be as evenly distributed as possible. This balancing can best be achieved by interpreting the indicator cards as taken for each power cylinder and at each power end, if the engine is of the double acting variety.

The accurately derived mean effective pressure is absolutely essential to the computation of horsepower (indicated) for the displacement type of prime mover. The indicated horsepower must first be obtained before the actual shaft horsepower is applied comparatively in making a determination of the engine's efficiency. As in all cases, the actual engine efficiency will equal the developed or actual shaft horsepower, divided by the theoretical horsepower: as

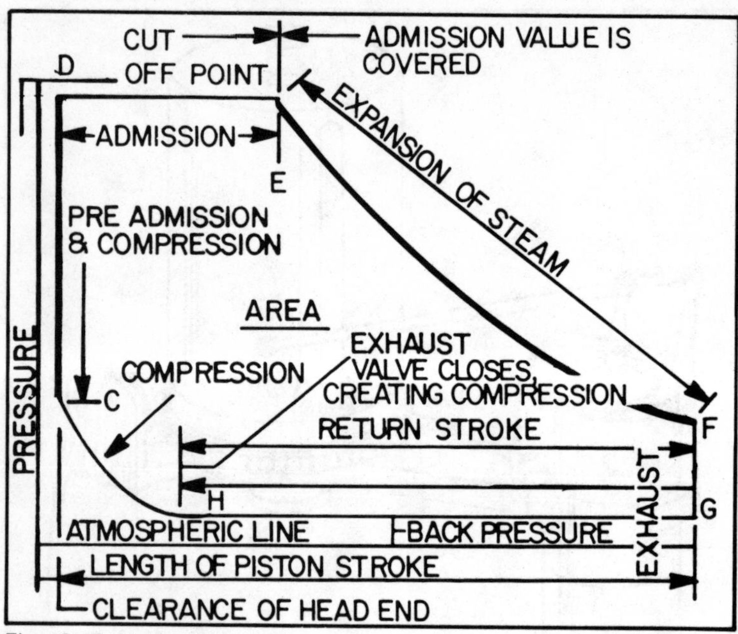

Fig. 10-17. Ideal engine card.

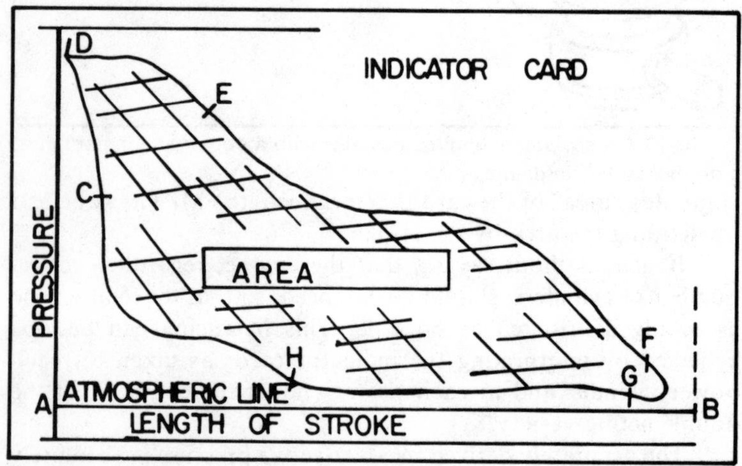

Fig. 10-18. Typical engine card.

$$\text{engine efficiency (percent)} = \frac{\text{actually developed shaft horsepower}}{\text{theoretical horsepower (IHP or thermal)}}$$

In acceleration type prime movers such as the steam and gas turbines which have no pistons, the following formula is used:

$$\text{acceleration efficiency} = \frac{\text{actually developed hp or torque}}{\text{heat delivered in Btus/min.} \ (\times)\ 0.0236}$$

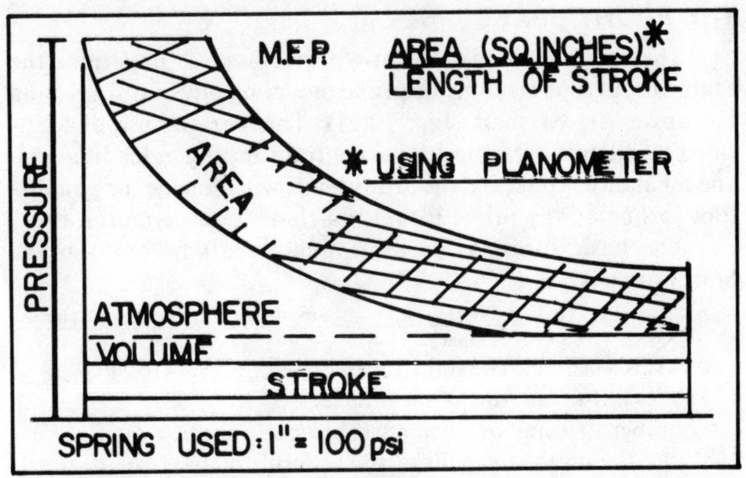

Fig. 10-19. Mean effective pressure graph.

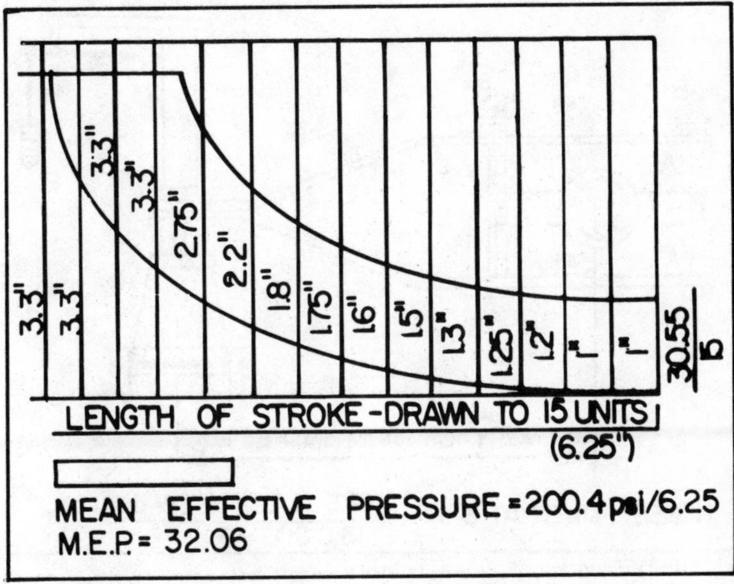

Fig. 10-20. Indicator card calculations for mean efffective pressure.

The displacement power (theoretical) is obtained by the indicated hp, which is the equivalent to thermal horsepower. This is basically the heat supplied, multiplied by 0.0236 for each Btu per minute supplied to the engine. Both variations are then compared rationally to the developed shaft hp.

THE PRONY BRAKE

The *prony brake* is a testing device used to determine the shaft output of a machine or prime mover by physically grasping the power-driven shaft (Fig. 10-21). Through the use of lubricated wooden clutches or grippers, a transmitting or loading arm, and a balance type scale, the prony blade will indicate the pounds-foot torque of any prime mover, machine, electric motor, etc.

The basic formula for determining shaft or prony brake horsepower is:

$$\text{shaft horse power} = \frac{6.28 \times L. \times N. \times (W\text{-}Wo)}{33{,}000} = \frac{\text{amount of lbs./foot torque/min.}}{33{,}000}$$

where: 6.28 = a fixed constant (2 times "Pi" or 3.1416)
L = the length of the torque arm in feet
N. = number of rpms or power strokes per minute
(W-Wo) = the no load weight of the arm subtracted from the full load, or gross weight of the arm, as loaded.

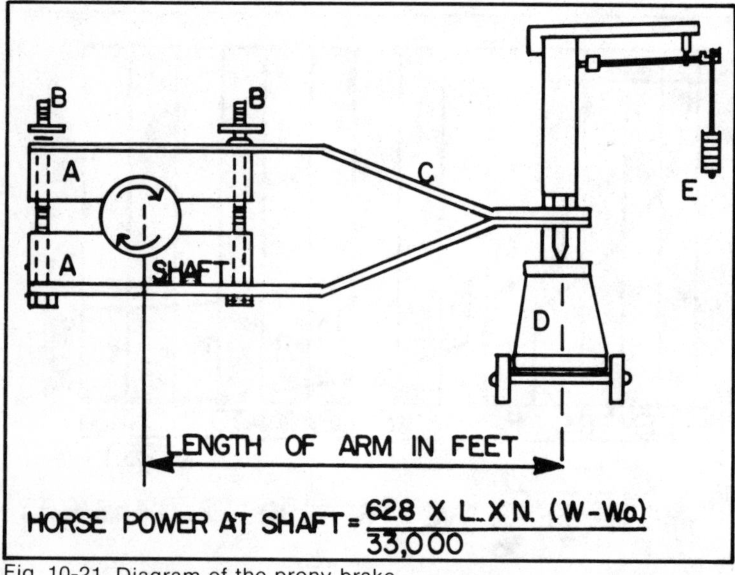

Fig. 10-21. Diagram of the prony brake.

With the brake assembly coupled about the selected point of shaft or journal contact, and the non-loaded arm resting upon the scale bed, the weight of the arm and assembly is noted (to be subtracted later, "Wo"). With the prime mover running at a predetermined point of rating, the brake tension adjusting wheels are evenly tightened to put tension on the shaft. The tension is increased to the point where the prime mover or engine is just about ready to stall or quit. The scale is constantly observed and the highest reading or weight balance is noted. The difference between this full load weight and the no-load weight of the arm is the actually developed shaft torque.

The distance of the arm from the center of the shaft to the weight bed of the scale will usually be found to be a simple even number. Since we are considering torque, we will be regarding it as so many pounds weight per 1 foot's length time. Thus, if our scale would indicate that we were developing a full load arm force or weight of 500 lbs. and we were using a 4-foot arm length, we would multiply the force, 500 lbs., by the 4-foot arm, giving us 2,000 pounds/foot torque..

The prony brake, in spite of its accuracy and simplicity, is being replaced by the *dynamometer*. There are two types of dynamometers—the electric type and the hydraulic type. The electric dynamometer is essentially an electric generator, while the hydraulic type is a large varrable output water pump.

DETERMINING INDICATED HORSEPOWER

The actual computation, by formula, of the engine's indicated horsepower is essential for the purposes of ascertaining the engine's mechanical efficiency. An explanatory step by step discussion is felt to be in order to acquaint the reader with the underlying factors and their concerted affects upon the final quotient of this formulation.

Basically, we will be collecting the data as required by the formula's factors: mean effective pressure, which can only be had by the application of an engine indicator to both power ends of the engine; the effective area of the piston face, or both areas if the engine is of the double acting variety, the effective length of the piston stroke in feet, which can be determined best by actually measuring the piston rods' movement or "crank throw", the number of power impulses or strokes occurring each minute, which will equal the number of revolutions counted per minute for a two cycle otto, diesel stirling engine or an single acting steam reciprocator (per the number of cylinders). The number will equal 0.5 for all four stroke engines; and the number will equal 2 for all double acting engines and for opposed piston type diesels. If the test is to be studied for the period of one full minute, then the divisor of 33,000 will be used; if the test is of one seconds' duration, then the figure 550 will be used. Referring directly to the formula:

$$\text{indicated horse power} = \frac{\text{MEP}(\times)\,\text{L}(\times)\,\text{A}(\times)\,\text{N}(\times)\,\text{k}}{33{,}000} \text{ or } \text{IHP} = \frac{\text{P.L.A.N.k}}{33{,}000}$$

The formula is rearranged as follows: $\text{IHP} = \frac{\text{P.}\times\text{A.}\times\text{L.}\times\text{N.}\times\text{k}}{33{,}000}$

Now, we may observe the intended sequence of operations as signified by our formula to construe the following logic. The pressure (P) which must be in its mean effective form is multiplied by the area (A) of the piston. Let's stop right here for some thought. When we multiply the pressure of force so acting on a body, such as the force of a working fluid as acted upon the area of a piston, the product of these factors will give us the total force or effort. Thus, with a pressure of say 100 psi MEP acting on a 6-inch diameter piston we will obtain the area of the piston—area= $D^2 \times .7854 = 6\times6=36; 36 \times .7854 = 28.3''$ or 28 square inches. The 100 psi MEP is now multiplied by the ara—100 psi×28 sq. in.=2,800 pounds total force created. What will this ineffective force by itself be doing? This otherwise static force will be moving through a distance as fixed by the stroke of the piston. Our formula is now shaping up as:

$$\text{IHP} = \frac{100\ \#\text{MEP} \times 28 \text{ sq. in.} \times \text{length (?)}}{33{,}000}$$

We measure the crank throw, or piston rod movement, and find a 4-foot distance as moved by the piston. Multiplying our static force of 2,800 pounds by the distance we will have this force move for each stroke, we have 2,800 lbs. × 4 ft. = 11,200 ft. lbs. Thus, we are now being told that we have a machine or prime mover that is capable, for each working stroke, of producing a calculated work equivalence of 11,200 ft. lbs. We know from physics that "work" can only be effected when a force actually and physically moves through a distance—such as 2,800 pounds force moving through the distance of 4 feet in the cylinder of our hypothetical engine.

Now that we have arrived at how much "work" we may accomplish per each cylinder, our formula is taking on this appearance:

$$\text{IHP} = \frac{100 \text{ psi MEP} \times 28 \times 4 \text{ ft.} \text{"P"} \times \text{"A"} \times \text{"L"}=}{33{,}000} = \text{work performed/cylinder}$$

Having obtained the work producing ability of one cylinder and only one power end impulse, we will go back to our hypothetical engine and find that it is of the four cylinder double acting type. Therefore, we must proceed further with our "work" consideration to include the number of cylinders (N) or number of power impulses per minute, and the "k" factor of how many power impulses occur for each stroke of the piston. In this case, because the engine is of the double acting variety, this figure will equal two.

$$\text{IHP} = \frac{"P" \times "A" \times "L" \times "N" \times "k"}{33{,}000} \qquad \text{IHP} = \frac{100 \times 28 \times 4 \times 40 \times 2}{33{,}000}$$

indicated horse power = 100 psi (×) 28 sq. in 40 power units 2 impulses

$$\text{indicated horse power} = \frac{\dfrac{100 \text{ psi}}{(\text{MEP})} (\times) \dfrac{29 \text{ sq. in}}{(\text{area-piston})} (\times) \dfrac{40 \text{ power units}}{\text{number-impulses}} (\times) \dfrac{2 \text{ impulses}}{\text{per stroke}}}{\#33{,}000 \text{ (conversion factor/min.)}}$$

We have, starting with the mean effective pressure, determined first the force, as "P" × "A", or MEP times the area of the piston. Next we obtained the work accomplished per cylinder, by multiplying the force by the distance moved, or in this cases the stroke of the piston—2,800 lbs. (×) 4 feet of travel= 11,200 ft. lbs. work performed/cylinder. The gross work performed by the entire 40 power strokes per minute (40 rpm) equals 11,200 ft. lbs. (work times 40 (work times one minute) times 2 number of acting impulses per stroke. Thus, 11,200 × 40 × 2 gives us a product of 896,000 ft. lbs of work performed per minute of operation.

To convert the figure of 896,000 foot pounds of work performed in each minute's operation of our engine to the preferred universally accepted horsepower equivalence, we must divide the work times duration of performance by an equal work/duration as advanced by Watt—which is 33,000 ft. lb/min. or 550 ft. lb/sec. Since we have counted the number (N) of the power impulses occuring each minute, our conversion constant must be selected as 33,000, or the equal amount of work performed in an equal amount of time (in this instance by the minute).

Applying the work performed by the engine in its entirety (by all cylinders and by all power ends of the cylinder) as a numerator, and Watt's conversion constant, for a minute's worth of work performed, as a denominator, we have:

$$\text{indicated horse power} = \frac{896{,}000 \text{ ft. lbs/min (from engine)}}{33{,}000 \text{ ft. lb/min (from Watt)}} = 27.15 \text{ IHP}$$

The indicated horsepower thus obtained will not in itself be indicative of the engine's true, or brake horsepower, or the actual usuable power at the drive shaft. Rather, the indicated horsepower will benefit the engineer with an initial index. From there the engineer may proceed to explore the progressively diminishing horsepower throughout its incurred deductions, as brought about by internal or frictional losses, design limitations and prevailing operational deficiencies. But the IHP is, nevertheless, the rightful

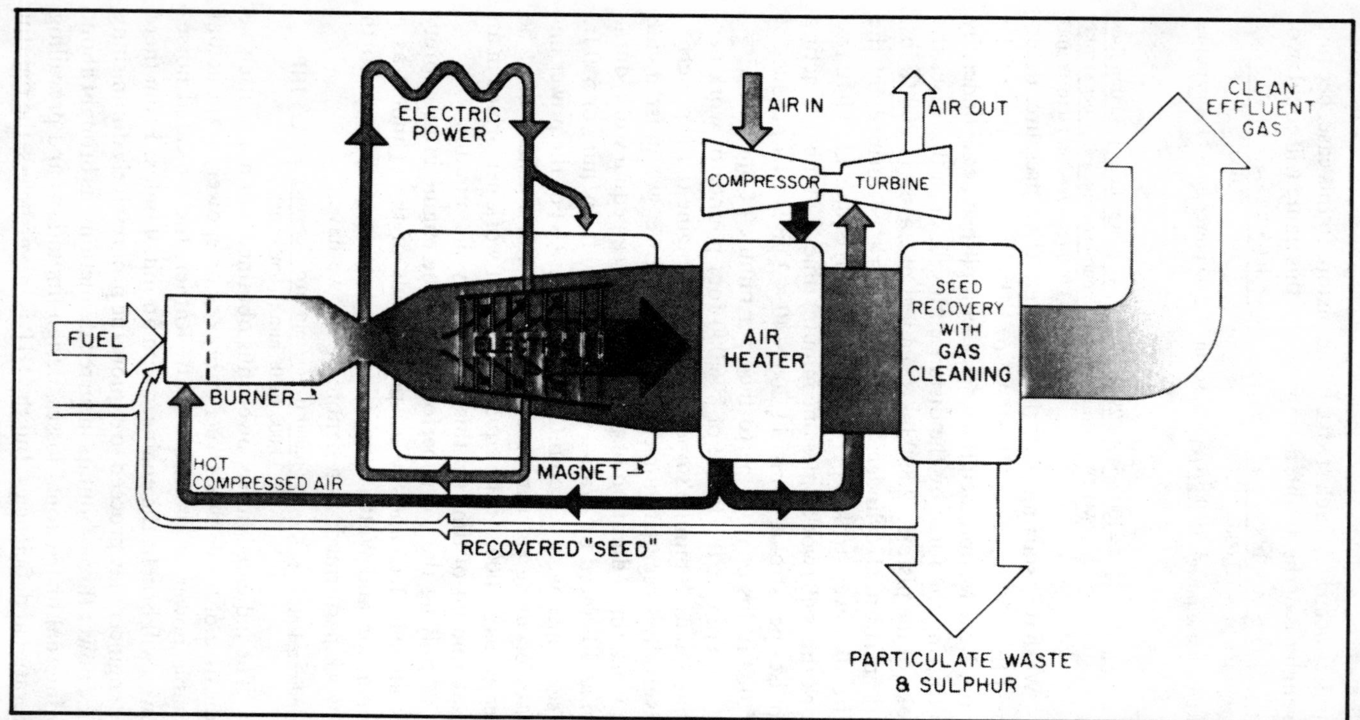

Fig. 10-22. Magnetohydrodynamics electric power generator flow diagram (courtesy of United States Department of Energy).

point of entry for the ultimate resolution of efficiency losses and their important identification in the power system as a whole.

SURVEY FORM: PART C

An examination of survey form part C will reveal that the prime movers included have been accordingly classified by their respective applications in regard to their driving force or energy origins. More specifically, the prime movers have been exclusively and independently categorized pursuant to behavioral "modes" such as thermal mechanical, thermal-mechanical or hydraulic. These categories will adequately subtend all known modes of power development or conversion, as will be encountered in the conventional fossilized fuel power plant.

It will be further noted that the majority of formulas given are so expressed or compounded to embrace the prime mover's efficiency in thermal/mechanical aspects. The subject of thermal horsepower has been covered earlier, but its significance will once again be underscored. In all cases of thermal to mechanical modes of energy conversion or transformation, this consideration will provide the most critical view of the "real" efficiency of the prime mover.

Fig. 10-23. Magnetohydrodynamics generator (courtesyEnergy Research and Development Administration).

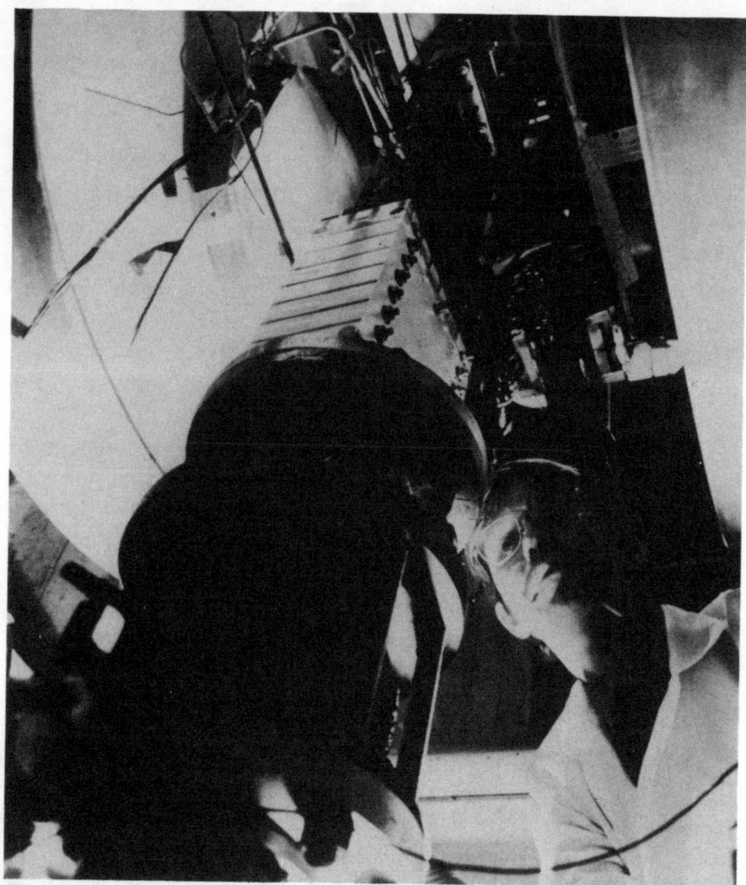

Fig. 10-24. A magnetohydrodynamics test section is fitted into a liquid metal loop at Argonne National Laboratory.

There is but one absolute indicator for determining the energy utilization index of the thermalprime mover—*heat in mechanical horsepower out*. This formula may be applied to any prime mover or engine of the thermal energy family such as all steam engines, the diesel and otto cycle engines, the stirling engine, the fuel cell, and the gas turbine and magnetohydrodynamic generator (MHD) (Figs. 10-22 through 10-24). The prime movers released from this consideration are naturally the hydraulic turbines, the Francis turbine, the Kaplan propeller, the Pelton wheel and the aerodynamic turbine (wind turbine).

Once more, all fossilized fuel-oriented prime movers may be analyzed under the thermal/mechanical formula. This formula not only offers an easily conducted survey, but provides the only true thermal/mechanical efficiency obtainable by any computational techniques. This formula will be developed as follows:

$$\text{T/M efficiency} = \frac{\text{mechanical horsepower as determined by shaft test (*)}}{\text{heat value of fuel required/min.(**)} \times 0.0236}$$

where: "T/M"= thermal/mechanical relationship

(*)-mechanical prony brake or dynamometer mechanism

(**)-calorific value of fuel (in Btus/lb.) times weight of fuel delivered

(This may also be abbreviated to "Btu heat release/minute).

It will be appreciated, then, that the above formula is in actuality the mechanical horsepower output divided by the thermal horsepower input. To refresh the reader's recollection of thermal horsepower, the absorption or release of 42.42 Btus of heat per minute would yield one mechanical horsepower if it were utilized by an engine of 100% efficiency. This conversion factor was obtained by expressing the following heat-work relationship:

$$\frac{\text{33,000-(horsepower)}}{\text{778-(work/heat)}} = \text{42.42 Btus/min./hp (theoretical)}$$

Conversely, should the horsepower be required from the heat release or Btus per minute input, we invert our expression to reformulate the function:

$$\frac{\text{778 (work/heat)}}{\text{33,000 (horsepower)}} = 0.0236 = \text{Number of horsepower in 1 Btu/min.}$$

This inverted relationship tells us that theoretically the maximum horsepower attainable from each Btu of heat released or delivered for conversion purposes could, under 100% efficiency conditions, achieve $\frac{236}{10,000}$ of a horsepower from that lone Btu of heat applied for one minute.

From our preceding discussions of the interconvertibility and conversion functions involving heat and its time release relationship to actual horsepower, we should be conducting our own formulas. We should be refurbishing our concepts and engineering attitudes to evaluate fuel supplies no longer as just so many gallons of oil or so high of a pile of coal, but rather as a proven equal and convertible quantity of actual horsepower units. In the final analysis, that is just exactly how the ultimate formula will be read:

$$\text{power plant efficiency} = \frac{\text{the actual usable horse power produced from our plant}}{\text{the tons of coal or gallons of oil needed for conversion}}$$

As practitioners of the engineering and scientific community, we must in the name of conservation adopt an unrelenting attitude toward acceptable efficiency criteria. Imperatively, then, our attitude should permit us to evaluate all conceivable energy conversion or utilization processes in their respective and appropriate

Table 10-2. Prime Mover Survey Form - Part C.

PART-"C"	PRIME SURVEY	PLANT CLASSIFICATION

Total H.P. Produced/Energy Consumed:________________ ________________
HP/BTU; KW/HP; KW/BTU; HP/Gal. passing Turbine, etc. Fossil, Fuel, Hydro, Diesel, etc.

Energy Conversion For Fossil Fuel : = $\dfrac{\text{Total Shaft Torque-FtLb-taken by Prony Brake}}{\text{Btus of Total Fuel Delivered to Prime Moverx778}}$ = Eff. ____ %

Turbine Efficiency Thermal-H.P. = $\dfrac{\text{Shaft HP Developed as taken by Prony Brake}}{\text{Lbs. Steam Delivered(x)"Q-1"-"Q-2" in BTU's(x).0236}}$ = ______ % Eff.

Energy Conversion For Mechanical Forms : = $\dfrac{\text{Total Output Shaft Horse Power}}{\text{Force(x) Distance Moved/min/33000 (X) 100=Mech.}}$ = ______ % Eff.

NOTE: "Q-1"=Enthalpy at admission presure(abs): "Q-2"=Enthalpy at Exhaust, psia. This data is provided in the Steam Tables, refer to Part-One. "Q"=BTU/Lb.

Theoretical Hydraulic *Water Turbine H.P.* := $\dfrac{QxWxH}{33{,}000}$ Where:Q=CuFt. Fluid Displacement/min/ W=Lbs.CuFt.Fluid; H=Head in Ft.= ______HP Maximum

Gas Turbine Eff. = $\dfrac{\text{Actual Shaft HP}}{\text{Calorific Value(x)Lbs.Fired(x).0236}}$ = ______ % Maximum

IHP-Reciprocating= $\dfrac{PxLxAxNxK}{33{,}000}$ Where:P=mep; L=Piston Stroke in Feet A=area of Piston; N=R.P.M.; K=Power Strokes

Brake HP at Shaft=$\dfrac{6.28xLxN(W\text{-}Wo)}{33{,}000}$ Where: "W"=/weight of gross Load in Lbs. "Wo"=Weight of Lever Arm-No Load

Engine Efficiency=$\dfrac{\text{Actual Brake HP}}{\text{IHP from mep card and formula}}$=Eff.______%

One Thermal Horse Power=The utilization of 42.42 BTU's/min=0.0236(x)BTU's fired.

context. Such future conceptual rationalizations must emanate from the all-important premise. How many shaft horsepower will effectively be delivered for any given quantity of heat release or amount of fuel consumption?

$$\text{conversion index} = \frac{\text{mechanical power output}}{\text{thermal power input}}$$

The higher this very real index figure is the more of an ideal prime mover or device we will have.

We must cease and desist the continued technological buffoonery that has thus far left us in the energy resource situation that we are in today. The diesel and the stirling cycles were already a known reality at the turn of the nineteenth century. They offer maximum attainable thermal efficiencies of 40 percent and 45 percent respectively, but still we have persisted in granting our engine manufacturers the economic luxuries of turning out reciprocating rejects that have efficiencies of less than 18 percent. (This figure, which is conservative, was obtained by me after taking the mean average horsepower/fuel consumption tested ratings of all major auto manufacturers—not the estimated ratings as offered.

I hope that appropriate legislation will some day be applied to the automotive industry to require them to furnish certified energy consumption/horsepower test results, as obtained by actual dynamometer determinations. When the Babcock and Wilcox Company or the Cleaver-Brooks Company spell out their boilers' ratings, they use the direct and unambiguous language as expected by the engineering profession—the proven pounds of steam output/per pounds of combustible fired. Perhaps this is why these two companies have been leaders in their product fields.

In part C will be formulated the most indicative and "absolute" computational instruments for purposes of ascertaining our prime movers' maximum attainable power, and the comparative portion as delivered at its drive shaft and known as its efficiency. In all thermal accelerator situations, we will be considering the thermal horsepower as provided for by either external or internal combustion. In respect to the displacement type prime movers, we will be applying the machines' indicated horsepower to its delivered shaft output. Once again, the true and only absolute parameters are the prime mover's mechanical shaft output as compared to its thermal horsepower input. Merely substitute the thermal hp (as obtained by the formula given) for whatever such consideration or aspect as specified. The revised formula will accordingly benefit the engineer with the machine's thermal/mechanical efficiency. See Table 10-2 for the part C survey form.

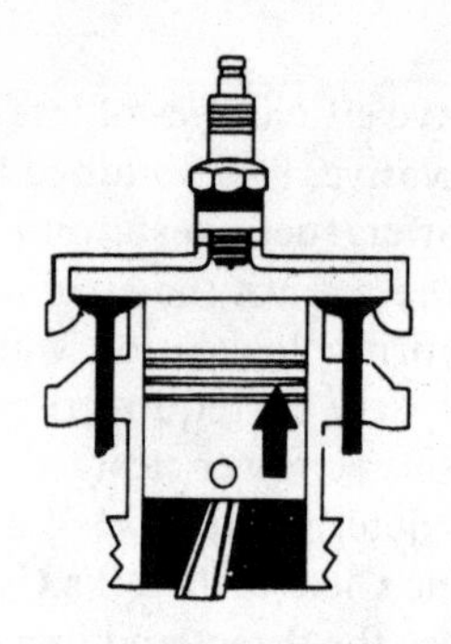

Chapter 11
Using The Hays-Republic Model Orsat Gas Analyzer

Since 1874 the *Orsat* method of measuring flue gases has been accepted as the standard of accuracy in industrial and scientific work. The orsat's reliability is best established by the fact that all other methods of combustion gas analysis are checked by this type of analyzer (Fig. 11-1).

The Hays-Republic model 621A orsat gas analyzers are designed for convenient fast and accurate analysis of flue gas constituents specifically: carbon dioxide ($C0^2$), oxygen (0^2), carbon monoxide (C0), and sulfur dioxide ($S0^2$).

UNPACKING THE INSTRUMENT

Unpack the orsat from its shipping case using care in handling the chemicals and the instrument. Check the packaging list to insure that all items listed are present. Report any missing items to Hays-Republic. If there is damage, notify the transportation firm. Any damage claims concerning items shipped f.o.b. the factory should be negotiated with the carrier responsible. In such cases it is advisable to retain the packing and carton for the claim adjustor's inspection. When the instrument is to be returned to the factory for repair, a good packing material to restrain movement and cushion shocks should be used in the shipping container.

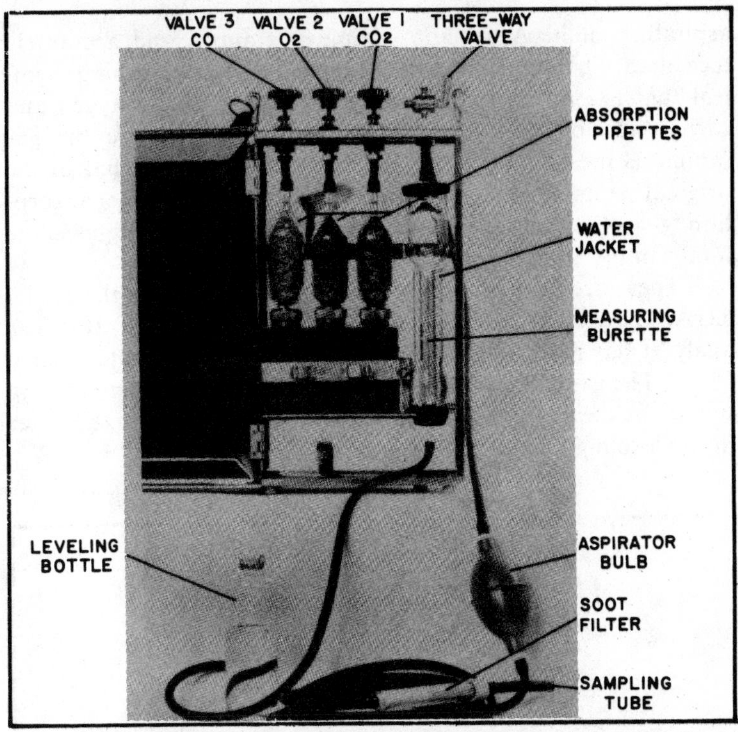

Fig. 11-1. The model 621A orsat gas analyzer (courtesy of Milton Roy Company, Hays-Republic Division).

THE PRINCIPLES OF OPERATION

The Hays-Republic model 621A orsat first measures a volume of gas, passes it through an absorbing chemical and then remeasures the sample residue. The measuring chamber or burette is furnished with a percentage scale which indicates the volume of sample absorbed.

The amount of excess air present under a given set of combustion conditions can be determined by employing the model 621A to measure carbon dioxide ($C0^2$), oxygen (0^2), and carbon monoxide (C0).

In an efficient combustion system the highest amount of carbon dioxide ($C0^2$) obtained using the smallest amount of excess air, without producing C0 is the primary objective. The percentage of carbon dioxide ($C0^2$) will vary with every fuel and condition. However, there is a specific percentage of $C0^2$ that will yield a maximum combustion efficiency for the prevailing firing conditions. The gas sample is drawn into the model 621A by the aspirating bulb. A specific sample is trapped and accurately measured. The sample is then passed through an absorbing chemical that has an affinity for carbon dioxide, then oxygen and carbon monoxide. After each absorption procedure the gas sample is measured and its volume is compared to that of the original sample. The *difference in volume* before and after absorption, gives the exact measurement of the percentage of gas in the total sample.

There are four simple steps in gas analysis using the model 621A analyzer. Through design and construction, this gas analyzer has made these steps easy, rapid, and accurate.

□ The gas sample to be analyzed is drawn into the measuring burette by means of the aspirator bulb, venting through the water in the leveling bottle (Fig. 11-2).

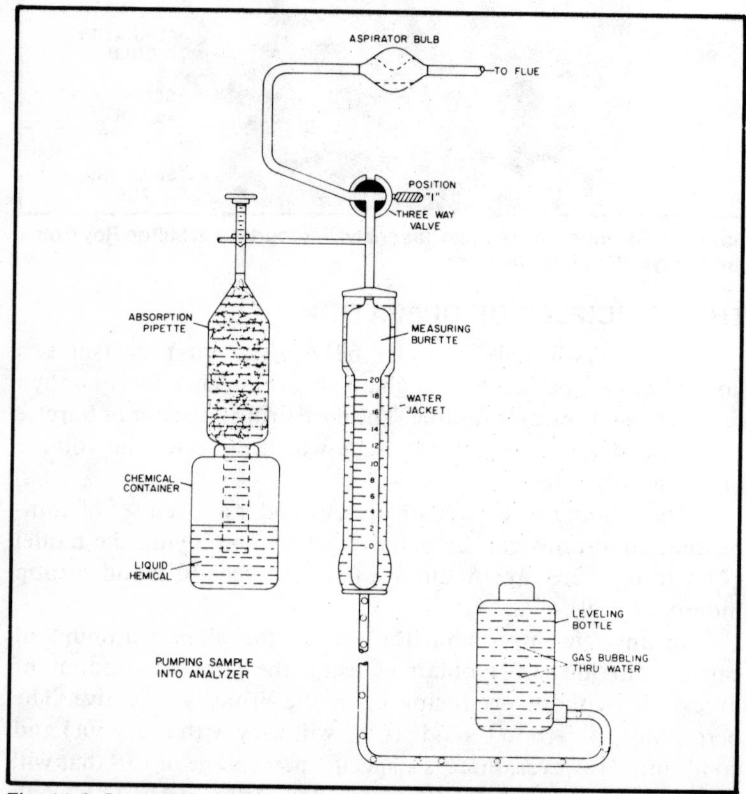

Fig. 11-2. Step one in gas analysis (courtesy of Milton Roy Company, Hays-Republic Division).

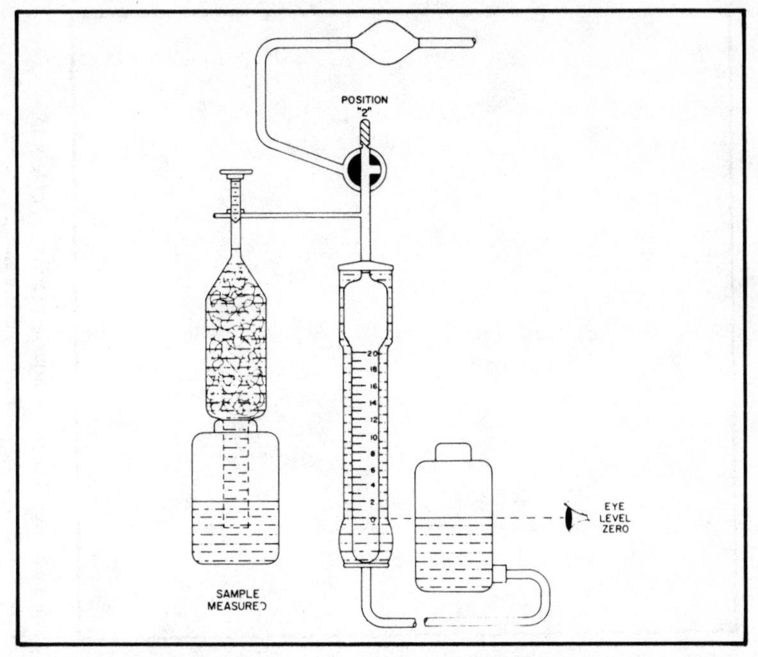

Fig. 11-3. Step two in gas analysis (courtesy of Milton Roy Company, Hays-Republic Division).

☐ The sample is measured by bringing the water level in the burette exactly to zero (0). The top of the burette is open to atmosphere so that the gas is at atmospheric pressure. This step is necessary for accuracy (Fig. 11-3).

☐ By raising the leveling bottle, water will rise in the burette forcing the gas sample into the selected absorption pipette. The pipette is filled with metallic material freshly saturated with chemicals which immediately absorb the sample constituent being measured (Fig. 11-4).

☐ The remainder of the sample is drawn back into the measuring burette by lowering the leveling bottle. The sample is remeasured and the reading from the burette scale is recorded (Fig. 11-5).

ANALYZER PARTS IDENTIFICATION

The Hays-Republic model 621A gas analyzers are provided in a light weight, easily portable case. They are designed to withstand active use in the boiler room or field, yet remain accurate and dependable.

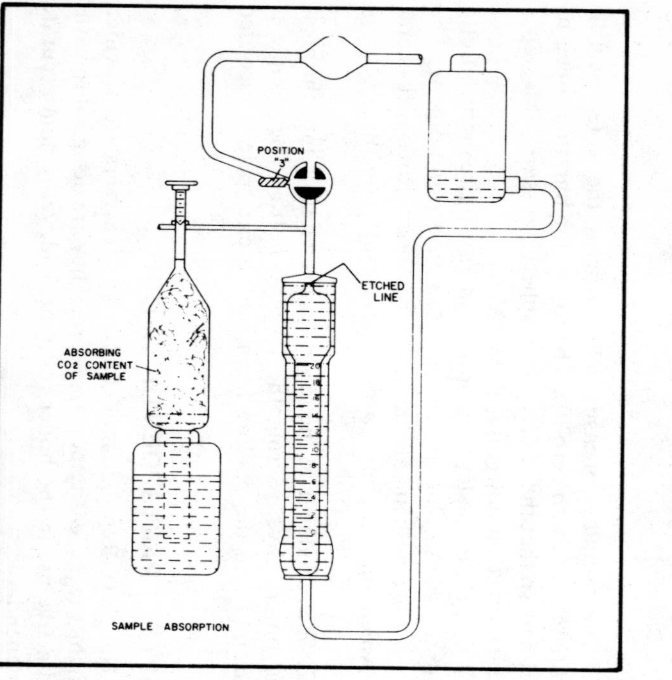

Fig. 11-4. Step three in gas analysis (courtesy of Milton Roy Company, Hays-Republic Division).

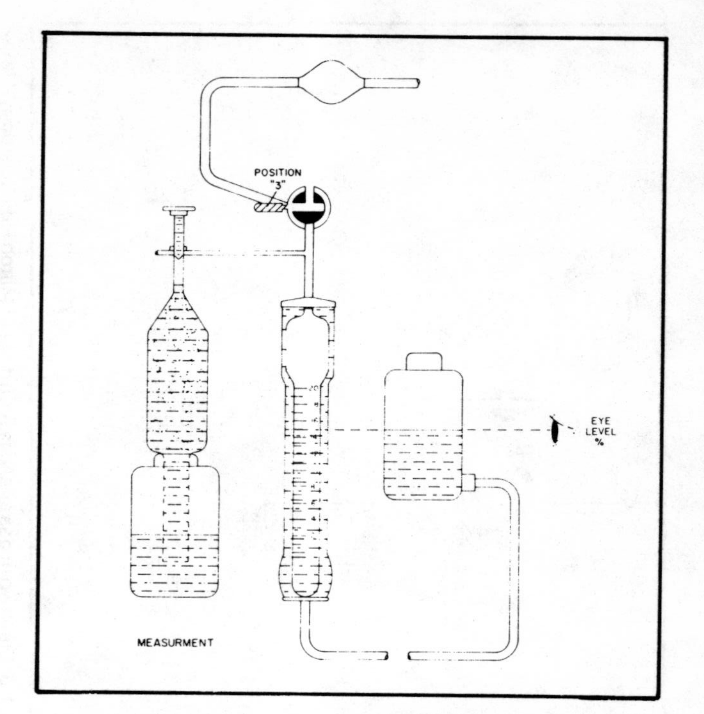

Fig. 11-5. Step four in gas analysis (courtesy of Milton Roy Company, Hays- Republic Division).

The water jacketed burette is individually calibrated and has an etched scale. The extra heavy glass tubing is encased and protected by a molded glass water jacket. The water jacket, which controls the temperature of the gas samples, is so shaped that it assists in the reading, acting as a magnifier.

The all metal three way valve speeds analysis. Position one is used when a gas sample is being drawn into the burette (Fig. 11-6).

Position two allows excess gas to be vented out of the burette as a measured sample is being trapped off at atmospheric pressure and stabilized temperature.

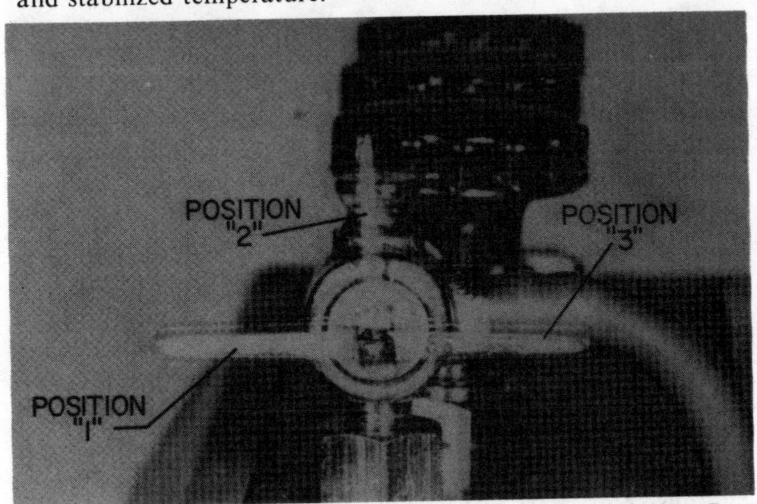

Fig. 11-6. Depiction of the various valve positions (courtesy of Milton Roy Company, Hays-Republic Division).

Position three allows the gas sample to be transported into the absorption pipettes and then drawn back into the burette by means of the leveling bottle position. In position three the gas sample is remeasured, and the percentage of absorption read directly on the burette scale.

The needle valves at the top of each absorption pipette are of a non corroding metal. Each absorbing pipette has its own needle valve, and closing a needle valve shuts off the passage way between that particular pipette and the measuring burette. The connecting header and the valve seats may be cleaned by passing a pipe cleaner through the sample header.

The absorption pipettes absorb because they are filled to the top with a metallic material which provides a greater absorption surface. These pipettes are molded to insure a rugged construction, and reduce breakage.

The rubber bag on the back of the analyzer, is used during operation and is connected to the oxygen (0^2) and the carbon monoxide (C0) chemical containers to prevent air exposure of these absorbing chemicals.

The chemical containers are employed to store the individual absorbing chemical. The leveling bottle is used as a vent when a sample is being drawn, and is the means of forcing a sample into the selected absorption pipette for determination of that constituent's volume. The leveling bottle, during use, is filled approximately three-fourths full of tap or distilled water.

The Hays-Republic model 621A gas orsat analyzer is easily dismantled. The tools required are a screwdriver and a pair of pliers. The absorption pipettes, rubber connectors, and chemical containers may be removed by releasing the slide bar and detaching the soft rubber connectors from their needle valve seats. The pipettes may be removed from the chemical containers by simply loosening the metal fittings.

ANALYZER PREPARATION

Three chemicals are furnished with each analyzer: cuprous chloride solution for carbon monoxide (C0), SEEZ 0^2 solution for oxygen (0^2) and cardisorber solution for carbon dioxide ($C0^2$). Use safe practices when handling chemicals. If a chemical is spilled on hands or clothing, apply vinegar and water immediately. Safety goggles should be worn when handling SEEZ and cuprous chloride. If SEEZ or cuprous chloride is splashed into the eyes, flood the eyes with water and apply a saturated solution of boric acid. Medical assistance is advised.

PLACING CHEMICALS IN THE MODEL 621A ORSAT GAS ANALYZER

☐ The analyzer should be placed on a firm, stable platform or table.
☐ Open both doors of the analyzer case.
☐ Remove the leveling bottle from the case. Connect the rubber tubing to the bottom of the leveling bottle, and the bottom of the burette. Place the leveling bottle on the table.
☐ Fill the leveling bottle three-fourths full of distilled or tap water.
☐ Fill the water jacket of the measuring burette by raising the top water jacket cap and carefully pour in enough water to fill the water jacket. To protect the contents of the analyzer from freezing, glycerine or a similar type of antifreeze solution may be used.

The solution used must be transparent and clean.

☐ The contents of the water jacket need not be changed and may be left in the water jacket indefinitely.

☐ Unsnap the leather carrying handle next to the three way valve.

☐ Remove the rubber tubing of each black chemical container from their respective plugs.

☐ A filling tube is furnished with the analyzer. Place the copper tip into the vent tubing of the chemical container to be filled. Select the chemical to be placed in the container and place the bottle (in its plastic base) on the table next to the analyzer.

☐ Carefully remove the cap of the chemical bottle and insert the rubber end of the filling tube into the chemical bottle solution. Caution: Do not allow the chemical to overflow.

☐ Select the needle valve at the top of the case related to the absorption pipette and its associated chemical container being filled. Open this needle valve with three or four full turns.

☐ Place the three way valve in position number three.

☐ Apply suction to the top of the leveling bottle. The chemical will appear in its absorption pipette and rise about half way. The chemicals are corrosive. Use care in handling and refer to warning label on bottle.

☐ Release the suction on the leveling bottle and allow the chemical in the pipette to drain into the chemical container.

☐ Repeat this operation until the chemical bottle is empty, or the chemical container is filled.

☐ Remove the filling tube from the chemical bottle and the chemical container vent tubing. Clean the filling tube with tap water. After cleaning the filling tube, store it for future fillings. Caution: Do not splash the chemical.

☐ Reconnect the chemical container vent tubing to the steel plug of the absorption pipette.

☐ Turn the three way valve to position one.

☐ Close the needle valve firmly. Do not tighten hard enough to damage the valve seat.

Each individual chemical container may be charged in turn, as described above. It is recommended that the sequence of carbon dioxide (cardisorber), oxygen (SEEZ) and carbon monoxide (cuprous chloride) be followed.

Empty the leveling bottle, place it in the analyzer case, close the doors of the case, and re-attach the leather strap. The analyzer may now be stored until an analysis is required.

PRELIMINARY PROCEDURES

Place the analyzer at a convenient point where the tests and measurements will be made. Place the analyzer on a stable table or hang on secure wall attachment. Open the case doors and place the burette so that it may be easily read. Remove the leveling bottle and fill with distilled or tap water.

Unsnap the leather handle by the three way valve. Remove the aspirator bulb assembly from its clamp and reel. Remove each of the vent tubings from their respective steel plugs. Connect the vent tubing for the oxygen and carbon monoxide chemical con-

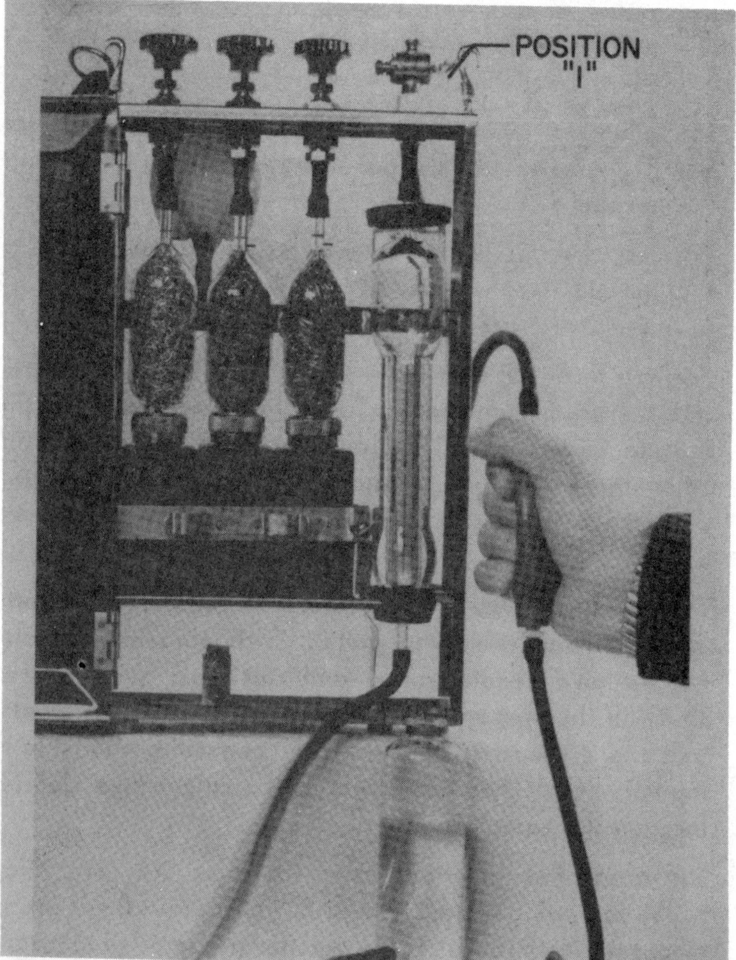

Fig. 11-7. Pump the aspirating bulb as shown (courtesy of Milton Roy Company, Hays-Republic Division).

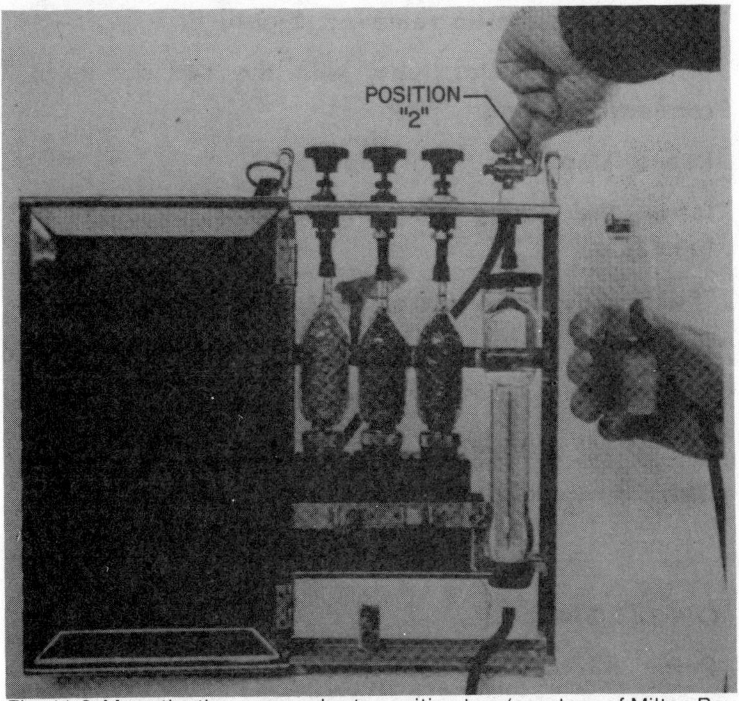

Fig. 11-8. Move the three way valve to position two (courtesy of Milton Roy Company, Hays-Republic Division).

tainers to the rubber bag manifold. The carbon dioxide tube should be left unattached. The chemical in each absorption pipette must be at the etched fill mark on the neck of the pipette.

If the chemical of any pipette is not at the etch mark proceed as follows:

☐ Place the three valve in position two.

☐ Raise the leveling bottle slightly above the top of the burette. The water level is controlled in the burette by clamping the rubber hose between it and the leveling bottle. As the water nears the top of the burette clamp the tubing tightly.

☐ Place the three way valve in position three.

☐ Open the needle valve at the top of the pipette at least one fourth of a turn.

☐ Lower the leveling bottle below the burette with the tubing clamped.

☐ Gradually release the clamp on the tubing until chemical begins to rise in the pipette.

☐ When the chemical reaches the etched fill line on the pipette clamp off the tubing and close the needle valve.

☐ If the chemical does not reach the etched fill line repeat the previous three steps as required.

GAS SAMPLING AND ANALYSIS

When sampling a gas select an entry point in the gas source. The sampling point must be truly representative of the gas sample, without air infiltration. It is recommended that a one-eighth inch black iron pipe be used as a sample probe. One end of the sample probe is attached to the soot filter tubing. The probe length must be determined by the sample location. In all cases it should extend approximately half way into the gas stream being analyzed. Place the leveling bottle, filled with distilled or tap water on the table or below the analyzer.

Place the three way valve in position one (Fig. 11-7) and pump the aspirating bulb to force a sample into the burette, and through the leveling bottle. Squeeze the aspirator bulb firmly seven or eight times with sufficient force to cause active bubbling of water in the leveling bottle.

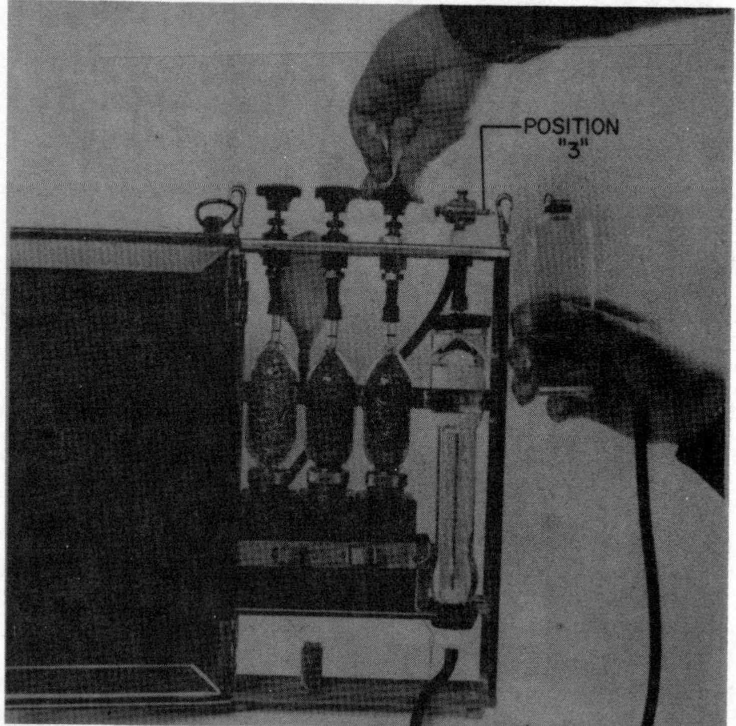

Fig. 11-9. The valve is moved to position three (courtesy of Milton Roy Company, Hays-Republic Division).

Slowly raise the leveling bottle until water shows near the bottom of the burette. Raise and lower the leveling bottle to work out entrapped gas bubbles. Be sure that the water level remains below zero.

When the entrapped gases are removed, clamp the rubber hose and raise the leveling bottle about 6 inches. Move the three way valve to position two (Fig. 11-8) and gradually release the clamp pressure on the leveling bottle tubing. When the water level reaches exactly zero (Fig. 11-9) clamp the tubing tightly and move three way valve to position three (Fig. 11-9).

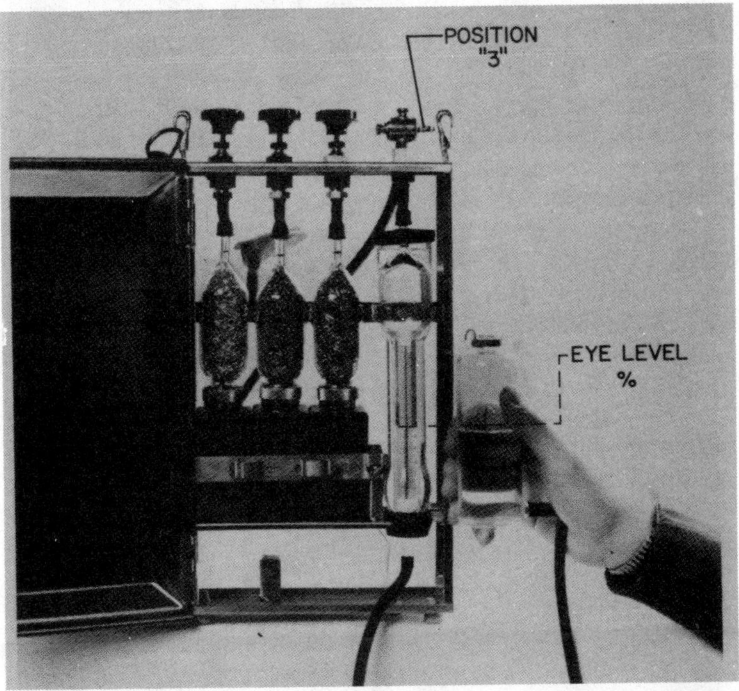

Fig. 11-10. With the two levels in line at eye level, read the burette scale (courtesy of Milton Roy Company, Hays-Republic Division).

Raise the leveling bottle until it is slightly above the burette. Select the absorption pipette for the sample to be measured and open this pipette needle valve three or four turns. The water will rise in the burette, pushing the gas into the absorption pipette.

As the water approaches the top of the burette, control its rise with hand pressure on the leveling bottle tubing. Be sure that the water moves up slowly. Stop the rising water when it reaches the small opening at the top of the burette, or the etch mark on the burette. This is accomplished by clamping the leveling bottle tubing tightly.

With the burette water level at the etch mark, close the absorption pipette's needle valve, and place the leveling bottle on the table or hang it from the bottom of the analyzer case. Wait approximately 30 seconds to insure complete absorption.

When absorption is complete, open the absorption needle valve. The absorbing chemical will rise in the pipette. As it approaches the neck of the pipete control its rise by clamping the rubber tubing of the leveling bottle. Stop the chemical exactly at the etch mark of the pipette, and close the absorption needle valve.

Align the liquid level of the leveling bottle with the liquid level in the burette. With the two levels in line (at eye level) (Fig. 11-10) read the burette scale. Record this reading. This is the measurement in percentage.

After this absorption and measurement repeat the procedure to insure that the same reading is obtained. If the amount of your second absorption increases, make a third absorption to insure a complete and correct analysis. It is essential that each constituent of the flue gas sample be completely absorbed before commencing the absorption of the next constituent.

Normally, a flue gas sample is measured in the model 621A gas analyzer as follows: $C0^2$, 0^2 and C0, in this order. The chemical used to absorb oxygen will absorb $C0^2$, and the chemical used to absorb C0 will absorb oxygen and carbon dioxide. For these reasons, carbon dioxide must be absorbed and measured first, followed by an oxygen measurement, and lastly by a carbon monoxide measurement. Note: The original sample must be retained throughout the analysis. Each individual component measurement is accomplished as outlined above. It is recommended that each measuring procedure be repeated at least twice to insure identical measurements.

The difference between the reading of carbon dioxide ($C0^2$) and oxygen (0^2) is the percentage of oxygen. The difference is the reading between oxygen (0^2) and carbon monoxide (C0) is the percentage of C0.

As an example, assume the absorption reading for $C0^2$, is 16 percent. An oxygen absorption measurement indicates 20 percent. The difference between the oxygen and $C0^2$ measurement (20 - 16) is 4 percent. This is the percent of oxygen. A C0 absorption is accomplished and a reading is 20.6 percent. The difference between the C0 absorption reading of 20.6 percent and oxygen of 20 percent (20.6 percent - 20) is 6 percent. This is the percentage of C0 in the gas sample.

MODEL 621AS-38.30 SULFUR DIOXIDE ANALYZER

The following instructions apply only to the analysis of sulfur dioxide (S0²) using the measuring accessories furnished with the standard 100CC model 621A-38.30. To measure sulfur dioxide the normal gas sample tubing is disconnected from the three way valve and the sulfur dioxide measuring system is attached to the three way valve.

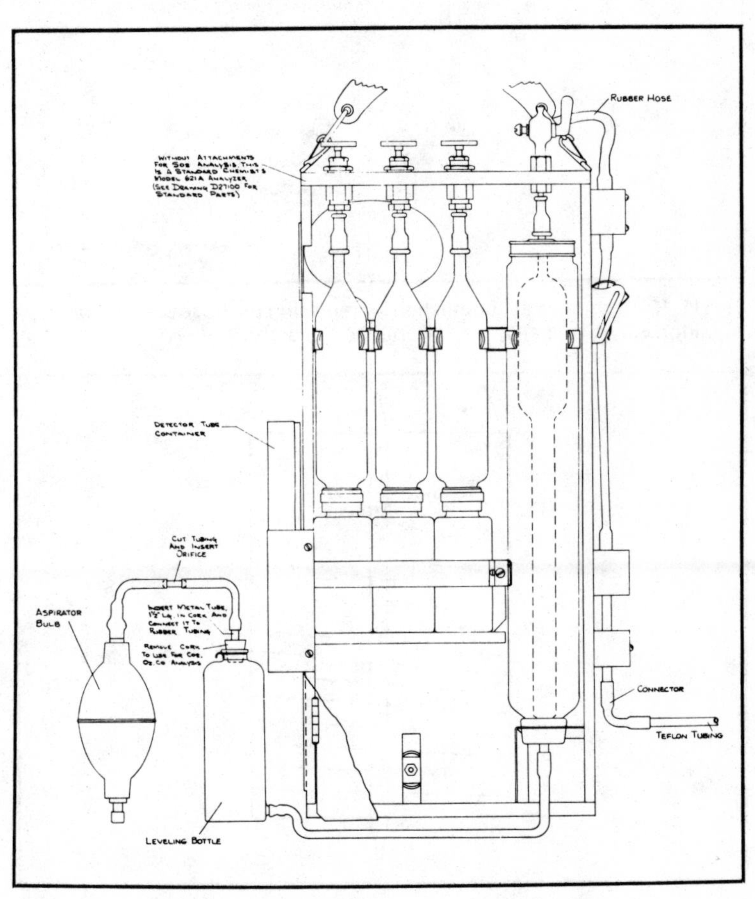

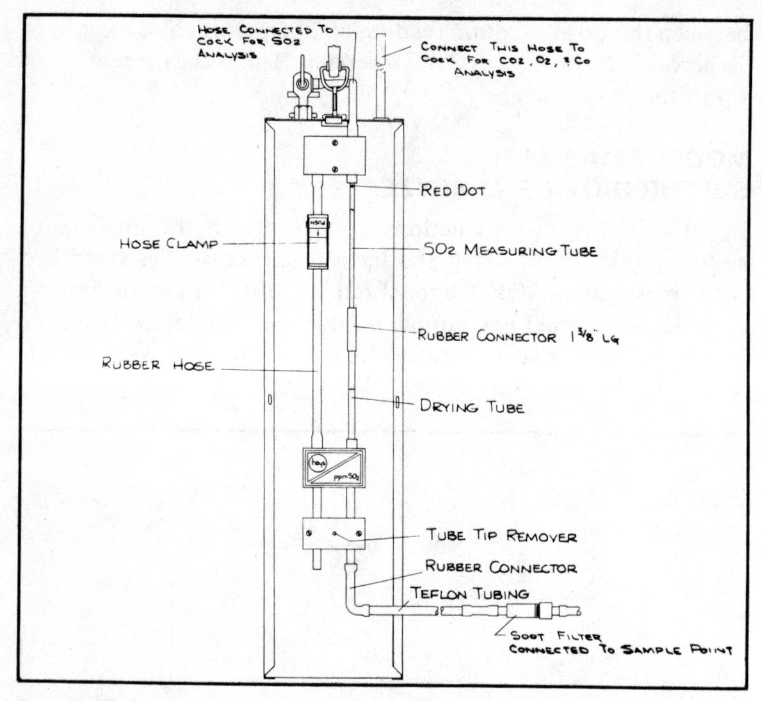

Fig. 11-12. Purging is accomplished by releasing the hose clamp (courtesy of Milton Roy Company, Hays-Republic Division).

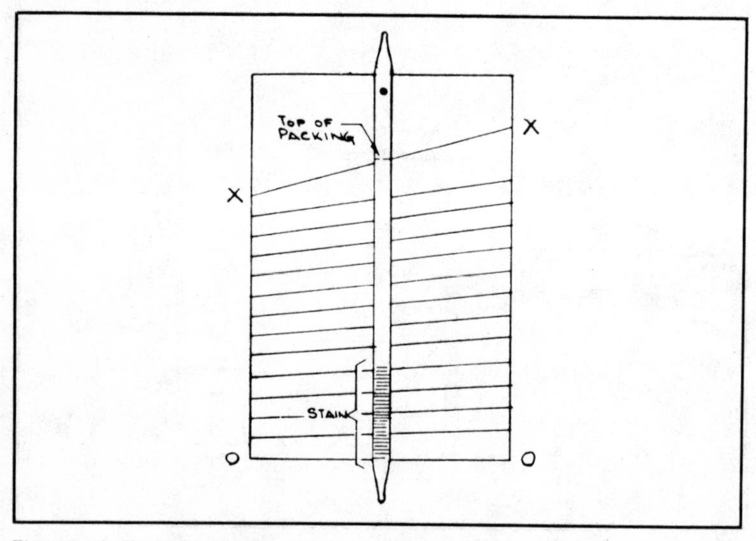

Fig. 11-13. Remove the detector tube and place it on the scale (courtesy of Milton Roy Company, Hays-Republic Division).

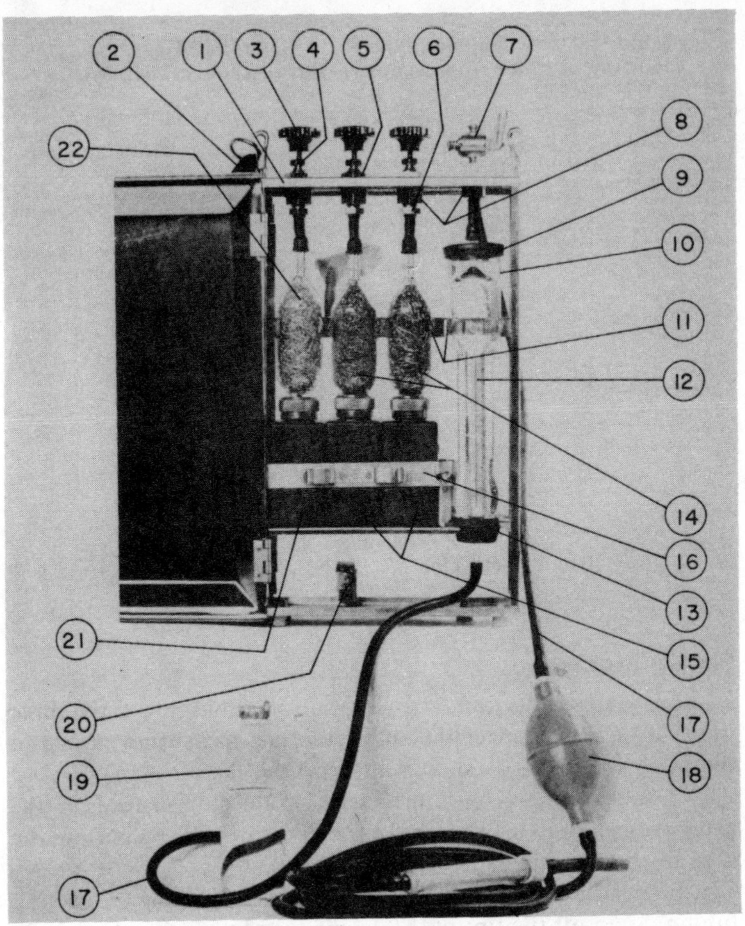

Fig. 11-14. Front view of the analyzer (courtesy of Milton Roy Company, Hays-Republic Division).

Principles of Operation

The analysis of sulfur dioxide is accomplished by drawing a 100cc sample of gas through drying and detector tubes. The detector tube contains a chemical which changes color in the presence of sulfur-dioxide.

The detector tube is removed from the analyzer and placed on a scale graduated 0-.30 percent. The vertical discoloration of the detector tube is indicative of the sulfur dioxide concentration.

The correct time for the gas sample to be exposed to the detector tube is approximately three minutes. A longer time will

Table 11-1. Information on the Parts Pictured in Fig. 11-14 (courtesy of Milton Roy Company, Hays-Republic Division).

		PROXIMATE ANALYSIS-%				ULTIMATE ANALYSIS-%					
RANK	STATE	MOIS-TURE	VOLA-TILE MATTER	FIXED CARBON	ASH	S.	H.	C.	N.	O.	HEATING VALUE-BTU's/LB.
Anthracite	PENNA.	4.4	4.8	81.8	9.0	0.6	3.4	80.	1.0	6.2	13,130
Semianthracite	ARK.	2.8	11.9	75.2	10.1	2.2	3.7	78.	1.7	4.0	13,360
Bituminous Coal:											
Low Volatile	MD.	2.3	19.6	65.8	12.3	3.1	4.5	75.	1.4	4.2	13,220
Medium Volatile	ALA.	3.1	23.4	63.6	9.9	0.8	4.9	77.	1.5	6.2	13,530
Medium Volatile	ALA.	3.1	23.4	63.6	9.9	0.8	4.9	77.	1.5	6.2	13,530
High Volatile-A	KY.	3.2	36.8	56.4	3.6	0.6	5.6	79.	1.6	9.2	14,090
High Volatile-B	OHIO	5.9	43.8	46.5	3.8	3.0	5.7	72.	1.3	14.	13,150
High Volatile-C	ILL.	14.8	33.3	39.9	12.	2.5	5.8	59.	1.0	20.	10,550
SEMIBITUMINOUS:											
Rank-A	WASH.	13.9	34.2	41.0	11.	0.6	6.2	58.	1.4	23.	10,330
Rank-B	WYO.	22.2	32.2	40.3	4.3	0.5	6.9	54.	1.0	33.	9,610
Rank-C	COLO.	25.8	31.1	38.4	4.7	0.3	6.3	50.	0.6	38.	8,580
Lignite	N. DAK.	36.8	27.8	30.2	5.2	0.4	6.9	41.	0.7	46.	6,960

S — SULFUR; H — HYDROGEN; C — CARBON; N — NITROGEN; O — OXYGEN

completely discolor the detector tube. A short sample time will result in a low concentration indication.

Set Up Procedure

Disconnect the standard gas analysis tubing from the three way valve, and connect the sulfur dioxide measuring tubing to the three way valve as shown in Fig. 11-11.

Install the ¼-inch stainless steel connector instead of the glass orifice. Snap off the tips of the detector tube by placing the tube in the tube tip remover (Fig. 11-12).

Install the detector tube, with the red dot up to the connecting tubing. Snap off the tips of the drying tube by placing the tube in the tube tip remover.

Install the drying tube with the drying agent of the bottom. The burette water jacket must be filled with water.

Connect the leveling bottle to the suction side of the aspirator bulb. See Fig. 11-11. Fill the leveling bottle two-thirds full of water. Close all needle valves on top of the model 621A.

Operation

Purge the system and check for moisture. Continue purging until all tubing is dry. Purging is accomplished by releasing the hose clamp (Fig. 11-12). Place the three way valve in position one and operate the aspirating bulb for approximately three minutes.

Table 11-2. Information on the Parts Pictured in Fig. 11-15 (courtesy of Milton Company, Hays-Republic Division).

PHOTO ITEM	DESCRIPTION	PART NUMBER	30 CC 621A 34:30	50 CC 621A 31:30	100 CC 621A 38:30	100 CC 621AS 38:30
·23	Connectors (4 required)	445-11	1	1	1	1
·24	Rubber Bag	445-34	1	1	1	1
25	CO Vent Tube	415-38	1	1	1	1
26	CO_2 Vent Tube	415-38	1	1	1	1
27	Knurled Nut (3 required)	961-160	1	1	1	1
	Rubber Gasket (3 required)	535-91	1	1	1	1
28	Soot Filter Assembly	C-00-192	1	1	1	1
29						
30						
31	Filling Tube Assembly	C-720-98	1	1	–	–
		C-720-145	–	–	1	1
32	Case Feet (4 required)	445-82	1	1	1	1
33						
33A						
34	Air Bag Manifold	C-720-16	1	1	1	1
35	Aspirator/Soot Filter Clamp	920-5	1	1	1	1
·	Compression Gasket (3 per set)	C-423-100	1	1	1	1
·	Clean Out Screw for Header	751-131	1	1	1	1
·	Valve Only for Aspirator Bulb	445-10	1	1	1	1
·	Gasket for Bottom of Burette	535-37	1	1	1	1
·	Steel Wool	920-33	1	1	1	1
·	Copper Wire	921-45	1	1	1	1
·	10 Detector Tubes, Drying Tubes and Tube Connectors	476-22	–	–	–	1
·	Orifice Assembly	C-720-757	–	–	–	1
·	Teflon Tubing	413-24-10′	–	–	–	1
·	Rubber Stopper	445-80	–	–	–	1
·	Package of Cleaning Wires	921-59	–	–	–	1
·	Aspirator Bulb Assembly	C-720-755	–	–	–	1
·	Connector for Rubber Stopper	762-61	–	–	–	1
·	Hose Clamp	245-50	–	–	–	1
·	Cardisorber	C-731-5-6	1	1	1	1
·	Seez O_2	C-731-4-6	1	1	1	1
·	Cuprous Chloride	C-731-7-6	1	1	1	1
·	Cleaning Solution	C-731-1-6	1	1	1	1

· Recommended Spare Parts

Replace the stainless steel connector with the glass orifice tube, remove the stopper from the leveling bottle, and tighten the hose clamp.

Place the three way valve in position two. Raise the leveling bottle until the water level reaches the etch mark on the burette. Control the water rise by clamping the leveling bottle connecting tubing.

With the water level at the burette etch mark and the tubing firmly clamped, place the three way valve in position one and insert the stopper in the leveling bottle.

Compress the aspirating bulb, and release the water bottle tube clamp. Maintain a steady suction by squeezing the aspirating bulb, and set the leveling bottle down.

The water level in the burette will fall slowly. In about three minutes it will reach zero (0).

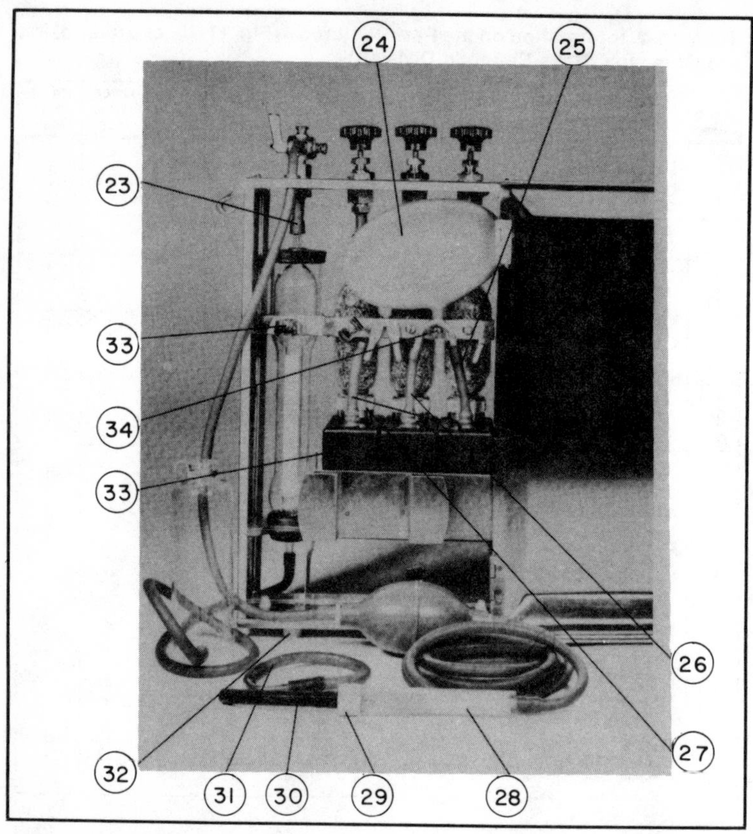

Fig. 11-15. Back view of the analyzer (courtesy of Milton Roy Company, Hays-Republic Division).

When the water level reaches zero (0) clamp the leveling bottle tubing, and place the three way valve in position two. Remove the detector tube immediately and place it on the scale as shown in Fig. 11-13. The sulfur dioxide concentration is at the highest point of color change. Discard the used drying and detector tubes.

Figures 11-14 and 11-15 show front and back view of the orsat gas analyzer. Tables 11-1 and 11-2 list parts.

Courtesy of Milton Roy Company, Hays-Republic Division

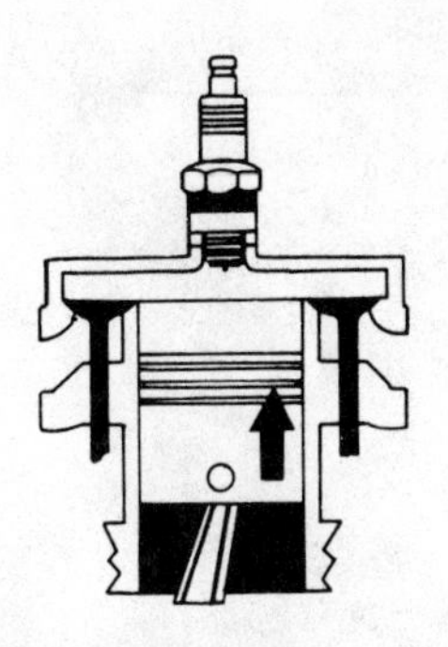

Chapter 12
Using the Keeler Model 59 Dual Flue Gas Analyzer

The dual flue gas analyzer utilizes two identical cells in which filaments possesing catalytic properties are housed. Each filament temperature is continuously measured and indicated on the meters (Fig. 12-1).

In one cell an atmosphere of butane is maintained around the filament which burns in ratio to the amount of oxygen present in the sample being drawn in by the instrument pump. Any change in filament temperature is indicated on the meter as a percentage of oxygen (Fig. 12-2).

Air is maintained in the other cell which burns in ratio to the amount of combustibles present in the sample. Any change in temperature is indicated on the meter as a percentage of combustibles.

Both oxygen and combustible readings are simultaneous with a response time of less than 30 seconds when using a sampling tube of 25 to 50 feet in length. A disposable cylinger of butane is mounted within the instrument to operate the oxygen cell, and can be expected to give 60 to 70 hours service.

OPERATING INSTRUCTIONS

☐ Plug into 115 volt, 60 cycle outlet.

☐ Switch "ON" allow short warm up as follows: Instrument temperature: @ 70 degrees 4 minutes; 40 degrees 15 minutes; and 0 degrees 30 minutes.

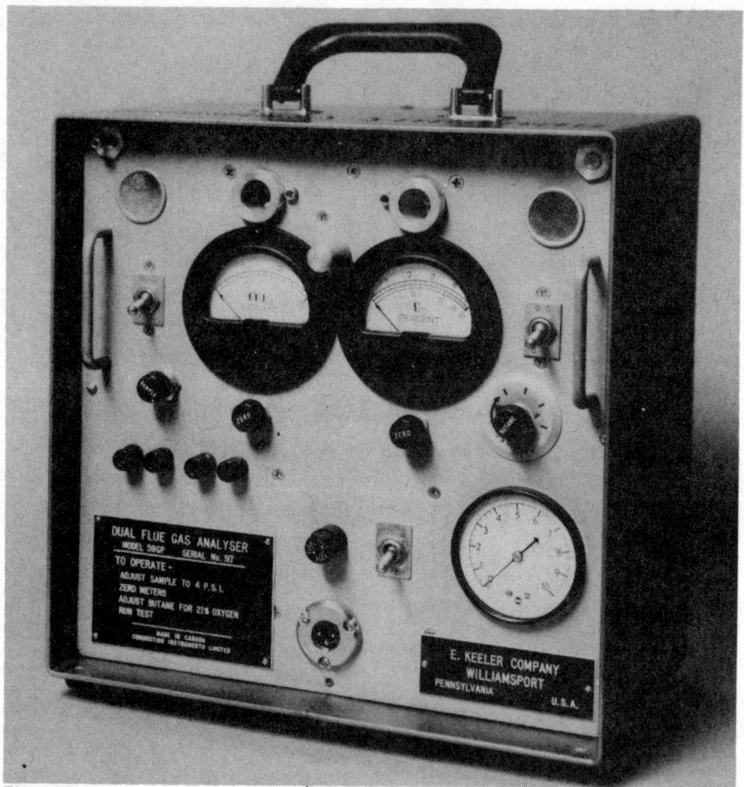

Fig. 12-1. Model 59GP dual flue gas analyzer (courtesy of E. Keeler Company).

☐ Adjust sample to 4 psi gauge pressure.

☐ Zero meters.

☐ Open and adjust butane valve for 21 percent or air reading. (Approximately two turns from closed position.)

☐ Insert probe in stack and proceed with tests.

It is not advisable to permit the unit to sample from combustion processes for extended periods of time when unattended, due to the rapid build up of condensation in the sample tube which, if not drained, may be drawn into the instrument in sufficient volume to overflow the water trap, saturate the filter materials, fill the pressure gauge and plug the sample control orificies in each cell with debris washed from the filter. Accumulated condensation in the plastic sampling hose may be drained by the operator by removing the hose end from the inlet fitting and shaking out the excess from the external trap and hose.

BUTANE SETTING

The oxygen calibration is not affected by rich or lean butane settings within the following limits. When sampling air (21 percent), the meter reads 23 percent. The instrument will indicate a 5 percent known sample gas as 5 percent, a 10 percent known sample gas as 10 percent, and a 15 percent known sample gas as 15 percent. There is no preceptible difference in the working range between 0 percent and 7 percent.

When sampling air (21 percent), the meter reads 15 percent due to lean butane valve setting. A 5 percent known sample gas is indicated as 5 percent. A 10 percent known sample gas is indicated as 10.25 percent. A 15 percent sample gas is indicated as 17 percent. There is no perceptible difference in the working range between 0 percent and 7 percent.

The 6 ounce butane refill lasts for 70 hours total use. A spare should be carried in the instrument cover pocket.

FILTRATION SYSTEM

Each unit is supplied with a 25-foot clear plastic sampling hose which forms part of the sample filtration system. It has been treated with a salt solution to aid the rapid condensation of water outside the instrument, where it can be readily observed and drained by the operator as it accumulates in the hose.

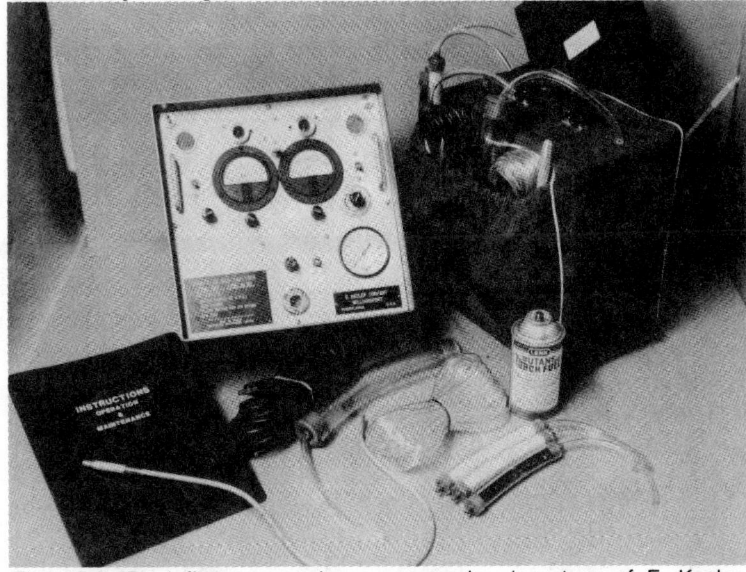

Fig. 12-2. Dual flue gas analyzer accessories (courtesy of E. Keeler Company).

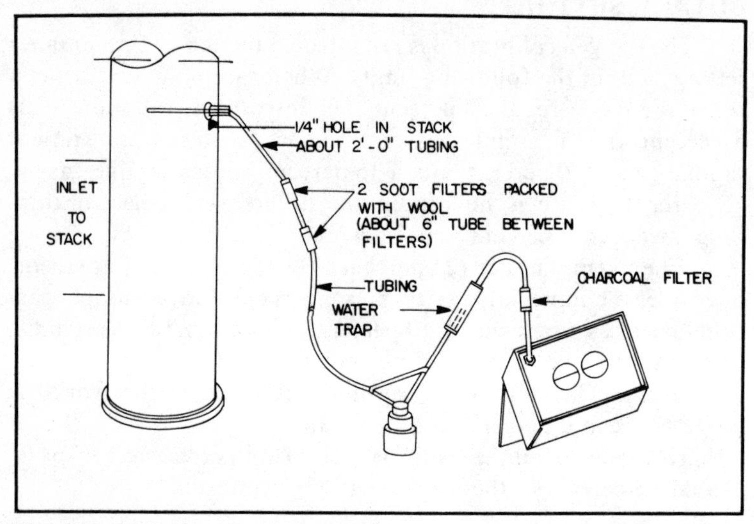

Fig. 12-3. Notice the filtration system.

Table 12-1. Data on Oxygen and Carbon Dioxide Percentages (courtesy of E. Keeler Company).

#4 FUEL OIL ULTIMATE CO_2 - 14.579% - DRY					
% EXCESS AIR	% CO_2	% O_2	% EXCESS AIR	% CO_2	% O_2
0	14.579	0.0	23	11.23	4.8
1	14.43	0.2	24	11.08	5.0
2	14.287	0.4	25	10.93	5.22
3	14.14	0.63	26	10.79	5.4
4	14.0	0.84	27	10.64	5.64
5	13.85	1.045	28	10.5	5.9
6	13.7	1.25	29	10.35	6.0
7	13.56	1.46	30	10.2	6.27
8	13.4	1.67	31	10.06	6.48
9	13.27	1.86	32	9.9	6.69
10	13.12	2.09	33	9.76	6.9
11	12.97	2.3	34	9.6	7.1
12	12.83	2.5	35	9.475	7.3
13	12.68	2.7	36	9.33	7.52
14	12.54	2.9	37	9.18	7.73
15	12.39	3.135	38	9.04	7.94
16	12.25	3.3	39	8.9	8.15
17	12.1	3.6	40	8.75	8.36
18	11.95	3.8	41	8.6	8.57
19	11.8	4.0	42	8.45	8.78
20	11.663	4.18	43	8.3	9.0
21	11.52	4.4	44	8.16	9.2
22	11.37	4.6	45	8.0	9.4

Should condensation be drawn into the unit, a water trap is provided to catch it. However, it has limited capacity and should be checked and drained. If the trap overflows, the double pass dirt filter will prevent the entry into the pump.

The filter material has the property of absorbing water, and if saturated can be regenerated by heating in a container at 600-650 degrees F. The filter material (absorbent virgin wool) should be replaced at least as often as every sixth butane cylinder. When sampling flue gases containing high sulfur percentages, the sample should be drawn through a flask containing a solution of water and sodium bicarbonate to prevent the erosion of the filter cotton, formation of alum in the combustion cells and possible low indications of oxygen and combustible (Figs. 12-3 and 12-4).

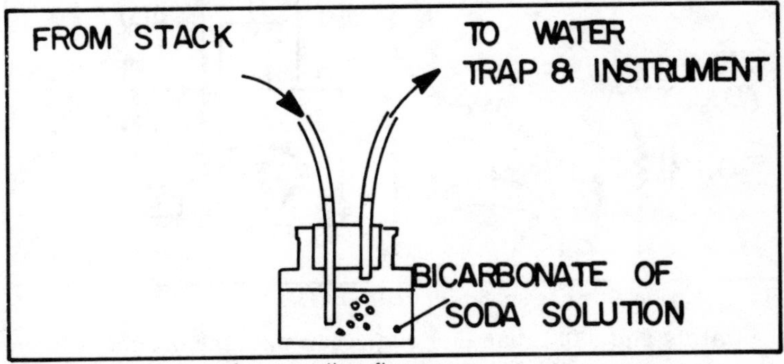

Fig. 12-4. Procedure for sampling flue gases.

PUMP

The pump requires a thorough cleaning when the pump sample pressure cannot be adjusted to 4 psi or when pressure fluctuations are indicated by the instrument pressure gauge. Completely disassemble the pump and clean all part of deposits. Be careful not to scratch the valve seats or check valve disc. Check the diaphragm for holes or tears.

Return the check valve discs and retaining screws to the same valve from which they were removed as the wear on each valve is different. They may not function correctly if mixed. If after cleaning the pump the sample pressure is still low, insert a pressure gauge in the plastic line which connects the instrument gauge to the sample adjusting needle valve to determine if the gauge is functioning correctly. If the instrument pressure gauge reads low it should be replaced or repaired.

The combustion cells have a glass window through which the active filaments can be observed. Both cells are equipped with

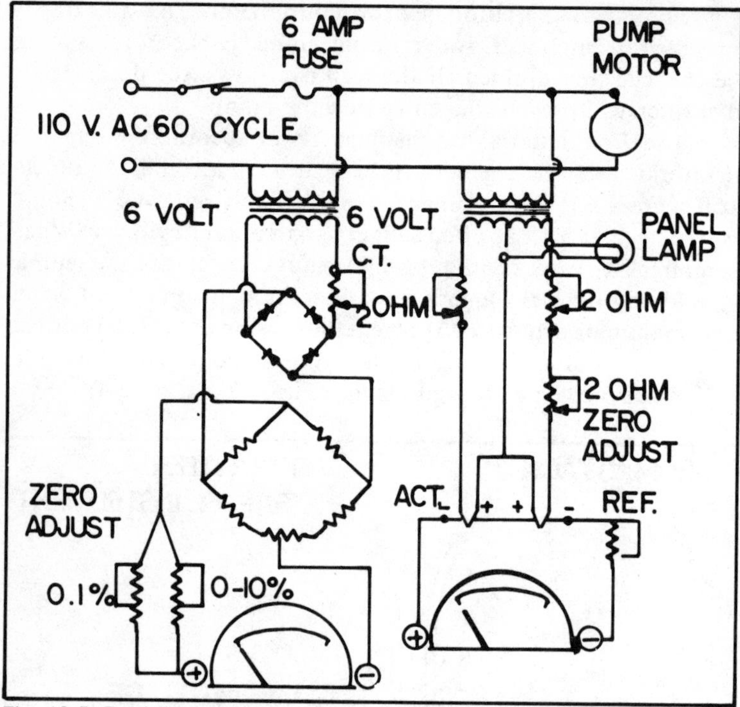

Fig. 12-5. Dual flue gas analyzer circuit.

aspirators and .008″ diameter orifices to pass the proper volume of sample gas at 4 psi operating pressure. The C0 cell aspirator introduces air into the sample stream to react any combustibles in the flue gases on the active filament. The oxygen cell aspirator mixes butane into the sample stream to react oxygen in the flue gases on the oxygen active filament.

FILAMENT CIRCUITS

Two types of measuring circuits are used both utilizing the principle of catalytic combustion for the indication of oxygen and combustible gases. The series opposing *thermocouple circuit* is two filaments with thermocouples in series connected with the polarity opposing so one thermocouple output is continuously bucking out the output of the other. Each thermocouple is inserted in a helical wound coil to measure the coil temperature. One coil is not catalytic (reference filament); the other has catalytic properties and heats up in the presence of a combustible gas. When the meters are zeroed, both filaments have the same output and cancel out reading zero on the meter. When a sample gas containing combustibles is passed through the cell, the catalytic filament heats up giving a higher output from the thermocouple and registering on the meter as a percentage of oxygen or C0 in the flue gases. The thermocouple type circuit is best suited to those applications where the atmosphere to be analyzed is normally reducing such as atmosphere generators, annealing furnaces and ovens with a combustibles content exceeding 1 percnt by volume.

The *wheatstone resistance bridge* circuit utilizes four balanced resistors in a net work arranged to indicate any change in the resistance of the bridge components. Two of the resistors are catalytic filaments. One is housed in a sealed well isolated from the sample. The other is inserted in the sample stream to react any combustibles present in the flue products. When C0, hydrogen or butane are encountered, the filament heats up and increases the filament resistance, which is indicated on the meter as a percentage of combustibles or oxygen. The wheatstone bridge circuit is best suited to those applications where the atmosphere to be analyzed is normally oxidizing or neutral, where indications of low percentages of oxygen or C0 are desired in the range of 0 to ½ of 1 percent full scale.

Oxidizing Atmospheres

The majority of applications are those involving the complete oxidation of fuel with approximately 5 percent excess oxygen being indicated in the flue gases (25 percent excess air) where the purpose is to heat air, make steam or heat water and not lose valuable Btu up the stack in the form of C0, soot or superheated air (Table 12-1). For these more common applications, the model 59SG is best suited with an oxygen indicating range of 0 to 21 percent and a double C0 indicating range of 0 to 1 or 2 percent and 0 to 10 percent. On the 0 to 1 percent C0 scale, readings of .04 percent are readily apparent. (.04 percent C0 is the maximum permissible).

Inert Atmospheres

Those oven or furnace atmospheres containing up to 2 percent combustibles are considered inert. The instrument is set and operated as per the instrument instruction plate, but the oven or furnace atmosphere should also be checked with the instrument butane supply turned "off." The oxygen calibration is accurate down to the stockhiometric point where there are no oxygen or combustibles indicated. As the percentage of combustibles increases, the oxygen indication becomes less sensitive and is the equivalent to having the butane mixture too rich.

If combustibles are indicated in the sample stream with no oxygen indicated, as a final check for oxygen turn the butane supply off. In this state the oxygen active filament will react the combustibles in the sample stream with any oxygen present and give a much amplified reading on low oxygen percentages if present. For continuous checking of inert atmosphere applications the model 59IA is best suited with a double C0 range of 0 to .5 percent and 0 to 5 percent full scale and 0 to 2 percent, on the low oxygen scale 0 to 10 percent high.

Reducing Atmospheres

Those oven or furnace atmospheres containing more than 2 percent combustibles should be sampled with the butane "OFF"; should the butane be left "ON", the indication may be zero oxygen, though the sample may contain ¼ to ½ percent and oxidize the product being processed.

CALIBRATION OF SAMPLE GASES

Calibration adjustments should only be attempted if pertinent known sample gases are available. We suggest the following:

2 percent oxygen, 8 percent carbon dioxide, Balance nitrogen; 5 percent oxygen, 8 percent carbon dioxide, Balance nitrogen; 5 percent carbon monoxide, 8 percent carbon dioxide, Balance nitrogen; 4 percent carbon monoxide, 8 percent carbon dioxide, and Balance nitrogen

The above sample gases are recommended for those general purpose model 59 units with C0 calibrations which are normally used for checking and setting oxidizing or inert atmospheres.

For the dual flue gas analyzer model 59RA supplied for use on reducing atmospheres with a combustible scale C0 plus hydrogen, we suggest the following sample gas: 4 percent carbon monoxide, 4 percent hydrogen, 1 percent oxygen, 8 percent carbon dioxide and the balance of nitrogen.

For general purpose applications other than reducing atmosphere, the combustible gas in the known sample should be specified as C0. This is the first gas to appear in the flue products and is the least reactive in the combustible cell. Units scales for C0 will read three times higher on the same percentage of hydrogen gas.

Do not use sample gases which have percentages of combustibles and oxygen in the same cylinder as units set to these samples will give erroneous indications on general purpose applications other than reducing atmospheres. The C0 cell is not a natural gas leak detector and will not indicate natural gas in air mixtures.

Setting the Calibration Thermocouple Circuits

For the C0 cell, using a millivolt meter with leads attached to the two thermocouple terminals on one filament, adjust the coil temperature via the resistor until the output is 9 MV on both the

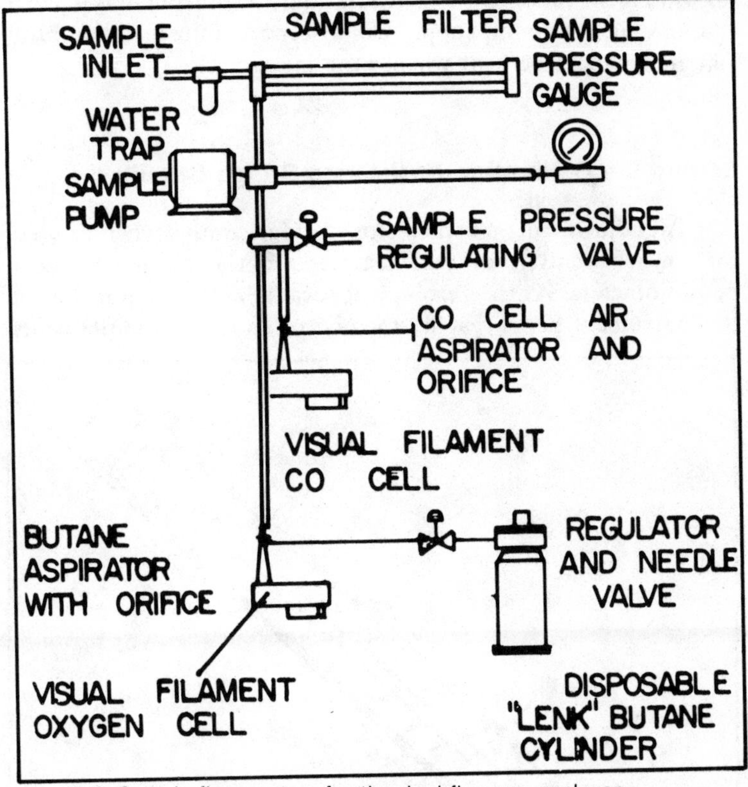

Fig. 12-6. Sample flow system for the dual flue gas analyzer.

active and reference filaments. Remove the millivolt meter and adjust the instrument meter to zero via the zero adjust on the panel. Pass a C0 sample gas through the instrument and adjust the reading to the sample via the meter adjusting potentiometer.

For the oxygen cell, adjust each filament output in 9 MV. Open the butane valve the required number of turns until the meter reads 21 percent. Let the unit run for three minutes; then pass the sample gas through the instrument and adjust the meter to the known sample via the proper meter adjusting potentiometer. If the disparity between the instrument reading and the sample gas is too great to be compensated for via the meter ad justing potentiometer, then reset the filament to a higher temperature (11 MV) if the reading is high, or a lower temperature value (8 MV) if the reading is low. Each time an adjustment is made in the circuit, the butan supply will also need adjustment to bring the meter to 21 percent when sampling air.

Setting the Calibration Resistance Bridge Circuits

All adjustments on wheatstone bridge circuits whether used for oxygen analysis or combustibles are via the meter adjust potentiometers. As the zero adjust located below each meter on the instrument panel reaches the end of its range, the filaments

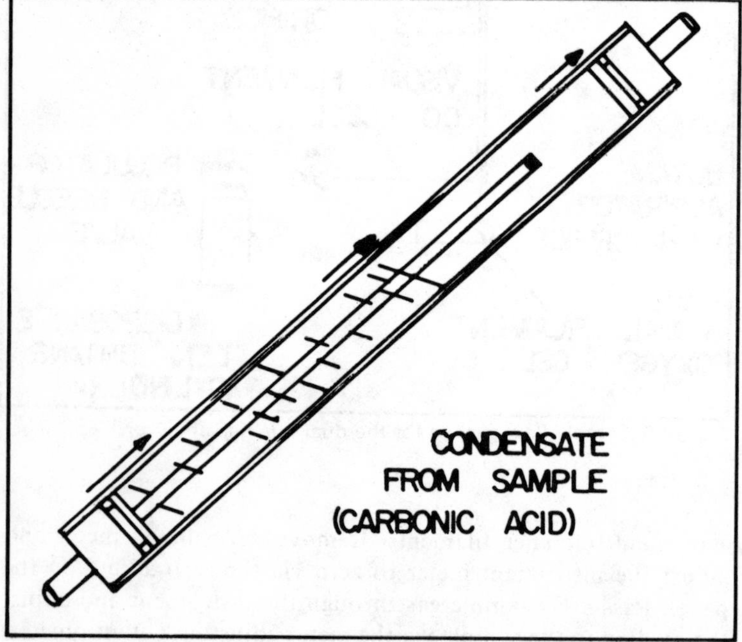

Fig. 12-7. External clear plastic water trap.

should then be reversed—the active placedin the reference well and the reference placed in the sample stream (Fig. 12-5). These filaments are interchangeable but if too far out of balance to be reversed, then replacement is advised.

Before making any major adjustments to the instrument, be sure all orifices are cleaned and free from partial obstruction by dirt or other solids. These are located in the brass fitting at the combustion cell sample inlet and at the C0 cell air aspirator (refer to Fig. 12-6). Remove the orifice and pass a .005″ diameter wire through the hole. The combustibles cell is subject to a smotering action if the air aspirator orifice is blocked or the combustibles percentage exceeds 15 percent by volume. This condition will cause the meter to rise quickly full scale and then settle back to zero or below zero.

SAMPLING PROBES AND HOSES

For most applications the stainless steel probe is adequate if the temperature does not exceed 1500 degrees F. At higher temperatures or when sampling directly in the flame, the probe begins to erode and the exposed nickle from the stainless steel cracks the carbon dioxide molecules into monoxide, giving a false combustibles indication on the C0 meter. For temperature exceeding 1500 degrees F., a ceramic probe is recommended (see Table 12-2).

Rubber tubing should not be used as it gives off combustibles when heated and absorbs condensible hydrocarbons when exposed to industrial solvents, releasing these combustibles later

Table 12-2. Parts List for the Dual Flue Gas Analyzer (courtesy of E. Keeler Company).

PARTS AND ACCESSORIES	
1 - 59-001	Sample pump, less mounting plate and motor.
1 - 59-002	Pump check valve discs.
1 - 59-004	Buna pump diaphragm (Pkg. of 5).
1 - 59-014	Pump Motor.
1 - 59-017	Sample Control Orifices, .008″ Dia.
1 - 59-018	0 to 10 PSI sample pressure gauge.
1 - 59-019	Sample pressure regulating needle valve.
1 - 59-028	25 ft. clear plastic sampling hose.
2 - 59-039	Internal filter pass packed with desiccant.
1 - 59-090	5 amp. fuse (Box of 5).
1 - 59-107	Active thermocouple type filament.
1 - 59-108	Reference thermocouple type filament.
1 - 59-119	Orifice for aspirator air inlet with plastic tube (0.15″).
2 - 59-163	Resistance bridge filaments for CO cell (Model 59SG).
1 - 59-165	Resistance bridge filaments for oxygen cell (Model 59IA).
1 - 59-167	External moisture scrubber charged with desiccant.
1 - 59-170	Lenk disposable butane cylinder.
1 - Linde' desiccant available in 1 lb. packages	
1 - Virgin wool filter material available in ¼ lb. packages.	

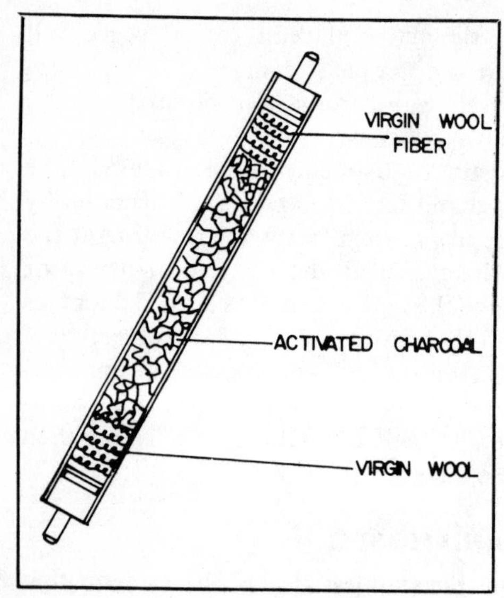

Fig. 12-8. Condensable hydrocarbon scrubber.

when the instrument is in use and the hose is warm. *Neoprene tubing* is the most serviceable, but is not clear or easily cleaned on the inside of the tube. *Tygen tubing* is recommended as it is a clear, non-absorbing, self-extinguishing plastic and permits cleanliness and condensates to be observed by the operator when in use (Fig. 12-7). In addition, the melting point is much lower than the temperature at which the material gives off erroneous combustibles, which lead to false settings of burners and other combustion equipment.

Plastic line filters charged with charcoal are available for use in the sampling line for scrubbing out high boiler point vaporized contaminants such as quench and cutting oils which tend to foul the internal filter passes, pump and analyzing filaments as they cool and condense inside the instrument flow system (Fig. 12-8). The internal instrument filter passes are packed with virgin wool and Linde desiccant which is available as an accessory in one pound packages.

Courtesy of E. Keeler Company

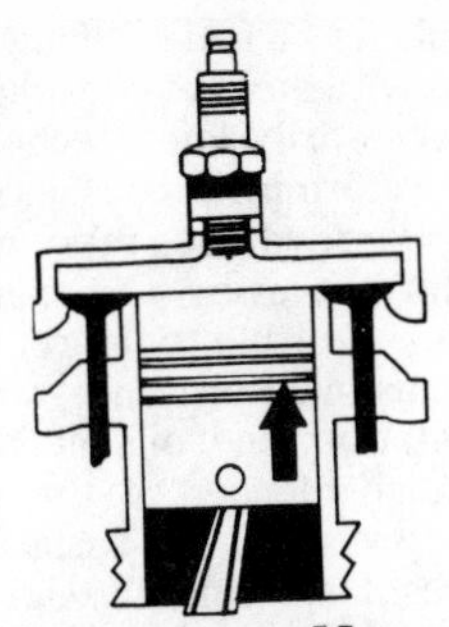

Appendix A
Heavy Fuel Oil Systems

This appendix provides information pertaining to the evolution of fuel oils, the effects shortages have had on fuel oil quality and consistency, the ways in which past and present system design practice differ, and some insight into what fuel conditions may be like a few years from now. It is impossible to provide specific design recommendations because every geographical area and every application will have its own peculiarities that dictate design features not universally acceptable, but a serious effort is made to provide information useful for basic considerations.

CHANGING FUEL OIL PROPERTIES

Over the past few decades, our nation's economy has developed to a point where the demand for gasoline has forced oil refineries to find new and better ways to refine crude oil. In the beginning of that period (1935), the refining process was referred to as a straight run process and only 10.45 percent of each barrel of crude was converted into distillate fuel oils. The heavy residual that remained, amounting to about 27 percent of each barrel, posed a disposal problem and was offered as a low cost fuel oil (Bunker C or 6).

By 1966, several additional crude oil refining processes were in use, and up to 89 percent of each barrel was taken off as gasoline, distillate oil, and other saleable products, with only 11 percent remeining for use as heavy fuel oil or as a blending base for heavy fuel oil. Oil producers know the chemical makeup of the products which are distilled from the crude, but there has been little interest in the chemical makeup of the residuals. Because of this, when a refiner has a petroleum waste disposal problem, he may blend it with the residual used as fuel oil. The rationale is that since no one knows what the chemical makeup is, it doesn't make any difference if the waste material is added.

From the standpoint of a burner manufacturer or a fuel user, changing from about 27 percent residual (from crude) to only 11 percent has resulted in today's residual oil being much heavier and more viscous than before. Some may even have an API gravity of less than 10 (heavier than water). The cracking processes used to increase the gasoline yield have, in effect, removed almost all of the lighter hydrocarbons from the crude. Because the residual that remains may have a pour point of 100 F and viscosities far higher than ASTM standards for 6 fuel oil, fuel oils that we buy today are made by blending distillates with the heavy residual.

Blending oils is not necessarily a new technique, but the components obtained from cracking and reforming processes used in producing today's blend may have a higher carbon/hydrogen ratio, which can lead to chemical instability and problems for the user. The methods of blending differ, depending upon the facilities available to the refiner or the distributor. At the refinery, blending is done under pressure, with components heated to various levels, and with agitation. Fuel oils blended in this manner tend to be more miscible than those blended at the distributor level, where preheated residuals will be poured on the surface of distillate oils and the blending occurs as the residuals sink through the distillate, and are hopefully dissolved. Low sulfur oils, forced into use by environmental considerations, have brought even more problems with the chemical and physical stability of the fuel.

To lower the sulfur content of a residual fuel, refiners "cut" it with a distillate, which not only has a lower gravity but most likely is a chemically unstable product of a thermal cracking process. Such blended low sulfur fuel oil has a pour point in the 0 to 30° F. range. To guarantee that this fuel oil remains in its blended state, it is necessary to provide heating and circulation within the storage tank.

Refined low sulfur fuel oil has a pour point of 90°F. or higher. The refinery process for producing low sulfur residuals tends to hydrogenate the oil (forming paraffin or long chain hydrocarbons). As a result the fuel has better chemical stability than those produced by the blending process. However, these oils containing long chain hydrocarbons are susceptible to sludge precipitation when stored in a tank without heat and/or circulation.

Oil experts contacted at five major suppliers and four distributors do not now know of any specification requirement that will assure a repeatable burning quality. They agree that viscosity and carbon residue are not good or reliable indicators of fuel oil quality. Several suspect that surfact tension is an important factor, but have no data to back it up. Some pointed out that in previous years, the refineries worked with only one or two basic crudes, but today they may be working with eight to 10 different crude oil supplies. Table A-1 shows the diversity of correct 6 oil properties.

The experts admit that blending these various oils may not be the most desirable practice, but that if the oil is properly stored and conditioned, the blending should not create any serious burning problems. When questioned

Table A-1. Some #6 Fuel Oil Analyses (courtesy of North American Manufacturing Company).

Gravity, °API	-2.4*	.30*	7.1	8.6	10.4	11.0	12.0	13.4	14.4	15.0	17.0	17.8*	25.0
Viscosity at 122 F, SSF	26	48	347	147	225	350	165	160	328	180	100	55	31
(Approx. SSU)	(235)	(450)	(3250)	(1450)	(2200)	(3400)	(1600)	(1550)	(3150)	(1750)	(950)	(525)	(280)
BSW, %	.60	.20	.20	.30	.40	.50	.60	.40	.80	.20	.80	.20	.20
Flash, P.M.C.C., F	148	182	245	190	190	230	235	230	230	245	190	230	220
Carbon Residue, %	2.80	7.25	11.80	11.25	11.60	9.50	10.00	8.90	8.80	10.36	7.40	1.00	2.24
Pour Point, F	+30	+65	+40	+30	+45	+50	+30	+40	+45	+35	+30	+90	+70
Sulfur, %	2.71	1.48	.51	.88	2.66	2.65	3.50	1.66	.83	.85	1.90	2.48	.74

* These are "odd-ball" fuels, sold as #6 fuel oil, that are produced in some refineries as by-products in the refining of lubricating oils and asphalts. As their supply is limited, they are usually obtained on a "spot purchase", and as "distress oil".

about the possibility of sludge formation, they indicated that while blending may not form sludge as we know it, the oils may be in contiguous contact rather than in solution; because of this, improper storage practice could accelerate the normal tendency for the heavier fluid components to separate from the blend.

At this time, North American Manufacturing Company is not involved with selection and design of in-tank heating systems. The above information is provided only to alert the reader to possible solutions to oil storage problems. North American system designs will begin at the pump suction point and will end at the return valve for the tank entry.

FUEL SUPPLY SITUATION

There is a strong possibility that natural gas will become unavailable for use in large industrial furnaces and boilers. It will probably be replaced by some forms of SNG, substitute or synthetic natural gas, made from coal. Gasification plants can be built at point of use, or a very large facility can be constructed, complete with a new piping system, to serve several industrial customers. There may be many of these plants in operation in the next 15 years.

Tremendous amounts of gasoline, jet, and diesel fuel are required to maintain the USA's transportation systems and our armed forces. Heating oils, 1 and 2 grade distillates, may be allocated or rationed for years to come and will be generally unavailable for boilers, industrial furnaces, ovens and kilns. Distillates now used as blending agents will be in short supply, or very expensive. Shortly after the loss of those distillates, we would expect 4 oil and the blended heavy oils also to disappear. (About 30 percent of every gallon of blended heavy oil is distillate). The only fuel oil remaining would be low sulfur residual.

Occasionally, straight run 3 or 4 oil becomes available in some localities. Reports imply that it will be available for at least a year or more, but it is usually difficult to get suppliers to provide a written contract stating the availability. North American Manufacturing Company recommends that oil storage, pumping, heating, and burning system designs be capable of handling a wide variety of fuels. If a system is designed for 2 oil, 6 cannot be accommodated; but a system designed for 6 oil can handle any oil from 2 through 6 and even some

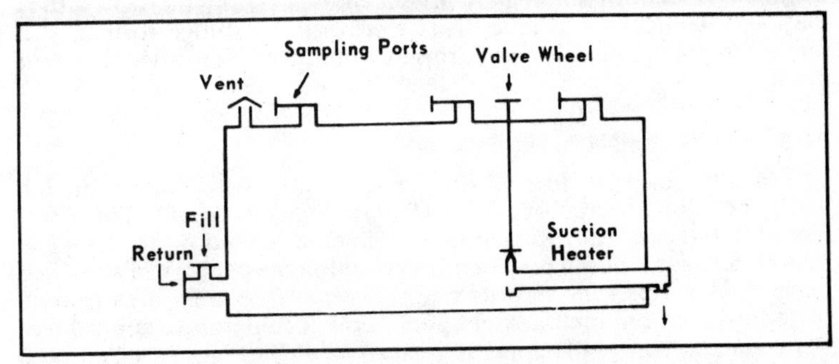

Fig. A-1. Side elevation view of an above-ground heavy oil storage tank (courtesy of North American Manufacturing Company).

Fig. A-2. Plan view of an above-ground heavy oil storage tank (courtesy of North American Manufacturing Company).

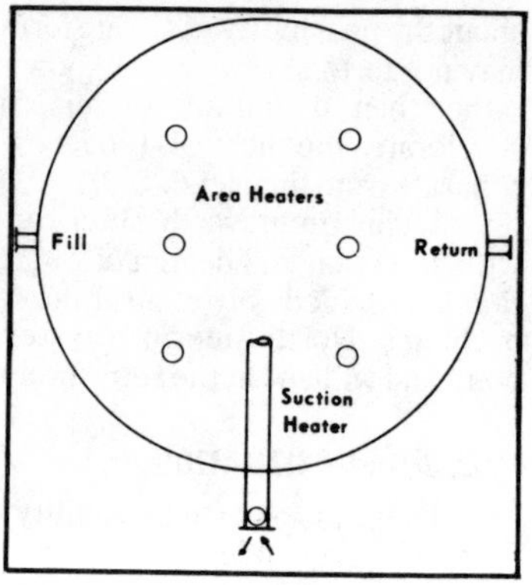

reclaimed oils. The investment requird to develop and install a new fuel oil system is such that the owner should provide himself with the flexibility required to permit use of any grade oil that might become available in the next 10 years.

FUEL OIL STORAGE

If the system designer/owner elects to compromise the total design to save a few dollars in the initial installation, he may experience difficulties in starting the system, or the daily operation may be fraught with nuisance type troubles, or major problems could arise that could interrupt plant production.

Possible chemical or physical instability of fuel oils, makes it necessary to consider a storage tank's environment. Unless a fuel oil tank is heated or stirred, the oil may stratify. The rate at which the heavier components will precipitate out of the blend is usually accelerated by low temperature. For blended oils, the pour point temperatures indicated in the specifications (usually zero to 30°F.) are true only when the oil is in the blended condition, before stratification. A blend may consist of 70 percent residual and 30 percent distillate oil.

The greater the difference in their specific gravities, the more serious the problems of separation stratification become. If the distillates used for blending are produced by a cracking process, they may be chemically unstable. This can result in sludge formation if temperatures are permitted to drop to within 20° of the specified pour point.

Bare Tanks for Straight Run Residual Oil

In the past, large fuel oil storage tanks were fitted with a suction heater positioned to remove fuel oil from the tank at a point 12 to 18 feet from the outside wall. Return lines delivered heated oil to the same region, forming a warm pool of oil deep within the center of a solidified mass of fuel. For very viscous oil, or a very large tank, area heaters were positioned to melt a larger pool of oil. Temperature control was set to maintain the pool temperature at about 80°F. Figures A-1 and A-2 show the general arrangement of large fuel oil tanks. The residual fuel oils made by the straight run process had 70°F. to 85°F. pour point.

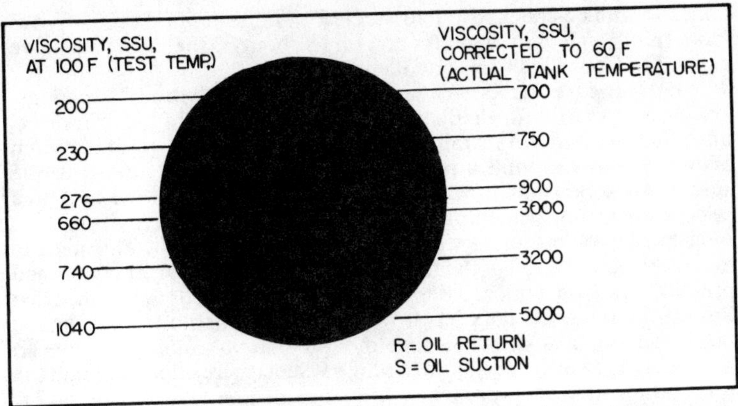

Fig. A-3. A sad case of stratification in a horizontal storage tank.

Apparently there was quite a temperature span between the point where the oil mass was in a solid state, to the point where fluidity returned; so a thick layer of oil would cling to the side of the tank, forming an excellent insulation for a warm pool of oil in the center. Under these conditions, it was not necessary to insulate fuel oil tanks.

Blended Heavy Oil Tanks are Not Self-Insulating

Fuel oil make-up has undergone quite a change and now most 4, 5 and 6 oils are blends of distillate and residuals or even crudes. Pour points of blended oils may range from 0 to 30°F. The temperature must not be permitted to fall within 20° of the pour point if the oil is to be kept in the blended state. If a blend cools too much, the residuals begin to separate out and / or sludge formation will begin. Both will slide to the bottom of the tank. If the oil is heated to maintain the temperature at 20°, above the pour point, all of the oil in the tank will be fluid. Steel sides of the tank will be exposed to warm liquid oil on the inside and cold wind on the outside. The heat loss must be maintained if oil fluidity is to be maintained when outside temperatures fall below the level of pour point ≤ 20°.

Low sulfur residual fuels as produced by a refinery process are the high pour point (high paraffin) oils that have everyone concerned. (In one case the pour point was 95°F and the viscosity at 122°F. was 44SSF. The oil regained its fluidity at about 100°
At that temperature level, there will be no doubt about the need for fuel oil tank insulation. Table A-2 gives useful information relative to fuel oil tanks, including heat loss information.

Because no one knows what type oil will be stored in a tank in the future, and because steam coils cannot be added after a tank is full, new fuel oil tanks should have steam coils covering the entire tank bottom. Insulation can be added when and if needed. Figure A-3 illustrates a typical tank of 6 fuel oil that has not been used or heated for a few months. Heavy oil systems should be started up well in advance of cold weather and left running until the warm season.

NFPA standards 30 and 31 serve as a good general guide for tank installation and location, but specific state, local, and insurance codes should be checked. Field erected oil tanks range from 20 feet to more than 100 feet in diameter and up to about 60 feet high, and have capacities from 40,000 to several million gallons.

Prefabricated tanks, available in sizes from 250 to 30,000 gallons, are used primarily for distillate products such as light fuel oil, diesel oil, gasoline, and are installed either above or below ground. When installed below ground a partially filled tank may be floated to the surface by subterranean water. To prevent this, anchor the tank to a concrete base, or place it far enough below the surface to provide an over-burden of sufficient weight to hold the tank down. This usually means placing the top of the tank 3 to 5 feet below grade, and it is good practice to set the tank at a depth where the top is at or below the frost line. If the oil in the tank has low viscosity (as distillates), lifting the fluid from the tank with a surface-mounted pump usually presents no difficulty. However, if a liquid has high viscosity (residuals) so that the total pump suction approaches 15 inches of mercury, the pump lift becomes critical, and pump cavitation mah be experienced.

The current fuel shortage has forced some plant operations to consider installing heavy fuel oil systems with real estate plot plans that cannot accommodate above-ground tankage. For underground heavy oil storage, construct a pumping and heating station on top of the tank, with the floor of the pump room below grade, resting on the

Table A-2. Recommended Heater Input for Balancing Normal Losses from Above-Ground Tanks (courtesy of North American Manufacturing Company).

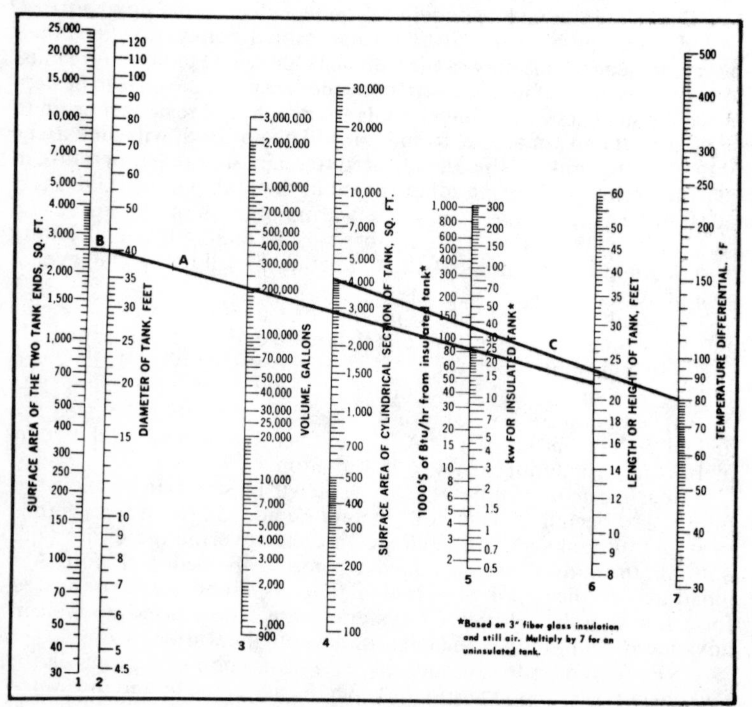

Table A-3. Tank Spacing (courtesy of North American Manufacturing Company).

Tank Spacing Capacity of tank, US gallons	Minimum distance, from property line which may be built upon, including the opposite side of a public way or from nearest important building or group of buildings*†	Minimum distance from nearest side of any public way*
275 or less	5 ft	5 ft
276 to 750	10 ft	5 ft
751 to 12 000	15 ft	5 ft
12 001 to 30 000	20 ft	5 ft
30 001 to 50 000	30 ft	10 ft

* If tank is equipped with an approved inerting or foam system, these distances may be halved to not less than five feet.

† Distances apply only if adjacent structures have fire protection. For unprotected structures, minimum distance must be twice that shown.

surface of the tank. This will usually limit the vertical lift to about 12 feet, requiring a pump vacuum of less than 10 inches of mercury. Very high viscosity, however, may necessitate use of a submerged pump or a pump at the level of the tank bottom in a dry well beside the tank.

Both above-ground and underground storage tanks are used for fuel oils. For practical reasons, underground tanks rarely exceed 20,000 gallons capacity. Tank location with regard to nearby structures and other tanks is dictated by applicable codes. See Table A-3.

NFPA codes require underground tanks to be at least one foot from the nearest wall of any basement, pit, or property line. Above-ground tanks over 550 U.S. gallons, must be equipped with emergency venting to relieve pressures over 2.5 psig, and must be located per the table below.

Adjacent tanks must be separated by no less than one-sixth the sum of their diameters, unless the diameter of one tank is less than half the diameter of the other. In this case, minimum separation is one half the diameter of the smaller tank. In the industry there has been some disagreement regarding the temperature at which oil may be returned to the tank.

ohot Oil Return—160°F to 250°F.

In many steel mills and glass plants, it is common to have oil recirculated at burning temperatures, with some of that hot oil returned directly to the storage tank. When circulating systems are stretched out to cover the distances required in large mills, the power consumption of the pumping system is an important design consideration. Pumping and circulating oil at 150 SSU instead of 1250 SSU requires less horsepower. If the fuel oil has reasonable flash point, if the tank is large, and if the return volume is small neither gasification of fuel within the tank nor sludge formation will be serious problems. Oil chemists claim that, contrary to popular opinion, sludge is not formed by overheating.

Warm Oil Return—100°F to 140°F

This is less likely to cause vaporization of fuel oil within the tank, a vapor pocket at pump suction and related pump cavitation problems, tank overheating and heat loss, boiloff of light ends reducing the calorific value of oil and increasing the handling problems with the remaining oil, or the possibility of venting an explosive vapor. With the fuel oils available today, the lower temperature is recommended.

Avoid Overheating

Some smaller installations will require a careful review of in-tank heating. Precautions should be taken to prevent overheating of the fuel oil caused by heating control system failure. If the electric immersion heater control switch were stuck in the on position in a small, insulated, prefabricated tank, the oil temperature might be raised to a point where distillation would begin. Some experts say that a dangerously explosive situation could develop. Installation of high tank temperature limit switches represents very low cost insurance. These automatic temperature switches could d-energize redundant electrical contactors to interrupt current flow to electric heaters or close a manual reset valve in the steam supply line to the fuel oil heater and activate tank steam flooding valves.

Restrictions on In-Tank Heating

Electric in-tank heating is very widespread, and a generally accepted practice, although FIA regulations seem to infer that electrical immersion heating is prohibited. The NFPA rule book recommends a maximum steam pressure of 15 psi for use with submerged oil heaters. This pressure was probably selected because it corresponds to a saturation temperature of about 250°F. However, most installations are operating at pressures and temperatures well above 15 psi and 250°F. (Throttling of steam through steam pressure regulators can result in localized superheating).

If it is found necessary to heat the entire contents of an oil tank, there are several options available. Steam coils can be placed to heat the entire tank bottom. Area steam heaters can be placed at strategic locations on the tank bottom as shown in Fig. A-2. Flexible electric strip type heaters of low watt density can be installed in a new tank. Return line oil heaters can be used to boost oil return temperature, permitting some control over tank temperature, whereas a simple *hot* oil return leaves tank temperature uncontrolled. Electrical tracing tape such as Chremelix electro-wrap can be applied to the tank exterior and covered with insulation.

All of these methods are currently in use. More than one may be employed. When sizing these systems, be sure to take advantage of the heat energy contained in reutrn oil.

In-Tank Circulation

It is not necessary to achieve complete temperature uniformity throughout the tank. If the oil temperature is slightly lower on the outside surface as compared to the center of the mass, some natural circulation will result. Positive circulation and agitation can be achieved if heat is applied at the bottom of the tank, and the oil return system discharges in an upward direction. Mechanical circulating devices are not in common practice. The main fuel oil pump and heater sets can be used to maintain tank heat and provie agitation during periods of zero oil consumption.

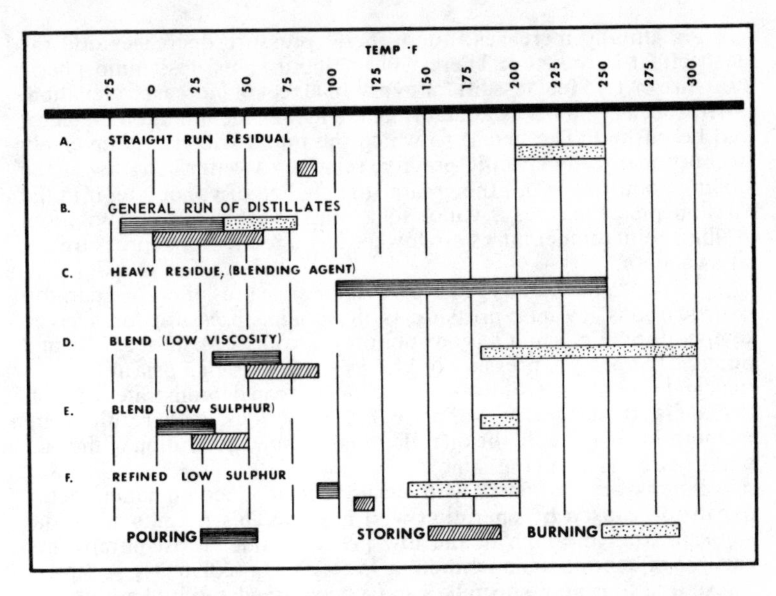

Fig. A-4. Variations in oil temperature requirements for pouring, storing and burning (courtesy of North American Manufacturing Company).

However, main fuel oil pump and heater sets are seldom of sufficient size to effectively agitate a tank. If a system is to be provided for that purpose, the fuel oil pump should be sized in a manner that would cause all of the oil stored, to pass through the pump at least once each week. The heaters in that system would be sized to provide sufficient heat to keep the oil at 20° above the pour point.

Summary of Recommendations on Oil Storage

For above-ground tanks, consider the possibility of installing tank insulation. (It must be waterproof). Fuel oil tanks should be constructed with provisions for heating the entire bottom of the tank. On larger tanks install steam coils at the time of construction. For smaller prefabricated tanks, provide for insertion of flexible electric heaters from the top of the tank.

Install ports to permit drawing oil samples from several locations within the tank. Install oil return systems so as to cause agitation and prevent BS&W (bottom sediment and water) from accumulating, especially near the suction connection. Check local, state, and federal codes governing tank installation. See Fig. A-4.

FUEL OIL HANDLING SYSTEMS

After a pump is started, the air is evacuated from the suction pipe and a vacuum is "pulled" upon the surface of the oil within the pipe. The pressure of the atmosphere on the surface of the oil in the tank then pushes the oil up to the pump. The theoretical maximum "lift" of a pump pulling a perfect vacuum (29.9" Hg) is 34 feet with a liquid having a specific gravity of 1.0. The practical lift we should expect a pumping system to overcome is limited by the following.

As altitude increases, atmospheric pressure decreases and the push effect is reduced. There will be about ½ psi less atmospheric pressure or 1.15 feet less lift for every 1000 feet of increased elevation. Also, specific gravity is a factor, as the lighter the liquid the higher it can be pushed. The height to which the liquid will rise is inversely proportional to its specific gravity, relative to water. The lower the boiling point of the oil, the greater the chance of vaporizing it in the suction pipe, causing a vapor lock and possible pump cavitation. Boiling point temperatures are lowered by a reduction in pressure i.e. by a vacuum.

The Cox chart (Fig. A-5) shows vapor pressures of normal paraffin hydrocarbons. (Vapor pressure is the boiling pressure for a given temperature, e.g. water has a vapor pressure of 14.7 psia at 212°F. and heptane has a vapor pressure of 14.7 psia at 200°F. or 3 psia at 120° lock and cavitation. (Air in an oil suction line can demonstrate some of the same symptoms as vapor lock. Vapor lock may result from evaporation of water in the oil rather than from vaporization of the oil, but the effect is about the same).

Pumps are designed with clearances to accommodate metal expansion caused by operation with hot oil. This permits some oil slippage from the high to the low pressure side. Most pumps are capable of continuous operation at 15 to 20″ Hg (2.036″ Hg = 1 psi). These first four points usually can be accommodated by limiting the system suction requirements to a maximum of 15″ Hg (7.37 psig).

Pressure drop in piping is directly proportional to the kinematic viscosity in SSU (seconds, Saybolt Un versal). Viscosity variation is best accommodated by a realistic pipe size selection and careful consideration of the following. How reliable is the viscosity information? Is the system designed for the lowest tank temperature that might occur? Provisions must be made to maintain design temperature. Considering the long term fuel supply, lighter grades may not always be available. Is the design based on the heaviest oil that may be

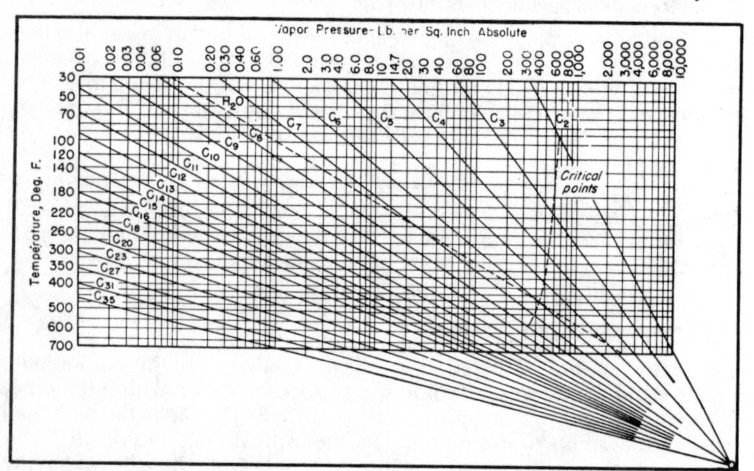

Fig. A-5. Cox chart for vapor pressures of normal paraffin hydrocarbons (courtesy of North American Manufacturing Company).

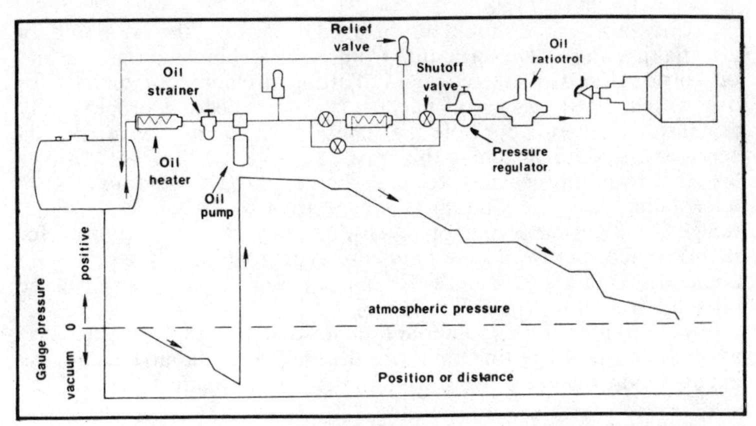

Fig. A-6. The total pressure boost caused by the pump equals the total system drop (courtesy of North American Manufacturing Company).

used? Suction piping size and pump motor hp must be selected to allow for the worst possible viscosity. (Quotations should indicate maximum viscosity, pressure and flow).

Allow for some extra drop, because strainers become dirty. Install vacuum gauges on both inlet and outlet of each oil strainer, to provide positive indication of a strainer-imposed flow restriction.

Pressure, flow, and viscosity determine a pump's horsepower requirement. The maximum pump capacity for pumps in North American's standard catalog is based on a viscosity of 2500 SSU.

Fuel oil pumps are normally of the positive displacement type. If such a pump is sized to move 14 gpm of fluid, 14 gpm must flow or the motor will stall. Outlet pressure builds up to whatever pressure is needed to overcome the resistance imposed upon the pump discharge i.e., a relief valve, pressure regulator, and a piping system. Inlet vacuum is a direct indication of the pressure drop imposed by the suction side piping system. Motor load is anfected by the total system pressure drop (outlet pressure + inlet suction), which in turn is a function of flow, and viscosity. See Fig. A-6.

The relief valve built into a pump is provided to protect the pump in the event of a temporary system flow blockage. If oil were to flow through this valve continuously, returning to the pump inlet, the oil would overheat and possibly cause mechanical damage to the pump. For control purposes, a separate back pressure regulator should be selected for continuous flow at a reasonable pressure drop, and positioned so that it will provide back pressure control at a strategic point is charging into the supply tank or pump suction.

Piping Material

This must be carefully selected if clogged strainers and atomizers are to be avoided. Years ago, pickled and oiled black steel pipe was standard for fuel oil systems because it was scale-free and pre-oiled. Pickled and oiled pipe is now difficult to obtain; so common practice is to select black iron pipe and acid flush the system prior to operation.

Galvanized pipe should not be used because of the possibility of zinc flaking and slidge formation from reaction between the zinc and oil-borne sulfur. Galvanized pipe and fittings are not recommended for use with natural gas or oil because the zinc coating process may temporarily cover a pin hole that could open up in service. For oil service, a welded system is the most desirable. Sonket weld fittings are the best and easiest to use; however butt weld joints are acceptable. Large oil piping systems are usually welded without flanges, except near components that may require removal for maintenance. Screwed type joints are acceptable, but the pipe joint compound must be suitable for use at 300°F. and must not be oil-soluble.

Teflon pipe joint compound is excellent and teflon tape is acceptable. Avoid getting tape over the end of the thread because the excess tends to break loose and plug small downstream orifices.

Pumping and Heating Systems

Two stage heating, pumping and circulating systems in the form of closed loops are the most desirable. See Figs. A-7 through A-10. When firing systems are required to modulate, ccurate temperature control from burner to burner is very important, and good control is the primary purpose of the loop arrangement.

First stage loops naturally are sized at 1.5 times th maximum burning rate, and the fuel oil heaters must be sized to heat the total pumping capacity. Oil in this first stage loop is normally heated from a low of about 50°F. to a high of 140°F.

Sizing of the second tage loop is based on two primary conditions: the burning rate and a circulating rate that will keep oil temperature drop to less than 10°. If the loop system is designed as shown in Fig. A-8, the fuel oil heater must be sized to heat all oil to be burned, from

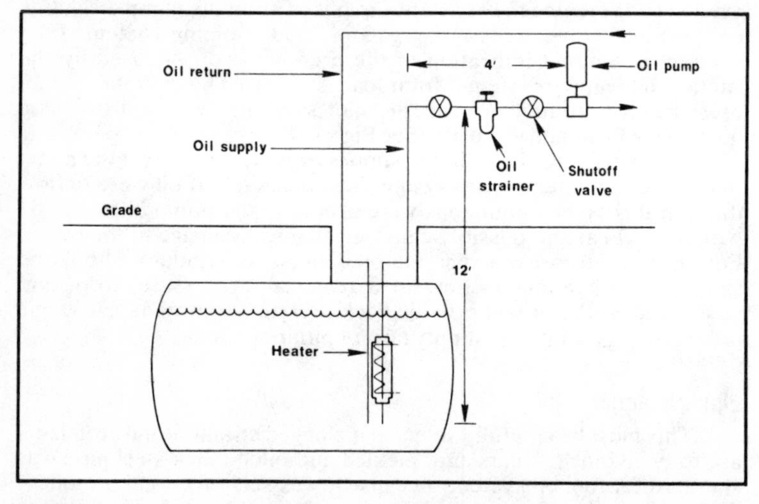

Fig. A-7. Diagram of a heating and pumping system (courtesy of North American Manufacturing Company).

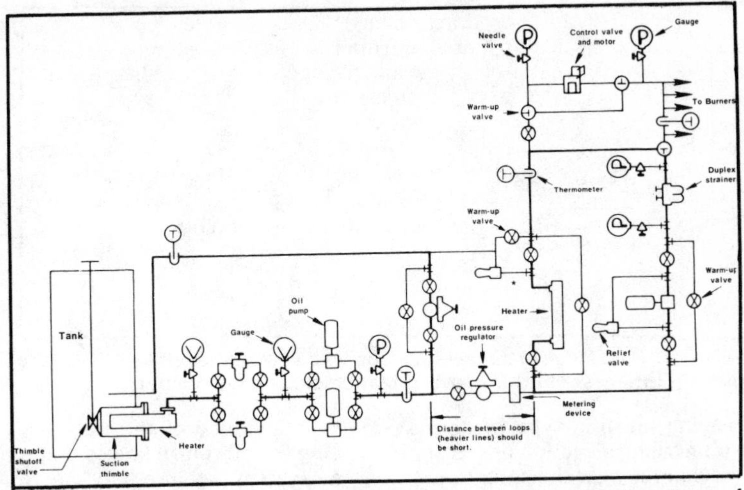

Fig. A-8. A two stage oil pumping and loop heating system (courtesy of North American Manufacturing Company).

loop entry temperature (140° F.) to atomizing temperature, plus all oil circulated (pump capacity) through a temperature rise equal to the loop temperature loss.

It will not always be possible to design a complete system without a few dead end sections. They can be accommodated if kept reasonably short, and will not cause a serious problem if properly heat traced and insulated. Consideration should be given to making provision for hot oil flushing and recirculation of dead end piping during system start up.

For large tank farms where recirculation over the long distance back to the tank is impractical, ooster pumps and day tanks are commonly used. See Fig. A-11. Where oil lines may be 6 inches or larger, steam tracers are commonly installed in the manner shown in Fig. A-12. Smaller lines are usually traced with steam, or a hot liquid, as shown in Fig. A-13, or traced electrically.

In-Tank Oil Heaters

Common design practice is to remove residual or blended fuel oil from a hot well (80°F.) located within the tank. A hot well is formed by

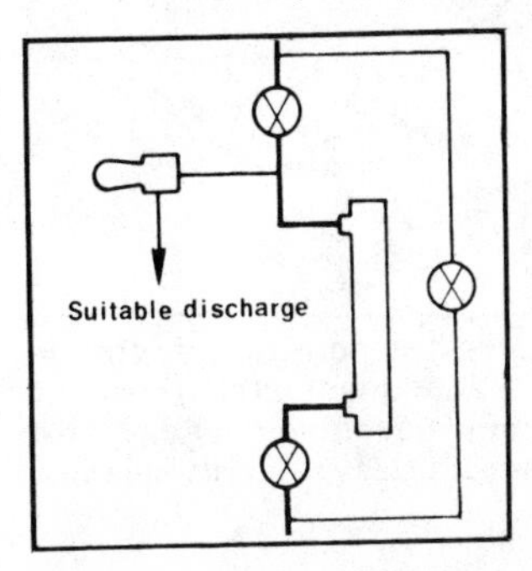

Fig. A-9. Alternate method for system temperature/pressure relief (courtesy of North American Manufacturing Company).

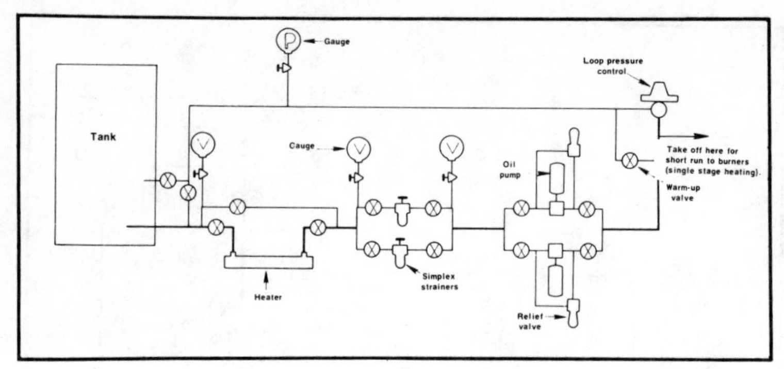

Fig. A-10. First stage or single-stage lop with suction heater external to the tank (courtesy of North American Manufacturing Company).

placing an area type heating system on the floor of the tank, by submerging a suction heater, or by placing the oil return line so that it discharges near the suction point. Some systems require only one of the above methods, others utilize all three. Before 1960, fuel oil storage systems for Bunker C or 6 oil, generally were not insulated because the fuel had a pour point between 60°F. and 80°F. It could be pumped at 100°F. and it was necessary to heat the oil above 200°F. for burning. bduring cold weather the pre-1960 Bunker C or 6 oil would solidify on the sides of the tank, and in effect, form an effective insulation barrier for a warm pool of oil that was formed inside by the heating system mentioned above.

The reader can appreciate that heavy fuel oil that must be stored today and in the future, could possibly have pour points that range all the way from 0°F. to 100°F.; some pour at 95°F. and are ready to burn at 120°F. If the tank temperatures are not maintained at some reasonable level, there is a very distinct possibility that as the oil cools, heavy liquid components could be formed because of chemical instability, or heavy components held in suspension (contiguous mixture) could precipitate out. These would then settle to the bottom of the storage tank.

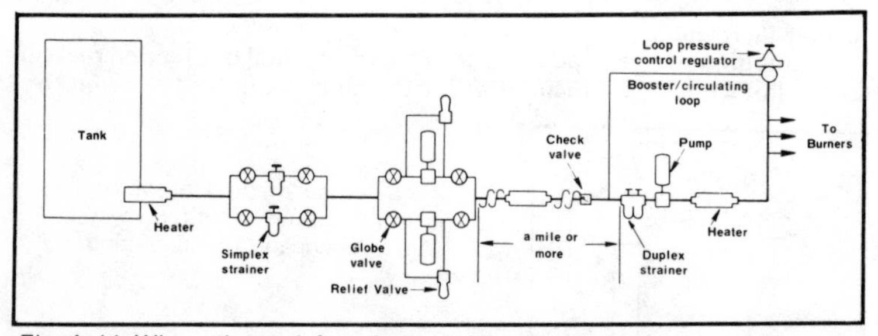

Fig. A-11. Where the tank farm is a considerable distance from point of use, a recirculating system is sometimes impractical. Higher first stage temperatures, booster heater and booster pumps may be required, or day tanks can be installed close to the point of final use (courtesy of North American Manufacturing Company).

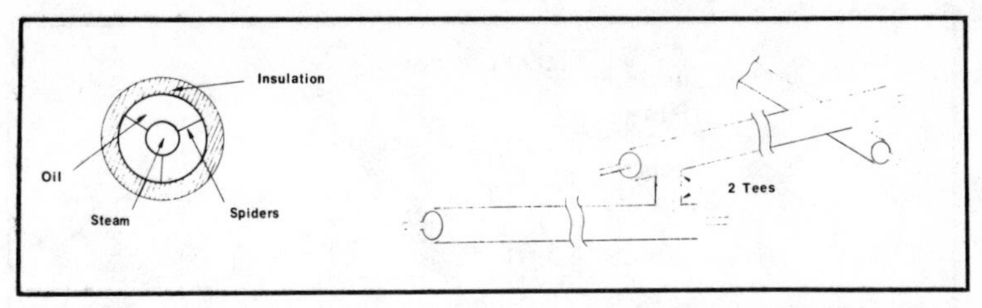

Fig. A-12. Internal steam tracing (courtesy of North American Manufacturing Company).

The entire tank heating system should b designed in a manner that will not permit the fuel to be heated, purposely or accidentally, to its flash point at atmospheric pressure. The designer of oil systems should keep in mind the fact that steam regulators, electrical contactors and thermostats do fail, and that they do not always fail safe (i.e. with heat input turned off). If in-tank heaters fail to shut off, tnc temperature could run away. Under some conditions, returning hot oil could raise tank temperature to a point where it would almost equal the return oil temperature. Positive steps shoutd be taken to minimize th hazards created by overheating fuel oils. With today's oils there is double justification for insulating storage tanks in cold climates to prevent oil separation that could be caused by low skin temperature and to prevent heat loss. The entire tank content may be liquid when the oil is held at temperature above ambient. There will be no insulating effect from solidified oil clinging to the walls. Any insulation added to outdoor tanks and piping must be weatherproof. Because of the current inconsistency of fuel oil blends, it is good practice to design in-tank heating systems that will provide either uniform heating or heating that will augment normal convection heat flow.

Suction heaters can be installed inside the tank and positioned to remove oil from a point near the center of the oil mass, or just outside the tank with the suction connection gathering oil from the tank periphery. A *submerged heater* is to be used on any grade oil where the viscosity could exceed a pumpable range if the tank were to be unheated during cold weather. These heaters are nrequently installed within a pipe casing fitted with a large gate valve at the inside end of that casing. The valve can be operated from outside the tank, thereby

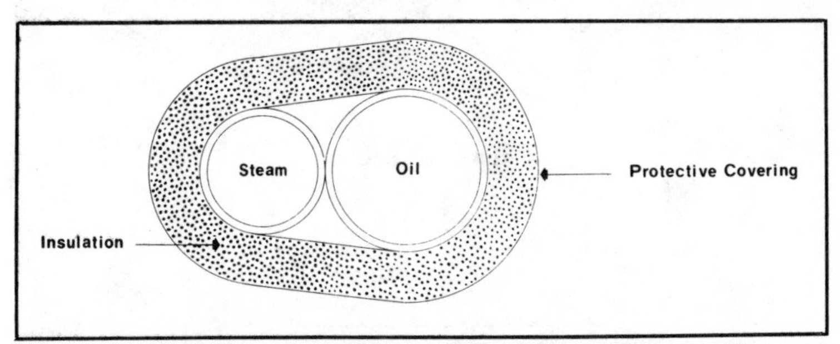

Fig. A-13. External steam tracing (courtesy of North American Manufacturing Company).

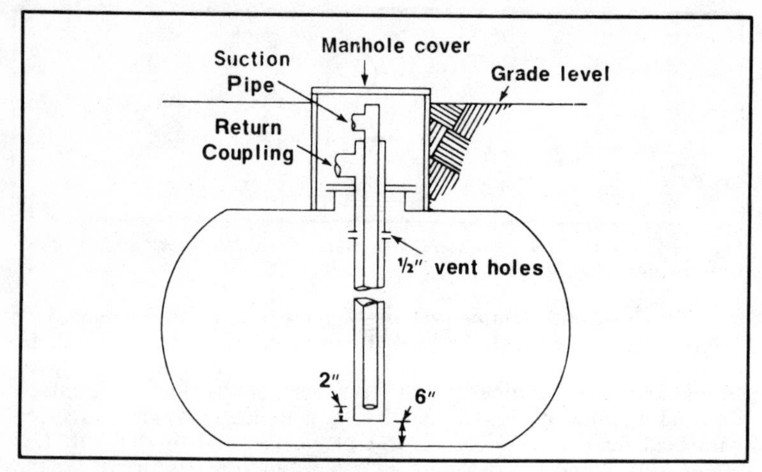

Fig. A-14. The use of return oil for preheating heavy oil underground is acceptable if the tank bottom is above the water table, and if the oil is a good quality that does not require agitation (courtesy of North American Manufacturing Company).

permitting heater removal for maintenance, without draining the tank. See Fig. A-8.

External heaters are to be used only when the viscosity of the oil is such that it could not cool to a point where pumpability is lost. When sizing piping connections for these particular heaters, it must be remembered that oil viscosity on the entry side of the heater, can be considerably higher than that at the discharge. See Fig. A-10.

Return line heaters are used only where the tanks are below ground and the oil from the heater is discharged into a hot well sometimes in the form of an inverted bucket suspended on the end of

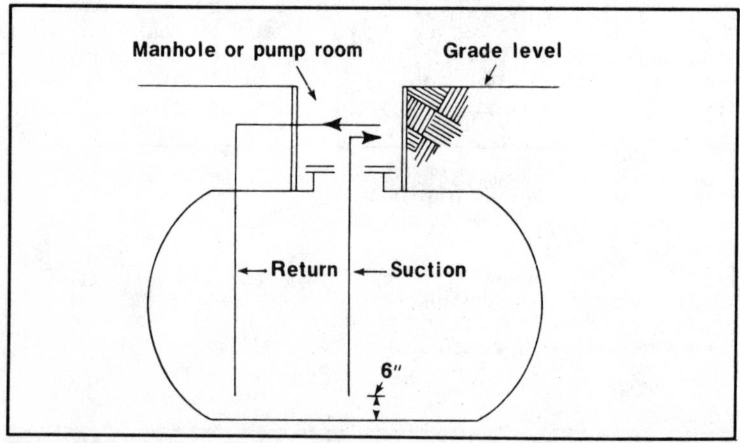

Fig. A-15. Underground tank piping arrangement for #6 oils that require agitation (courtesy of North American Manufacturing Company).

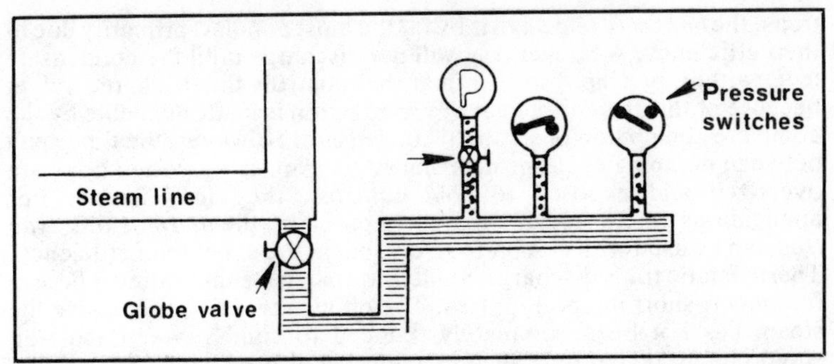

Fig. A-16. A trap is a U-section that acts as a condensate well to form a protective barrier between live stream and temperature-sensitive gauges and pressure switches. Where there is danger of freezing, the siphon should be filled with glycol. If a hand valve is installed to permit switch replacement, it should be of a type that provides for key locking in the open position (courtesy of North American Manufacturing Company).

the return and suction piping. This arrangement was developed primarilh to eliminate the need to place steam or electric heaters in a tank below ground. Storing heavy fuel oil below ground requires supervision of oil temperatures when the fuel oil is not in use. The main pump and heater set must be operated, as required, to maintain oil temperature at a point 20° above the pour point. Typical return line arrangements are shown in Figs. A-14 and A-15.

Line heaters are the most commonly used type of fuel oil heater. Their application is simple and straightforward. Typical examples of installations are shown in Figs. A-8 through A-11.

Steam heaters are the most effective method for heating oil where precise temperature control is desired. Steam temperature/pressure regulators modulate the steam rate as required to maintain oil temperature within a narrow range.

Every steam pressure gauge and pressure switch should be fitted with a siphon (pigtail), and every gauge should have a manual shutoff vatve. A siphon is an ordinary ¼-inch pipe nipple formed with a 360° for gauge and switch internals. It is good practice to install siphons on oil pressure switches also, because most switches available today are not suitable for continuous operation when exposed to fluids above 150°F.

When used outdoors, freezing may be a problem. In case it may be possible to build a single trap (filled with ethylene glycol) that would be common to a multiple switch/gauge combination. Figure A-16 illustrates this.

When steam is used as an oil heating or tracing medium, the energy available is equal to the value of the heat of vaporization at a given saturated temperature and pressure. Steam will, by the condensation process, give up this energy without a reduction in temperature or pressure.

Steam Traps

The basic purpose of the trap is to drain condensate from the steam lines, without blowing live steam to the atmosphere. In selecting steam traps, the *bucket type* design is by far the most popular, primarily due to their efficiency. A bucket trap will not discharge until the condensate level within the trap, raises a float that controls the discharge valve; because of this the entire heat of vaporization is made available by the complete conversion of steam to condensate. However, the time span between dumping cycles is determined by trap sizing. When traps are oversized and exposed to cold climates, they may freeze. For applications where freezing may be a problem, the *thermostatic type trap* can be used with complete safety, but at the expense of efficiency. Thermostatic traps discharge condensate laden steam to atmosphere in frequenct, short interval, pulses. The efficiency is lower because the steam has not been completely reduced to condensate. Each trap should be protected by a steam strainer, and fitted with a blow down valve.

Condensate from steam traps, used in conjunction with fuel oil heaters, may be oil contaminated. Be aware of the potentially serious problem of delivering oil-contaminated condensate to a boiler. The amount of condensate flowing from the traps can be considerable and its disposal should not be indiscriminate. Oil separators can be used in condensate return lines if the value of the condensate justifies their cost.

Electric Fuel Oil Heaters

These are rated at 12 watts per square inch and are available up to 60 kW per unit. Submersible suction heaters are available up to 60 kW.

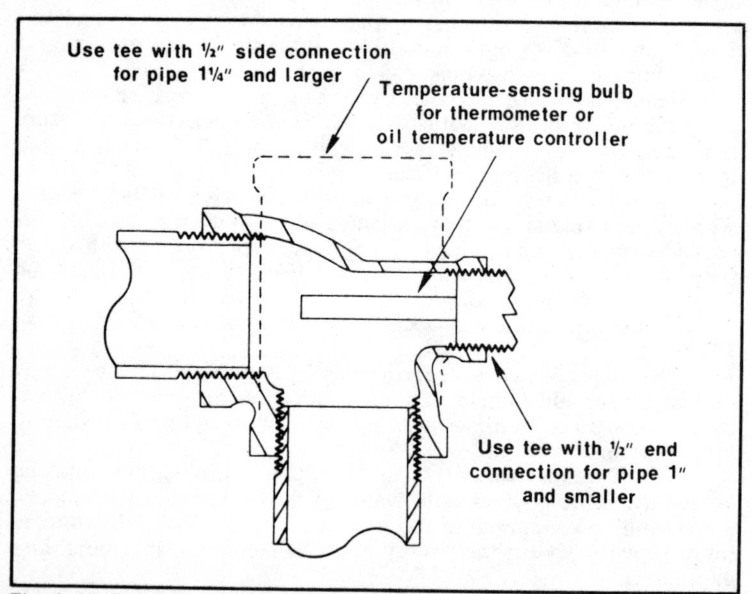

Fig. A-17. Recommended temperature bulb or thermometer installation in a heavy oil line. Oil flow may be in either direction. Steam or electric tracing must not contact the tee or the thermometer housing (courtesy of North American Manufacturing Company).

Line heaters can be installed in series or in parallel where it is necessary to have larger capacities than listed. For sizing electric heaters, the following equation includes a 30 percent safety factor:

$$\text{watts} = \text{required gph} + \text{required temp rise} + 1.3.$$

The majority of electric oil heaters used in industrial systems are required to operate on three phase current. Electric heaters do not modulate heat energy input as do steam heaters; the heat is either on or it is off. For that reason, electric heating systems must rely heavily upon temperature differential ranges that are built into a snap action thermostat that actuates an electrical contactor, which in turn applies or disconnects the three phase current. This action can in some cases, represent a heavy, cycling electrical load.

North American catalog heaters are fitted with simple snap action thermostats that have proven to be fairly reliable. They are factory installed within the heater, and are difficult to service or replace. Snap action thermostats of a higher quality with a closer differential are available. These units have capilliary bulbs which are fitted into piping wells as shown in Fig. A-17.

Stepping type thermostatic controls are available, but at present, they are very expensive and do not provide true modulating control. Application of a variac is impractical at this time.

The watt density is a factor that drastically affects the cost of the heaters. Our experience shows that 12 watts per square inch is a reasonably conservative rating that will not cause serious problems. There will be occasions where a watt density of 18 is specified, requested or offered. The hazard with the higher rating is that the surface temperature of the element can run very hot, and if the oil flow through the heater is low, improperly balanced, or if air is trapped within the heater shell, the element could burn out, or coke up. Heaters for heating the entire contents of tanks are designed in a slightly different form and are usually rated at only 5 watts per square inch. Table A-4 provides contactor size selection for various heater sizes and voltage conditions.

The oil heating and tracing system requirements should be carefully reviewed before selecting the heat source. If steam is unavailable, the tendency is to use electric heat. During early design stages, the total connected electric power requirements, including tracing, should be estimated and reviewed with the user. Installation of a small packaged boiler may prove to be the best choice and should not be overlooked, because steam is the most economical selection from the standpoint of long term operating cost, and it provides better control.

Heat Tracing

Heat tracing is used in several ways depending upon the duty intended. Listed below are two of the types that are generally available.

Steam tracing is applied in two methods Figs. A-12 and A-13. Internal tracing lines (usually black iron pipe) are normally used in oil transportation systems requiring pipe larger than 4-inch. External tracers are by far the most commonly used with piping systems for transportation, circulation, and burner connections. Black iron pipe or copper tubing is used for external tracing.

Table A-4. Electric Fuel Oil Heater Current Ratings (courtesy of North American Manufacturing Company).

Heater kW	208/3/50-60 Amps	230/3/50-60 Amps	460/3/50-60 Amps
5	13.9	12.5	6.25
6	16.6	15.0	7.5
8	22.2	20.0	10.0
10	27.7	25.0	12.5
12	33.2	30.0	15.0
15	41.5	37.5	18.8
16	44.4	40.0	20.0
18	49.9	45.0	22.5
20	56	50	25
24	67	60	30
30	83	75	38
36	100	90	45
42	117	105	53
48	134	120	60
60	166	150	75

Tracing steam supply pressures range from 5 psig to 150 psig and there is considerable controversy regarding the best pressure level, with that controversy hinging upon the effect high steam temperature may have on fuel oil. When fuel oil is not moving, or it is recirculated in a loop having zero or very low consumption, the oil may be overheated by the steam and "coke up" within the pipe. To prevent this from happening, steam pressure is controlled to keep steam temperature from exceeding the desired oil temperature.

Steam is not desuperheated as it flows through a pressure reducing regulator. For example, if 150 psig steam (366°F. saturation temperature) is reduced to 15 psig through a regulator, the exit steam will be superheated and very near the 366°F. level—not the desired 250°F, corresponding to saturated steam at 15 psig. To alleviate this situation, it would be necessary to desuperheat the steam or to use other temperature controlling schemes.

North American's engineering group believes that the coking problem is not as serious as it once was, and that as long as the oil is kept moving and the steam temperature does not exceed 365°F., coking will not be a problem. The tracing on outdoor piping systems should operate at all times when the temperature is at or below freezing.

On indoor recirculated systems, the tracing should be used only when starting a cold system or on piping sections exposed to frequently opened doors or windows. Refer to the section on steam trap selection. The most effective way to maintain heat in an operating oil system is by recirculation. Standard procedure for sizing steam tracing lines is to determine the maximum heat loss based upon ambient temperature, oil temperature, pipe size, insulation type and thickness. To accommodate irregular maintenance of insulation and steam traps, it is suggested that the steam system heating capacity should be approximately three times that of the heat loss figure. In sizing the steam tracers, the pressure drop allowance per hundred feet of pipe should not exceed 5 psi.

Electric tracing is available in the form of tape that clings to the piping walls or as cable that is clipped to the pipe. Suppliers provide charts for determining the heat requirement, oil thickness of insulation.

Some of the tracing tapes available are self-limiting and require no thermostatic controls. Their electric resistance increases with pipe temperature, thereby allowing less current flow, and reducing wattage output. Theoretically, the heat output of the tape reaches a point where it is in equilibrium with the heat losses.

The cable form of tracing is controlled by surface contacting thermostats that are placed at intervals along the piping system. These thermostats actually turn the power on and off, operating very much like an electric fuel oil heater.

This form of tracing is simple to design and easy to install, but very expensive. Because the control is automatic, electric tracing can remain on stream at all times.

Manual Valves

The following is a brief review of the various types of hand valves used in fuel oil systems. Valves that are strictly for occasional shutoff service and never for control or throttling are normally *gate valves*. The primary advantage of this type of valve is low pressure drop and low cost. Except for repacking, they cannot be field-repaired. A *rising stem gate valve* has threads in the valve bonnet and on the valve stem. The valve stem rises out of the valve bonnet as the valve is opened. On valves up to 4", the valve wheel is fastened directly to the stem and rises with it. On larger rising stem valves, the valve wheel is clipped to the valve bonnet and the stem appears to screw out of the wheel. This type of valve is not always a good choice for outdoor service or where the threaded stem could be covered with dirt or ice, which could prevent or delay closing.

Non-rising stem gate valves have internal threads in the gate and threads on the stem. The valve wheel and stem are clipped to the valve bonnet. When this valve is opened, the disc rises on the stem threads, and the valve wheel does not rise. The stem threads are always being flushed clean by the fluid in the pipe. A real advantage here is space in that the valve has no external dimensional change when opened or closed. The disadvantage is that visual observation will not indicate if the valve is open or closed.

Gate valves are intended for use primarily in piping systems carrying water, oil and other liquids. They are not normally recommended for use in compressed air or steam lines, although there may be exceptions.

Globe valves are intended for use with gas, air, steam or liquids where frequent and tight shutoff service is required. They have reasonably good throttling characteristics and are often used for manual control. Globe valves are of the rising stem and valve wheel design because the total movement of valve internals is considerably less than that of full-ported, low-drop gate valves. Globe valves have higher pressure drop than gate valves of the same size. Although more costly, field-repair is possible. All globe valve discs can be replaced or resurfaced and lapped to correct minor leaking. Some of the better quality valves provide for complete seat and disc replacement. Valve discs are available for a wide variety of services—resilient materials like rubber, neoprene, and teflon; soft metals like brass; and very hard alloys for high pressure steam service. Valve seats vary too, depending upon the service required. To take full advantage of the wide selection of globe valve types and materials, it is important to make a careful choice.

When a hand valve is required to provide fine tuning or throttling, a *needle valve* should be selected. These are also well suited for tight shutoff, but are usually too expensive to use for that service alone. Needle valves are frequently used for pressure gauge shutoff service because when partially throttled they are capable of dampening gauge fluctuations.

Check Valves

A swing-check or a ball-check valve is commonly used on the suction side of an oil pump to prevent loss of prime when the pump is above the oil level. If the pump packing gland is not leaking, if there are no open valves in the system, and if there are no air leaks, a check valve is not needed. They are of some use, however, when starting a system where it is necessary to prime the pump. They are not intended for low pressure tight shutoff; so they may not prevent the pump from losing its prime during an extended down time. *Foot valves* were originally designed for use at the bottom of a water well to prevent drying of the pump leathers and resultant loss of prime.

Apparently some years ago, someone thought it would be a good idea to install foot valves in oil piping systems, but it is North American's opinion that their use causes more trouble than they are worth. Check valves and foot valves are widely used to minimize priming loss on intermittant operations. In large industrial applications, where the fuel oil pumps are running continuously, the use of a check valve in the suction line is limited TO original priming. Where industrial fuel pumps have flooded suctions, suction line check valves are not required.

Relief Valves

The styles, service and application of relief valves as applied to oil systems is not too involved, but it is important to understand their application prior to selection. For oil line pressure control, a diaphragm type or pilot operated relief valve will provide the best performance. For a system or pump protection, the poppet type, is usually adequate, and will often cost less money than the same size diaphragm-operated unit. In oil systems there are several basic applications, all of which are described below.

A relief valve can be used as a light oil loop pressure controller, frequently referred to as a back pressure regulator. It is usually installed in the piping system immediately after the last burner takeoff. This application can be handled by a simple and inexpensive poppet type relief valve that has a wide swing in controlled pressure level from the no flow condition to maximum flow. Long lines, wide turndown and high pressure drop require the addition of a pressure regulator in every fuel takeoff line when the pressure in the supply line is controlled by a poppet type relief valve.

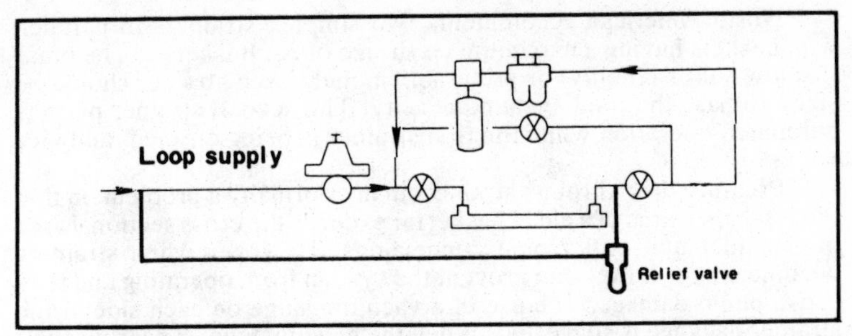

Fig. A-18. Preferred relief valve piping arrangement for protection of an oil heater from excessive pressure caused by failure to open valves or by overheating (courtesy of North American Manufacturing Company).

It is good practice to install a relief valve to protect every positive displacement pump from overpressure that could be accidentally imposed. A poppet type relief valve can serve this function, but it should not be expected to provide outlet pressure control. If this valve is not built into the pump head by the manufacturer, a pump protection relief valve must be placed in the piping system in a manner that will relieve the pump outlet and return it immediately to the inlet.

These valves are usually sized to relieve full pump capacity over a 40 pound range. For example, if the relief valve is set at 100 psi, the pump outlet pressure will reach 140 psi before the relief valve passes its full volume. Because of the possibility of overheating, it is not good practice to permit a pump to operate for extended periods of time with this valve open wide.

For oil heater protection, there are at least two philosophies regarding relief valve application and either of the two mentioned are acceptable as long as their purpose is understood. The preferred method for this application is to apply a full pump capacity diaphragm or poppet type relief valve as shown in Fig. A-18. This arrangement protects against the possibility of a heater outlet valve being left in the closed position, heater temperature control failure and general system overpressure. One drawback for this method is that some relief valves tend to leak and once operated, it is difficult to determine if the valve returned to the full closed position. To protect against this possibility, select a high quality valve. An alternate method of applying relief valves is to use a poppet type valve installed as shown in Fig. A-19. This valve provides some protection, but discharge does not return to tank. The potential oil discharge is both a nuisance and a hazard.

Oil Strainers

The primary purpose of these strainers is to protect the pump from damage by granular foreign matter in the oil. Simplex strainers are preferred (over duplex strainers) on the suction side of the pump to avoid the pressure loss of the duplex strainer's internal three-way valve, and to avoid the possibility of an air leak through the packing gland of that valve. Suction line strainers may be under a vacuum of 12 to 14 inches of mercury and air leaking in at this point could create a problem in the pumping system. The presence of air in the suction line will cause the same type of pump noise experienced with cavitation.

North American recomments two simplex strainers in parallel, with baskets having a maximum mesh size of 20. Baskets can be brass (for low sulfur oil only) or preferably monel. Each strainer should be sized to pass the total system capacity. The second strainer permits continued operation while the first strainer is being cleaned, and vice versa.

Pressure drop through strainers in not normally a problem, in that the total open area of a clean basket far exceeds the cross sectional area of the inlet and outlet pipe connections. However when strainers become clogged, they can prevent the system from operating and may cause pump damage. The use of a vacuum gauge on each side of the strainer package is suggested. When the vacuum level is nearly equal, the basket is clean. As strainer baskets become clogged, the vacuum on the pump side of the strainer will rise noticeably above the vacuum on the tank side. If there is a problem is starting oil flow, the extra gauges may help locate the problem.

SUMMARY

To design an oil handling system, all the following items of information must be accumulated and analyzed:

☐ Oil specifications including viscosity, sulfur, flash point and pour point. If possible, determine if the oil is a blend, or a straight residual.
☐ Maximum burning rate required.
☐ What viscosity oil will be circulated in the first stage loop, and what temperature is required to achieve that viscosity?
☐ Determine oil flow requirement for the first stage loop. A good figure to use is 1.5 times the burning rate. Pump selection will be based on standard capacities nearest to that required. The oil heaters must be sized to heat all of the oil the pump will move.
☐ Is the system to operate year-round or seasonally? If seasonally, how long are the down periods likely to be? Can the system be filled with light oil prior to shutdown, or must provisions be made to reheat the oil within the pipes?
☐ If the tank is to be above grade, what is the maximum pressure head? If below grade, what is the tank diameter, and where is the top of the tank relative to the normal grade line?
☐ How many tanks will the system draw from?

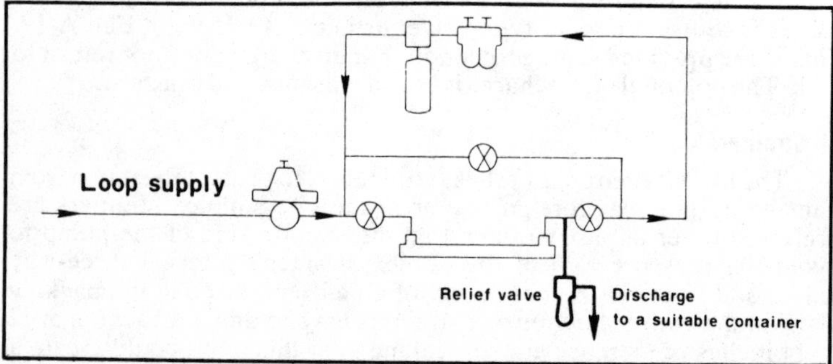

Fig. A-19. Alternate arrangement for over-pressure protection of an oil heater (courtesy of North American Manufacturing Company).

☐ Will the oil temperature in the tank fall to less than 20° above the pour point? If so, and the oil solidifies, how will it be liquefied?
☐ Will the suction heater be submerged in the oil, or will it be a line heater placed outside of the tank?
☐ If submerged, will the suction line be naturally flooded, or will the oil need to be mechanically drawn from the tank? If flooded, has provision been made to remove the heater in the event of a failure?
☐ At design viscosity and rated flow, what is the pressure drop through the heater? Through what temperature range must the heater raise the oil? Normally these heaters are sized to raise the oil from 50°F. to about 130 or 140°F. or whatever temperature will result in a viscosity acceptable for fuel pump motor size. Raising the outlet temperature above 140°F. may cause pump cavitation.
☐ What pipe size should be used at the entry of the heater? Low temperature oil entering the heater may have a viscosity that is very high. Remember that 15" Hg vacuum (7.4 psg) is the maximum usable suction on the pump inlet.
☐ What pipe size is needed for the hot oil outlet?
☐ How is the heat to be provided—electric or steam?

Courtesy of North American Manufacturing Company

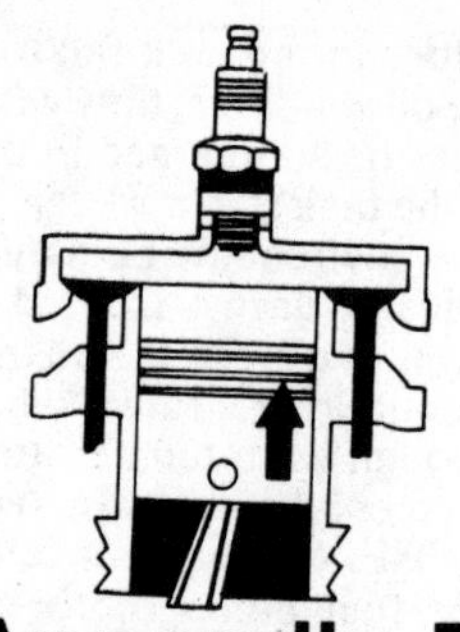

Appendix B
Availability of Smaller Size Coal Boiler Plants

The energy crisis that occurred during the 1974 oil embargo highlighted the fact that supplies of oil and natural gas were limited and that government agencies would even further control their use by allocation, tax incentives and/or price controls. Companies and institutions that have purchased industrial boilers since the embargo have had to make some very difficult evaluations and the results have been that more new plants are going to coal firing.

Research and development on fluidized bed combustion, coal gasification and liquefaction have intensified, but widespread commercial use is still many years away. Therefore, coal fired boilers purchased in the near future will be equipped with essentially the same type of firing equipment that has been used in the past.

Although the energy crisis has drastically changed the industrial boiler market, it is interesting to note that many of the boilers purchased in the pre-crisis period were selected to provide flexibility in the type of fuel that could be burned. A large percentage of the industrial boilers purchased form the E. Keeler company during the 1970, 1971 and 1972 period were designed for either initial or future coal firing.

Not only are more smaller coal fired boilers being purchased, but many customers are insisting on the flexibility to burn other fuels. Most of the coal fired units purchased during the last few years have had auxiliary gas and/or oil burners or at least provisions were made for their future installation. In addition, practically all industrial customers are considering plant wastes, whether solid, liquid, or gaseous, as possible supplemental fuels.

Before discussing the availability of industrial coal fired boilers, it should be explained that this paper will cover only stoker fired, water tube boilers with a maximum steam capacity of about 250,000 lbs./hr. Although there have been many pulverized coal and cyclone fired boilers built for capacities of less than 250,000 lbs/hr., it is generally agreed that the economic breakpoint is in the 200,000 to 250,000 lbs./hr. range.

Practically all modern stoker fired boilers are of the two drum type. The furnace and convection bank arrangements vary among boiler manufacturers but two basic designs are the "long drum" and "cross drum" as shown on Figs. B-1 through B-3.

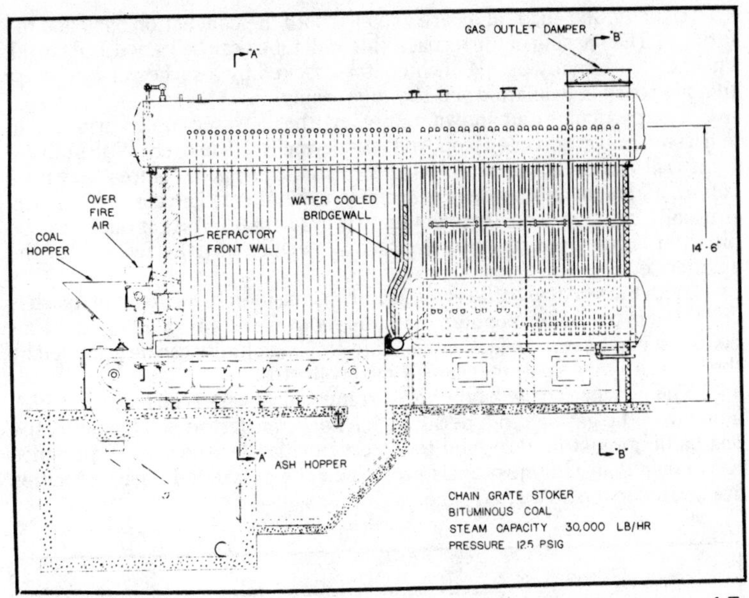

Fig. B-1. A "long drum" boiler with a chain grate stoker (courtesy of E. Keeler Company).

The terms "long drum" and "cross drum" have not been used extensively relative to bent tube boilers. They are, however, descriptive of modern designs and their use will probably become more common. On a "long drum" boiler, the flue gas flows lengthwise to the drums, whereas on a "cross drum" it flows across or perpendicular to the drums.

"LONG DRUM" BOILERS

Figure B-1 shows a "long drum" boiler with a chain grate stoker. The steam drum extends the full length of the setting while the lower drum extends only the length of the convection bank. This type of unit is particularly suitable where head room is limited, and its low profile permits shop assembly in the smaller sizes. Considerable standardization has been accomplished on these units and engineering costs can be minimized.

The side wall headers extend the full length of the setting and are supplied with water from the lower drum by short tubes in the last gas pass. There are no external supply tubes. To provide optimum circulation characteristics, the furnace side wall tubes terminate in the steam drum without intermediate headers. The furnace side wall tubes which are "heavy steamers" enter the drum on the horizontal center line. This, along with a liberal steam drum size, assures a stable water level and dry steam can be obtained without complicated steam drum internals.

This type of unit is top supported from a structural steel frame with columns at the four corners and saddles under each end of the steam drum. The lower drum is suspended from the steam drum by the bank tubes. Horizontal buckstays and intermediate columns are supplied as necessary to provide a substantial casing frame between the front and rear columns.

Usually, 2½-inch tubes are used for both the convection bank and the furnace. The spacing of the furnace side wall tubes can be varied as desired, but they are usually on 4½-inch centers backed by 2-inches of refractory tile, appropriate insulation and 10 gauge casing.

The particular unit shown on Fig. B-1 has a water cooled bridge wall and a similar water cooled front wall can be applied if desired. With both the front wall and the bridge wall water cooled, the furnace requires very little refractory and maintenance is reduced to a minimum. The convection bank is usually baffled vertically to obtain four gas passes. The gas outlet can be directed either upward or to the rear as required by the location of the auxiliary equipment.

With the vertical baffle arrangement a considerable amount of flyash is collected in the boiler hoppers, and this reduces the dust loading to the air pollution equipment. In fact, some state emission limits can be achieved by the use of a single stage mechanical dust collector.

The convection bank tubes are arranged in a staggered pattern, which along with the gas baffling, permits high heat transfer rates. Of course, the gas baffling must be designed to keep velocities reasonable. Experience has shown that with most coals erosion can be prevented if gas velocities are limited to the 30-40 fps range.

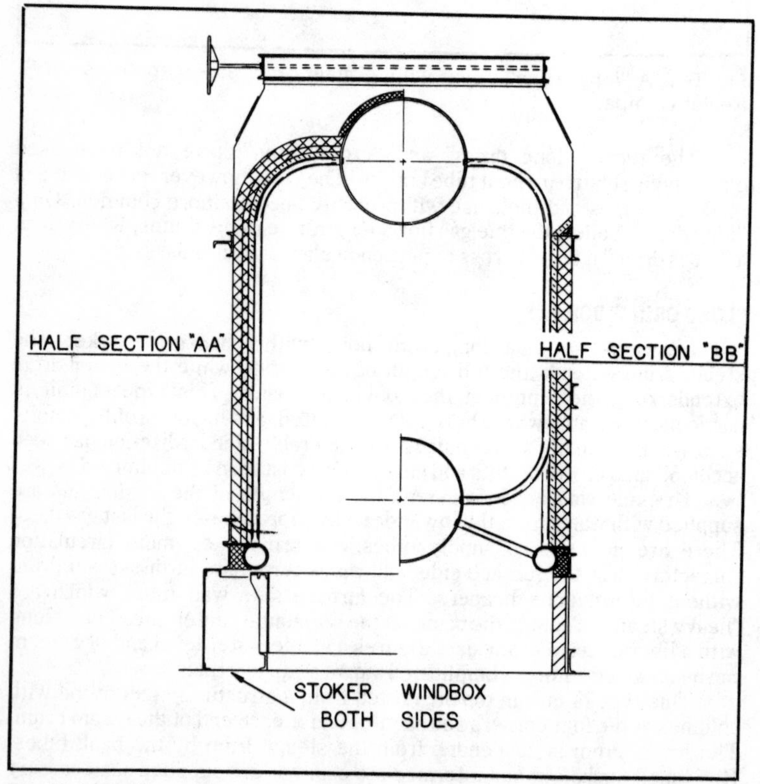

Fig. B-2. Stoker and windbox detail (courtesy of E. Keeler Company).

Sootblower operation and maintenance is minimized on the "long drum" boiler because usually only one blower is required to clean the entire convection bank as shown on Fig. B-1.

The "long drum" boiler can be shop assembled for steam capacities up to about 300,000 lbs/hr. The only field work required is the installation of the stoker and furnace refractory. Shop assembly not only reduces the use of costly field labor, but also permits better quality control.

For steam capacities greater than 30,000 lbs./hr. field erected units can be supplied. Their application is usually limited to those installations where head room or equipment arrangement problems exist.

The "long drum" boiler can be fitted with almost any type of stoker, but the more usual application is either the underfeed or chain grate.

Although coal may not be the initial fuel, many customers prefer the "long drum" boiler because it has the flexibility for future stoker firing. In these cases, the oil and/or gas burners are installed in the front wall and the stoker grate can be installed initially and bricked over or provisions can be made for installing it in the future.

The "long drum" type boiler is especially suitable for institutional or small industrial plants where reliability, simplified operation and ease of maintenance are desired. The casing is usually bolted and the convection bank tubes are on alternate front to rear spacing to facilitate tube replacement. All normal maintenance, such as tube replacement, can be done without welding or other costly special procedures.

"CROSS DRUM" BOILERS

The boiler shown in Fig. B-3 is a typical "cross drum" unit of the type usually recommended for the larger steam capacities. This type of unit is completely field erected and since there are no shipping limitations, it can be custom designed for a particular fuel, space consideration and/or arrangement of auxiliary equipment.

For capacities up to about 175,000 lbs/hr. a bottom-supported unit with a multi-pass convection bank is usually recommended. For higher capacities, it is often desirable to use a top supported unit with a single pass convection bank.

The unit shown is bottom-supported at two elevations; the furnace and stoker loads are taken on a structural frame at the operating floor, and the convection bank loads are taken on a structural frame under the lower drum.

Basically, the convection bank of this unit consists of three cross flow and one parallel flow gas passes with a top gas outlet. All of the tubes in the last parallel flow pass are downcomers. Gas baffles are formed with refractory tile and castable. The bank tubes are usually 2½ inches on alternate side to side spacing of 4⅛ inch and 5¼ inch. This permits replacement of individual tubes without disturbing others for access.

The convection bank tubes are in line rather than staggered and this permits higher gas velocities without tube erosion. On multipass banks, gas velocities are usually limited to the 40-50 fps range while on single pass banks, they can be as high as 70 fps.

The furnace tubes can be either 2½ inch or 3¼ inch. Normally 2½ inch tubes are used on 5 inch centers, but some customers prefer centers as close as 3½ inch. The furnace tubes are backed by 2 inches of refractory tile, appropriate insulation and 10 gauge casing.

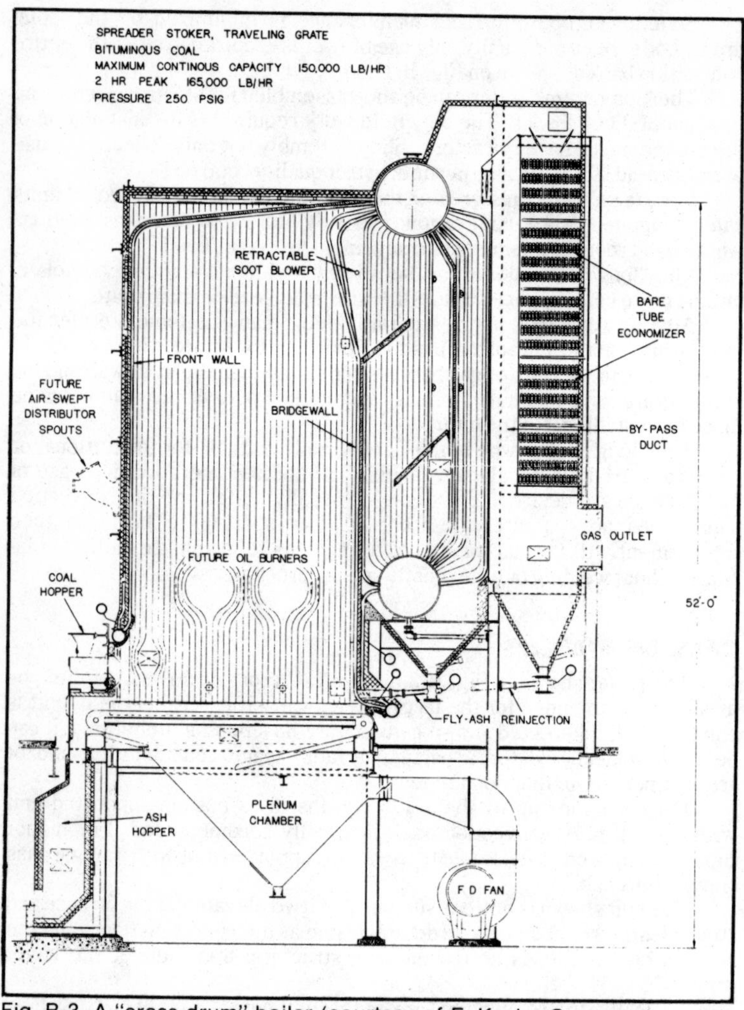

Fig. B-3. A "cross drum" boiler (courtesy of E. Keeler Company).

Although not usually recommended in this size range, tangent tube construction with skin casing can be used on all four furnace walls. The heat absorption characteristics are nearly the same for both the closely spaced and tangent tube construction, and the increased cost associated with tangent tube construction usually cannot be justified.

The method of supplying water to the lower water wall headers is not specifically shown on Fig. B-3, but it is usually done by individual supply tubes from either the top or bottom drum. All supply tubes are inside the boiler setting and do not require external insulation or lagging.

The casing is supported from a structural frame completely independent of the pressure parts. Welded casing is preferred to keep air infiltration to a minimum, but bolted casing is sometimes supplied.

The unit shown in Fig. B-3 is typical of those fitted with provisions for future auxiliary gas and/or oil burners. The burners are usually placed in one side wall and they must be high enough to permit operation on auxiliary fuel without overheating the grates, but low enough to assure that all furnace circuits receive heat over enough height to maintain positive circulation.

To prevent overheating, the grate must be covered with several inches of ash or bricked over when the auxiliary burners are used. Usually the ash cover is preferred unless there are extended periods where the stoker is not used. The burner diffuser must be pulled back when the stoker is in operation, and experience has shown that with a small amount of air leakage through the burner, no further protection from radiant heat is required. If it is anticipated that there will be extended periods when the auxiliary burners will not be used, refractory burner throat plugs may be used to eliminate the air infiltration. The refractory plugs can be installed without entering the furnace.

The particular unit shown on Fig. B-3 was fitted with a bare tube economizer arranged for down flow gas and up flow water. An integral economizer gas bypass duct with associated dampers was supplied to control the gas temperature to the air pollution control equipment.

The unit has a spreader stoker with a traveling grate. The spreader stoker is probably the most popular in the 30,000 to 250,000 lbs/hr. range since it will handle a wide variety of coals and provides excellent response to load changes. In the 30,000 to 70,000 lbs/hr. range, spreader stokers are often fitted with reciprocating, oscillating, or dump grates rather than traveling grates. The same type of boiler is also often used with bin feed (cross feed) chain grate or vibra-grate stokers.

The unit shown in Fig. B-3 has provisions in the front wall for the future installation of air-swept distributor spouts to handle solid industrial waste. A separate fuel preparation and handling system is required for the waste fuel, but due to the increasingly difficult problem of disposing of industrial waste, many customers are either installing this equipment initially or making provisions fo future installation by bending the tubes and arranging the casing accordingly.

TYPICAL CHAIN GRATE STOKER

Figure B-4 shows a typical chain grate stoker application. This unit was designed for bituminous coal with a maximum continuous capacity of 40,000 lbs/hr. at 100 PSIG and a peak capacity of 44,000 lbs/hr. The tubes and the casing on one side wall are arranged for an oil burner capable of the peak capacity.

The unit shown has a small refractory ignition arch below the front wall header. In some cases, the front wall is completely water cooled, but in these instances the front wall header and lower few feet of tubes are studded and covered with 3 or 4 inches of castable. The castable in this area then acts as an ignition arch and provides radiant heat to the incoming coal. Although some type of small ignition arch is usually used with chain grate stokers burning bituminous coal, very few stoker or boiler manufacturers recommend long front or rear arches that were so common a few years ago before overfire air was used extensively.

On a unit of this type with a chain grate stoker, the furnace heat release should not exceed 30,000 Btu/cu. ft. Grate heat release rates are usually based on past experience with a particular coal but generally the maximum

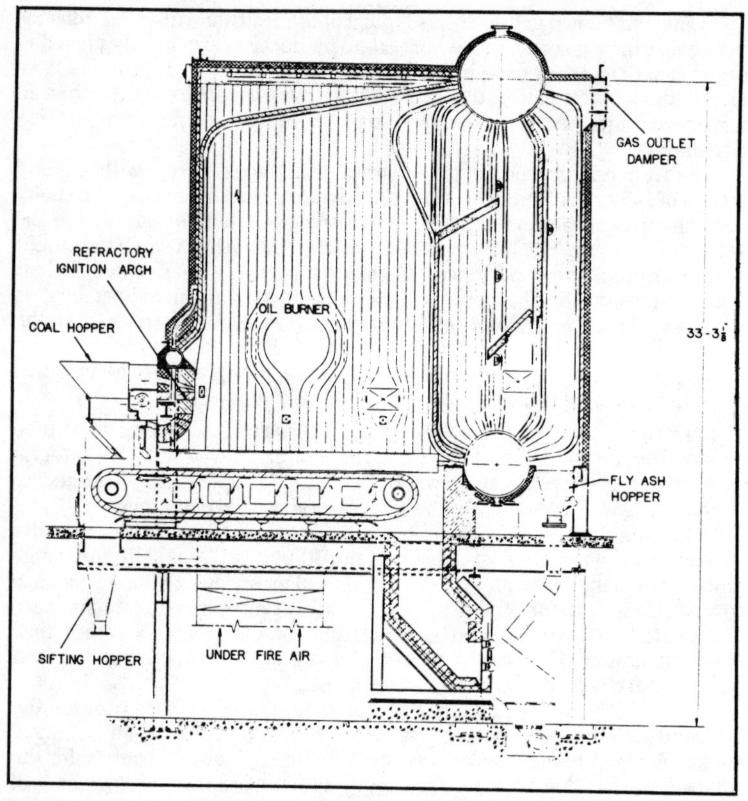

Fig. B-4. Typical chain grate stoker application (courtesy of E. Keeler Company).

with a high ash and high moisture bituminous coal is 375,000 Btu/sq. ft. With the better grades of coal, the grate heat release can often be increased to as much as 450,000 Btu/ sq. ft.

This particular unit has 2½ inch tubes in both the furnace and convection bank. The front wall and bank tubes are an alternate 4⅛ inch and 5¼ inch side to side spacing. The furnace side wall tubes are on 5 inch centers.

TYPICAL SPREADER STOKER

The spreader stoker installation shown in Fig. B-5 was designed for a maximum continous capacity of 50,000 lbs./hr. at 610 PSIG and 700° F. on bituminous coal. It is also suitable for a 4 hour peak of 57,500 lbs./hr.

A convection type superheater was used with a retractable soot blower in the cavity between the screen tubes and the superheater. Generally, retractable soot blowers are not necessary on stoker fired units unless as in this case, the first blower at the furnace exit is exposed to furnace radiation.

With spreader stoker firing, 25-50 percent of the coal is burned in suspension and this increases the amount of ash carryover to the convection

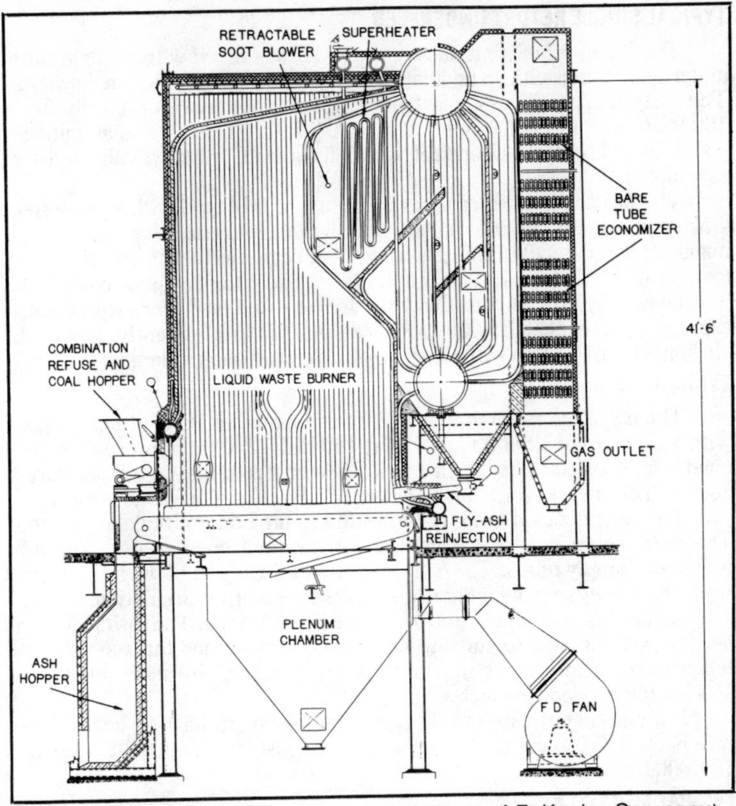

Fig. B-5. Spreader stoker installation (courtesy of E. Keeler Company).

passes. To keep carbon loss reasonable, flyash re-injection is usually utilized. On this particular installation, only the flyash from the boiler hopper is re-injected. In some cases where permitted by the type of air pollution equipment, the flyash collected in the economizer hoppers may also be re-injected. It is now very unusual to re-inject from any pick-up point beyond the economizer such as from mechanical dust collector hoppers.

All boiler bank tubes are 2½ inches on alternate side to side spacing of 4⅛ inches and 5¼ inches. The furnace and superheater tubes are 2½ inch and 2 inch, repectively. The bridge wall, front wall and superheater tubes are on side to side spacing of 4⅞ inches. The furnace side wall tubes are on 3½ inch centers with 1 inch clear between tubes.

This unit was designed to burn two types of industrial wastes. A burner in the side wall handles a liquid waste and the coal hoppers and feeders are of a combination type that will handle both coal and solid industrial wastes such as wood, paper, etc.

With spreader stokers, the furnace heat release rate should not exceed 30,000 Btu/cu. ft. The grate heat release rate can vary from about 450,000 Btu/sq. ft. for the dump grate to about 750,000 Btu/sq. ft. for the traveling grate.

TYPICAL SINGLE RETORT UNDERFEED

The "long drum" unit shown on Fig. B-6 is fitted with a single retort underfeed stoker with an undulating grate arranged for side ash discharge. The maximum continous capacity firing bituminous coal is 26,000 lbs/hr. at 100 PSIG with a two hour peak of 30,000 lbs./hr. This particular unit has a water cooled bridge wall and refractory front wall, and it was shipped shop assembled.

The furnace heat release rate for units of this type with an underfeed stoker is usually in the 35,000 to 45,000 Btu/cu. ft. range. Stoker manufacturers usually limit the grate heat realease to 425,000 Btu/sq.ft. for average coals and 475,000 Btu/sq. ft. for exceptionally good coals.

Generally, single retort stokers are not used on boilers with a steam capacity greater than 30,000 to 35,000 lbs./hr. Consequently, practically all "long drum" boilers with this type of stoker are shop assembled.

TYPICAL VIBRA-GRATE

The unit illustrated in Fig. B-7 was designed to burn bituminous coal with a capacity of 120,000 lbs./hr. at 450 PSIG and for a future superheat condition of 100,000 lbs/hr. at 450 PSIG with a total steam temperature of 700° F. This unit is typical of the larger vibragrate stoker installations.

The water cooled grate is connected to the boiler circulation system. The rear stoker header is supplied with water from the lower drum by individual supply tubes. The front riser tubes form part of the furnace front wall where they receive radiant heat to assure positive circulation.

All furnace and convection bank tubes are 2½ inch. The lower side wall headers are inclined to suit the water cooled grate and furnace side wall tubes are on 5 inch centers. The bank tubes are on alternate side to side spacing of 4 inch and 5⅛ inch.

With an installation of this type, the maximum furnace heat release rate is 30,000 Btu cu. ft. The grate heat release rate is usually limited to 400,000 Btu/sq. ft.

TYPICAL MULTIPLE RETORT

The "long drum" unit in Fig. B-8 is fitted with a multiple retort underfeed stoker. It was designed for a steam capacity of 42,000 lbs./hr. at

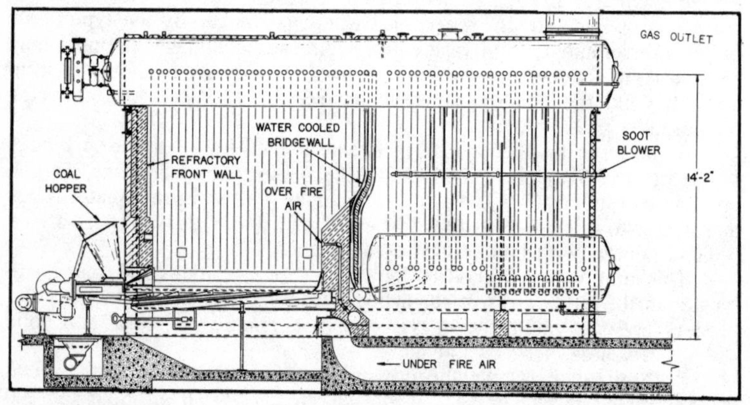

Fig. B-6. "Long drum" unit is fitted with a single retort underfeed stoker (courtesy of E. Keeler Company).

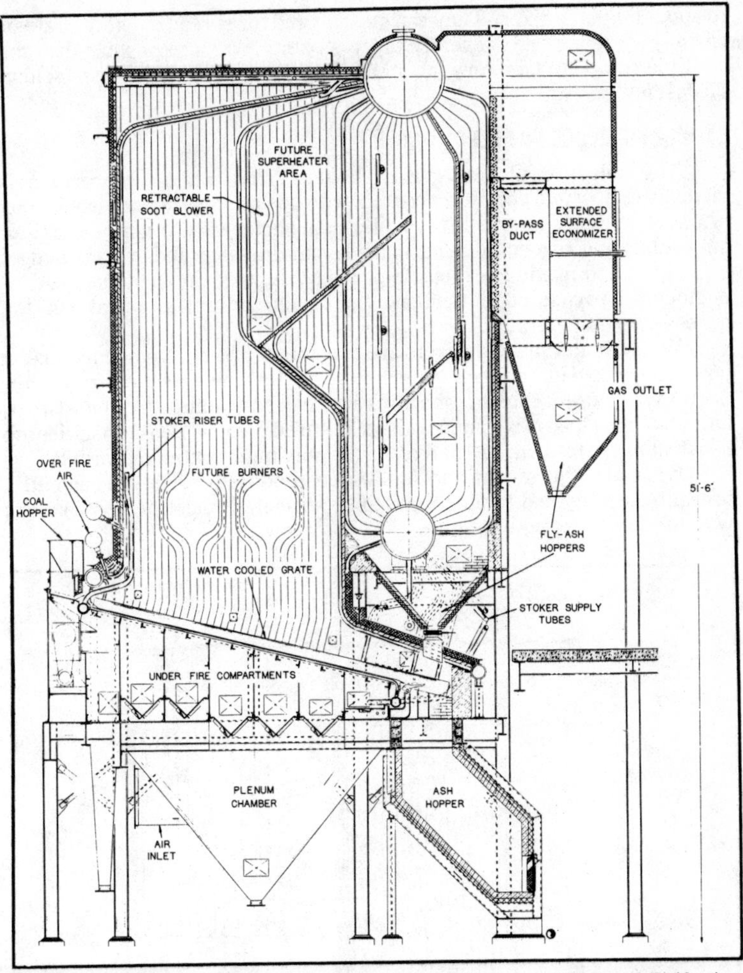

Fig. B-7. A typical vibra-grate stoker installation (courtesy of E. Keeler Company).

150 PSIG when burning bituminous coal. Due to the size, this unit was completely field erected.

In general, this type of unit with a multiple retort stoker is designed for a maximum furnace heat release of 35,000 Btu/cu. ft. but under certain circumstances, it may be increased to 45,000 Btu/cu. ft. Stoker manufacturers usually limit the grate heat release to 600,000 Btu/sq. ft.

The lower section of the water cooled bridge wall is covered with armor block to protect the tubes from erosion and clinkers. The side wall headers and tubes are usually studded and refractory covered in the immediate grate area.

As indicated previously in this paper, field erected "long drum" boilers similar to the one shown in Fig. B-8 can be supplied for steam capacities up to about 60,000 lbs./hr. They can be arranged for spreader, vibra-grate or chain grate firing and presently these stokers are more popular than the multiple retort. In fact, very few multiple retort stokers have been sold in the last few years.

STOKER FIRED BOILER SELECTION

The selection of stoker fired boilers from catalog information is virtually impossible and direct contact with one or more boiler manufacturers is usually necessary. Most boiler manufacturers are quite willing to help select and size equipment, make layout drawings and provide budget pricing, but to maximize their help, it is vital that the customer and/or engineers provide good fundamental design information and set the necessary parameters.

Before boiler manufacturers are consulted, it is recommended that at least some consideration be given to the following specific areas. How many units and what steam capacities are desired? Remember that a two or four hour peak load is common with stoker fired boilers. Peak capacities are usually limited to about 110 percent of the maximum continuous rating.

What are the steam and feedwater conditions? The steam outlet pressure, quality and temperature (if superheated) must be known along with the feedwater temperature.

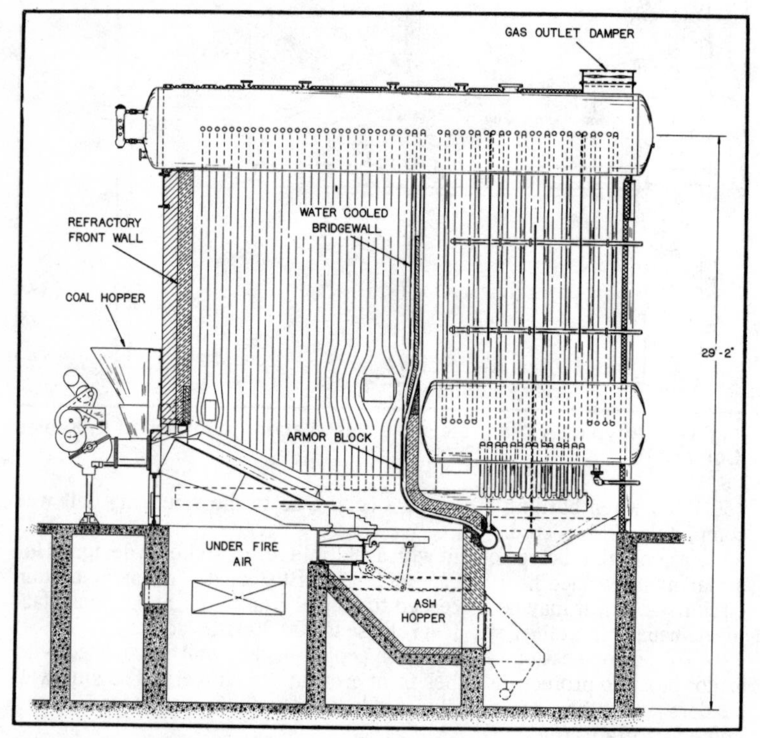

Fig. B-8. This unit is fitted with a multiple retort underfeed stoker (courtesy of E. Keeler Company).

What type of coal will be used? Usually a particular coal is selected for design, but it is advisable to have one or more alternate sources and they should also be considered when selecting and sizing equipment.

The boiler manufacturer must have the proximate analysis, sulphur content and higher heating value for the design coal and each alternate. In addition, the ultimate analysis of the design coal is necessary for combustion and efficiency calculations. The free swelling index of the design coal and alternate coals should be known to help determine burning charactertistics.

The oxidizing and reducing ash softening temperatures are necessary to help select the stoker and determine the maximum furnace exit gas temperature. Several stoker manufacturers also desire an ash analysis. Both the softening temperature and analysis should be available for the design coal and each alternate.

Is heat recovery planned? With current fuel prices, heat recovery is usually considered for all except the very small units. Economizers are more popular than air heaters except when the "as fired" moisture content of the coal exceeds 30 percent. Heated combustion air can present problems with equipment arrangement and often increases stoker maintenance and operating problems. When the extra duct work and larger fans necessary with an air heater are considered, the initial cost is usually less for an economizer.

How much condensate makeup is anticipated and will it be softened? What type of boiler water treatment, phosphate or chelant, is planned? How much continuous blowdown will be required?

What are the load charactertistics? Such items as average, minimum and maximum load should be established. The type of load is important in the boiler and stoker design. For example, are load swings expected and, if so, what are the magnitudes and rates of change.

What type of air pollution equipment is anticipated? Will the design be based on state regulations or federal guidelines? What are the maximum and minimum gas temperatures to the control equipment?

Where will the boilers be located—elevation, etc? Will the boiler be totally or partially enclosed? Will there be a basement for ash handling equipment?

What type of coal and ash handling equipment will be used? Should the boiler be equipped with auxiliary gas and/or oil burners? Burners can be supplied initially or if this area is uncertain, provisions can be made to make future installation much easier.

Some thought should be given to the current or future availability of waste fuels. Liquid, gaseous or solid wastes can often be used as supplemental fuels. Of course, the boiler manufacturer must have an elemental analysis and the physical characteristics to determine whether they are suitable for burning and how they should be introduced into the furnace.

Are there any preferences as to the type of controls? What are the electric current characteristics?

What philosophy will be used in selecting the number and quality of operating personnel? Will the plant people do their own maintenance? Are there any space limitations or preferred auxiliary equipment arrangements?

Courtesy of E Deeler Company from a paper written by Dan McCoy, Chief Engineer.

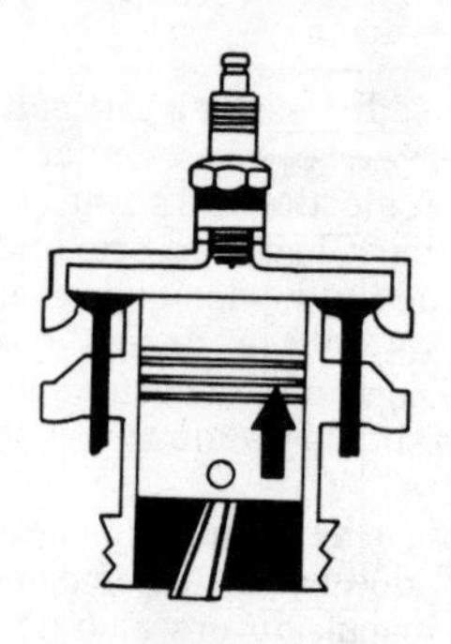

Appendix C
Charts & Tables

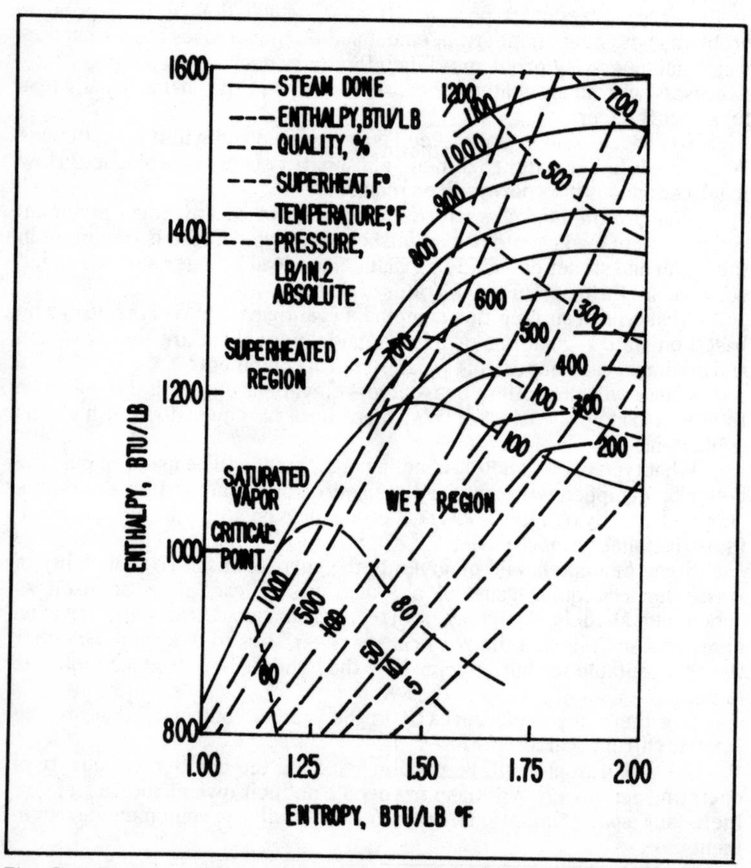

Fig. C-1. Enthalpy and entropy graph.

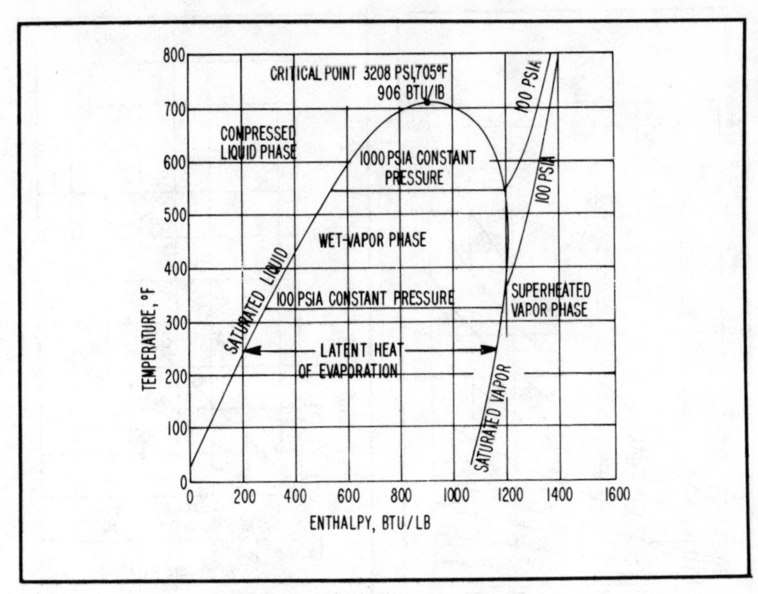

Fig. C-2. Temperature and enthalpy graph.

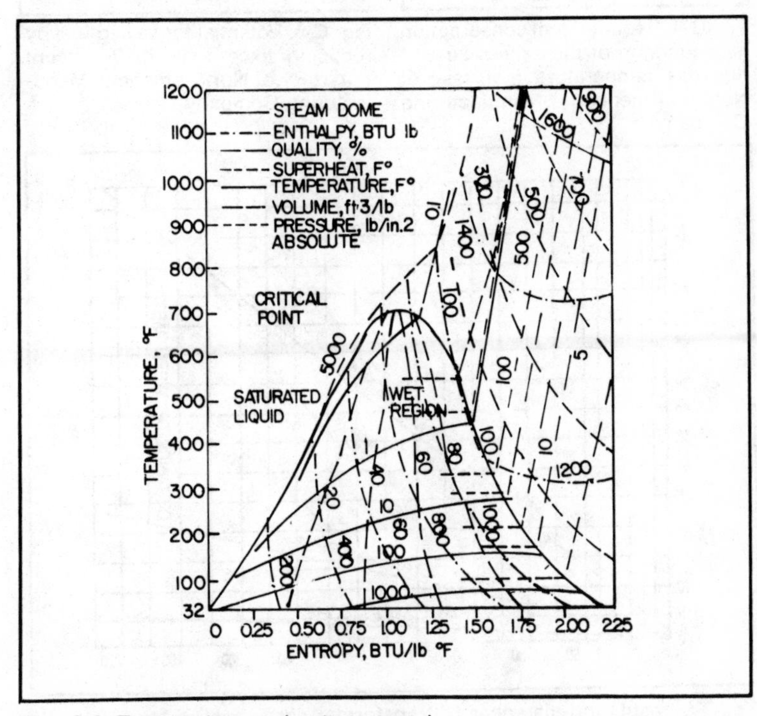

Fig. C-3. Temperature and entropy graph.

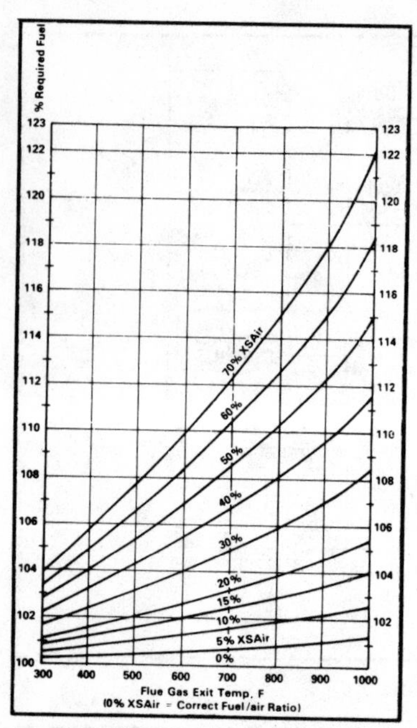

Fig. C-4. Required fuel consumption as a function of the excess air and flue gas temperature (courtesy of North American Manufacturing Company).

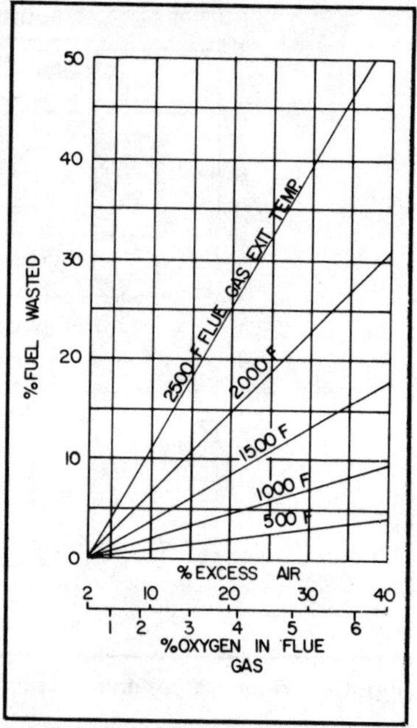

Fig. C-5. Potential for saving fuel by reducing excess air fo 2 percent (courtesy of North American Manufacturing Company).

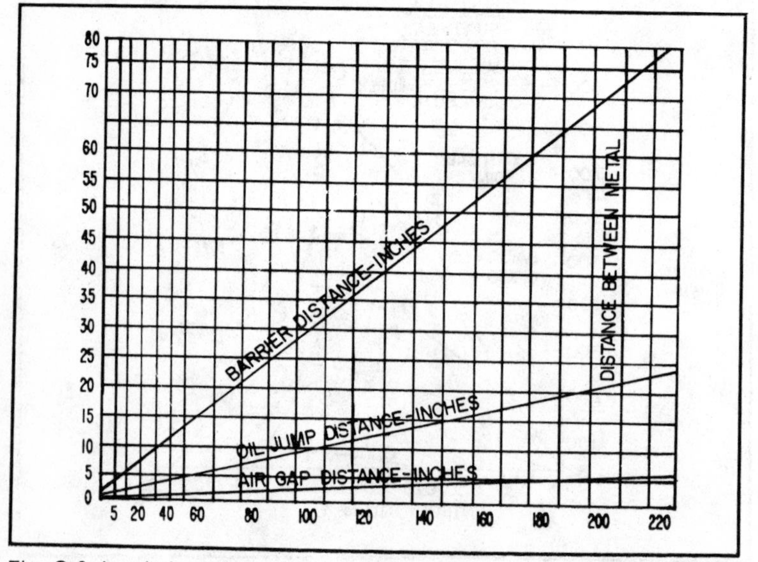

Fig. C-6. Insulating distances of transformer parts.

Table C-1. Electrical Formulas (courtesy of Caterpillar Tractor Company).

To Obtain	Alternating Current Single-Phase	Alternating Current Three-Phase	Direct Current
Kilowatts	$\frac{V \times I \times pf}{1{,}000}$	$\frac{1.732 \times V \times I \times pf}{1{,}000}$	$\frac{V \times I}{1{,}000}$
kV•A	$\frac{V \times I}{1{,}000}$	$\frac{1.732 \times V \times I}{1{,}000}$	
Horsepower Required To Drive Known kW Generator	$\frac{kW}{.746 \times \text{Eff. (Gen.)}}$	$\frac{kW}{.746 \times \text{Eff. (Gen.)}}$	$\frac{kW}{.746 \times \text{Eff. (Gen.)}}$
kW Input To a Motor of Known hp	$\frac{hp \times .746}{\text{Eff. (Motor)}}$	$\frac{hp \times .746}{\text{Eff. (Motor)}}$	$\frac{hp \times .746}{\text{Eff. (Motor)}}$
Full Load Amperes When Motor Horsepower is Known	$\frac{hp \times 746}{V \times pf \times \text{Eff.}}$	$\frac{hp \times 746}{1.732 \times V \times \text{Eff.} \times pf}$	$\frac{hp \times 746}{V \times \text{Eff.}}$
Amperes When kW Is Known	$\frac{kW \times 1{,}000}{V \times pf}$	$\frac{kW \times 1{,}000}{1.732 \times V \times pf}$	$\frac{kW \times 1{,}000}{V}$
Amperes When kV•A Is Known	$\frac{kV{\bullet}A \times 1{,}000}{V}$	$\frac{kV{\bullet}A \times 1{,}000}{1.732 \times V}$	
Frequency (cps)	$\frac{\text{Poles} \times rpm}{120}$	$\frac{\text{Poles} \times rpm}{120}$	
Reactive kV•A (kV•AR)	$\frac{V \times I \times \sqrt{1 - (pf)^2}}{1{,}000}$	$\frac{1.732 \times V \times I \times \sqrt{1 - (pf)^2}}{1{,}000}$	
% Voltage Regulation	$\frac{100\,(V_{NL} - V_{FL})}{V_{FL}}$	$\frac{100\,(V_{NL} - V_{FL})}{V_{FL}}$	$\frac{100\,(V_{NL} - V_{FL})}{V_{FL}}$

Where: V = Line-To-Line Volts
I = Line Current Amperes
pf = Power Factor

Table C-2. Approximate Efficiencies for Squirrel Cage Induction Motors (courtesy of Caterpillar Tractor Company).

Horsepower	Full-Load kW Required	Full-Load Efficiency
1/2	0.6	68%
3/4	0.8	71
1	1	75
1-1/2	1.5	78
2	1.9	80
3	2.7	82
5	4.5	83
7-1/2	6.7	83
10	8.8	85
15	13	86
20	16.8	89
25	21	89
30	24.9	90
40	33.2	90
50	41.5	90
60	49.2	91
75	61.5	91
100	81.2	92
125	101.5	92
150	122	92
200	162.5	92
250	203	92
300	243	92
350	281	93
400	321	93
450	362	93
500	401	93
600	482	93

Table C-3. Direct Current Motors Full Load Current in Amperes (courtesy of Caterpillar Tractor Company).

HORSEPOWER	CURRENT AT 115 V	CURRENT AT 230 V	CURRENT AT 550 V
1/4	3	1.5	
1/3	3.8	1.9	
1/2	5.4	2.7	
3/4	7.4	3.7	1.6
1	9.6	4.8	2
1-1/2	13.2	6.6	2.7
2	17	8.5	3.6
3	25	12.5	5.2
5	40	20	8.3
7-1/2	58	29	12
10	76	38	16
15	112	56	23
20	148	74	31
25	184	92	38
30	220	110	46
40	292	146	61
50	360	180	75
60	430	215	90
75	536	268	111
100		355	148
125		443	184
150		534	220
200		712	295

Table C-4. Typical Equipment Power Tolerances (courtesy of Caterpillar Tractor Company).

Device	Voltage		Frequency	Harmonics and Noise	Remarks
	Variation	Duration of Interruption	Variation		
NEMA Induction Motors	± 10%	Varies With Load 30 Cycle Reclosure Usually Acceptable	± 5%	Increases Heat	Sum of Voltage and Frequency Not to Exceed ± 10%
NEMA AC Control Relays	± 10% Continuously Pickup On — 15% Hold In — 25% (Approximate)	Drops Out In One Cycle or Less	± 5%	Insensitive	
Solenoids-Valves, Brakes, Clutches	± 30% to 40%	1/2 Cycle			
Starter Coils, Motor Contactors					
AC Dropout	− 30% to − 40%	2 Cycles			
AC Burnout	− 15% to 10%	Continuous			
DC Dropout	− 30% to − 40%	5 to 10 Cycles			
Fluorescent Lights	− 10%				Erratic Start
Incandescent Lights	− 25% to + 15%				Short Life
Mercury Vapor Lights	− 50%	2 Cycles			Extinguished
Communications Radio, TV, Telephone	± 5%			Variable Sensitive to Spike	
Computers	± 10% − 8%	1 Cycle	+ 1/2 Hz	5%	
Electronic Tubes	± 5%			Variable	
Inverters	+ 5% at Full Load		± 2 Hz	2% Sensitive to Spikes	May Require Isolating Transformer, Filters
Thyristor (SCR)	+ 10% at No Load, − 10% Transient			Sensitive	
Recitifiers, Solid-State Diode	± 10%			Sensitive	

NOTE: Final Determination of Power Requirements Must Result From Equipment Supplier's Specific Recommendations

Table C-5. Conversion Formulas (courtesy of Westinghouse Electric Corporation).

CONVERSION FORMULAS

TO FIND	DIRECT CURRENT	ALTERNATING CURRENT — SINGLE PHASE	ALTERNATING CURRENT — † THREE PHASE
Amperes When Horse Power is Known	$\frac{H.P. \times 746}{Volts \times Efficiency}$	$\frac{H.P. \times 746}{Volts \times Efficiency \times P.F.}$	$\frac{H.P. \times 746}{Volts \times 1.73 \times Efficiency \times P.F.}$
Amperes When Kilowatts is Known	$\frac{KW \times 1000}{Volts}$	$\frac{KW \times 1000}{Volts \times P.F.}$	$\frac{KW \times 1000}{Volts \times 1.73 \times P.F.}$
Amperes When kva is Known		$\frac{kva \times 1000}{Volts}$	$\frac{kva \times 1000}{Volts \times 1.73}$
Kilowatts	$\frac{Amperes \times Volts}{1000}$	$\frac{Amps. \times Volts \times P.F.}{1000}$	$\frac{Amps. \times Volts \times 1.73 \times P.F.}{1000}$
kva		$\frac{Amps. \times Volts}{1000}$	$\frac{Amps. \times Volts \times 1.73}{1000}$
Power Factor		$\frac{Kilowatts \times 1000}{Amps. \times Volts}$ or $\frac{KW}{kva}$	$\frac{KW \times 1000}{Amps. \times Volts \times 1.73}$ or $\frac{KW}{kva}$
Horse Power (Output)	$\frac{Amps. \times Volts \times Efficiency}{746}$	$\frac{Amps. \times Volts \times Efficiency \times P.F.}{746}$	$\frac{Amps. \times Volts \times 1.73 \times Efficiency \times P.F.}{746}$

Power Factor and Efficiency when used in above formulas should be expressed as decimals.
† For 2-phase, 4-wire substitute 2 instead of 1.73.
† For 2-phase, 3-wire substitute 1.41 instead of 1.73.

Table C-6. Full Load Currents of Motors (courtesy of Westinghouse Electric Corporation).

AMPERES—FULL-LOAD CURRENT

HP OF MOTOR	DIRECT-CURRENT MOTORS			ALTERNATING-CURRENT MOTORS: SINGLE-PHASE MOTORS		SQUIRREL-CAGE INDUCTION MOTORS: TWO-PHASE					SQUIRREL-CAGE INDUCTION MOTORS: THREE-PHASE					SLIP-RING INDUCTION MOTOR: TWO-PHASE					SLIP-RING INDUCTION MOTOR: THREE-PHASE				
	115-VOLT	230-VOLT	550-VOLT	110-VOLT	220-VOLT	110-VOLT	220-VOLT	440-VOLT	550-VOLT	2300-VOLT	110-VOLT	220-VOLT	440-VOLT	550-VOLT	2300-VOLT	110-VOLT	220-VOLT	440-VOLT	550-VOLT	2300-VOLT	110-VOLT	220-VOLT	440-VOLT	550-VOLT	2300-VOLT
¼				4.8	2.4																				
½	4.5	2.3		7	3.5	4.3	2.2	1.1	.9		5.0	2.5	1.3	1.0											
¾	6.5	3.3	1.4	9.4	4.7	4.7	2.4	1.2	1.0		5.4	2.8	1.4	1.1		6.2	3.1	1.6	1.3		7.2	3.6	1.8	1.5	
1	8.4	4.2	1.7	11	5.5	5.7	2.9	1.4	1.2		6.6	3.3	1.7	1.3		6.7	3.4	1.7	1.4		7.8	3.9	2.0	1.6	
1½	12.5	6.3	2.6	15.2	7.6	7.7	4.0	2	1.6		9.4	4.7	2.4	2.0		11.7	5.9	3.0	2.3						
2	16.1	8.3	3.4	20	10	10.4	5	3	2.0		12.0	6	3.0	2.4		12.5	6.3	3.1	2.5		14.4	7.2	3.6	2.9	
3	23	12.3	5.0	28	14		8	4	3.0			9	4.5	4.0			8.7	4.3	3.5		20.2	10	5.0	4	
5	40	19.8	8.2	46	23		13	7	6			15	7.5	6.0			13.0	6.5	5.2			15	7.5	6	
7½	58	28.7	12	68	34		19	9	7			22	11	9.0			20.0	10.0	7.6			25	13	10	
10	75	38	16	86	43		24	12	10			27	14	11			24.3	12.1	10.0			28	14	11	
15	112	56	23				33	16	13			38	19	15			39	19.5	15.6			45	23	18	
20	140	74	30				45	23	19			52	26	21	5.7		49	24.7	19.8			56	28	22	
25	185	92	38				55	28	22	6		64	32	26	7		60	30.0	24.0	6.4		67	34	27	7.5
30	220	110	45				67	34	27	7		77	39	31	8		72	36.0	28.8	7.8		82	41	33	9
40	294	146	61				88	44	35	9		101	51	40	10		93	46.5	37.3	9.5		106	53	42	11
50	364	180	75				108	54	43	11		125	63	50	13		113	57	45	12.1		128	64	51	14
60	436	215	90				129	65	52	13		149	75	60	15		135	68	54	14.0		150	75	60	16
75	540	268	111				156	78	62	16		180	90	72	19		164	82	65	17.3		188	94	75	19
100		357	146				212	106	85	22		246	123	98	25		214	108	87	21.7		246	123	99	25
125		443	184				268	134	108	27		310	155	124	32		267	134	108	27		310	155	124	31
150			220				311	155	124	31		360	180	144	36		315	158	127	32		364	182	145	37
175																									
200			295				415	208	166	43		480	240	195	49		430	216	173	44		490	245	196	52

Table C-7. Selecting Wire and Fuse Sizes for Motor Branch Circuits (courtesy of Westinghouse Electric Corporation).

(Based on Room Temperature 30° C. 86° F.)

FULL-LOAD CURRENT RATING OF MOTOR	MINIMUM ALLOWABLE SIZE OF COPPER WIRE, A. W. G. OR MCM NATIONAL ELECTRIC CODE			FOR RUNNING PROTECTION OF MOTORS		MAXIMUM ALLOWABLE RATING OF BRANCH-CIRCUIT FUSES WITH CODE LETTERS			
AMPERES	RUBBER TYPES R. RW. RU (14-6)	TYPE RH HEAT-RESISTANT GRADE RUBBER	TYPES TA V. AVB	MAX. RATING OF N.E.C. FUSES AMPERES	MAX. SETTING OF TIME-LIMIT PROTECTIVE DEVICE AMPERES	SINGLE-PHASE and SQUIRREL CAGE FULL VOLTAGE, RESISTER AND REACTOR STARTING CODE LETTERS F TO V, INCL.	SINGLE-PHASE and SQUIRREL CAGE FULL VOLTAGE, RESISTOR OR REACTOR STARTING CODE LETTERS B TO E, INCL. Auto-transformer STARTING F TO V	SQUIRREL-CAGE, AUTO-TRANSFOMER STARTING. CODE LETTERS B TO E INCL.	ALL MOTORS CODE LETTER A. D. C. AND WOUND-ROTOR MOTORS
1	14	14	14	2	1.25	15	15	15	15
2	14	14	14	3	2.50	15	15	15	15
3	14	14	14	4	3.75	15	15	15	15
4	14	14	14	6	5.0	15	15	15	15
5	14	14	14	8	6.25	15	15	15	15
6	14	14	14	8	7.50	20	15	15	15
8	14	14	14	10	10.0	25	20	20	15
10	14	14	14	15	12.50	30	25	20	15
12	14	14	14	15	15.00	40	30	25	20
14	12	12	14	20	17.50	45	35	30	25
16	12	12	14	20	20.00	50	40	35	25
18	10	10	14	25	22.50	60	45	40	30
20	10	10	14	25	25.0	60	50	40	30
24	10	10	14	30	30.0	80	60	50	40
28	8	10	12	35	35.0	90	70	60	45
32	8	8	10	40	40.0	100	80	70	50
36	6	8	10	45	45.0	110	90	80	60
40	6	6	10	50	50.0	125	100	80	60
44	6	6	8	60	55.0	125	110	90	70
48	4	6	8	60	60.0	150	125	100	80
52	4	6	6	70	65.0	175	150	110	80
56	4	4	6	70	70.0	175	150	120	90
60	3	4	6	80	75.0	200	150	120	90
64	3	4	6	80	80.0	200	175	150	100
68	2	4	6	90	85.0	225	175	150	110
72	2	3	4	90	90.0	225	200	150	110
76	2	3	4	100	95.0	250	200	175	125
80	1	3	4	100	100.0	250	200	175	125
84	1	2	4	110	105.0	250	225	175	150
88	1	2	4	110	110.0	300	225	200	150
92	0	2	3	125	115.0	300	250	200	150
96	0	1	3	125	120.0	300	250	200	150
100	0	1	3	125	125.0	300	250	200	150
110	00	0	2	150	137.5	350	300	225	175
120	000	0	2	150	150.0	400	300	250	200
130	000	00	1	175	162.5	400	350	300	200
140	0000	00	1	175	175.0	450	350	300	225
150	0000	000	0	200	187.5	450	400	300	225
160	250	000	00	200	200.0	500	400	350	250
170	250	000	00	225	213	500	450	350	300
180	300	0000	00	225	225	600	450	400	300
190	300	250	000	250	238	600	500	400	300
200	350	250	000	250	250	600	500	400	300
220	400	300	0000	300	275	......	600	450	350
240	500	350	250	300	300	......	600	500	400

Wire sizes shown in this table are for single motor, for short distances from feeder center to motor, therefore the wire sizes are tabulated as minimum. Where a group of mo tors are involved, special consideration must be given in selecting proper wire size.

Wire sizes are based on not more than three conductors in raceway or cable.

TAble C-8. Transformer KVA Ratings and Sound Levels in Decibels.

KVA Rating-Transformer	Typical Sound Levels in Decibels
50	48
100	51
300	55
500	56
1,000	60
5,000	67
10,000	70
50,000	77
100,000	82
200,000	88
400,000	91
600,000	94
800,000	96
1,000,000	98

Table C-9. English to Metric Conversion Factors (courtesy of Caterpillar Tractor Company).

SYMBOL	WHEN YOU KNOW	MULTIPLY BY	TO FIND	SYMBOL
BTU	BRITISH THERMAL UNIT	1055.0	JOULE	J
BTU/HP-HR	BRITISH THERMAL UNIT/ HORSEPOWER-HOUR	0.001 415	MEGAJOULES/KILOWATT-HOUR	MJ/KW-HR
BTU/HR	BRITISH THERMAL UNIT/ HOUR	1055.0	JOULES/HOUR	J/HR
BTU/MIN	BRITISH THERMAL UNIT MINUTE	0.017 584	KILOWATT	KW
°C	CELSIUS (DEGREES)	[(1.8 C) + 32]	FAHRENHEIT (DEGREES)	°F
CU FT	CUBIC FEET	0.0283	CUBIC METER	M^3
CU FT/HR	CUBIC FEET/HOUR	0.0283	CUBIC METER/HOUR	M^3/HR
CFM	CUBIC FEET/MINUTE	0.0283	CUBIC METER/MINUTE	M^3/MIN
CU IN	CUBIC INCH	0.016 387 1	LITER	L
CU IN	CUBIC INCH	0.000 016 387 1	CUBIC METER	M^3
°F	FAHRENHEIT (DEGREES)	[0.5555 (F-32)]	CELSIUS (DEGREES)	°C
FT/MIN	FEET/MINUTE	0.3048	METER/MINUTE	M/MIN
FT	FEET	0.3048	METER	M
FT H_2O	FEET OF WATER	2.986 08	KILOPASCAL	KPA
GPH	GALLON/HOUR	3.7854	LITER/HOUR	L/HR
GPM	GALLON/MINUTE	3.7854	LITER/MINUTE	L/MIN
HP	HORSEPOWER	0.7457	KILOWATT	KW
IN HG	INCH OF MERCURY	3.3768	KILOPASCAL	KPA
IN	INCH	25.4	MILLIMETER	MM
IN H_2O	INCH OF WATER	0.248 84	KILOPASCAL	KPA
KW	KILOWATT	56.869 88	BRITISH THERMAL UNIT/MINUTE	BTU/MIN
L	LITER	61.0236	CUBIC INCH	CU IN
μ	MICRON	1.0	MICROMETER	μM
LB	POUND	0.4536	KILOGRAM (MASS)	KG
LB	POUND	4.448 22	NEWTON (FORCE)	N
LB FT (FT-LB)	POUND FOOT	1.355 818	NEWTON METER	N•M
LB IN (IN-LB)	POUND INCH	0.112 985	NEWTON METER	N•M
LB/IN	POUNDS/INCH	0.175 126 8	NEWTON/MILLIMETER	N/MM
LB/IN	POUNDS/INCH	175.1268	NEWTON/METER	N/M
LB/HP-HR	POUND/HORSEPOWER-HOUR	608.28	GRAM/KILOWATT HOUR	G/KW-HR
LB/HR	POUND/HOUR	453.6	GRAM/HOUR	G/HR
M^3	CUBIC METER	61 023.61	CUBIC INCH	CU IN
PSI	POUNDS/SQUARE INCH	6.894 75	KILOPASCAL	KPA
QT	QUART	0.946 35	LITER	L
SQ FT	SQUARE FEET	0.0929	SQUARE METER	M^2
SQ IN	SQUARE INCH	6.4516	SQUARE CENTIMETER	CM^2
U.S. GAL	U.S. GALLON	3.7854	LITER	L

Table C-10. Volume and Capacity Equivalents (courtesy of Caterpillar Tractor Company).

UNIT	CU. IN.	CU. FT.	CU. YD.	CU. CM	CU. M	U.S. LIQUID GALLONS	IMPERIAL GALLONS	LITERS
1 CU. IN.	1	.000579	.0000214	16.39	.0000164	.004329	.00359	.0164
1 CU. FT.	1728	1	.03704	28,317	.028	7.481	6.23	28.32
1 CU. YD.	46,656	27	1	764,600	.765	202	167.9	764.6
1 CU. CM	.061	.0000353	.00000131	1	.000001	.000264	.00022	.001
1 CU. M	61,020	35.31	1.308	1,000,000	1	264.2	220.2	1000
1 U.S. LIQUID GAL.	231	.1337	.00495	3785	.003785	1	.833	3.785
1 IMPERIAL GAL.	277.42	.16	.00594	4545.6	.004546	1.2	1	4.546
1 LITER	61.02	.03531	.001308	1000	.001	.2642	.22	1
1 ACRE FT.	—	43,560	1613.33	—	1233.5	325,850	271,335	—

(There is no standard liquid barrel; by trade custom, 1 BBL. of petroleum oil, unrefined — 42 gallons)

Table C-13. Head and Pressure Equivalents (courtesy of North American Manufacturing Company).

HEAD, FEET OF OIL [6]	HEAD, FEET OF WATER[7]	PRESSURE, psi	PRESSURE, "Hg[7]
1	0.900	.390	0.794
1.260	1.132	.491	**1**
2	1.800	.780	1.588
2.52	2.26	.982	**2**
2.56	2.31	**1**	2.04
3	2.70	1.170	2.38
3.78	3.39	1.473	**3**
4	3.60	1.560	3.18
5	4.50	1.950	3.97
5.13	4.62	**2**	4.08
6	5.40	2.34	4.76
6.30	5.63	2.44	**5**
7	6.30	2.73	5.56
7.56	6.75	2.93	**6**
7.69	6.92	**3**	6.12
8	7.20	3.12	6.35
8.82	7.92	3.43	**7**
9	8.10	3.51	7.15
10	9.00	3.90	7.94
10.05	9.07	3.92	**8**
10.26	9.23	**4**	8.16
11.34	10.18	4.41	**9**
12.60	11.30	4.89	**10**
12.82	11.54	**5**	10.19
13.86	12.44	5.39	**11**
15.12	13.56	5.87	**12**
15.38	13.85	**6**	12.23
16.38	14.72	6.38	**13**
17.64	15.85	6.87	**14**
17.96	16.15	**7**	14.27
18.90	16.98	7.36	**15**
20	18.00	7.80	15.88
20.5	18.46	**8**	16.31
23.1	20.8	**9**	18.35
25.2	22.5	9.76	**20**
25.6	23.1	**10**	20.3
27.7	24.9	10.80	**22**
28.2	25.4	**11**	22.4
30	27.0	11.70	23.8
30.2	27.1	11.75	**24**
30.8	27.7	**12**	24.4
32.8	29.4	12.74	**26**
33.3	30.0	**13**	26.5
35.3	31.7	13.75	**28**
35.9	32.3	**14**	28.5

[6] 0.9 sp gr (about a No. 4 oil)
[7] at 39.2 F (4 C)

HEAD FEET OF OIL[6]	HEAD, FEET OF WATER[7]	PRESSURE, psi	PRESSURE, "Hg[7]
37.7	33.9	14.696	29.922
37.8	34.1	14.74	**30**
38.5	34.6	**15**	30.6
40	36.0	15.60	31.8
50	45.0	19.50	39.7
51.3	46.2	**20**	40.7
60	54.1	23.4	47.6
64.2	57.7	**25**	51.0
70	63.1	27.35	55.6
76.9	69.2	**30**	61.2
80	72.1	31.2	63.5
90	80.8	**35**	71.5
100	90.1	39.0	79.4
102.5	92.3	**40**	81.4
110	99.1	42.9	87.3
120	108.1	46.8	95.3
128.3	115.4	**50**	102.0
130	117.1	50.7	103.2
140	126.1	54.6	111.2
150	135.1	58.5	119.1
154.0	138.5	**60**	122.3
160	144.1	62.4	127.0
170	153.1	66.3	135.0
179.5	161.6	**70**	142.5
190	171.2	74.2	150.9
200	10.2	78.0	158.8
205	184.6	**80**	163.1
225	203	88.0	178.7
230	208	**90**	183
250	225	97.5	198
256	231	**100**	203.8
275	248	107.5	218.8
300	270	117.0	238.2
320	288	**125**	255
325	293	127.0	258
350	315	136.5	278
385	346	**150**	306
400	360	156.0	318
500	450	195.0	397
513	462	**200**	407.7
600	541	234	476
700	631	273	556
770	692	**300**	611.6
800	721	312	635
900	811	351	715
100	901	390	794

Table C-11. Head and Pressure Equivalents (courtesy of North American Manufacturing Company).

UNIT	MM HG (0° C)	IN. HG (0° C)	IN. WATER (39° F)	FT. WATER (39° F)
MM HG	1	0.03937	0.53526	0.0446
IN. HG	25.4	1	13.5955	1.13296
IN. WATER	1.86827	0.07355	1	0.08333
FT. WATER	22.4192	0.88265	12	1
LBS. PER SQ. IN.	51.7149	2.03602	27.6807	2.3067
KILOGRAMS PER SQ. CM	735.559	28.959	393.71171	32.80931
BAR	750.062	29.530	401.4742	33.45618
ATMOSPHERES	760	29.9213	406.79375	33.89948
KILO PASCAL	7.500 62	.295 30	4.014 742	.334 5618

	LB PER SQ. IN.	KILOGRAMS PER SQ. IN.	BAR	ATMOSPHERES (14.7 PSI)	KILO PASCAL
MM HG	0.01934	0.00136	0.00133	0.001315	—
IN. HG	0.49115	0.03453	0.03386	0.03342	—
- WATER	0.03613	0.00254	0.00249	0.00246	0.249
FT. WATER	0.43352	0.030479	0.02989	0.02950	2.989
LBS. PER SQ. IN.	1	0.07031	0.06895	0.06805	6.895
KILOGRAMS PER SQ. CM	14.2233	1	0.98067	0.96784	98.067
BAR	14.504	1.01972	1	0.98692	100.
ATMOSPHERES	14.6959	1.03323	1.01325	1	101.325
KILO PASCAL	0.145 038	0.010 1972	0.010000	0 009 86.920	1

Table C-13. Viscosity Conversion (courtesy of North American Manufacturing Company).

v IN SSU_{100}	v IN CENTI-STOKES	v IN SSU_{100}	v IN SSF_{122}	v IN CENTI-STOKES	v3 IN SSU_{100}	v IN SSF_{122}	v IN CENTI-STOKES
32	1.81	250	27.8	53.7	2500	254.0	539.0
35	2.70	300	32.7	64.6	3000	305.0	647.0
40	4.24	400	42.5	86.2	4000	407.0	863.0
50	7.36	500	52.3	107.8	5000	509.0	1079.0
60	10.32	600	62.2	129.4	6000	610.0	1294.0
70	13.08	700	72.0	151.0	7000	712.0	1510.0
80	15.66	800	82.2	172.6	8000	814.0	1726.0
90	18.12	900	92.0	194.2	9000	916.0	1942.0
100	20.55	1000	102.3	216.0	10000	1018.0	2158.0
150	31.90	1500	153.0	324.0	1500	1526.0	3236.0
200	43.00	2000	204.0	432.0	2000	2039.0	4315.0

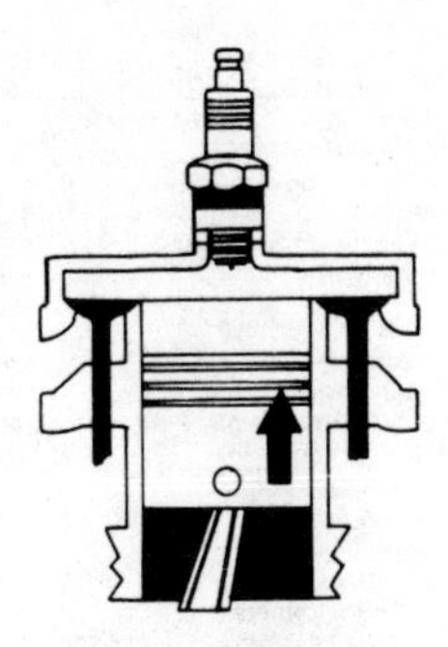

Index